Lecture Notes in Computer Science

Founding Editors

Gerhard Goos
Juris Hartmanis

Editorial Board Members

Elisa Bertino, *Purdue University, West Lafayette, IN, USA*
Wen Gao, *Peking University, Beijing, China*
Bernhard Steffen, *TU Dortmund University, Dortmund, Germany*
Moti Yung, *Columbia University, New York, NY, USA*

The series Lecture Notes in Computer Science (LNCS), including its subseries Lecture Notes in Artificial Intelligence (LNAI) and Lecture Notes in Bioinformatics (LNBI), has established itself as a medium for the publication of new developments in computer science and information technology research, teaching, and education.

LNCS enjoys close cooperation with the computer science R & D community, the series counts many renowned academics among its volume editors and paper authors, and collaborates with prestigious societies. Its mission is to serve this international community by providing an invaluable service, mainly focused on the publication of conference and workshop proceedings and postproceedings. LNCS commenced publication in 1973.

Michael Wand · Kristína Malinovská ·
Jürgen Schmidhuber · Igor V. Tetko
Editors

Artificial Neural Networks and Machine Learning – ICANN 2024

33rd International Conference on Artificial Neural Networks
Lugano, Switzerland, September 17–20, 2024
Proceedings, Part IX

Editors
Michael Wand
IDSIA USI-SUPSI
Lugano, Switzerland

MeDiTech, SUPSI
Lugano, Switzerland

Jürgen Schmidhuber
KAUST Center of Generative AI
Thuwal, Saudi Arabia

IDSIA USI-SUPSI
Lugano, Switzerland

Kristína Malinovská
Comenius University
Bratislava, Slovakia

Igor V. Tetko
Helmholtz Zentrum München
Neuherberg, Germany

BigChem GmbH
Unterschleißheim, Germany

ISSN 0302-9743   ISSN 1611-3349 (electronic)
Lecture Notes in Computer Science
ISBN 978-3-031-72355-1   ISBN 978-3-031-72356-8 (eBook)
https://doi.org/10.1007/978-3-031-72356-8

© The Editor(s) (if applicable) and The Author(s), under exclusive license
to Springer Nature Switzerland AG 2024

This work is subject to copyright. All rights are solely and exclusively licensed by the Publisher, whether the whole or part of the material is concerned, specifically the rights of translation, reprinting, reuse of illustrations, recitation, broadcasting, reproduction on microfilms or in any other physical way, and transmission or information storage and retrieval, electronic adaptation, computer software, or by similar or dissimilar methodology now known or hereafter developed.
The use of general descriptive names, registered names, trademarks, service marks, etc. in this publication does not imply, even in the absence of a specific statement, that such names are exempt from the relevant protective laws and regulations and therefore free for general use.
The publisher, the authors and the editors are safe to assume that the advice and information in this book are believed to be true and accurate at the date of publication. Neither the publisher nor the authors or the editors give a warranty, expressed or implied, with respect to the material contained herein or for any errors or omissions that may have been made. The publisher remains neutral with regard to jurisdictional claims in published maps and institutional affiliations.

This Springer imprint is published by the registered company Springer Nature Switzerland AG
The registered company address is: Gewerbestrasse 11, 6330 Cham, Switzerland

If disposing of this product, please recycle the paper.

Preface

In recent years, Machine Learning has become more important than ever before. Large Language Models have revolutionized language-based tasks, with an impact far beyond the research community and IT-related industries: Artificial Intelligence for solving day-to-day tasks has become available for a wide range of end users across the world.

Machine Learning not only influences our daily lives, but also many fields of science and technology. As a specific example, we present Artificial Intelligence in organic chemistry and pharmaceutical research: a variety of tasks in this field are tackled with state-of-the-art Neural Network methods, leading to improved design and higher security of medical drugs, and to better solutions for chemical tasks in general, improving the quality of life of a large number of persons across the globe.

It is in this context that we proudly present the Proceedings of the 33th International Conference on Artificial Neural Networks (ICANN 2024). ICANN is the annual flagship conference of the European Neural Network Society (ENNS). This edition was co-organized by Istituto Dalle Molle di studi sull'intelligenza artificiale (IDSIA USI-SUPSI https://www.idsia.usi-supsi.ch) and by the Marie Skłodowska-Curie (MSC) Innovative Training Network European Industrial Doctorate "Advanced machine learning for Innovative Drug Discovery" (AIDD https://ai-dd.eu), supported by the MSC Doctoral Network "Explainable AI for Molecules" (AiChemist https://aichemist.eu). After two years of on-line and two years of hybrid conferences, ICANN 2024 was again organized as an in-person event, held on the premises of Università della Svizzera italiana (USI) and Scuola Universitaria Professionale della Svizzera italiana (SUPSI) in Lugano from September 17 to September 20, 2024.

ICANN 2024 featured three main conference tracks, namely Artificial Intelligence and Machine Learning, Bio-inspired Computing, and an Application Track. Dedicated members of the ICANN community also organized three workshops:

- AI in Drug Discovery
- Explainable AI in Human-Robot Interaction
- Reservoir Computing

   as well as three special sessions:

- Spiking Neural Networks and Neuromorphic Computing
- Accuracy, Stability, and Robustness in Deep Neural Networks
- Neurorobotics.

   Two tutorial sessions

- FEDn – A scalable federated machine learning framework for cross-device and cross-silo environments
- TSFEL - A Hands-on Introduction to Time Series Feature Extraction

were likewise proposed and organized by the community, as well as the

- Tox24 Challenge (prediction of toxicity of chemical compounds).

The proceedings of the conference are published as Springer volumes belonging to the Lecture Notes in Computer Science series. The conference had a total of 764 articles submitted to it. The papers went through a double-blind peer-review process supervised by experienced Area Chairs who suggested decisions to Program Chairs. In total, 564 Area Chairs, Program Committee (PC) members, and reviewers participated in the review process. The reviewers were on average assigned 3–4 articles each and submissions received on average 2.03 reviews each. A list of reviewers/PC Members who agreed to publish their names is included in the proceedings.

Based on the Area Chairs' and reviewers' comments, 310 articles (40.5% of initial submissions) were accepted, including 180 manuscripts selected for oral presentations. Out of the total number of accepted articles the majority (285 papers) were full articles with an average length of 15 pages, 20 manuscripts were short articles with an average length of 10 pages, and 5 were abstracts with an average length of 3 pages.

The accepted papers of the 33rd ICANN conference are published as 11 volumes, including one open-access volume with papers supported by the AIDD project.

The authors of accepted articles came from 29 different countries. As indicated by first author affiliation the largest number of articles came from China, followed by Germany, Japan, and Italy. While the majority of the articles were from academic researchers, the conference also attracted contributions from many industries including large pharmaceutical companies (Pfizer, Bayer, AstraZeneca, Johnson & Johnson), information and communication technology companies (Fujitsu and Baidu inc.), as well as multiple startups. This speaks to the increasing use of artificial neural networks in industry. Four keynote speakers were invited to give lectures on the timely aspects of advances in understanding the brain (Michael Reimann); new insights into cortical attention mechanisms and context-dependent gating and how they might inspire future developments in AI (Walter Senn); the current state of cognitive systems and how the full range of bio-signals can be utilized to further enhance human-robot interactions (Tanja Schultz); and a general overview of the past, present and future of machine learning (Jürgen Schmidhuber).

These proceedings provide comprehensive and up-to-date coverage of the dynamically developing field of Artificial Neural Networks. They are of major interest both for theoreticians as well as for applied scientists who are looking for new innovative approaches to solve their practical problems. We sincerely thank the Program and Steering Committee, Area Chairs, and the reviewers for their invaluable work.

September 2024

Michael Wand
Kristína Malinovská
Jürgen Schmidhuber
Igor V. Tetko

Organization

General Chairs

Jürgen Schmidhuber   KAUST Center of Generative AI, Saudi Arabia, and IDSIA USI-SUPSI, Switzerland

Igor V. Tetko   Helmholtz Munich, Germany and BigChem GmbH, Germany

Program Chairs

Michael Wand   IDSIA USI-SUPSI, Switzerland and MeDiTech, SUPSI, Switzerland

Kristina Malinovska   Comenius University Bratislava, Slovakia

Honorary Chair

Stefan Wermter   University of Hamburg, Germany

Organizing Committee Chairs

Katya Ahmad   Helmholtz Munich, Germany
Alessandra Lintas   University of Lausanne, Switzerland

Local Organizing Committee

Stefano van Gogh   IDSIA USI-SUPSI, Switzerland
Qinhan Hou   IDSIA USI-SUPSI, Switzerland
Nicolò La Porta   SUPSI, Switzerland
Alessandro Giusti   IDSIA USI-SUPSI, Switzerland
Vittorio Limongelli   USI, Switzerland
Cesare Alippi   IDSIA USI-SUPSI, Switzerland
Elena Invernizzi   IDSIA USI-SUPSI, Switzerland
Alessia Gianinazzi   IDSIA USI-SUPSI, Switzerland

Communication Chairs

Sebastian Otte University of Lübeck, Germany
R. Omar Chavez-Garcia IDSIA USI-SUPSI, Switzerland

Steering Committee

Stefan Wermter University of Hamburg, Germany
Angelo Cangelosi University of Manchester, UK
Igor Farkaš Comenius University Bratislava, Slovakia
Chrisina Jayne Teesside University, UK
Matthias Kerzel University of Hamburg, Germany
Alessandra Lintas University of Lausanne, Switzerland
Kristína Malinovská Comenius University Bratislava, Slovakia
Alessio Micheli University of Pisa, Italy
Jaakko Peltonen Tampere University, Finland
Brigitte Quenet ESPCI Paris, France
Ausra Saudargiene Lithuanian University of Health Sciences, and Vytautas Magnus University, Lithuania
Roseli Wedemann Rio de Janeiro State University, Brazil
Sebastian Otte University of Lübeck, Germany

Area Chairs

Alessandro Antonucci IDSIA USI-SUPSI, Switzerland
Alessandro Facchini IDSIA USI-SUPSI, Switzerland
Alessio Micheli University of Pisa, Italy
Anthony Cioppa University of Liège, Belgium
Ausra Saudargiene Lithuanian University of Health Sciences, and Vytautas Magnus University, Lithuania
Brigitte Quenet ESPCI Paris PSL, France
Chen Zhao King Abdullah University of Science and Technology, Saudi Arabia
Daniele Palossi IDSIA USI-SUPSI, Switzerland
Davide Bacciu University of Pisa, Italy
Fabio Rinaldi IDSIA USI-SUPSI, Switzerland
Felix Putze University of Bremen, Germany
Francesca Faraci MeDiTech/BSP SUPSI-DTI, Switzerland
Gabriela Andrejková P. J. Šafárik University in Košice, Slovakia
Hui Liu University of Bremen, Germany

Igor Farkaš Comenius University Bratislava, Slovakia
Kevin Jablonka Friedrich Schiller University Jena, Germany
Marcello Restelli Politecnico di Milano, Italy
Marco Forgione IDSIA USI-SUPSI, Switzerland
Matthias Karlbauer University of Tübingen, Germany
Michela Papandrea ISIN, DTI, SUPSI, Switzerland
Mihai Andries IMT Atlantique, France
Oleg Szehr IDSIA USI-SUPSI, Switzerland
Rafael Cabañas de Paz University of Almería, Spain
Silvio Giancola King Abdullah University of Science and Technology, Saudi Arabia
Thang Vu University of Stuttgart, Germany
Yibo Yang King Abdullah University of Science and Technology, Saudi Arabia
Zuzana Černeková Comenius University Bratislava, Slovakia

Workshop and Special Session Chairs

Workshop: AI in Drug Discovery

Djork-Arné Clevert Pfizer GmbH, Germany
Igor Tetko Helmholtz Munich, Germany

Workshop: Explainable AI in Human-Robot Interaction

Stefan Wermter University of Hamburg, Germany
Angelo Cangelosi University of Manchester, UK
Igor Farkaš Comenius University Bratislava, Slovakia
Theresa Pekarek-Rosin University of Hamburg, Germany

Workshop: Reservoir Computing

Alessio Micheli University of Pisa, Italy
Gouhei Tanaka Nagoya Institute of Technology, Japan
Claudio Gallicchio University of Pisa, Italy
Benjamin Paassen University of Bielefeld, Germany
Domenico Tortorella University of Pisa, Italy

Special Session: Spiking Neural Networks and Neuromorphic Computing

Sander Bohté	CWI Amsterdam, Netherlands
Sebastian Otte	University of Lübeck, Germany

Special Session: Accuracy, Stability, and Robustness in Deep Neural Networks

Vera Kurkova	Institute of Computer Science of the Czech Academy of Sciences, Prague Czech Republic
Ivan Tyukin	King's College, London, UK

Special Session: Neurorobotics

Igor Farkaš	Comenius University Bratislava, Slovakia
Kristína Malinovská	Comenius University Bratislava, Slovakia
Andrej Lúčny	Comenius University Bratislava, Slovakia
Pavel Petrovič	Comenius University Bratislava, Slovakia
Michal Vavrečka	Czech Technical University in Prague, Czechia
Matthias Kerzel	University of Hamburg, Germany
Hassan Ali	University of Hamburg, Germany
Carlo Mazzola	Italian Institute for Technology, Italy

Program Committee

Abraham Yosipof	CLB, Israel
Adam Arany	KU Leuven, Belgium
Adrian Mirza	Helmholtz Institute for Polymers in Energy Applications, Germany
Adrian Ulges	RheinMain University of Applied Sciences, Germany
Alan Anis Lahoud	Örebro University, Sweden
Albert Weichselbraun	University of Applied Sciences of the Grisons (FHGR), Switzerland
Alessandra Roncaglioni	Istituto di Ricerche Farmacologiche Mario Negri, Italy
Alessandro Giusti	IDSIA USI-SUPSI, Switzerland
Alessandro Manenti	USI, Switzerland

Alessandro Trenta	University of Pisa, Italy
Alessio Gravina	University of Pisa, Italy
Alex Doboli	Stony Brook University, USA
Alex Shenfield	Sheffield Hallam University, UK
Alexander Schulz	Bielefeld University, Germany
Alexandra Reichenbach	Heilbronn University of Applied Sciences, Germany
Ali Rodan	University of Jordan, Jordan
Alireza Raisiardali	Pragmatic Semiconductor Limited, UK
Aliza Subedi	Tribhuvan University, Nepal
Amir Mohammad Elahi	EPFL, Switzerland
Ana Claudia Sima	SIB Swiss Institute of Bioinformatics, Switzerland
Ana Sanchez-Fernandez	Johnson & Johnson Innovative Medicine, Belgium/JKU Linz, Austria
Andrea Licciardi	ICAR-CNR, Italy
Andreas Mayr	Johannes Kepler University Linz, Austria
Andreas Plesner	ETH Zurich, Switzerland
Andrej Lucny	Comenius University Bratislava, Slovakia
Aneri Muni	University of Montreal and Mila AI Institute, Canada
Angeliki Pantazi	IBM Research - Zurich, Switzerland
Angelo Moroncelli	IDSIA USI-SUPSI, Switzerland
Anne-Gwenn Bosser	Lab-STICC, ENIB, France
Anthony Strock	Stanford University, USA
Antonio Liotta	Free University of Bozen-Bolzano, Italy
Aparna Raj	BITS Pilani, Dubai Campus, United Arab Emirates
Ardian Selmonaj	IDSIA USI-SUPSI, Switzerland
Arnaud Gucciardi	University of Ljubljana, Slovenia
Artur Xarles	Universitat de Barcelona, Spain
Asma Sattar	University of Pisa, Italy
Aurelio Raffa Ugolini	Politecnico di Milano, Italy
Baohua Zhang	Beijing Institute of Technology, China
Baojin Huang	Wuhan University, China
Barbara Hammer	Bielefeld University, Germany
Bikram Kumar De	Texas State University, USA
Blerina Spahiu	University of Milan-Bicocca, Italy
Bo Li	Baidu Inc., China
Bogdan Kwolek	AGH University of Krakow, Poland
Bojian Yin	Innatera B.V., Netherlands
Brian Moser	German Research Center for Artificial Intelligence, Germany

Bulcsú Sándor	Babeş-Bolyai University, Romania
Cesare Donati	Politecnico di Torino, Italy
Chengeng Liu	Wuhan University, China
Chenxing Wang	Beijing University of Posts and Telecommunications, China
Chi Xie	Tongji University, China
Chong Zhang	Xi'an Jiaotong-Liverpool University, China
Chrisina Jayne	Teesside University, UK
Christoph Reinders	Leibniz University Hannover, Germany
Chrysoula Kosma	École Normale Supérieure Paris-Saclay, France
Cleber Zanchettin	Universidade Federal de Pernambuco, Brazil
Congcong Zhou	Sir Run Run Shaw Hospital, Zhejiang University, China
Coşku Can Horuz	University of Lübeck, Germany
Cunjian Chen	Monash University, Australia
Cyril Zakka	Stanford University, USA
Dania Humaidan	University Hospital Tübingen and Hertie Institute for Clinical Brain Research, Germany
Daniel Frank	University of Stuttgart, Germany
Daniel Nissani (Nissensohn)	Independent Research, Israel
Daniel Ortega	University of Stuttgart, Germany
Daniel Rose	University of Vienna, Austria
Daniele Angioletti	Università della Svizzera italiana, Switzerland
Daniele Castellana	Università degli Studi di Firenze, Italy
Daniele Malpetti	IDSIA USI-SUPSI, Switzerland
Daniele Zambon	IDSIA USI-SUPSI, Switzerland
Darío Ramos López	University of Almería, Spain
Davide Borra	University of Bologna, Italy
Dehui Kong	Sanechips; ZTE, China
Denis Kleyko	Örebro University, Sweden
Diana Borza	Babeş-Bolyai University, Romania
Dinesh Kumar	Bennett University, India
Dirk Väth	University of Stuttgart, Germany
Dongmian Zou	Duke Kunshan University, China
Doreen Jirak	Istituto Italiano di Tecnologia, Italy
Douglas McLelland	BrainChip, France
Duarte Folgado	Fraunhofer Portugal AICOS, Portugal
Dulani Meedeniya	University of Moratuwa, Sri Lanka
Dumitru-Clementin Cercel	Politehnica University of Bucharest, Romania
Dylan Muir	SynSense, Switzerland
Dylan R. Ashley	IDSIA USI-SUPSI, Switzerland

E. J. Solteiro Pires	Universidade de Trás-os-Montes e Alto Douro, Portugal
Elena Šikudová	Charles University, Czech Republic
Elia Cereda	IDSIA USI-SUPSI, Switzerland
Elia Piccoli	University of Pisa, Italy
Emmanuel Okafor	King Fahd University of Petroleum and Minerals, Saudi Arabia
Evaldo Mendonça Fleury Curado	Centro Brasileiro de Pesquisas Físicas and National Institute of Science and Technology for Complex Systems, Brazil
Evgeny Mirkes	University of Leicester, UK
Farhad Nooralahzadeh	Zurich University of Applied Sciences, University of Zurich, Switzerland
Fatemeh Hadaeghi	University Medical Center Hamburg-Eppendorf (UKE), Germany
Fatima Ezzeddine	Università della Svizzera italiana, Switzerland
Federico Errica	NEC Laboratories Europe, Germany
Fedor Scholz	University of Tübingen, Germany
Filipe Miguel Cardoso Micu Menezes	Helmholtz Munich, Germany
Flávio Arthur Oliveira Santos	Universidade Federal de Pernambuco, Brazil
Florian Lux	University of Stuttgart, Germany
Francesco Faccio	IDSIA USI-SUPSI, Switzerland/KAUST AI Initiative, Saudi Arabia
Francesco Landolfi	Università di Pisa, Italy
Francis Colas	Inria, France
Frédéric Alexandre	Inria, France
Gabriel Haddon-Hill	Keio University, Japan
Gabriela Sejnova	Czech Technical University in Prague, Czech Republic
Gabriele Lagani	ISTI-CNR, Italy
Gerrit A. Ecke	Mercedes-Benz AG, Germany
Gianvito Losapio	Politecnico di Milano, Italy
Giorgia Adorni	IDSIA USI-SUPSI, Switzerland
Giovanni Dispoto	Politecnico di Milano, Italy
Giovanni Donghi	University of Padua, Italy
Giuliana Monachino	University of Applied Sciences and Arts of Southern Switzerland (SUPSI), Switzerland
Gugulothu Narendhar	TCS Research, India
Guillaume Godin	BigChem, Switzerland
Habib Irani	Texas State University, USA
Hanno Gottschalk	TU Berlin, Germany
Haoran Yang	Sichuan University, China

Hasby Fahrudin	AIBrain, South Korea
Hicham Boudlal	Mohammed First University of Oujda, Morocco
Hitesh Laxmichand Patel	Oracle/New York University, USA
Houssem Ouertatani	IRT SystemX & Univ. Lille, CNRS, Inria, France
Huang Yifan	Northeast Electric Power University, China
Hubert Cecotti	California State University, Fresno, USA
Hugo Cesar de Castro Carneiro	Universität Hamburg, Germany
Huifang Ma	Northwest Normal University, China
Igor Tetko	Helmholtz Munich, Germany
Ivor Uhliarik	Comenius University Bratislava, Slovakia
Jan Kalina	Czech Academy of Sciences, Institute of Computer Science, Czech Republic
Jan Niehues	KIT, Germany
Jan Prosi	University of Tübingen/International Max Planck Research School for Intelligent Systems, Germany
Jan Wollschläger	Bayer Pharmaceuticals, Germany
Jannis Vamvas	University of Zurich, Switzerland
Jérémie Cabessa	University of Versailles Saint-Quentin, France
Jia Cai	Guangdong University of Finance and Economics, China
Jiahui Chen	Xiamen University, China
Jialiang Xu	Soochow University, China
Jian Zhang	Zhejiang University, China
Jing Han	University of Cambridge, UK
Jingzehua Xu	Tsinghua University, China
Jinlai Ning	King's College London, UK
Jiong Wang	Beijing Normal University, China
Jiwen Yu	Peking University, China
Jizhe Yu	Dalian University of Technology, China
João Ricardo Sato	Universidade Federal do ABC, Brazil
Johannes Kriebel	University of Münster, Germany
Johannes Zierenberg	Max Planck Institute for Dynamics and Self-Organization, Germany
Jorge Lo Presti	University of Pavia, Italy
Julian Cremer	Pfizer, Germany
Julie Keisler	EDF R&D, Inria, France
Julien Marteen Akay	Bielefeld University of Applied Sciences and Arts, Germany
Jun Zhou	Wuhan University, China
Junjie Zhou	Nanjing University of Aeronautics and Astronautics, China
Junzhou Chen	College of William and Mary, USA

Kai Mao	Xi'an Jiaotong University, China
Kevin Scheck	University of Bremen, Germany
Keyan Jin	Macao Polytechnic University, Macao SAR, China
Khoa Phung	University of the West of England, UK
Kiran Lekkala	University of Southern California, USA
Kohei Nakajima	University of Tokyo, Japan
Konstantinos Chatzilygeroudis	University of Patras, Greece
Krechel Dirk	RheinMain University of Applied Science, Germany
Kristína Malinovská	Comenius University Bratislava, Slovakia
Lapo Frascati	ODYS, Italy
Laura Azzimonti	IDSIA USI-SUPSI, Switzerland
Laurent Larger	FEMTO-ST Institute, Université Bourgogne-Franche-Comté, France
Laurent Mertens	KU Leuven, Belgium
Laurent Udo Perrinet	Institut des Neurosciences de la Timone, Aix Marseille Univ - CNRS, France
Lazaros Iliadis	Democritus University of Thrace, Greece
Lea Multerer	IDSIA USI-SUPSI, Switzerland
Lei Li	University of Copenhagen, Denmark
Lenka Tetkova	Technical University of Denmark, Denmark
Leon Scharwächter	University of Tübingen, Germany
Leonardo Olivetti	Uppsala University, Sweden
Lewis Mervin	AstraZeneca, UK
Lina Humbeck	Boehringer Ingelheim Pharma GmbH & Co. KG, Germany
Lindsey Vanderlyn	University of Stuttgart, Germany
Logofatu Doina	Frankfurt University of Applied Sciences, Germany
Lu Yang	Wuhan University, China
Lubomir Antoni	Pavol Jozef Šafárik University in Košice, Slovakia
Luca Butera	IDSIA USI-SUPSI, Switzerland
Luca Sabbioni	ML cube, Italy
Luís Gonçalves	Universidade Federal de Pernambuco, Brazil
Lyra Puspa	Vanaya NeuroLab, Indonesia and Canterbury Christ Church University, UK
Maëlic Neau	ENIB, France/Flinders University, Australia
Mahsa Abazari Kia	Northeastern University London, UK
Maksim Makarenko	Saudi Aramco, Saudi Arabia
Manas Mejari	IDSIA USI-SUPSI, Switzerland
Manon Dampfhoffer	Univ. Grenoble Alpes, CEA, List, France
Manuel Traub	University of Tübingen, Germany

Marco Paul E. Apolinario	Purdue University, USA
Marco Podda	University of Pisa, Italy
Marco Tarabini	Politecnico di Milano, Italy
Marcondes Ricarte da Silva Júnior	Federal University of Pernambuco, Brazil
Marek Suppa	Comenius University Bratislava, Slovakia
Marina Garcia de Lomana	Bayer AG, Germany
Markus Heinonen	Aalto University, Finland
Marta Lenatti	Consiglio Nazionale delle Ricerche, Italy
Martin Lefebvre	Université catholique de Louvain, Belgium
Martin Ritzert	Georg-August Universität Göttingen, Germany
Masanobu Inubushi	Tokyo University of Science, Japan
Matej Fandl	Comenius University Bratislava, Slovakia
Matej Pecháč	Tachyum s.r.o., Slovakia
Matteo Rufolo	IDSIA USI-SUPSI, Switzerland
Matthias Kerzel	Universität Hamburg, Germany
Matthias Rupp	Luxembourg Institute of Science and Technology, Luxembourg
Matus Tuna	Comenius University Bratislava, Slovakia
Maximilian Kimmich	University of Stuttgart, Germany
Maynara Donato de Souza	Federal University of Pernambuco, Brazil
Mengdi Li	University of Hamburg, Germany
Mengjia Zhu	IMT School for Advanced Studies Lucca, Italy
Michal Bechny	UNIBE/SUPSI, Switzerland
Michal Burgunder	Università della Svizzera italiana, Switzerland
Michal Vavrecka	CIIRC CTU, Czech Republic
Michela Sperti	Politecnico di Torino, Italy
Michele Fontanesi	University of Pisa, Italy
Mikhail Andronov	Università della Svizzera Italiana, Switzerland
Mingyang Li	Stanford University, USA
Mingyong Li	Chongqing Normal University, China
Miroslav Strupl	IDSIA USI-SUPSI, Switzerland
Moritz Wolter	Rheinische Friedrich-Wilhelms-Universität Bonn, Germany
Muhammad Arslan Masood	Aalto University, Finland
Muhammad Burhan Hafez	University of Southampton, UK
Mykhailo Sakevych	Texas State University, USA
Nabeel Khalid	German Research Center for Artificial Intelligence, Germany
Navdeep Singh Bedi	IDSIA USI-SUPSI, Switzerland
Nicolò La Porta	Università della Svizzera Italiana, Switzerland
Niklas Beuter	Technische Hochschule Lübeck, Germany
Niko Dalla Noce	University of Pisa, Italy

Oh-hyeon Choung	dsm-firmenich, Switzerland
Olivier J. M. Béquignon	Leiden University, The Netherlands
Omran Ayoub	University of Applied Sciences and Arts of Southern Switzerland, Switzerland
Oscar Mendez Lucio	Recursion, Spain
Osvaldo Simeone	King's College London, UK
Otto Brinkhaus	Spleenlab GmbH, Germany
Pascal Tilli	University of Stuttgart, Germany
Paul Czodrowski	JGU Mainz, Germany
Paul Kainen	Georgetown University, USA
Paula Štancelová	Comenius University Bratislava, Slovakia
Paula Torren-Peraire	Johnson & Johnson Innovative Medicine, Belgium
Pavel Denisov	University of Stuttgart, Germany
Pavel Kordík	Czech Technical University in Prague, Czech Republic
Pavel Petrovič	Comenius University Bratislava, Slovakia
Peiyu Liang	Temple University, USA
Peng Qiao	NUDT, China
Pengjie Liu	Southern University of Science and Technology, China
Pengyu Li	Yanshan University, China
Petia Koprinkova-Hristova	Institute of Information and Communication Technologies, Bulgarian Academy of Sciences, Bulgaria
Petra Vidnerová	Institute of Computer Science, Czech Academy of Sciences, Czech Republic
Philipp Allgeuer	University of Hamburg, Germany
Plinio Moreno	Instituto Superior Técnico/University of Lisbon, Portugal
Qinhan Hou	IDSIA USI-SUPSI, Switzerland
Quentin Jodelet	Tokyo Institute of Technology, Japan
Raphael Yokoingawa de Camargo	Universidade Federal do ABC, Brazil
Răzvan-Alexandru Smădu	National University of Science and Technology POLITEHNICA Bucharest, Romania
Reyan Ahmed	University of Arizona, USA
Ricardo O. Chávez García	IDSIA USI-SUPSI, Switzerland
Riccardo Massidda	Università di Pisa, Italy
Riccardo Renzulli	University of Turin, Italy
Robert Legenstein	Graz University of Technology, Austria
Robertas Damaševičius	Kaunas University of Technology, Lithuania
Robin Winter	Pfizer, Germany

Rodolphe Vuilleumier	École normale supérieure-PSL, Sorbonne Université, CNRS, France
Rodrigo Braga	NOVA School of Science and Technology, Portugal
Rodrigo Clemente Thom de Souza	Federal University of Paraná, Brazil
Roseli S. Wedemann	Universidade do Estado do Rio de Janeiro, Brazil
Roxane Jacob	University of Vienna, Austria
Ru Zhou	RuiJin Hospital LuWan Branch, Shanghai Jiaotong University School of Medicine, China
Ruinan Wang	University of Bristol, UK
Ruixi Zhou	Beijing University of Posts and Telecommunications, China
Rupesh Raj Karn	New York University Abu Dhabi, United Arab Emirates
Samuel Genheden	AstraZeneca R&D, Sweden
Sandra Mitrovic	IDSIA USI-SUPSI, Switzerland
Sankalp Jain	NCATS-NIH, USA
Sara Joubbi	University of Pisa, Italy
Seema Dilipkumar Aswani	BITS Pilani, Dubai Campus, UAE
Seiya Satoh	Tokyo Denki University, Japan
Semih Beycimen	Cranfield University, UK
Senhui Qiu	Ulster University, UK
Sergei Katkov	Free University of Bozen-Bolzano, Italy
Sergio Mauricio Vanegas Arias	LUT University, Finland
Shangchao Su	Fudan University, China
Sheng Xu	Chinese University of Hong Kong, Shenzhen, China
Shenyang Liu	University of Central Florida, USA
Sherjeel Shabih	Humboldt University, Germany
Shi Haoran	China Water Northeastern Investigation, Design & Research Co., Ltd., China
Shingo Murata	Keio University, Japan
Shinnosuke Matsuo	Kyushu University, Japan
Shiyao Zhang	University of Bremen, Germany
Sho Shirasaka	Osaka University, Japan
Simiao Zhuang	TUM Beijing, China
Simon Heilig	Ruhr University Bochum, Germany
Šimon Horvát	Slovakia
Simone Bonechi	University of Siena, Italy
Simone Lionetti	Hochschule Luzern, Switzerland
Siyu Wu	Central South University of Forests and Technology, China
Stefano Damato	IDSIA USI-SUPSI, Switzerland

Stéphane Meystre	MeDiTech/SUPSI, Switzerland
Steve Azzolin	University of Trento, Italy
Sudip Roy	Indian Institute of Technology Roorkee, India
Sujala D. Shetty	BITS Pilani, Dubai Campus, United Arab Emirates
Taoran Fu	Hunan University & Hunan Institute of Engineering, China
Teste Olivier	Université Toulouse 2, IRIT (UMR5505), France
Thierry Viéville	Inria, France
Tianyi Wang	Nanyang Technological University, Singapore
Tim Schlippe	IU International University of Applied Sciences, Germany
Tingyu Lin	TU Wien, Austria
Tuan Le	Pfizer, Germany
Valerie Vaquet	Bielefeld University, Germany
Vangelis Metsis	Texas State University, USA
Vani Kanjirangat	IDSIA USI-SUPSI, Switzerland
Varun Ojha	Newcastle University, UK
Veronica Lachi	Fondazione Bruno Kessler, Italy
Viktor Kocur	Comenius University Bratislava, Slovakia
Vincenzo Palmacci	University of Vienna, Austria
Wei Dai	Robo Space, China
Weiqi Li	Peking University, China
Weiran Chen	Soochow University, China
Wenjie Zhang	Shandong University, China
Wenwei Gu	Chinese University of Hong Kong, China
Wolfram Schenck	Bielefeld University of Applied Sciences and Arts, Germany
Xavier Hinaut	Inria, France
Xi Wang	National University of Defense Technology, China
Xiangxian Li	Shandong University, China
Xiangyuan Peng	Technical University of Munich, Germany
Xiaochen Yuan	Macao Polytechnic University, Macao SAR, China
Xiaomeng Fu	University of Chinese Academy of Sciences, China
Xiaowen Sun	University of Hamburg, Germany
Xiaoxiao Miao	Singapore Institute of Technology, Singapore
Xingda Yao	Zhejiang University of Technology, China
Xinxin Luo	Southeast University, China
XinZhi Lin	Beihang University, China
Xun Lin	Beihang University, China

Yan Jiang	Nanjing University of Information Science and Technology, China
Yang Cao	Shanghai University of Finance and Economics, China
Yangfan Zhou	Southwest University of Science and Technology, China
Yangxun Ou	East China Normal University, China
Yao Du	Beihang University, China
Yaxin Hu	University of Lübeck, Germany
Ye Hu	Pfizer, Germany
Yi Li	Lancaster University, UK
Yichi Zhang	Fudan University, China
Yiming Tang	Shanghai Lixin University of Accounting and Finance, China
Ying Tan	Key Laboratory for Computer Systems of State Ethnic Affairs Commission, Southwest Minzu University, China
Yiqing Shen	Johns Hopkins University, USA
Yixuan Xiao	University of Stuttgart, Germany
Yong Luo	Wuhan University, China
Yongtao Tang	National University of Defense Technology, China
Yuankun Chen	University of Science and Technology, China
Yuansheng Ma	Soochow University, China
Yuchen Guo	Institute of Information Engineering, Chinese Academy of Sciences, China
Yuichi Katori	Future University Hakodate, Japan
Yuji Kawai	Osaka University, Japan
Yusen Wu	Sichuan University, China
Yutaka Nakamura	Riken, Japan
Yuya Okadome	Tokyo University of Science, Japan
Zdravko Marinov	Karlsruhe Institute of Technology, Germany
Zeyao Liu	Key Institute of Information Engineering, Chinese Academy of Sciences, China
Zhang Ke	China University of Petroleum (Beijing), China
Zhenjie Yao	Institute of Microelectronics of the Chinese Academy of Sciences, China
Zheyan Gao	Tianjin University, China
Zhiheng Qiu	City University of Macau, China
Zhihuan Xing	Beihang University, China
Zuzana Berger Haladova	Comenius University Bratislava, Slovakia

Plenary Talks

Past, Present, Future, and Far Future of Machine Learning

Jürgen Schmidhuber

IDSIA USI-SUPSI, Switzerland, and KAUST AI Initiative, Saudi Arabia

I'll discuss modern Artificial Intelligence and how the principles of the G, P and T in Chat GPT emerged in 1991. I'll also discuss what's next in AI, and its expected impact on the future of the universe.

Dendritic Computations and Deep Learning in the Brain

Walter Senn

University of Bern, Institut für Physiologie, Computational Neuroscience Lab, Switzerland

Artificial Intelligence, through its working horse of neural networks, is inspired by the biological example of the brain. The unprecedented success of AI in modeling cognitive processes, in turn, inspires functional models of the brain. Yet, when looking into the brain, additional biological structures become apparent, such as dendritic morphologies, interneuron circuits, recurrent connectivity, error representations, top-down signaling and various gating hierarchies. I will give a review on these biological elements and show how they may integrate in an energy-based theory of cortical computation. Dendrites and cortical microcircuits turn out to implement a real-time version of error-backpropagation based on prospective errors. The theory is inspired by the least-action principle in physics from which all dynamical equations of motions are derived. We likewise derive the neuronal dynamics, including the synaptic dynamics with gradient-descent learning, from our Neuronal Least-Action (NLA) principle. The principle tells that the cortical activities and the real-time learning follows a path that minimizes prospective errors across all neurons of the network. Prospective errors in output neurons relate to behavioral errors, while prospective errors in deep network neurons relate to errors in the neuron-specific dendritic prediction of somatic firing. I will explain how these ideas relate to cortical attention mechanisms and context-dependent gating that link to, and potentially inspires, recent developments in AI.

Biosignal-Adaptive Cognitive Systems

Tanja Schultz

University of Bremen, Fachbereich 3 - Mathematik und Informatik, Cognitive Systems Lab, Germany

I will describe technical cognitive systems that automatically adapt to users' needs by interpreting their biosignals: Human behavior includes physical, mental, and social actions that emit a range of biosignals which can be captured by a variety of sensors. The processing and interpretation of such biosignals provides an inside perspective on human physical and mental activities, complementing the traditional approach of merely observing human behavior. As great strides have been made in recent years in integrating sensor technologies into ubiquitous devices and in machine learning methods for processing and learning from data, I argue that the time has come to harness the full spectrum of biosignals to understand user needs. I will present illustrative cases ranging from silent and imagined speech interfaces that convert myographic and neural signals directly into audible speech, to interpretation of human attention and decision making in human-robot interaction from multimodal biosignals.

A Model of Neocortical Micro- and Mesocircuitry and Its Applications

Michael Reimann

Blue Brain, Swiss Federal Institute of Technology Lausanne, Switzerland

We present a large-scale, biophysically detailed model of rat non-barrel somatosensory regions. Building upon an earlier version of such a model, we increased the spatial scale of the model and enhanced its biological realism. The most salient improvements are: First, construction of realistic synaptic connectivity as the union of two algorithms, one for local connections, and another for long-range connections. Second, introduction of methods to build a model inside a standardized voxel atlas. This, combined with the connectivity algorithms allows models of brain regions to be developed separately and then easily integrated. Third, improvements in the methods to compensate for missing extrinsic inputs and to validate an in-vivo-like activity regime.

We demonstrate several applications of the model that make use of its specific advantages over more simplified models: First, studying the rules of synaptic plasticity at the population level. Second, studying the effect of heterogeneous and non-random connectivity on circuit function and reliability. Third, studying the accuracy and biases inherent in spike sorting algorithms.

Contents – Part IX

Human-Computer Interfaces

Combining Contrastive Learning and Sequence Learning for Automated Essay Scoring 3
   XiaoYi Wang, Jie Liu, Jianshe Zhou, and Wang Jiong

PIDM: Personality-Aware Interaction Diffusion Model for Gesture Generation 19
   Takahiro Shibasaki, Yutaka Nakamura, and Yuya Okadome

Prompt Design Using Past Dialogue Summarization for LLMs to Generate the Current Appropriate Dialogue 33
   Yuya Okadome, Akishige Yuguchi, Ryota Fukui, and Yoshio Matsumoto

Recommender Systems

Click-Through Rate Prediction Based on Filtering-Enhanced with Multi-head Attention 45
   Meihan Yao, Shuxi Zhang, Lang lv, Jianxia Chen, Mengyu Lu, Gaohang Jiang, Liang Xiao, and Zhina Song

Enhancing Sequential Recommendation via Aligning Interest Distributions 60
   Yiyuan Zheng, Beibei Li, Beihong Jin, and Rui Zhao

LGCRS: LLM-Guided Representation-Enhancing for Conversational Recommender System 74
   Ruobing Wang, Xin He, Hengrui Gu, and Xin Wang

Multi-intent Aware Contrastive Learning for Sequential Recommendation 89
   Junshu Huang, Zi Long, Xianghua Fu, and Yin Chen

Subgraph Collaborative Graph Contrastive Learning for Recommendation 105
   Jie Ma, Jiwei Qin, Peichen Ji, Zhibin Yang, Donghao Zhang, and Chaoqun Liu

Time-Aware Squeeze-Excitation Transformer for Sequential Recommendation 121
   Hongwei Chen, Luanxuan Liu, Zexi Chen, and Xia Li

Environment and Climate

Carbon Price Forecasting with LLM-Based Refinement
and Transfer-Learning 139
  Haiqi Jiang, Ying Ding, Rui Chen, and Chenyou Fan

Challenges, Methods, Data–A Survey of Machine Learning in Water
Distribution Networks 155
  Valerie Vaquet, Fabian Hinder, André Artelt, Inaam Ashraf,
  Janine Strotherm, Jonas Vaquet, Johannes Brinkrolf,
  and Barbara Hammer

Day-Ahead Scenario Analysis of Wind Power Based on ICGAN
and IDTW-Kmedoids 171
  Yun Wu, Wenhan Zhao, Yongbin Zhao, Jieming Yang, Diwen Liu,
  Ning An, and Yifan Huang

Enhancing Weather Predictions: Super-Resolution via Deep Diffusion
Models 186
  Jan-Matyáš Martinů and Petr Šimánek

Hybrid CNN-MLP for Wastewater Quality Estimation 198
  Marco Cardia, Stefano Chessa, Alessio Micheli,
  Antonella Giuliana Luminare, and Francesca Gambineri

Short-Term Forecasting of Wind Power Using CEEMDAN-ICOA-GRU
Model 213
  Yun Wu, Wei Zheng, Yongbin Zhao, Jieming Yang, Ning An, and Dan Feng

City Planning

Predicting City Origin-Destination Flow with Generative Pre-training 233
  Mingwei Zhang, Lizhong Gao, Qiao Wang, and Weihao Gao

Vehicle-Based Evolutionary Travel Time Estimation with Deep Meta
Learning 246
  Chenxing Wang, Fang Zhao, Haiyong Luo, Yuchen Fang,
  Haichao Zhang, and Haoyu Xiong

Machine Learning in Engineering and Industry

APF-DQN: Adaptive Objective Pathfinding via Improved Deep
Reinforcement Learning Among Building Fire Hazard 265
  Ke Zhang, Dandan Zhu, Qiuhan Xu, Hao Zhou, and Xuemei Peng

DDPM-MoCo: Enhancing the Generation and Detection of Industrial
Surface Defects Through Generative and Contrastive Learning 280
   Xiaozong Yang, Huailiang Tan, and Xinyan Wang

Detecting Railway Track Irregularities Using Conformal Prediction 295
   Andreas Plesner, Allan P. Engsig-Karup, and Hans True

Identifying the Trends of Technological Convergence Between Domains
Using a Heterogeneous Graph Perspective: A Case Study of the Graphene
Industry 310
   Shan Jiang, Yuan Meng, and Danni Zhou

Machine Learning Accelerated Prediction of 3D Granular Flows in Hoppers ... 325
   Duy Le, Linh Nguyen, Truong Phung, David Howard,
   Gayan Kahandawa, Manzur Murshed, and Gary W. Delaney

RD-Crack: A Study of Concrete Crack Detection Guided by a Residual
Neural Network Improved Based on Diffusion Modeling 340
   Yubo Huang, Xin Lai, Zixi Wang, Muyang Ye, Yinmian Li, Yi Li,
   Fang Zhang, and Chenyang Luo

Applications in Finance

Anomaly Detection in Blockchain Using Multi-source Embedding
and Attention Mechanism 357
   Ao Xiong, Chenbin Qiao, Baozhen Qi, and Chengling Jiang

Beyond Gut Feel: Using Time Series Transformers to Find Investment Gems ... 373
   Lele Cao, Gustaf Halvardsson, Andrew McCornack,
   Vilhelm von Ehrenheim, and Pawel Herman

MSIF: Multi-source Information Fusion for Financial Question Answering 389
   Man Lin, Delong Zeng, Jiarui Ouyang, and Ying Shen

Artificial Intelligence in Education

A Temporal-Enhanced Model for Knowledge Tracing 407
   Shaoguo Cui, Mingyang Wang, and Song Xu

Social Network Analysis

Position and Type Aware Anchor Link Prediction Across Social Networks 425
   Dongwei Zhu, Yongxiu Xu, Hongbo Xu, Hao Xu, Qi Wang,
   and Wenhao Zhu

Artificial Intelligence and Music

LSTM-MorA: Melody-Accompaniment Classification of MIDI Tracks 443
  Hui Liu, Leon Flaack, Shiyao Zhang, and Tanja Schultz

Software Security

Ch4os: Discretized Generative Adversarial Network
for Functionality-Preserving Evasive Modification on Malware 461
  Christopher Molloy, Furkan Alaca, and Steven H. H. Ding

SSA-GAT: Graph-Based Self-supervised Learning for Network Intrusion
Detection 476
  Qian Liu, Hui Zhang, Youpeng Zhang, Lin Fan, and Xue Jin

Author Index 493

Human-Computer Interfaces

Combining Contrastive Learning and Sequence Learning for Automated Essay Scoring

XiaoYi Wang[1], Jie Liu[1,2], Jianshe Zhou[1(✉)], and Wang Jiong[1]

[1] Capital Normal University, Beijing, China
1441425098@qq.com
[2] North China University of Technology, Beijing, China

Abstract. The objective of automated essay scoring (AES) is to employ artificial intelligence techniques to automate the scoring process and minimize the impact of subjective factors on grading. Previous works tend to treat it solely as a regression or classification task, without considering the integration of both. Additionally, neural networks trained on limited samples often exhibit poor performance in capturing the deep semantics of texts. To enhance the performance of AES, we propose a novel approach that combines contrastive learning with sequence learning, effectively integrating regression loss and classification loss. This paper employs a variety of data augmentation techniques to construct negative samples suitable for contrastive learning, aiming to alleviate the inherent sample imbalance issue in essay datasets. Additionally, we propose to utilize sequence learning for essay scoring, incorporating empirical distribution based on the general distribution characteristics of the essay dataset to address the issue of unbalanced prediction results caused by sample imbalance. Experimental results demonstrate that the proposed multi-task learning framework outperforms the single-task learning framework in enhancing the effectiveness of automatic essay scoring.

Keywords: Automated essay scoring · Data augmentation · Contrastive learning

1 Introduction

Automated essay scoring technology involves the prediction of scores by computers after providing a certain amount of samples. An effective AES system can save teachers a significant amount of teaching costs and provide timely feedback on the quality of student essays.

Existing AES models can be broadly categorized into two training frameworks. One is the single-task learning framework, where the scoring model is trained to predict scores for specific features of essays, such as overall features, coherence scores, and thematic consistency scores [3,10,16,17,22]. Another is the multi-task learning framework, where the overall score and feature scores of

---

X. Wang and J. Liu—Contributed equally.

essays are combined for prediction simultaneously [7,8,13]. For the latter, the model predicts feature scores while the learned feature distribution contributes to predicting overall scores, thus providing greater interpretability from the scoring perspective. However, existing multi-task learning frameworks have high annotation requirements for training data [12], which not only involve annotating overall scores but also predicting feature scores, leading to increased manpower and resource costs. To address these issues, this paper proposes a novel multi-task training framework for evaluating the overall scores of essays. This framework does not require the annotation of feature scores but instead combines contrastive learning with sequence learning tasks to enhance the robustness of the scoring model.

Constructing appropriate positive and negative examples for contrastive learning is challenging. Obtaining essay samples suitable for contrastive learning is difficult due to the uncertainty of sample quality and the significant manpower and time costs involved. To address the problem of sample size in training datasets, data augmentation techniques have been proposed. However, when these data augmentation methods are applied to essays, they may produce many results that do not align with reality. If dirty data generated by these methods is mixed into the dataset, it not only fails to improve the predictive accuracy of the model through expanding data samples but may also decrease model accuracy. Therefore, the goal of automatic essay scoring is to better supplement data samples while improving model performance.

To achieve this objective, we have addressed three key questions:

(1) Adaptability of Essay Data and Data Augmentation Methods.
    Whether it is Chinese or English essays, ensuring a fundamental rule for data augmentation is essential: the generated results should constitute a coherent essay. Many scholars have not been stringent in their selection of data augmentation methods, which is evidently flawed. For example, employing methods like back translation, where a Chinese essay is first translated into a second foreign language and then back to Chinese, although such augmentation methods may enhance the model's accuracy in random learning instances, they lack interpretability. Moreover, there's no evidence to suggest that such methods can be generalized across different topics or genres and remain effective.
(2) Lack of Prominent Essay Scoring Features.
    Currently, much of the research on essay scoring focuses on combining models, adjusting losses, and updating encoding methods. However, the overall score of an essay often changes due to specific features that are easily overlooked during the scoring process. For a computer, minor changes in features may not significantly alter the overall score, but in real evaluation, they accurately decrease the score. This situation should be observed by the model, yet existing methods are not sensitive to it.
(3) Task Division in Automatic Essay Scoring.
    Existing research divides essay scoring into two tasks: linear regression, where the model is expected to directly provide an accurate score, and

sequence regression, similar to multi-task classification, where the model treats each score as a separate class and ultimately outputs the essay category. However, both tasks have their shortcomings. Linear regression fails to account for the holistic nature of essay scoring, which is not simply a sum of individual scores but a comprehensive assessment. On the other hand, in multi-class classification, when score ranges are large, small differences between categories can lead to significant errors.

Our main contributions can be summarized as follows.

(1) We propose multiple data augmentation schemes and validated their effectiveness through extensive experiments.
(2) We introduce the use of contrastive learning methods by constructing positive-negative-original sample pairs from both original and augmented data.
(3) We transform automatic essay scoring from a multi-classification task into a sequence regression task, and we design corresponding loss functions for this sequence regression problem to enhance its effectiveness in the scoring process.

2 Related Work

2.1 Automatic Essay Scoring Technology

The use of computer-assisted essay scoring systems has a long history of research, with each system employing different criteria for score evaluation. Traditional AES often relies on regression or ranking systems with complex handcrafted features to assess essay scores. Larkey [9] employed manually crafted features based on linguists' prior knowledge, yielding good performance even with limited data but exhibiting poor portability. Ridley et al. [15] predicted trait scores of essays across different domains using CNN and attention mechanisms. Marek Rei [1] proposed using LSTM to represent textual meaning and learning the contribution of specific words to text scores for automated text scoring. Masaki Uto [17] introduced a model combining deep neural network AES with manually crafted document-level features. Yang et al. [24] applied a method combining ranking loss with regression loss to the BERT model. Wang et al. [19] employed multiple BERT models for multi-faceted essay scoring, combining document feature scores, vocabulary feature scores, and segment feature scores to obtain the final essay prediction score. Li et al. [10] proposed an AES method based on multi-scale features, which utilized Sentence-BERT (SBERT) to vectorize sentences and connect them to the DNN-AES model.

2.2 Contrastive Learning Technology

For contrastive learning, there are generally two types of methods for constructing positive and negative samples. One type generates new samples through data augmentation methods, with the new samples serving as positive or negative samples paired with original samples. In the SimCSE model proposed by Gao et al. [4], it was found that two rounds of drop-out as a method of positive sample enhancement. Yan et al. [23] proposed to combine four different data augmentation strategies to complete contrastive learning. The second type selects positive and negative samples from the original data, without concerns about the reliability of the data. Giorgi et al. [5] proposed selecting distant sentences as negative samples because the semantics of sentences farther apart in the same article are more difficult to match. Unlike the typical paradigm of constructing sample pairs through data augmentation, Xie et al. [21] treated every two essays in the entire dataset as sample pairs, globally ranking all essays, effectively avoiding the possibility of incorporating dirty data and ensuring more comprehensive construction of sample pairs.

Therefore, this paper employs multiple data augmentation methods that align with essay writing habits. By utilizing contrastive learning techniques, the model can capture specific scoring features. Additionally, a combination of multiple losses is designed to correct results, thereby enhancing the accuracy of the model in essay scoring.

3 Method

This section presents an overview of the proposed automatic essay scoring method. Figure 1 depicts the overall framework, and this section will outline the entire process and the details of the model.

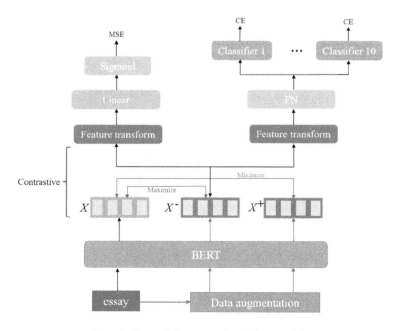

Fig. 1. Overall framework of the model

3.1 Overall Architecture

Figure 1 illustrates the complete framework model. The positive and negative sample construction module employs multiple construction strategies and incorporates a wealth of external knowledge to enrich data augmentation. The scoring module consists of two parts: one is the regression task scoring module, and the other is the sequence regression task scoring module tailored to the characteristics of essay data. Lastly, contrastive learning involves comparing the positive and negative samples constructed by data augmentation with the original samples, thereby enabling the model to better capture essay features.

3.2 Positive and Negative Sample Construction

For essay data, considering the original essay as the anchor sample, the corresponding positive samples are essays with the same score, while the negative samples are essays with scores different from the anchor sample.

Positive Sample Data Augmentation. To construct positive sample data, we design three augmentation methods. Firstly, we utilize the large-scale pre-trained language model (PLM) Roberta [11] for data augmentation. We randomly select K words, generate their vector representations using the PLM, and find the word vectors with the smallest distance to these representations by searching for similar words. These words are then swapped with the original ones, and the modified text is converted back to its textual form. Secondly, we rely on external knowledge sources for text augmentation. Similarly, we randomly select K words and replace them with synonyms from WordNet. Additionally, we construct an abbreviation-expansion table [14] applicable to English sentences. Expressions such as "she's" are expanded to random variations like "she is" or "she has", or "she is" is abbreviated as "she's". Since these modifications do not alter the semantic meaning of the text and do not affect the sentence structure, they are suitable for creating positive samples. Thirdly, we modify the original samples based on model mechanisms to achieve data augmentation. Inspired by the unsupervised learning approach in SimCSE [4], we leverage the random dropout process during text serialization. This stochastic process ensures that even when the same sentence is input into the BERT model, it generates two different text vectors. By harnessing this property, we can create diverse positive samples, enriching our dataset and improving the model's generalization ability.

Negative Sample Data Augmentation. The construction of negative samples is mainly achieved from two perspectives. Firstly, word-level replacement is conducted based on the common habits of handwritten English word errors [18]. This involves randomly replacing words based on a compiled list of error words. Since some English words have very similar spellings, and it is easy to make letter writing errors or omissions while handwriting, similar words are summarized

Table 1. Examples of partially incorrect word expressions

Example	Replace
swim	sw**in**
staff	st**uff**
Row	R**aw**
expl**ain**	explan
conf**ir**med	conf**or**med

and replaced. The error word list contains a total of 13,551 words, with some examples shown in Table 1.

The second method involves character perturbation. It involves randomly shuffling, inserting, or deleting characters within words [20]. The usage is illustrated in Table 2. The key point of these two construction methods lies in whether other words with different meanings are used to replace the current word, rather than just altering characters.

Table 2. Examples of partially incorrect word expressions

Method	Example	Replace
Insert	Clear	Clea**u**r
Delete	Problem	Probem
Swap	Victory	Vic**ot**ry
Substitution	M**ot**ivation	M**or**ivation

3.3 Contrastive Learning

Due to the necessity of applying contrastive learning to both scoring subtasks simultaneously, we design three different application methods. First, we apply a nonlinear transformation to the text vectors generated by the encoder to obtain new text features, decoupling them from the features in the scoring task, and then calculate the distances between the transformed text features. Second, we simultaneously use contrastive learning in both the regression task and the sequence regression task, allowing each subtask to correct itself independently rather than as a whole. Additionally, we directly apply the feature vectors obtained from the BERT model to the generated text feature vectors. However, since both subtasks undergo nonlinear transformations to obtain text score features, indirect contrastive learning may not necessarily correct the text vectors to perform better in both scoring subtasks. Treating the contrastive learning portion as a third subtask alongside the other two subtasks is theoretically infeasible for overall scoring. Hence, it is more appropriate to directly compute contrastive learning on the feature vectors, as this approach can be applied to both subtasks simultaneously and is not affected by nonlinear transformations. After experimental validation, we found that directly computing contrastive learning on the feature vectors yields better results. Therefore, we choose to compute contrastive learning on the feature vectors obtained from the BERT model.

The essay's original data and data augmented data are passed through an encoder to generate essay feature vectors, which are then subjected to contrastive learning. We use KL divergence to calculate the distance between vectors:

$$KL(P \parallel Q) = \sum P(x) log \frac{P(x)}{Q(x)} \quad (1)$$

where P and Q are two probability distributions, with P representing the true distribution and Q representing the predicted distribution. KL divergence is used to correct the predicted distribution with reference to the true distribution during the final correction. Once the distances between samples are obtained, the focus of contrastive learning is to minimize the distance between anchor samples and positive samples, while maximizing the distance between anchor samples and negative samples to achieve the goal of contrastive learning. The formula for the final objective loss calculation is shown in Eq. (2):

$$CL = \frac{KL(X, X^+)}{KL(X, X^-)} \quad (2)$$

As the objective loss needs to be minimized, we aim to reduce the distance between the anchor sample and positive sample while increasing the distance between the anchor sample and negative sample. This approach ensures that the distance between positive and negative samples in the vector space becomes increasingly distant, while the distance between positive samples becomes closer. This process ultimately leads to the desired outcome.

3.4 Scoring Module

The scoring module mainly consists of the regression task and the sequence regression task. The model diagram is shown in Fig. 2.

The upper part of the model corresponds to the regression task, while the lower part represents the sequence regression task. We utilize the BERT model to encode the essay data into vectors, obtaining the text features of the essay, which are then separately transmitted to the two scoring tasks to obtain the final output features.

Regression Task Scoring. Essay scores are not simple categorical labels but rather a comprehensive consideration of multiple features. In actual teacher grading, there is a tendency toward certain scores, typically within a range of one point. If there is a greater tendency toward positive scoring features, the integer part of the score will be increased, otherwise, it will remain unchanged. In

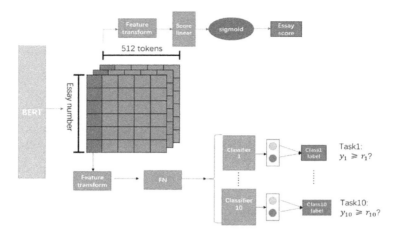

Fig. 2. Overall scoring module

classification tasks, only probability predictions for different score categories are made, making it difficult to reflect scoring tendencies. Therefore, we design the automated essay scoring task primarily as a regression task. The result is passed to a linear layer to obtain the final score:

$$f_1 = Tanh\left(W_1 X_{CLS} + b_1\right) \tag{3}$$

where X_{CLS} represents the [CLS] feature vector obtained by slicing the output of BERT. W_1 represents the weights, and b_1 is the bias. Nonlinear transformation is performed using the $Tanh$ activation function, and some features are discarded through dropout. This process results in separated score features.

$$f_2 = Sigmoid\left(W_2 f_1 + b_2\right) \tag{4}$$

Due to the normalization of the scores, normalization is also applied during the prediction stage. W_2 represents the weights of the second linear layer, and b_2 is the bias. During the training stage, the Mean Squared Error (MSE) is computed as the loss function:

$$Loss_{MSE} = \frac{1}{n}\sum_{i=1}^{n}(Y_i - \widehat{Y_i})^2 \tag{5}$$

where Y_i represents the predicted score of the i-th sample, and $\widehat{Y_i}$ represents the true score of the i-th sample.

Sequence Regression Task for Scoring. Consider the realistic scoring scenarios, teachers grade essays based on specific scoring criteria when assigning scores. Therefore, each score given to an essay is justified by these criteria. Consequently, we propose to view the process of scoring based on these criteria as a sequential regression scoring process. When grading an essay, we start from the lowest score and evaluate it based on a series of classifiers. If the essay meets the criteria for a particular score, it is judged as true, the score is incremented by one, and the essay proceeds to the next classifier for further evaluation. If it does not meet the criteria, it is judged as false, and the score from the previous classifier is output as the final result. Since scoring points are cumulative, if the features represented by the scoring points satisfy the criteria for the current score, it is judged as true, and the essay is accordingly graded. However, the imbalance in the essay dataset can affect the sequential regression process. Therefore, we employ the sequential regression task as an auxiliary task to correct scoring biases, enabling the regression task to accurately assign scores.

Initially, the features are processed through a fully connected network (FN). The output [CLS] token is then mapped to another text vector using another FN to decouple the features used by the two tasks, before being fed into the FN.

$$f_c = FN(X_{CLS}) \tag{6}$$

$$f_N = ReLU\left(W_n f_C + b_n\right) \tag{7}$$

Finally, a combination of multiple linear and nonlinear layers is used to obtain a feature vector slightly larger than the score range.

$$f_{classifier} = (W_C f_N + b_C) \tag{8}$$

To accomplish sequence regression scoring, we design multiple classifiers corresponding to each score. The extracted features are transformed into two features representing whether they belong to the current score value using linear layers.

$$SR = \sum_{i=1}^{n} CE = \sum_{i=1}^{n} -\left[y_i log p + (1-y_i) log (1-p)\right] \tag{9}$$

where y_i represents the true value of the i-th sample, the target loss results of each classification are summed up to form the final sequence regression loss. Due to the characteristics of essay scores, such a sequence regression method is inevitably imbalanced. Specifically, scores tend to concentrate at both ends, leading to inaccurate results. Therefore, we design a balance mechanism to serve as an auxiliary correction for the regression task scores, forming the second task. Each classification task for each score is treated as a sub-task, combined into a complete task, and the sum of the target losses should overall be 1. Weight distributions are assigned to each sub-task's loss based on the distribution of the training set data samples, and we use an empirical distribution to correct the target loss.

$$\lambda_t = \frac{\sqrt{N_k}}{\sum_{k=1}^{K} \sqrt{N_k}} \tag{10}$$

where N represents the number of samples, and k represents the rank. Originally, the empirical distribution was simply based on scores. Here, adding the square root to the numerator and denominator ensures that categories with fewer samples are not overly penalized, thus balancing the impact of samples on classification more effectively. To better influence the target loss, we multiply each λ of the empirical distribution as a weight with the cross-entropy loss results for each score, the final multi-task scoring part is completed.

4 Experimental Result

4.1 Dataset

We utilize the ASAP dataset from Kaggle competition [6], which comprises English essays written by students of various ages, covering multiple topics and writing styles. To explore the performance of automatic scoring in capturing scoring features, we select the first subset of scores with scientifically reasonable score ranges and divisions as our dataset. The dataset consists of 1783 essays, divided into training, validation, and test sets in an 8:1:1 ratio.

4.2 Settings

The training batch size was set to 3, including one positive sample, one negative sample, and one original sample, to facilitate contrastive learning within the same batch. To match the batch size of the test set, gradient accumulation was performed, resulting in an effective batch size of approximately 15. The AdamW optimizer was employed with a learning rate of 1e-5.

4.3 Compared Models

To demonstrate the effectiveness of adding contrastive learning and sequence regression, we select the following models for comparison:

CNN-LSTM-ATT [2]: It adopts hierarchical modeling of word-sentence-document. It obtains the representation of sentences by convolving words and then uses LSTM to obtain the representation of the text at the sentence level. Attention pooling is applied between levels instead of simply averaging the results.

R^2BERT [24]: It combines regression loss and ranking loss to generate the final scores from the features extracted by BERT's [CLS] tokens through a Dense layer. It is trained using the combined loss of regression and ranking.

BERT: The original BERT model differs from this study in that it did not undergo data augmentation, resulting in differences in the data samples.

4.4 Evaluation Indicators

We choose Quadratic Weighted Kappa (QWK) to evaluate the models. Its scoring range is [0, 1], where a value closer to 1 indicates better results. The calculation formula for QWK is as follows:

$$QWK = 1 - \frac{\sum_{i,j} w_{i,j} O_{i,j}}{\sum_{i,j} w_{i,j} E_{i,j}} \quad (11)$$

where $O_{i,j}$ represents the count of instances where the i-th category is classified as the j-th category, $E_{i,j}$ represents the expected agreement based on the true contingency table, and $w_{i,j}$ denotes the weight term. Higher QWK values indicate that predicted scores closely align with the actual scores in essay grading, reflecting the accuracy of the model's predictions.

4.5 Results of Different Data Augmentation Methods

Since the data augmentation in the experiment includes four types of positive sample and two types of negative sample augmentation methods, testing was conducted to determine which combination of samples could better improve the results. Initially, experiments were conducted on the results of single-sample augmentation, as shown in Table 3.

Table 3. Data augmentation results

Negative	Positive			
	Two dropouts	Word Embedding	Wordnet	Checklist
CharSwap	0.750	0.721	0.724	0.684
SpellingError	0.647	0.724	0.660	0.657

The enhancement method with two dropouts has the least impact on semantics, indicating that using it as a combination for positive sample enhancement can achieve better results. The results of the Checklist are not satisfactory, suggesting potential inaccuracies in the abbreviation and expansion items. The other two positive sample enhancement methods both involve the introduction of external knowledge and word replacement. However, complete synonym replacement is not always feasible for the same word, for instance, replacing "right" with "correct". In essays, inappropriate word replacements can lead to inaccuracies, hence these two methods may not yield desirable outcomes. Regarding negative sample enhancement methods, directly replacing with misspelled words yields better results than replacing with other words, indicating that the model can recognize the deduction for spelling errors and make correct score judgments accordingly.

Theoretically, the best negative sample strategy among the four positive sample strategies is SpellingError, while among the negative sample enhancement strategies, the combination of CharSwap and Wordnet demonstrates the best theoretical effectiveness, followed by two dropouts. As different strategies yield varying degrees of changes in the model, we adopt a complementary approach. After combining the positive and negative sample enhancement methods, the predicted results are compared again with those of single positive sample enhancement, with most results showing improvements compared to before combining.

4.6 Essay Scoring Experiment Results

We conduct experiments with other models under the same conditions, and the results are presented in Table 4.

Table 4. Overall scoring results

Models	QWK
CNN-LSTM-ATT	0.755
R^2BERT	0.760
BERT	0.678
Ours	0.759

In the current scenario of sufficient training samples, we expand the size of the essay dataset through data augmentation. Positive samples were included in the dataset with the same scores, while scarce low-score essays were supplemented with additional data. As shown in Table 4, our method showed improvement for some models, with performance comparable to the best-performing R^2BERT. This indicates that a larger training dataset generally leads to better performance, but introducing noisy data may degrade model performance.

4.7 Ablation Study

We conduct experiments after removing different components. Specifically, we tested four scenarios: without contrastive learning (w/o Contrastive learning), without sequence regression tasks (w/o Sequence Regression), without sequential regression with experience (w/o Sequential regression with experience), and a scenario where everything except regression tasks was removed. The results are presented in Table 5.

Table 5. Results of ablation experiments

Models	QWK
Ours	0.759
Regression-only	0.727
w/o Sequence Regression	0.707
w/o Contrastive learning	0.728
w/o Sequential regression with experience	0.749

The objective of the empirical distribution is to correct the final results that may be biased towards either extreme due to imbalances in the training samples. To investigate the impact of the experience distribution on scoring, apart

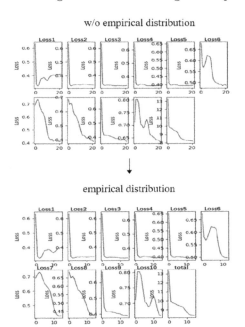

Fig. 3. Empirical distribution before and after use-loss comparison

from calculating the overall QWK, we also examined the loss in the sequence regression component. The loss visualization is presented in Fig. 3. The performance of the model decreased to varying degrees when different modules were removed, indicating that each module is indispensable. When only data augmentation methods were used to alter the input essay texts, the performance improved by 4.9% points compared to the original BERT model, demonstrating that the enhancement of positive and negative samples is beneficial for scoring on the current dataset. The removal of sequence regression tasks led to the most significant performance decrease, highlighting the necessity of integrating classification and regression tasks through sequential regression for essay scoring. After removing contrastive learning, the model performance decreased by 4.8% points, indicating the crucial importance of introducing contrastive learning. Finally, removing the sequence regression task with experience distribution reduced the model's performance by 1.1% points, suggesting that this task plays a certain role in refining the regression task.

5 Conclusion

This paper introduces an automatic essay scoring method that integrates contrastive learning with sequence learning. To construct negative samples suitable for contrastive learning, we employ diverse data augmentation techniques. These techniques alleviate the inherent sample imbalance in essay datasets, enhancing the model's feature capture capabilities through the interaction between augmented and original samples. Additionally, we propose utilizing sequence learning for essay scoring, effectively integrating classification and regression tasks. To address the issue of unbalanced prediction results caused by sample imbalance, we incorporate empirical distribution based on the general distribution characteristics of the essay dataset. Experimental results demonstrate that the multi-task learning framework outperforms the single-task learning framework in enhancing automatic essay scoring, providing a foundation for future research advancements. In the future, we plan to explore the design of high-quality negative samples for essay data, as well as further improvements in contrastive loss.

Acknowledgments. This work is supported partially by National Key Research and Development Program of China (2020AAA0109703), National Natural Science Foundation of China (62076167, U23B2029), and the general project of the 14th Five-Year Scientific Research Plan of the National language commission (YB145-16).

References

1. Alikaniotis, D., Yannakoudakis, H., Rei, M.: Automatic text scoring using neural networks. In: Proceedings of the 54th Annual Meeting of the Association for Computational Linguistics (Volume 1: Long Papers), pp. 715–725 (2016)
2. Dong, F., Zhang, Y., Yang, J.: Attention-based recurrent convolutional neural network for automatic essay scoring. In: Proceedings of the 21st conference on computational natural language learning (CoNLL 2017), pp. 153–162 (2017)
3. Farag, Y., Yannakoudakis, H., Briscoe, T.: Neural automated essay scoring and coherence modeling for adversarially crafted input. In: Proceedings of the 2018 Conference of the North American Chapter of the Association for Computational Linguistics: Human Language Technologies, Volume 1 (Long Papers), pp. 263–271 (2018)
4. Gao, T., Yao, X., Chen, D.: SimCSE: simple contrastive learning of sentence embeddings. In: Proceedings of the 2021 Conference on Empirical Methods in Natural Language Processing, pp. 6894–6910 (2021)
5. Giorgi, J., Nitski, O., Wang, B., Bader, G.: DeCLUTR: deep contrastive learning for unsupervised textual representations. In: Proceedings of the 59th Annual Meeting of the Association for Computational Linguistics and the 11th International Joint Conference on Natural Language Processing (Volume 1: Long Papers), pp. 879–895 (2021)
6. Hamner, B., Morgan, J., Lynnvandev, M., Ark, T.: The hewlett foundation: Automated essay scoring. Kaggle (2012)
7. Hussein, M.A., Hassan, H.A., Nassef, M.: A trait-based deep learning automated essay scoring system with adaptive feedback. Int. J. Adv. Comput. Sci. Appl. **11**(5) (2020)

8. Kumar, R., Mathias, S., Saha, S., Bhattacharyya, P.: Many hands make light work: using essay traits to automatically score essays. In: Proceedings of the 2022 Conference of the North American Chapter of the Association for Computational Linguistics: Human Language Technologies, pp. 1485–1495 (2022)
9. Larkey, L.S.: Automatic essay grading using text categorization techniques. In: Proceedings of the 21st Annual International ACM SIGIR Conference on Research and Development in Information retrieval, pp. 90–95 (1998)
10. Li, F., Xi, X., Cui, Z., Li, D., Zeng, W.: Automatic essay scoring method based on multi-scale features. Appl. Sci. **13**(11), 6775 (2023)
11. Liu, Y., et al.: Roberta: A robustly optimized BERT pretraining approach (2019). arXiv preprint arXiv:1907.11692
12. Mathias, S., Bhattacharyya, P.: ASAP++: enriching the ASAP automated essay grading dataset with essay attribute scores. In: Calzolari, N. (eds.) Proceedings of the Eleventh International Conference on Language Resources and Evaluation (LREC 2018). European Language Resources Association (ELRA), Miyazaki, Japan (2018). https://aclanthology.org/L18-1187
13. Mathias, S., Bhattacharyya, P.: Can neural networks automatically score essay traits? In: Proceedings of the Fifteenth Workshop on Innovative Use of NLP for Building Educational Applications, pp. 85–91 (2020)
14. Morris, J., Lifland, E., Yoo, J.Y., Grigsby, J., Jin, D., Qi, Y.: TextAttack: a framework for adversarial attacks, data augmentation, and adversarial training in NLP. In: Proceedings of the 2020 Conference on Empirical Methods in Natural Language Processing: System Demonstrations, pp. 119–126 (2020)
15. Ridley, R., He, L., Dai, X.Y., Huang, S., Chen, J.: Automated cross-prompt scoring of essay traits. In: Proceedings of the AAAI Conference on Artificial Intelligence. vol. 35, pp. 13745–13753 (2021)
16. Uto, M., Aomi, I., Tsutsumi, E., Ueno, M.: Integration of prediction scores from various automated essay scoring models using item response theory. IEEE Trans. Learn. Technol. **16**(6), 983–1000 (2023)
17. Uto, M., Xie, Y., Ueno, M.: Neural automated essay scoring incorporating handcrafted features. In: Proceedings of the 28th International Conference on Computational Linguistics, pp. 6077–6088 (2020)
18. Vajjala, S., Majumder, B., Gupta, A., Surana, H.: Practical natural language processing: a comprehensive guide to building real-world NLP systems. O'Reilly Media (2020)
19. Wang, Y., Wang, C., Li, R., Lin, H.: On the use of BERT for automated essay scoring: joint learning of multi-scale essay representation. In: Proceedings of the 2022 Conference of the North American Chapter of the Association for Computational Linguistics: Human Language Technologies, pp. 3416–3425 (2022)
20. Wei, J., Zou, K.: Eda: Easy data augmentation techniques for boosting performance on text classification tasks. In: Proceedings of the 2019 Conference on Empirical Methods in Natural Language Processing and the 9th International Joint Conference on Natural Language Processing (EMNLP-IJCNLP), pp. 6382–6388 (2019)
21. Xie, J., Cai, K., Kong, L., Zhou, J., Qu, W.: Automated essay scoring via pairwise contrastive regression. In: Proceedings of the 29th International Conference on Computational Linguistics, pp. 2724–2733 (2022)
22. Yamaura, M., Fukuda, I., Uto, M.: Neural automated essay scoring considering logical structure. In: Wang, N., Rebolledo-Mendez, G., Matsuda, N., Santos, O.C., Dimitrova, V. (eds.) Artificial Intelligence in Education. AIED 2023. LNCS(), vol. 13916. Springer, Cham (2023). https://doi.org/10.1007/978-3-031-36272-9_22

23. Yan, Y., Li, R., Wang, S., Zhang, F., Wu, W., Xu, W.: ConSERT: a contrastive framework for self-supervised sentence representation transfer. In: Proceedings of the 59th Annual Meeting of the Association for Computational Linguistics and the 11th International Joint Conference on Natural Language Processing (Volume 1: Long Papers), pp. 5065–5075 (2021)
24. Yang, R., Cao, J., Wen, Z., Wu, Y., He, X.: Enhancing automated essay scoring performance via fine-tuning pre-trained language models with combination of regression and ranking. In: Findings of the Association for Computational Linguistics: EMNLP 2020, pp. 1560–1569 (2020)

PIDM: Personality-Aware Interaction Diffusion Model for Gesture Generation

Takahiro Shibasaki[1(✉)], Yutaka Nakamura[2], and Yuya Okadome[1,2]

[1] Department of Information and Computer Technology, Tokyo University of Science, Tokyo, Japan
reserve.shiba@gmail.com, okadome@rs.tus.ac.jp
[2] Information R&D and Strategy Headquarters, RIKEN, Kyoto, Japan

Abstract. In a dyadic conversation, the behaviors of one participant, such as nodding and smiling, are influenced by those of the conversation partner. The velocity and magnitude of motion during conversation are also affected by the personality traits of each participant. In this paper, we propose the personality-aware interaction diffusion model (PIDM) for a dyadic conversation. PIDM generates interaction behaviors based on the masking features of all participants and participants' personalities. We apply PIDM to the motion generation during a dyadic conversation, and the differences in the generated results according to the personality are investigated. The results suggested that PIDM can change the distribution of generated behaviors by adjusting the extraversion which is the one parameter of the Big Five.

Keywords: Diffusion model · Human-Human Interaction · Personality

1 Introduction

Motion generation for an interaction situation is essential in the development of "natural" communication agents [3]. Behaviors of attendances of conversation affect each other, and it is necessary to simultaneously consider all participant's behavior for the communication agent. In a dyadic conversation situation, behaviors of both participants tend to synchronize [2], *i.e.*, nodding and smiling occur simultaneously. It is necessary to consider both participants' behaviors at the same time.

In addition to the synchronization between participants, the physical properties of gestures, such as velocity and magnitude, are also affected by personality traits [14,17]. Lippa et al. [14] assess body movements and personality and demonstrate that extroverts' gestures are faster. Nakano et al. [17] show that gesture patterns differ according to personality traits. Expressed gestures which depend on one's temperament are observed and interpreted by other participants during the conversation. In the human-agent communication, we consider how behaviors of communication agents are objectively perceived.

In this paper, we propose a personality-aware generation model of interaction based on the diffusion model to generate interaction gestures for a dyadic conversation as an extension of the interaction diffusion model (IDM) [21] to handle the personality. IDM generates behaviors during conversation based on the diffusion model, which is a deep generative model. While IDM only considers another participant's behavior, the proposed model, PIDM, is further conditioned by the personality traits for motion generation. We assume the distribution of possible behaviors is structured by the personality traits, and thus the personality controls the generation result of PIDM. In our approach, personality traits, especially the Big Five personality traits [5], are evaluated by a third party and used as one of the input features. PIDM reflects another participant's behaviors and the participants' personalities.

The Big Five measures the variables that constitute personality. This approach captures personality in five traits: Extraversion - Agreeableness - Conscientiousness - Neuroticism - Openness. Naumann et al. [18] demonstrate that subjective and objective evaluations of personality tend to be consistent. Due to the difficulty of using subjective personality in real situations, this insight suggests that the estimated personality is useful as the input feature for the actual conversation task.

PIDM is applied to the behavior generation task of a dyadic conversation. Differences in the tendency of the generated behaviors according to the personality are evaluated. In this experiment, the extraversion which is the one parameter of the Big Five is changed. The experimental results suggest that PIDM adjusts behavior generation during conversation according to the personality parameter.

2 Related Work

In this section, we describe the relationship between personality and human behaviors, and human motion generation based on generative models. Several studies investigate the effect of the Big Five traits on gesture [1,11,24]. Breil et al. [1] show that gestures serve as one of the cues to evaluate the Big Five personality traits. Particularly, Smith et al. [24] found that fast and energetic gestures are positively correlated with extraversion. Ishii et al. [11] investigated that participants' gestures differ based on their Big Five traits during face-to-face conversations.

Deep generative models approximate the probability distribution of the observed data and are expected to generate diverse samples according to the modeled distribution. Several deep generative models, *e.g.*, Variational AutoEncoders (VAE) [12], Generative Adversarial Networks (GAN) [6], and Diffusion Models [10,25] are often used for motion generation. Li et al. [13] developed a method for generating diverse gestures from speech audio based on VAE. The generation method for interaction behaviors using GAN is proposed by Nishimura et al. [20].

Diffusion models [10,25] are able to produce high-quality samples by iterative processes and stable training [29]. Diffusion probabilistic models, such as denoising diffusion probabilistic models (DDPM) [10], are utilized to generate human

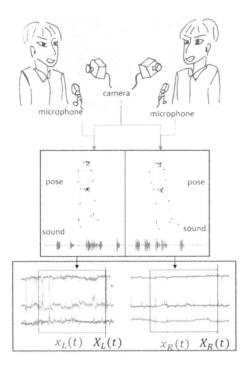

Fig. 1. Relationship between features $X(t)$ and features $x(t)$ in this paper

motion from text [26, 28]. IDM [21] generates consistent human-like behavior by using the mask in dyadic conversations, and the backend of IDM is a denoising diffusion implicit models (DDIM) [25] that require fewer iterative calculations than DDPM.

3 Personality-Aware Interaction Diffusion Model

In this section, we describe PIDM, the behavior generation model that is aware of personality based on objective evaluation during conversation.

3.1 Features

The pose and voice features of two participants at time t are set to $X_L(t)$ and $X_R(t)$. T time-step features of the two participants are also defined as $x_L(t) = [X_L(t-i)|i = 0, \ldots, T-1], x_R(t) = [X_R(t-i)|i = 0, \ldots, T-1]$. The features of two participants must be handled simultaneously to generate interaction behaviors. The feature at t is designed $x(t) = [x_L(t), x_R(t)]$. Figure 1 shows the relationship between $X(t)$ and $x(t)$ at time t.

The Big Five traits are used to assess the participants' personalities. Each personality trait of the Big Five is scored from 1 to 7. In this paper, the author

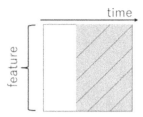

Fig. 2. Example of mask shapes used during training. The white and gray regions represent the features considered as conditions and the features to be masked, respectively. (Color figure online)

assesses participants' personalities. In the proposed method, the Big Five scores calculated from the personality questionnaire are used in the motion generation model, and the personality information of two participants during conversation is represented as ψ.

3.2 Denoising Diffusion Implicit Models

In this paper, the generation processes of our proposed model are based on DDIM [25]. DDIM is composed of a forward process to add noise to the original data and a reverse process to denoise.

In the forward process, Gaussian noise is gradually added to the original data x_0. Data x_k is obtained after adding noise to x_0 for k times. From $q(x_k|x_0) = \mathcal{N}(\sqrt{\alpha_k}x_0, (1-\alpha_k I))$, x_k is calculated by

$$x_k = \sqrt{\alpha_k}x_0 + \sqrt{(1-\alpha_k)}\epsilon, \tag{1}$$

where $\alpha_K \in [0,1)$ is a decreasing sequence of constant and $\epsilon \sim \mathcal{N}(0,1)$ is the noise.

In the reverse process, initial Gaussian noise is gradually denoised. The function $f_\theta(x_s)$ that estimates x_0 from x_s is calculated by $f_\theta(x_s) = x_0 = (x_s - \sqrt{1-\alpha_s}\epsilon_\theta(x_s))/\sqrt{\alpha_s}$, where s is the iteration step, $s \in S$ and $\epsilon_\theta(x_s)$ represents the function in DDIM that estimates the added noise. x_{s-1} for $s > 1$ is

$$x_{s-1} = \sqrt{\alpha_{s-1}}f_\theta(x_s) + \sqrt{1-\alpha_{s-1}}\frac{x_s - \sqrt{\alpha_{s-1}}f_\theta(x_s)}{\sqrt{1-\alpha_s}}. \tag{2}$$

By repeating equation (2), the data x_0' can be generated.

3.3 PIDM

PIDM ignores some features of $x(t)$ by applying a mask for training steps, where the mask is a binary matrix. PIDM then generates the features of the masked region by a diffusion model. Figure 2 shows an example of the mask M.

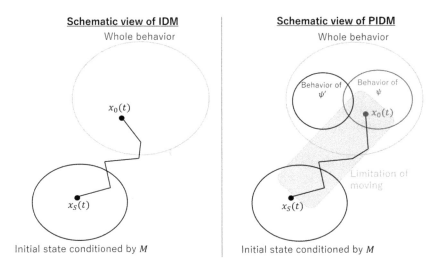

Fig. 3. The calculation scheme of IDM and PIDM. The black circle, black line, and Gray circle in the left and right figures are initial state with the mask, trajectory of iterative generation, and manifold of whole behavior, respectively. The orange shade in the right figure is the area limiting the trajectory during the iterative process. (Color figure online)

In this study, we assume the manifold of interaction behaviors of the trained model is structured by the personality. Personalities of third-party evaluation are associated with behaviors during conversation in the forward process of PIDM to structure the manifold. For the reverse process of the model, the direction of perturbation in the iterative process is limited by the personalities.

The difference between IDM and PIDM is briefly shown in Fig. 3. The left figure is the calculation of IDM conditioned only by the mask. The right figure is the calculation of PIDM conditioned by the mask and the personalities of third-party evaluation. In PIDM, the direction of trajectory is limited by personality ψ. The whole behavior of the right figure, PIDM, contains the set of behaviors for each personality trait.

Forward Process of PIDM. The conditional probability of noised data $x_k(t)$ at iteration step k is represented using $x_0(t), M,$ and ψ; $q(x_k(t)|x_0(t), M, \psi)$. $x_k(t)$ is calculated by

$$x_k(t) = M \otimes x_0(t) + (1 - M) \otimes (\sqrt{\alpha_k} x_0(t) + \sqrt{(1-\alpha_k)}\epsilon), \tag{3}$$

where $\otimes$ is the Hadamard product. For equation (3), noise is added only to the "masked" region without personality in the forward process.

In the proposed method, x_0 is directly estimated by f_θ, which is similar to the approach of Ramesh et al. [23]. The loss function with mask information $L(x_0, \hat{x}_0, M, \psi)$ is calculated as the L1 loss;

$$L(x_0, \hat{x}_0, M, \psi) = E[||(1-M) \otimes x_0 - (1-M) \otimes \hat{x}_0||_1], \quad (4)$$

where $\hat{x}_0 = f_\theta(x_k; \psi)$, i.e., $\hat{x}_0$ is conditioned by the personality.

Note that the size and aspect ratio of the mask are randomly determined during training. Personality information is assigned by setting full personality ψ for 70% of the data, setting the personality information of one participant to 0 for 20%, and setting $\psi = 0$ for the remaining 10%. This approach is to enhance the generalization ability of PIDM.

Reverse Process of PIDM. In the PIDM, features in the masked region $(1-M) \otimes x_s(t)$ are the generation target, and $M \otimes x_0(t)$ is known information. $x_{s-1}(t)$ with M and ψ becomes

$$x_{s-1}(t) = M \otimes x_0(t) + (1-M) \otimes \bar{x}_{s-1}(t; \psi), \quad (5)$$

and, $\bar{x}_{s-1}(t; \psi)$ is calculated using f_θ;

$$\bar{x}_{s-1}(t; \psi) = \sqrt{\alpha_{s-1}} f_\theta(x_s; \psi) + \sqrt{1-\alpha_{s-1}} \frac{x_s - \sqrt{\alpha_{s-1}} f_\theta(x_s; \psi)}{\sqrt{1-\alpha_s}}. \quad (6)$$

By repeating this procedure until $s = 1$, behaviors in dyadic conversation are generated.

Note that PIDM can generate behaviors even if one or two participants' personality is set to 0, i.e., $\psi = 0$ if two participants' personality is not considered. PIDM can generate human-like behaviors even when the personality is unknown.

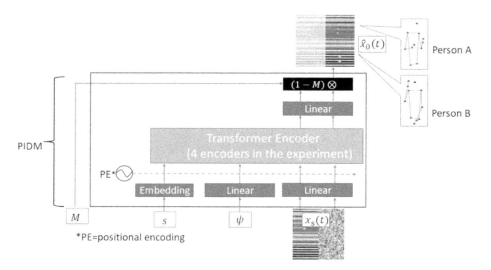

Fig. 4. PIDM network architecture. PIDM receives M, s, ψ, and $x_s(t)$, and the model outputs the estimated result of $x_0(t)$

3.4 Network Architecture

Figure 4 shows PIDM network architecture. PIDM receives $x_s(t)$, ψ, s, and M, and then the generation result of the masked part $(1 - M) \otimes \hat{x}_0$ is output. The behavior of the two people is generated simultaneously in the results.

The embedding layer converts s to 512-dimensional features. The personality values of two people ψ and the noised signal $x_s(t)$ are also converted to 512-dimensional features by a full connection layer. Ordering information is added to these converted features by the positional encoder, and then features are sent to the transformer encoder. In the PIDM, four layers of the transformer encoder [26] model are utilized as the network architecture. The encoder result is reproduced to the original data dimensions by a full connection layer and serves as $\hat{x}_0$. $x_{s-1}(t)$ is generated using this output and equations (5) and (6).

4 Experiments

PIDM is applied to the interaction data and the behavior generation task in a dyadic conversation. In this experiment, we investigate the distribution of generated results of PIDM changes according to Big Five personality traits.

4.1 Dataset Construction

In this experiment, features are extracted from the videos which recorded dyadic conversations, *i.e.*, conversations between two participants. We gathered the video data, and the conversation pairs are collected by offering a company. The ratio of male to female participants is 3 to 5 and each pair knew each other in advance. Each pair agrees to use the video data for the study. The conversation is an open-domain conversation with no special instructions regarding gestures.

The videos are recorded by using the 4K resolution camera, and microphones are attached to each person for recording voice signals. Three 10-minute sessions are conducted for each pair, and conversations of four pairs are recorded. As a result, the total amount of video is approximately two hours.

The pose features are extracted for each frame of the video data using MediaPipe [15]. Because the participants sat facing each other during conversation, in the behavior generation task, 11 points of joint information are used: nose, left shoulder, right shoulder, left elbow, right elbow, left wrist, right wrist, left hip, right hip, left index, and right index (Fig. 5).

Each pose feature is down sampled from 30 to 5 fps to smooth the signals. Because the sampling rate of the speech signal is 48Khz, the features using the values measured in the last 0.2 s are calculated by $\hat{A}(t) = \max_{t \in (t-0.2, t]} A^2(t)$, where $A(t)$ represents the sound signal. After these processes, the pose and audio features are combined.

To measure personality traits, ten item personality inventory (TIPI) [7] is a widely used as the approximation of the Big Five traits [19,24,27]. Oshio et al. [22] have developed a Japanese version of the TIPI (TIPI-J), which showed

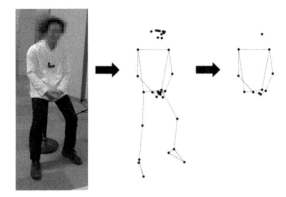

Fig. 5. Eleven points extracted from the video and MediaPipe.

acceptable levels of reliability and validity. Because questionnaire items of TIPI-J can be answered from the appearance and impression, the annotator uses TIPI-J to assess eight participants' personalities in this experiment. The personality information and down-sampled pose and audio features are constructed as the dataset for the behavior generation task.

4.2 Experimental Settings

PIDM is trained on two hours of the conversation data except for the test data. Two-minute samples from each pair are divided into test data. The total length of the test dataset is eight minutes. The time length of data T is set to $T = 50$. The number of iteration steps in the reverse process is set to 10. PIDM is trained using a Laptop computer with Nvidia GeForce RTX 3070 laptop, core i7 cpu, and 16 GB memory. For the test, all features after $T/2 = 25$ are removed by masking and then generated, that is, the model forecasts future behavior.

Fréchet Inception Distance(FID) [9,26] and multimodality [8,13] are calculated for the generated behaviors of the test data, and the generated results are evaluated. FID is a criterion that determines the similarity of the distributions of the generated results and original data. A smaller FID means that the distributions of the two data are similar.

The multimodality is a criterion that calculates the variety of the motions generated for certain clip data. In this experiment, the multimodality is calculated using the nose and right and left wrist features because the shoulders and waist do not move significantly.

The eight participants are divided into two groups: One is "H" with an extraversion score of 4.5 or higher, and the other is "L" with an extraversion score of less than 4.5. Because the velocity, magnitude, and frequency of the gesture are positively correlated with the extraversion score [14,17], it is expected that there are differences in behavior between the two groups. In the experiment, an extraversion score is changed for each group, and it is evaluated whether PIDM

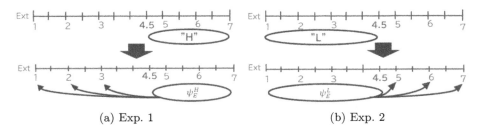

Fig. 6. Extroversion scores set in the generation of behaviors. The label 'Ext' means extraversion score. In exp. 1, the extraversion scores are set ψ_E^H, 3, 2, and 1. In exp. 2, the extraversion scores are set ψ_E^L, 5, 6, and 7.

can generate various and consistent behaviors. The following is a summary of the experiment for each group. Note that the extraversion scores of one participant are described as ψ_E^H and ψ_E^L, respectively. FID_H means FID score calculated between the generated result for each extraversion score and the original data in group H, and FID_L is also calculated between the generated result and the original data in group L.

Exp. 1 In the generation of behaviors in group H, the extraversion score is set to ψ_E^H, 3, 2, and 1. Behaviors are generated by assigning a lower extraversion score than the original score. Figure 6(a) shows the setting of exp. 1. Behaviors are also generated with $\psi = 0$, *i.e.*, the personality of each participant is set to zero.

Exp. 2 In the generation of behaviors in group L, the extraversion score is set to ψ_E^L, 5, 6, and 7. Behaviors are generated by assigning a higher extraversion score than the original score. Figure 6(b) shows the setting of exp. 2. Behaviors are also generated with $\psi = 0$, *i.e.*, the personality of each participant is set to zero.

4.3 Experimental Results

Result: exp. 1 Fig. 7 shows the example of the behavior generation results in exp. 1. The original data (Fig. 7(r)) shows a smooth hand-raising motion. The result generated with ψ_E^H (Fig. 7(o)) shows arm-crossing behavior. On the other hand, the generated result that the extraversion score is 1 (Fig. 7(a)) shows slight behavior from the start. Because the extraversion score is low, it is suggested that large behaviors are less likely to be generated.

The scores of the generation results, FID_H, FID_L, and multimodality, in exp. 1, are listed in Table 1. FID_H of the generation result with the extraversion score of ψ_E^H is the smallest. FID_H increases when the extraversion score is decreased from ψ_E^H to 3, 2, and 1 in the behavior generation. On the other hand, FID_L decreases when the extraversion score is decreased from ψ_E^H to 3, 2, and 1. It is considered that the distribution of generated behaviors is close to the distribution

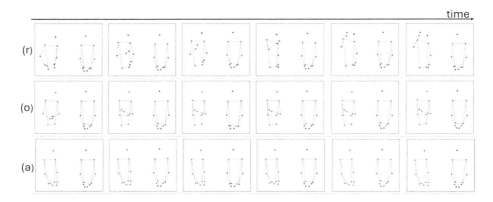

Fig. 7. Generation results in exp. 1. (r) Original signal, (o) result generated with an extraversion score of ψ_E^H, (a) result generated with an extraversion score of 1. Black, blue, and red shapes indicate the original participant, conversation partner, and generated poses, respectively. (Color figure online)

Table 1. FID_H, FID_L and Multimodality for each extraversion score including $\psi = 0$ in exp. 1. The mean and standard deviation are calculated by 10 times generation.

H	ψ_E^H	3	2	1	$\psi = 0$
FID_H	0.66 ± 0.27	0.68 ± 0.32	0.78 ± 0.32	1.03 ± 0.50	0.72 ± 0.26
FID_L	1.92 ± 0.26	1.59 ± 0.29	1.37 ± 0.35	1.26 ± 0.33	1.89 ± 0.31
Multimodality	0.18 ± 0.03	0.19 ± 0.04	0.19 ± 0.04	0.19 ± 0.04	0.19 ± 0.03

in group L by decreasing the extraversion score from ψ_E^H. Furthermore, FID_H of the generation result with $\psi = 0$ is higher than with the extraversion score of ψ_E^H. On the other hand, there is no difference in the multimodality of generation results of ψ_E^H, 3, 2, and 1.

Result: exp. 2 Fig. 8 shows the example of the behavior generation results in exp. 2. The original data (Fig. 8(r)) shows the wrist moving with both hands down. The result generated with ψ_E^L (Fig. 8(o)) shows little behavior from the start. On the other hand, the generation result that the extraversion score is 7 (Fig. 8(a)) shows both hands raising gradually from the start. Because the extraversion score is high, it is suggested that large behavior tends to be obtained.

The scores of the generation results, FID_H, FID_L, and multimodality, in exp. 2, are listed in Table 2. FID_L of the generation result with the extraversion score of ψ_E^L is the smallest. FID_L increases when the extraversion score is increased from ψ_E^L to 5, 6, and 7 in the behavior generation. On the other hand, FID_H decreases when the extraversion score is increased from ψ_E^L to 5, 6, and 7. FID_L of the generation result with $\psi = 0$ is higher than with the extraversion score of ψ_E^L.

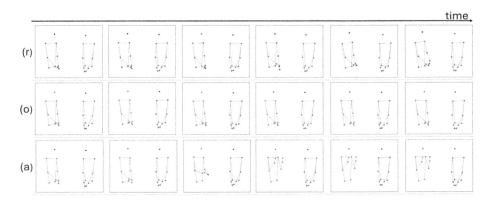

Fig. 8. Generation results in exp. 2. (r) Original signal, (o) result generated with an extraversion score of ψ_E^L, (a) result generated with an extraversion score of 7. Black, blue, and red shapes indicate the original participant, conversation partner, and generated poses, respectively. (Color figure online)

Table 2. FID_H, FID_L and multimodality for each extraversion score including $\psi = 0$ in exp. 2. The mean and standard deviation are calculated by 10 times generation.

L	ψ_E^L	5	6	7	$\psi = 0$
FID_H	1.29 ± 0.12	1.05 ± 0.15	1.03 ± 0.16	0.99 ± 0.17	1.22 ± 0.11
FID_L	0.52 ± 0.17	0.73 ± 0.27	0.87 ± 0.29	1.05 ± 0.43	0.62 ± 0.30
Multimodality	0.20 ± 0.03	0.21 ± 0.03	0.22 ± 0.03	0.22 ± 0.03	0.21 ± 0.03

The multimodality of the generation result increases by setting the extraversion raised from ψ_E^L to 5, 6, and 7. This shows that PIDM can generate more various behaviors by increasing the extraversion score. The multimodality of the behavior generation with $\psi = 0$ is higher than with ψ_E^L. This result also shows that personality traits control the probability of expressing gestures.

5 Discussion

In this section, we discuss the experimental results and limitations.

5.1 Discussion of Experimental Results

The multimodality with the extraversion score ψ_E^H in exp. 1 is smaller than with the extraversion score ψ_E^L in exp. 2. While previous researches show that gestures' velocity, magnitude, and frequency are positively correlated with the extraversion score, the relationship between gesture variation and the extraversion score [14,17] is not mentioned. If PIDM has the ability to learn the relationship between extraversion and not only the physical property but also variations of gestures, this model is expected to associate personality traits with behavior. The relationship between personality traits and generated results needs continued investigation.

The questionnaire used for personality assessment should be considered. TIPI-J [22] examined Big Five traits in relation to existing questionnaires. For the extraversion, correlations between TIPI-J and other Big Five questionnaires ranged from $r = 0.67$ to $r = 0.85$. On the other hand, for the traits of agreeableness and openness, the correlations between TIPI-J and the other Big Five questionnaires showed about $r = 0.50$. It is necessary to investigate other questionnaires to measure personality traits.

5.2 Limitation

In this experiment, the data obtained from participants is insufficient to cover the variation in personality and behavior. In addition to each person's personality, it is necessary to consider the combination of personalities of pairs in a dyadic conversation situation [16]. Increasing the amount of the conversation data is necessary to investigate the robustness and effectiveness of the experimental results.

In this experiment, because we aim to obtain preliminary insight, one annotator evaluates the personalities of the participants. To improve the quality of the annotations, we make as impartial a judgment as possible, referring to various information from the video itself, such as the gestures of the participants. However, To stabilize the estimation of personality evaluation, multiple annotators need to assess personality. For instance, a corpora annotated by multiple people is created for emotion estimation in the natural language processing field [4].

6 Conclusion

In this paper, we propose a personality-aware interaction diffusion model (PIDM) for modeling behavior in a dyadic conversation that reflects personality by objective evaluation. PIDM is a model for generating behaviors during a conversation which simultaneously deals with the personalities and features of participants. PIDM is applied to the behavior generation during a dyadic conversation, and the generated results with different personalities are evaluated. The results suggest that PIDM can generate different behaviors by adjusting the personality information.

Generating and evaluating behaviors based on other personality traits, *e.g.*, Agreeableness and Conscientiousness, is future work. Implementing the generated behavior by PIDM to CG agents and communication robots and evaluating human impressions are also our future tasks.

Acknowledgment. This work was supported by JST Moonshot R&D Grant Number JPMJMS2011 (Development of Semi-autonomous CA), JSPS KAKENHI Grant Numbers 23K16977.

References

1. Breil, S.M., Osterholz, S., Nestler, S., Back, M.D.: 13 contributions of nonverbal cues to the accurate judgment of personality traits. The Oxford Handbook of Accurate Personality Judgment, pp. 195–218 (2021)
2. Delaherche, E., Chetouani, M., Mahdhaoui, A., Saint-Georges, C., Viaux, S., Cohen, D.: Interpersonal synchrony: a survey of evaluation methods across disciplines. IEEE Trans. Affect. Comput. **3**(3), 349–365 (2012)
3. Forlizzi, J.: How robotic products become social products: an ethnographic study of cleaning in the home. In: Proceedings of the ACM/IEEE International Conference on Human-Robot Interaction, pp. 129–136 (2007)
4. Gaillat, T., Zarrouk, M., Freitas, A., Davis, B.: The SSIX corpora: three gold standard corpora for sentiment analysis in English, Spanish and German financial microblogs. In: LREC: Language Resources and Evaluation Conference, pp. 2671–2675. European Languages Resources Association (ELRA) (2018)
5. Goldberg, L.R.: The development of markers for the big-five factor structure. Psychol. Assess. **4**(1), 26 (1992)
6. Goodfellow, I., et al.: Generative adversarial nets. In: Advances in Neural Information Processing Systems, vol. 27 (2014)
7. Gosling, S.D., Rentfrow, P.J., Swann, W.B., Jr.: A very brief measure of the big-five personality domains. J. Res. Pers. **37**(6), 504–528 (2003)
8. Guo, C., Zou, S., Zuo, X., Wang, S., Ji, W., Li, X., Cheng, L.: Generating diverse and natural 3D human motions from text. In: Proceedings of the IEEE/CVF Conference on Computer Vision and Pattern Recognition, pp. 5152–5161 (2022)
9. Heusel, M., Ramsauer, H., Unterthiner, T., Nessler, B., Hochreiter, S.: GANs trained by a two time-scale update rule converge to a local Nash equilibrium. In: Advances in Neural Information Processing Systems, vol. 30 (2017)
10. Ho, J., Jain, A., Abbeel, P.: Denoising diffusion probabilistic models. Adv. Neural. Inf. Process. Syst. **33**, 6840–6851 (2020)
11. Ishii, R., Ahuja, C., Nakano, Y.I., Morency, L.P.: Impact of personality on nonverbal behavior generation. In: Proceedings of the 20th ACM International Conference on Intelligent Virtual Agents, pp. 1–8 (2020)
12. Kingma, D.P., Welling, M.: Auto-encoding variational bayes (2013). arXiv preprint arXiv:1312.6114
13. Li, J., et al.: Audio2Gestures: generating diverse gestures from speech audio with conditional variational autoencoders. In: Proceedings of the IEEE/CVF International Conference on Computer Vision, pp. 11293–11302 (2021)
14. Lippa, R.: The nonverbal display and judgment of extraversion, masculinity, femininity, and gender diagnosticity: a lens model analysis. J. Res. Pers. **32**(1), 80–107 (1998)
15. Lugaresi, C., et al.: MediaPipe: A framework for building perception pipelines (2019). arXiv preprint arXiv:1906.08172
16. Moon, Y., Nass, C.: How "real" are computer personalities? Psychological responses to personality types in human-computer interaction. Commun. Res. **23**(6), 651–674 (1996)
17. Nakano, Y., Oyama, M., Nihei, F., Higashinaka, R., Ishii, R.: The generation of agent gestures expressing personality traits. Hum. Interface Soc. **23**(2), 153–164 (2021). (in Japanese)
18. Naumann, L.P., Vazire, S., Rentfrow, P.J., Gosling, S.D.: Personality judgments based on physical appearance. Pers. Soc. Psychol. Bull. **35**(12), 1661–1671 (2009)

19. Neff, M., Wang, Y., Abbott, R., Walker, M.: Evaluating the effect of gesture and language on personality perception in conversational agents. In: Allbeck, J., Badler, N., Bickmore, T., Pelachaud, C., Safonova, A. (eds.) IVA 2010. LNCS (LNAI), vol. 6356, pp. 222–235. Springer, Heidelberg (2010). https://doi.org/10.1007/978-3-642-15892-6_24
20. Nishimura, Y., Nakamura, Y., Ishiguro, H.: Human interaction behavior modeling using generative adversarial networks. Neural Netw. **132**, 521–531 (2020)
21. Okadome, Y., Nakamura, Y.: Generating interaction behavior during a dyadic conversation using a diffusion model. In: IEEE International Conference on Computer and Automation Engineering, pp. 152–157 (2024)
22. Oshio, A., Shingo, A., Cutrone, P.: Development, reliability, and validity of the Japanese version of ten item personality inventory (TIPI-J). Jpn. J. Pers./Pasonariti Kenkyu **21**(1), 40–52 (2012)
23. Ramesh, A., Dhariwal, P., Nichol, A., Chu, C., Chen, M.: Hierarchical text-conditional image generation with clip latents. arXiv preprint arXiv:2204.06125 **1**(2), 3 (2022)
24. Smith, H.J., Neff, M.: Understanding the impact of animated gesture performance on personality perceptions. ACM Trans. Graph. (TOG) **36**(4), 1–12 (2017)
25. Song, J., Meng, C., Ermon, S.: Denoising diffusion implicit models (2020). arXiv preprint arXiv:2010.02502
26. Tevet, G., Raab, S., Gordon, B., Shafir, Y., Cohen-Or, D., Bermano, A.H.: Human motion diffusion model (2022). arXiv preprint arXiv:2209.14916
27. Wang, Y., Tree, J.E.F., Walker, M., Neff, M.: Assessing the impact of hand motion on virtual character personality. ACM Trans. Appl. Percept. (TAP) **13**(2), 1–23 (2016)
28. Zhang, M., Cai, Z., Pan, L., Hong, F., Guo, X., Yang, L., Liu, Z.: MotionDiffuse: text-driven human motion generation with diffusion model. IEEE Trans. Pattern Anal. Mach. Intell. **46**(6), 4115–4128 (2024)
29. Zhu, W., et al.: Human motion generation: a survey. IEEE Trans. Pattern Anal. Mach. Intell. **46**(4), 2430–2449 (2023)

Prompt Design Using Past Dialogue Summarization for LLMs to Generate the Current Appropriate Dialogue

Yuya Okadome[✉][iD], Akishige Yuguchi[iD], Ryota Fukui, and Yoshio Matsumoto[iD]

Tokyo University of Science, Tokyo, Japan
okadome@rs.tus.ac.jp

Abstract. Recent technological innovations in large language models (LLMs) produce incredible performance. This also has a similar impact on dialogue systems. However, following fluently current dialogue from the past dialogue is crucial, especially for chat-oriented dialogue systems, which are difficult for only LLMs to handle. In this paper, we propose a prompt design using a method summarizing dialogue for LLMs to generate the current appropriate dialogue in chat-oriented dialogue systems. For dialogue summarization, we first use a hand-crafted dialogue summarization corpus and two other corpora, and then a language model that summarizes dialogue in several sentences is fine-tuned on the combined corpora. We conducted two experiments for the performance evaluation of the proposed method. One is to evaluate how much the constructed model summarizes dialogue in some patterns. Another is to evaluate a performance predicting the current dialogue by prompting an LLM using the summarization model in contrast to the whole past dialogue. Through all the evaluation, the results suggest that the proposed prompt design is useful for dialogue generation using LLMs.

Keywords: Large Language Model · Prompt · Summarization

1 Introduction

The generation of natural sentences has been achieved by the technological innovations in large language models (LLMs) [1,11,15]. Chat-oriented dialogue systems that fluently reply to human utterances are expected to be developed using LLMs. LLMs can generate replies of the dialogue system from the given text instructions.

Recent conventional chat-oriented dialogue systems have adopted end-to-end deep learning approaches [4,17]. The text generation rules are trained using a huge amount of data, and the trained model is implemented into the dialogue systems [7,12]. Especially in LLMs, "natural" sentence generation can be realized because extremely large parameters and data and human-in-loop training methods are used [11].

Since an LLM is trained as a conditioning model with a huge amount of data, the model can generate natural sentences from added conditions, *i.e.*, prompts [11]. Hence, the effective prompt design is crucial for LLMs to demonstrate high-performance language generation [5,8]. The current conversation sentence is generated by following the previous context in the chat-oriented dialogue. If all context information is maintained and added to the prompt, increasing cost by a large number of tokens and truncation of long-term context occurs. In addition, the performance degradation is reported when the prompt with long context is input to an LLM [10]. These insights are critical for chat-oriented dialogue systems that have to handle long-term context. Besides, in human dialogue tasks, text summarization is effective for information compression of long-term context because the summarization is known to be useful for the task of dialogue hand-over [16].

In this paper, we propose a prompt design with past dialogue summarization to an LLM for constructing chat-oriented dialogue systems. In our proposed method, a hand-crafted dialogue summarization corpus and two types of corpora (a document type and a summary type) are combined, and a model summarizing dialogue in several sentences is fine-tuned on the combined corpora. The aim of using two types of corpora is the generalization ability and availability of the dataset. The output of the fine-tuned summarization model for past dialogue is fed into a prompt for LLM, and then, the current dialogue is estimated.

To validate the proposed method, the following two experiments are conducted.

1. The inspection of the performance of the fine-tuned Japanese dialogue summarization model
2. Predicting the current dialogue by prompting an LLM using past dialogue summarization

In the first experiment, the effect of the number of samples in the dialogue summarization dataset on fine-tuning the language model is inspected. From this experiment, the "decent" quality of summarization is obtained if the number of data is small. In the second, the difference in the generated dialogue between prompts with full context (whole past dialogue) and context summarization is investigated. Although the number of input tokens is reduced, similar outputs are obtained from both full context and summarization in the current dialogue prediction task. The results of these experiments suggest that the proposed method is useful for the prompt to an LLM.

2 Proposed Method

Our proposed prompt design is based on dialogue summarization. The transformer-based language model (*e.g.*, T5 model [13]) that can summarize the Japanese sentences is used for fine-tuning the Japanese dialogue summarization. We adopt that the summary style is several sentences [6] for a dialogue summarization.

Furthermore, the hand-crafted summarization corpus is prepared because there are only few Japanese daily dialogue summarization corpora. To improve the generalization ability of the dialogue summarization model, we combine different corpora such as a news summarization corpus.

2.1 Japanese Dialogue Summarization Corpus

To construct the hand-crafted summarization corpus, the dyadic conversation data is extracted from the Nagoya University Conversation Corpus (NUCC) [3], and a part

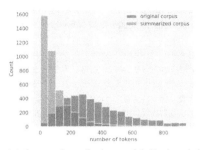
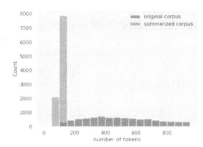

(a) the number of tokens of dolly-ja original and summarization texts.

(b) the number of tokens of livedoor original and summarization texts.

Fig. 1. The number of tokens of each corpus. Blue and red histograms are the length of the original and summarized tokens, respectively. (Color figure online)

of the extracted conversation is summarized. The NUCC is built by transcription of recorded spoken dialogue, and filler and ambiguous utterances (*i.e.*, backchannel information) are also preserved. It is expected that the corpus is consistent with the result of speech recognition which the actual dialogue system receives.

On-the-fly transcribing this information for the text input of the dialogue system is difficult. In our summarization corpus, backchannel information included in NUCC is removed, and then, conversation data is summarized.

Because one dialogue is created from one hour of speech information, summarizing the whole conversation is hard. Therefore, about 20 utterances are extracted and summarized. Participants' IDs in the original corpus are changed to A and B. To show the quality of the fine-tuning model on a quite small number of data, 45 summarized dialogues are extracted and the Japanese dialogue summarization corpus (JDSC) is constructed.

2.2 Other Corpora

Because 45 summarized dialogues are not enough for fine-tuning the language model, the quality of summarization becomes low. To overcome this problem, two corpora, Livedoor news summarization[1] and dolly-15k-ja[2] are used for fine-tuning. The amount of pairs of documents and summaries from the Livedoor corpus is about 200k, and that of dolly-ja is about 4k. To close the amount of data of each corpus, 10k data is randomly sampled from the Livedoor corpus.

Figure 1 shows the histograms of the token count of sentences in each corpus. In the Livedoor corpus, the tokens count of summarization is much smaller than the original sentences, and the variance of the count of summarization is large in the dolly corpus. It is expected that corpora with different properties are combined to prevent over-fitting on a certain corpus.

---

[1] https://github.com/KodairaTomonori/ThreeLineSummaryDataset.
[2] https://huggingface.co/datasets/kunishou/databricks-dolly-15k-ja.

Table 1. Results of ROUGE-1, -2, and -L, BERT score and Sentence-BERT for each summarization model and ChatGPT versus target summary. A higher value is better for each score.

		ROUGE-1	ROUGE-2	ROUGE-L	BERT score	Sentence-BERT
Summary Type	ALL	**0.464 ± 0.078**	**0.166 ± 0.086**	**0.286 ± 0.076**	**0.744 ± 0.031**	**0.953 ± 0.012**
	HF	0.422 ± 0.062	0.136 ± 0.061	0.255 ± 0.053	0.725 ± 0.023	0.943 ± 0.016
	TEN	0.382 ± 0.066	0.115 ± 0.062	0.245 ± 0.056	0.715 ± 0.025	0.939 ± 0.015
	W/O	0.320 ± 0.073	0.071 ± 0.052	0.211 ± 0.055	0.694 ± 0.027	0.929 ± 0.019
	ChatGPT	0.417 ± 0.090	0.133 ± 0.074	0.273 ± 0.059	0.725 ± 0.027	0.947 ± 0.017

3 Experiment

Two experiments are conducted to investigate the performance of the proposed method. One experiment is the inspection of the performance of the dialogue summarization model, and another experiment is to verify the performance of the current dialogue prediction on LLM by different patterns of prompt design. In the performance investigation of the summarization model, the amount of dialogue summarization data is changed for examining the effect of summarization data. In the availability test, dialogue prediction using prompts of each summary sentence and whole contexts are performed, and how "similar" they are investigated.

3.1 Performance of the Summarization Model

The number of data of JDSC, Livedoor, and dolly-ja is 45, 10000, and 4172, respectively. In this experiment, two corpora, Livedoor and dolly-ja are fixed, and 5-fold cross-validation is performed on JDSC, *i.e.*, the training data of conversation in each validation is 36. In each validation, the fine-tuning models of all conversation data (ALL), eighteen conversation data of JDSC in each validation (HF), ten conversation data of JDSC in each validation (TEN), and using only Livedoor and dolly data (W/O) are compared.

In this experiment, the T5 model [13] is used as the language model for the summarization task, because T5 model shows the high performance on the abstraction summarization task [2] and the Japanese pretrained model is recently launched. The T5 model which is trained on Japanese sentences[3] is used for the fine-tuning.

Results. Table 1 shows the scores of ROUGE-1, -2, and -L [9], BERT score [18], Sentence-BERT [14] between outputs of each summarization model and the target summary. These scores are frequently used as evaluation criteria of the abstraction summarization task [2]. In addition to fine-tuning models, the summarization result by the LLM (ChatGPT, gpt-3.5-turbo) is also shown as one of the references.

The score of TEN, HF, and ALL is higher than W/O from Table 1. The performance of summarization is improved by combining several corpora even if the amount of dialogue data is less than 40. From the scores of TEN, HF, and ALL, if only a few task-related data, *i.e.*, JDSC are used, the performance is much improved.

---
[3] https://huggingface.co/retrieva-jp/t5-base-long.

Table 2. Prompt design using summarization (ALL, W/O) and whole context (FULL). Original prompts in Japanese and translated prompts in English are shown.

Context	Japanese (original)	English (translate)
ALL, W/O	{participant A}と{participant B}が会話をしています。ここまでの会話の内容を{summary}と要約できるとき、以下の{participant A}の発言{sentence}に対し、{participant B}の返答を一つ出力してください。	{participant A} and {participant B} are having a conversation. When the conversation up to this point can be summarized as {summary}, please output one response from {participant B} to the following {sentence} from {participant A}.
FULL	{participant A}と{participant B}が会話をしています。ここまでの会話の内容が{context}であるとき、以下の{participant A}の発言{sentence}に対し、{participant B}の返答を一つ出力してください。	{participant A} and {participant B} are having a conversation. When the conversation up to this point is {context}, please output one response from {participant B} to the following {sentence} from {participant A}.

3.2 Dialogue Prediction

The performance of the task of current dialogue prediction is investigated using a prompt to an LLM which includes summarization sentences that are generated using a fine-tuning language model on whole JDSC (ALL), *i.e.* proposed prompt design is based on the dialogue summarization. In this experiment, prompts including whole context (FULL) and summarization output by the language model trained without JDSC (W/O) are compared with a prompt with ALL.

Prompts with summarization model (ALL, W/O) and whole context (FULL) are designed as in Table 2. {participant A} and {participant B} assume the roles of a user and a dialogue system. In this experiment, these prompts are input to the LLM (ChatGPT, gpt-3.5-turbo), and one reply is predicted. Outputs of the LLM of FULL and ALL are similar when the summarization sentence is efficiently compressing original contexts because both prompts are almost the same.

In this experiment, the Japanese Daily Dialogue Corpus[4] is used for dialogue prediction tasks to prevent information leakage. The sentences in this corpus are close to the actual chat-oriented dialogue. In the corpus, about a thousand dialogue is recorded in each five conversation contents. Five dialogues are sampled from each content, and the last utterance is predicted.

Dialogue System Overview. The overview of the dialogue system used in this study is shown in Fig. 2. The past dialogue is input to the fine-tuned summarization model, and the summarized sentences are output. The current dialogue is predicted by the LLM with the prompt including the summarization.

---
[4] https://github.com/jqk09a/japanese-daily-dialogue.

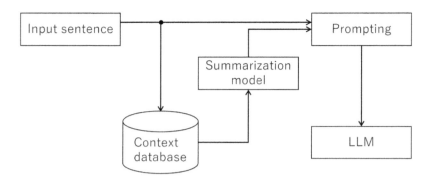

Fig. 2. The scheme of the LLM-based dialogue system used in this study.

Table 3. Results of ROUGE-1, -2, and -L, BERT score and Sentence-BERT for the prompt design with each summarization model (ALL, W/O) versus a prompt design with FULL. A higher value is better for each score.

	ROUGE-1	ROUGE-2	ROUGE-L	BERT score	Sentence-BERT
ALL v.s. FULL	**0.535 ± 0.238**	**0.328 ± 0.302**	**0.499 ± 0.237**	**0.818 ± 0.082**	**0.928 ± 0.037**
W/O v.s. FULL	0.452 ± 0.194	0.232 ± 0.202	0.416 ± 0.198	0.800 ± 0.062	0.910 ± 0.073

The summarization is used as the past dialogue in this system. If the amount of information is equivalent between the summarization and the whole past dialogue, it is expected that the predicted result by the LLM becomes close.

Results. Table 3 shows the scores of ROUGE-1, -2, and -L [9], BERT score [18], Sentence-BERT [14] between ALL and W/O, and FULL. The amount of the input token of FULL, W/O, and ALL is 478 ± 107, 196 ± 20.1, and 192 ± 23.7, respectively. The token count is dramatically reduced by using a prompt with context summarization.

The scores of ALL versus FULL are high even if the task is the prediction from Table 3. It is expected that the outputs of ALL and FULL are similar although the number of input tokens for text generation is reduced. However, there are "dissimilar" prediction results because the variances of the ROUGE scores are large. Investigating the factor of "dissimilar" including the format of summary is one of the future works.

The scores of W/O versus FULL are lower than ALL. The past dialogue is not efficiently compressed using W/O because the performance of summarization is not good. This result means the performance of the summarization affects the output quality of the LLM.

Table 4 shows the examples of prediction results of the current dialogue. From the table, the outputs of FULL, ALL, and W/O differ from the target. Person A talks about the experience of going to the "road" in this conversation. However, the results of FULL and ALL ignore the A's experience, and the output of W/O asks the impression of the road. The dialogue prediction is difficult for LLMs even if the whole context is input.

Table 4. Examples of dialogue prediction. Original Japanese sentences and translated English sentences are shown in the table. The two previous sentences in English* are "A: Yes, I have. Have you been there too? B: I have been there on a road trip. The number of curves exceeds my expectations."

Method	Predict (Japanese)	Predict (English)
True	A: 一般人が運転するには困難な道路です。高速道路の上信越自動車道が作られた理由もよく分かります。	A: It is a difficult road for the ordinary person to drive. I can understand why the Jyoshinetsu Expressway was built.
FULL	A: 本当ですか？私も行ってみたいですね。カーブがたくさんあるということは、ドライブが楽しめそうですね。	A: Really? I would like to go there too. It would be a fun drive because there are so many curves.
ALL	A: そうですか、それは楽しそうですね。私もドライブが好きなので、いつか行ってみたいです。	A: That sounds like fun. I also like driving, so I would like to go there someday.
W/O	A: 本当ですか？それは興味深いですね。どの道が一番印象に残りましたか？	A: Really? That's interesting. Which road impressed you the most?

* Original sentences in Japanese are "A: そうです。あなたも行ったことがありますか？ B: 私もドライブで行ったことがありますよ。聞きしに勝るカーブの数でした。".

The meanings of outputs of FULL and ALL are close to each other, and the sentences of FULL and W/O are not similar. By improving the text summarization model, it is suggested that the prompt with summarization generates similar sentences to FULL.

4 Conclusion

In this paper, we proposed the prompt design for an LLM to efficiently compress the context information. The dialogue summarization model is fine-tuned on a small number of dialogue summarization corpus and other two corpora. The generated context summary is implemented to a prompt to an LLM, and the current dialogue is output. We evaluated the proposed model through two experiments, the quality of summarization and a prompt with the summary to an LLM. The results of the two experiments are that a small dialogue corpus improves the output quality of the summarization model, and the outputs from an LLM are similar between prompts with whole and summarized context. From these experiments, it is suggested that the proposed prompt design for an LLM is useful for constructing a dialogue system.

We only discussed the performance of the summarization model in this study. In the experiment of dialogue prediction, the performance of the proposed method for the long-term dialogue (*e.g.*, more than a few thousand tokens) was not investigated in this paper due to the length limitation of the dialogue in the corpus. Our future work is to develop an actual dialogue system with the proposed method for more long-term dialogue and investigate the human impression.

The other limitation is that we only discussed the score of the current dialogue prediction in the experiment. To validate the effectiveness of a dialogue system with the proposed method, it is necessary that generated utterances are evaluated by human participants.

Acknowledgement. This work was supported by JSPS KAKENHI Grant Number 23K16977 and 21H04418.

References

1. Anil, R., et al.: PaLM 2 technical report (2023). arXiv preprint arXiv:2305.10403
2. Fabbri, A.R., Kryściński, W., McCann, B., Xiong, C., Socher, R., Radev, D.: SummEval: re-evaluating summarization evaluation. Trans. Assoc. Comput. Linguist. **9**, 391–409 (2021)
3. Fujimura, I., Chiba, S., Ohso, M.: Lexical and grammatical features of spoken and written Japanese in contrast: exploring a lexical profiling approach to comparing spoken and written corpora. In: Proceedings of the VIIth GSCP International Conference. Speech and Corpora, pp. 393–398 (2012)
4. Gao, J., Galley, M., Li, L.: Neural approaches to conversational AI: Question answering, task-oriented dialogues and social chatbots. Now Foundations and Trends (2019)
5. Hoang, D.N., Cho, M., Merth, T., Rastegari, M., Wang, Z.: (dynamic) prompting might be all you need to repair compressed LLMs (2023). arXiv preprint arXiv:2310.00867
6. Kodaira, T., Komachi, M.: The rule of three: abstractive text summarization in three bullet points. In: Proceedings of the 32nd Pacific Asia Conference on Language, Information and Computation (2018)
7. Lei, W., Jin, X., Kan, M.Y., Ren, Z., He, X., Yin, D.: Sequicity: simplifying task-oriented dialogue systems with single sequence-to-sequence architectures. In: Proceedings of the 56th Annual Meeting of the Association for Computational Linguistics (Volume 1: Long Papers), pp. 1437–1447 (2018)
8. Li, L., Zhang, Y., Chen, L.: Prompt distillation for efficient LLM-based recommendation. In: Proceedings of the 32nd ACM International Conference on Information and Knowledge Management, pp. 1348–1357 (2023)
9. Lin, C.Y.: ROUGE: A package for automatic evaluation of summaries. In: Text summarization branches out, pp. 74–81 (2004)
10. Liu, N.F., et al.: Lost in the middle: How language models use long contexts (2023). arXiv preprint arXiv:2307.03172
11. Ouyang, L., et al.: Training language models to follow instructions with human feedback. Adv. Neural. Inf. Process. Syst. **35**, 27730–27744 (2022)
12. Park, Y., Ko, Y., Seo, J.: BERT-based response selection in dialogue systems using utterance attention mechanisms. Expert Syst. Appl. **209**, 118277 (2022). https://doi.org/10.1016/j.eswa.2022.118277, https://www.sciencedirect.com/science/article/pii/S0957417422014166
13. Raffel, C., et al.: Exploring the limits of transfer learning with a unified text-to-text transformer. J. Mach. Learn. Res. **21**(140), 1–67 (2020). http://jmlr.org/papers/v21/20-074.html
14. Reimers, N., Gurevych, I.: Sentence-BERT: sentence embeddings using Siamese BERT-networks. In: Proceedings of the 2019 Conference on Empirical Methods in Natural Language Processing and the 9th International Joint Conference on Natural Language Processing (EMNLP-IJCNLP), pp. 3982–3992. Association for Computational Linguistics, Hong Kong, China (2019)
15. Touvron, H., et al.: LLaMA: Open and efficient foundation language models (2023). arXiv preprint arXiv:2302.13971

16. Yamashita, S., Higashinaka, R.: Optimal summaries for enabling a smooth handover in chat-oriented dialogue. In: Proceedings of the 2nd Conference of the Asia-Pacific Chapter of the Association for Computational Linguistics and the 12th International Joint Conference on Natural Language Processing: Student Research Workshop, pp. 25–31 (2022)
17. Yamazaki, T., Yoshikawa, K., Kawamoto, T., Mizumoto, T., Ohagi, M., Sato, T.: Building a hospitable and reliable dialogue system for android robots: a scenario-based approach with large language models. Adv. Robot. **37**(21), 1364–1381 (2023). https://doi.org/10.1080/01691864.2023.2244554
18. Zhang, T., Kishore, V., Wu, F., Weinberger, K.Q., Artzi, Y.: BERTScore: evaluating text generation with BERT. In: Proceedings of the 2020 International Conference on Learning Representations (2020)

Recommender Systems

Click-Through Rate Prediction Based on Filtering-Enhanced with Multi-head Attention

Meihan Yao(✉), Shuxi Zhang, Lang lv, Jianxia Chen, Mengyu Lu, Gaohang Jiang, Liang Xiao, and Zhina Song

School of Computer Science, Hubei University of Technology, Wuhan 430068, China
202211215@hbut.edu.cn

Abstract. Click-Through Rate (CTR) prediction is a crucial task in recommend systems (RSs), particularly in large-scale industrial applications. CTR models based on graph neural networks (GNNs) is currently the mainstream technology, however, it also encounters challenges in terms of feature interactions and user interests. In the context of RSs, the individual attribute information associated with each entity holds significant importance beyond the inherent interactions between users and items. Rational utilization and capturing of dependencies among these attributes can substantially enhance the predictive accuracy of the model. Therefore, we propose two attention mechanisms of interactions, including internal interaction and external interaction, to handle the information exchange within the attributes of either users or items individually, as well as the interactions between the attributes of users and items. Simultaneously, we introduce a multi-head attention (MHA) mechanism for the interaction selection, to capture features from different dimensions. Furthermore, the behavioral data reflecting user interests unavoidably contains noise, we attenuates noise by utilizing fast Fourier transform (FFT) filtering algorithms to transfer information from the spatial domain to the frequency domain. Experimental results on three benchmarks show that our model outperforms existing SOTA models.

Keywords: Click-Through Rate · Filtering Algorithm · Graph Neural Networks · Feature Interaction

1 Introduction

Currently, recommendation systems (RSs) are widely employed to predict users' potential interests in products, often situated within large online platforms, such as Taobao and Facebook. Click-through rate (CTR) prediction is an essential metric for assessing the performance of recommendation models, which defines the accuracy of RS model by predicting whether or not a user will click on a special item.

Supported by the National Natural Science Foundation of China (No.42301434).

CTR prediction methods primarily include techniques such as logistic regression (LR), matrix factorization (MF), deep neural networks (DNNs) and so on. As an early CTR prediction model, LR possesses some advantages such as speed, lightweight structure, and strong interpret-ability. With the increasing complexity of data, however, simple LR-based models [1,2] are insufficient to support intricate feature extraction. Subsequent enhanced models introduce non-linear expressions to address this limitation. MF is recognized as a benchmark solution for CTR prediction [3], in which users and items representations have been captured from their interactions (e.g., behaviors like favorites, clicks, and purchases). Moreover, since users' attributes significantly influence users' preferences (e.g., gender, age), while items' attributes (e.g., price, color) also play a crucial role in determining their attractiveness for users, factorization machines (FM) [4,5] are utilized to solve CTR predictions with attribute embeddings, to capture more fine-grained collaborative filtering (CF) information. This results in a new challenge of attribute feature interactions for CTRs.

Unlike traditional RS models, DNNs-based RS models can automatically capture complicated data relationships and nonlinear user-item interactive information, to obtain more complexly interactive feature representations. Many researchers utilize FM for feature interactions as well as DNNs for relationships between features. For instance, NFM [6] and DeepFM [7] introduce hidden layers of multi-layer perceptrons (MLP) to FM models to non-linearly capture structural information. However, numerous results of these models indicate that it is less effective for them to solely rely on MLP for feature interactions. This is because not all feature interactions are beneficial for predictions, thus capturing unrelated feature interactions is akin to capturing noise. To identify the functions of feature interactions, some models incorporate selection mechanisms to select features information [8,9]. However, these models have high computational complexity for high-order interaction calculations. Afterward, graph neural networks (GNNs) have been demonstrated to have advanced performance in capturing high-order features [10–12]. Nevertheless, existing GNNs-based CTR models often treat all attribute node interactions equally, overlooking the differences between different types of attributes and thus unable to make decision effectively. In contrast to the one-graph construction strategy adopted by most GNN recommendation models, the model proposed in this paper introduces a graph construction strategy involving two types of graphs, aiming to capture the internal correlations among attributes of the same entity and the correlations between attributes of different entities.

Mining users interests is another critical goal of CTR prediction, since noise is inevitably present in the users behavioral data [13,14]. Meanwhile, noise will improve a large number of parameters in the GNNs to lead to over-fitting, thus significantly reducing CTR prediction accuracy. Therefore, we utilize a noise filtering algorithm based on fast Fourier transform (FFT) [15,16], which is variant of Discrete Fourier transform (DFT). DFT is a fundamental method in the field of digital signal processing (DSP), transforming discrete signals from time to frequency domains [17]. Afterward, FFT is a more efficient algorithm for computing DFT, recursively re-expressing a sequence of length N as DFT [18], to reduce the time complexity from $O(N^2)$ to $O(NlogN)$ with improvement of its scalability as well. Because FFT can transform signals into frequency domains with more periodic features, it is good at filtering out noise signals in DSPs. Hence, FFT is gradually introduced into the recommendations to reduce the noise impact between features. For instance, FMLP-Rec [19] and FMAS model [20] propose a learnable filter to handle noisy signals in the original embeddings for sequence recommendations.

Therefore, we propose a novel GNNs-based recommendation model focus on feature interaction and users behavioral noise. The proposed model utilizes Filtering-enhanced with Multi-Head Attention, namely FMHA. First, we construct two internal users or items graphs and one user-item interactive graph, to connect relations between users or items features. Moreover, we utilize DFT as a filter strategy to filter noise and rich feature information. Specifically, we utilize FFT to transform input representations from the spatial domain to the frequency domain, and then restored the denoised representations to the spatial domain through the inverse FFT process. Afterward, we design a dual mechanisms incorporating the interaction path selection and the MHA. Finally, we obtain the probability prediction of whether a user will click on a particular item. Experiments indicate that our model effectively enhances recommendation performance. In summary, our main contributions are described as follows:

1) Propose two graph construction strategies. The first involves inter-user or inter-item graphs by incorporating the inherent attributes and nodes of users or items. The second entails an interactive graphs, which interconnects user and item attributes to create a user-item graph.
2) Devise a dual attention mechanisms consisting of the interaction selection and the MHA function. The interaction selection mechanism assigns larger message weights to nodes with closer relationships. The MHA mechanism extracts the feature interaction information from multiple dimensions.
3) Employee a FFT-based filtering algorithm to attenuate noise information. We integrate a filtering component into each stacked block.

2 Related Work

2.1 Feature Interaction for CTRs

In recent years, DNNs methodologies have been harnessed to enhance existing models. Google introduced the Wide & Deep architecture [21], amalgamating the "Wide" segment for memorization and the "Deep" segment for generalization to enhance model efficacy. However, this architecture requires a lot of effort on feature engineering and is unable to generalize feature combinations that have not appeared before. For example, DeepFM [7] incorporates a wide component to emulate feature interactions, integrating a MLP with hidden layers into traditional FM models. However, it relies solely on MLP for feature interactions, yields sub-optimal results. DCN [22] and xDeepFM [23] explicitly capture high-order interactions through interactive networks, but they incur notably high computational complexity.

Discovering features interaction is a challenge for CTR, as manually combining features is labor-intensive, time-consuming, and unable to identify potential effective feature interactions. To address it, AFM [24] utilizes an attention mechanism to assign different importance weights for various interaction features. However, scalarization operations weaken the expressive power of feature vectors. AutoInt [8] employs self-attention [25] to automatically capture high-order features by stacking multiple attention layers, nevertheless, overlooking the interactions between low-order features.

2.2 GNNs-Based CTRs

GNNs have demonstrated excellent potential for high-order feature interactions in CTR prediction models. The reason is that GNNs-based CTR models proposes user-item interactions as a bipartite graph, alleviating sparsity and cold start issues, such as NGCF [9] and LightGCN [26]. However, these models focus on the user-item interactions ignoring the attributes interactions. Afterward, Fi-GNN [27] leverage GNNs to learn node relationships for attribute interaction aggregation. However, it overlooks implicit collaborative signals in user behaviors. To capture both high-order feature interactions and collaborative signals, DG-ENN [28] designs attribute graphs and user-item collaborative graphs. Furthermore, to alleviate sparsity of CTRs, it utilizes users similarity relations and items transitions to enrich the original user-item interaction relationships. Nevertheless, these models treat all attribute node interactions equally, neglecting the differences between different types of attributes, thus failing to make decisions more effectively.

3 The FMHA Framework

In this section, we describe our innovative FMHA framework. We begin by defining the problem this framework aims to solve, followed by a general overview of the model (Fig. 1). Afterward, we describe the five layers that consist the core of our model.

- Input and Embedding Layer: input pre-treatment nodes from datasets and maps nodes from high-dimensional sparse vectors to low-dimensional dense vectors.
- Graph Construction Layer: establishes connections between nodes, including an internal fully connected graph for users or items with their own attributes and an external connected graph for interacting users and items.
- Feature Interaction Layer: conducts message propagation and aggregation based on two graphs, enabling each node to aggregate higher-order representation information. Subsequently, through gated recurrent unit (GRU), the representations of nodes at different stages are fused to obtain the final node representation.

- Filter Enhancement Layer: filters out noise during the interaction between raw data and features. We use FFT to transform feature signals from spatial domain to spectral domain and then back to spatial domain.
- Fusion and Prediction Layer: combines features from different interaction stages and ultimately outputting the predicted probability of interaction between users and items. A dot product calculation is then performed on the overall features of users and items, resulting in the model's predicted probability of interaction between the user and item.

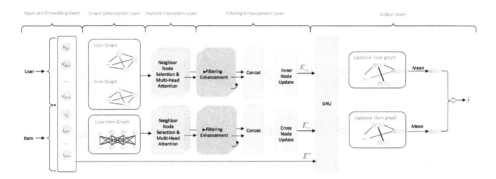

Fig. 1. Overall architecture of FMHA model.

3.1 Problem Statement

We define our training set as D, consisting of input-output pairs (X_n, Y_n): $D = (X_n, Y_n)_{1 \leq n \leq N}$, where N is the sample size. Here we use Y_n to present the predicted results, and formulate X_n as: $X_n = \{U_{id}, I_{id}, U_{att}, I_{att}\}$. The value of X_n denotes user and item along with their attribute features. In this context, U_{id} and I_{id} denote a user and an item node, respectively. U_{att} and I_{att} represent users and items attributes, respectively, such as $U_{att} = u_{attr1}, ... u_{attri}, ..., u_{attrn\ 1 \leq i \leq n}$, $I_{att} = \{i_{attr1}, ..., i_{attrj}, ..., i_{attrm}\}_{1 \leq j \leq m}$, in which u_{attri} and i_{attrj} stand for the node of attributes, n and m are the number of attributes for the user and item in this sample. Therefore, the input graph g is comprised of users, items, their attributes, and their interaction information. In results of our model, y is the true label of the data sample, and $\hat{y}$ is the predicted result of our model in this context.

3.2 Input and Embedding Layer

Afterward, we utilize one-hot encoding to represent nodes as vectors. Subsequently, a trainable matrix $V \in \mathbb{R}^{d \times n}$ is employed to transform sparse one-hot vectors from the high-dimension into low-dimensional dense vectors $e_i \in \mathbb{R}^{d \times 1}$ as (1):

$$e_i = V\ node_i \qquad (1)$$

where 'node' is used to represent, and there are n nodes in total, $node_i \in \mathbb{R}^{n \times 1}$ denotes the i-th node vector. $\mathbb{R}$ means the values of the vector are real number. d represents the dimensionality of the low-dimensional dense vector. Hence, a piece of input data is defined as $e = [e_1, e_2, ..., e_{m+n+2}]$ responding a label Y_n.

3.3 Graph Construction Layer

In this module, we construct three graphs, including two internal graphs: the user's internal graph G_u which include labeled nodes and attributes of users, the item's internal graph G_i that include labeled nodes and attributes of items, user-item interaction graph G_{ui}, in which all nodes are fully connected, indicating that each node is directly linked to every other node in the graph. In the interaction graph G_{ui}, the nodes of users (u_{id} and u_{attr}) are connected to the nodes of items (i_{id} and i_{attr}), but their own nodes are not connected. It's worth noting that these three graphs together form a fully connected graph G.

3.4 Feature Interaction Layer

To capture the correlation between features, feature interaction is indispensable in this model (Fig. 2). Feature interaction is performed on the three graphs described above. In the graph, if there is an edge between two nodes, it indicates a correlation between these two nodes, and the model learns these correlations. However, not all connections are beneficial. Additionally, the importance of different features for prediction varies; for example, the price of a item is obviously more important than its color. Therefore, it is necessary to selectively choose neighbour nodes and assign different weights to different features.

Neighbor Node Selection Mechanism. Firstly, we design a mechanism for selecting neighbour nodes, retaining nodes with a stronger relationship with the central node. Specifically, we utilized a MLP within a hidden layer to calculate the weight of the edge between node pairs through dot product. The formula is in (2):

$$p_{ij} = \sigma \left(W_2 \delta \left(W_1 \left(H_i \odot H_j \right) + b_1 \right) + b_2 \right) \quad (2)$$

where the low-dimensional dense vector representation of the node is represented as H, H_i and H_j represent the vectors of two adjacent nodes. $\odot$ is the element-wise Hadamard product. $W_1 \in R^{e \times d \times hidden}$ and $W_2 \in R^{e \times d \times hidden}$ denotes the weights of the first and second linear layer in the MLP, respectively. Meanwhile, $b_1 \in R^{e \times 1}$ and $b_2 \in R^{e \times 1}$ denotes the bias of the first and second linear layer in the MLP respectively. e is the number of edges in the current batch graph, $hidden$ is the hidden layer's size. δ is the activation function ReLU for the first layer, and σ is the sigmoid activation function. p_{ij} is the result obtained through the calculation in formula (3), where the values are normalized by the sigmoid activation function to be within the range (0, 1).

Additionally, the sum of all edge weights of the node's adjacent edges is equal to 1, representing the strength of the connections between nodes. After obtaining the edge weights p we retain the top-k proportion of edges in each graph, and remaining edges' weights are set to zero. The calculation is in (3), (4):

$$id_k = \text{argtop}_k \ p_{ij} \tag{3}$$

$$p_{ij}\left[-id_k\right] = 0 \tag{4}$$

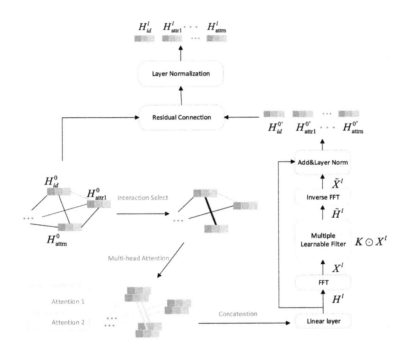

Fig. 2. Structure of Interaction and Filtering Enhancement Layer.

Here, $argtop_k$ is used to select the top-k scores p_{ij} in each graph, id_k is the index of the selected maximum k scores, and $-id_k$ is the index of the edges in the graph that were not selected. The weights of these non-selected edges are then set to 0, meaning only edges with weights large enough are retained. In the process, the second attention calculation is performed as (5):

$$c_{ij} = \text{LeakyReLU}\left(W_a\left(H_i \odot H_j\right)\right) \tag{5}$$

Here, W_a is a trainable weight matrix, $(H_i, \ H_j)$ of is a pair of neighboring nodes, $\odot$ denotes a dot product, LeakyReLU is the activation function, c_{ij} represents the attention score, which can be interpreted as the edge weight. Softmax function normalizes the attention scores, as shown in (6):

$$\alpha_{ij} = \frac{\exp(c_{ij})}{\sum_{j' \in N_i} \exp(c_{ij'})} \tag{6}$$

where α_{ij} is the attention score obtained through computation. The formula for updating node representation is as (7):

$$H_i^o = \sigma \left(\sum_{j \in \mathcal{N}} \alpha_{ij} p_{ij} W_b (H_i \odot H_j) \right) \tag{7}$$

W_b represents the trainable linear transformation matrix, p_{ij} and α_{ij} are the attention scores calculated in the first and second computations, respectively. These scores are combined to calculate the feature interaction weight between node feature vectors. σ is the sigmoid activation function, H_i^o represents updated node features.

Multi-head Attention Mechanism. As aforementioned, we utilize a MHA to obtain polysemy of features interactions. In particular, H represents the number of heads, and feature vectors are mapped into H heads, in which every head denotes a latent dimension. Each head has its own attention scores and independently executes (7). Notably, we split the feature vectors into H parts, and use a linear layer in advance to change the dimension of the feature vectors to $H \times d$. Finally, the updated vectors from each head are concatenated, as follows (8):

$$H_i^o = ||_{h=1}^{H} \sigma \left(\sum_{j \in \mathcal{N}} \alpha_{ij}^h p_{ij} W_b^h (H_i \odot H_j) \right) \tag{8}$$

where $||$ represents the connection, α_{ij}^h is the attention calculation for the h-th head, and W_b^h is the trainable linear transformation matrix for the h-th head. For convenience in subsequent calculations, H_i^o is transformed into $H_i^o \in \mathbb{R}^d$ through a linear layer.

3.5 Filtering Enhancement Layer

Through a learnable filtering layer, we perform filtering operations on each dimension of the frequency domain features, followed by skip connections and layer normalization. Given the input representation matrix for the l-th layer as H^l, H^l is transformed into the frequency domain in (9):

$$X^l = F(H^l) \in C^{n \times d} \tag{9}$$

In this context, F is a one-dimensional FFT. X^l is a multi-dimensional vector, It represents the spectral representation of H^l. Next, we modulate the spectrum by multiplying it with a learnable filter, denoted as $W \in C^{n \times d}$. The formula is as (10):

$$\tilde{X}^l = W \odot X^l \tag{10}$$

Here, ⊙ represents matrix element-wise multiplication. Finally, the modulated signal $\tilde{X}^l$ is transformed from the spectral domain back to the time domain with inverse FFT as (11):

$$\tilde{H}^l \leftarrow F^{-1}(\tilde{X}^l) \in R^{n \times d} \tag{11}$$

F^{-1} is the inverse one-dimensional FFT. With FFT and inverse FFT, noise is filtered. Finally, to alleviate the issues of gradient vanishing and training instability, we employ operations such as skip connections and layer normalization, as shown in (12):

$$\tilde{H}^l = Layernorm(F^l + Dropout(\tilde{F}^l)) \tag{12}$$

3.6 Output Layer

We represent the output of the embedding layer as E^0, which is also the input for the interaction layer $E^0 = H^0$. Let $E^l_{uu,ii}$ denote the representation of nodes in the internal graph after interaction layer updates. E^l_{ui} represents the representation of nodes in the external graph after updating cross-interaction. To integrate updates from different graphs and different layers for reliable joint decision-making, we use a gated recurrent unit (GRU) to fuse E^0, $E^l_{uu,ii}$ and E^l_{ui}. The formula is as (13):

$$\mathcal{F}_g = \text{GRU}\left(E^0, E^l, E^l_{ui}\right)(g \in V) \tag{13}$$

$\mathcal{F}_g$ represents the final feature set for all nodes, $V = U + I$ denotes the set of all nodes, U is the set of nodes in the user graph, and I is the set of nodes in the item graph. Finally, we extract the user feature set $\mathcal{F}_U$ as well as the item feature set $\mathcal{F}_I$ from $\mathcal{F}_g$. We then calculate the average of feature sets to get overall features representations for the user graph E^F_u and the item graph E^F_i, The prediction $\hat{y}$ is obtained by taking the dot product of these representations:

$$\hat{y} = \text{sigmoid}(\text{sum}(E^F_u \odot E^F_i)) \tag{14}$$

Here, E^F_u, $E^F_i \in \mathbb{R}^{b \times d}$, $\hat{y} \in \mathbb{R}^b$, b represents the batch size. Through the sigmoid function, the range of values for $\hat{y}$ is constrained to (0, 1), indicating the probability of CTR by the system.

We employ a binary cross-entropy loss function predicting whether a user will click on an item:

$$L = -\left(y \cdot \log \hat{y} + (1 - y) \cdot \log (1 - \hat{y})\right) \tag{15}$$

where y is the true label of the data sample, and $\hat{y}$ is the predicted result of our model in this context.

4 Experimental Results and Analysis

In this section, empirical studies were conducted to validate the effectiveness of the FMHA model.

Table 1. COMPARISON RESULTS

Models	Metrics	MovieLens 1M	Book-Crossing	AliEC
FM	AUC	0.8761	0.7417	0.6171
	NDCG@5	0.8761	0.7616	0.0812
	NDCG@10	0.8761	0.8029	0.112
NFM	AUC	0.8985	0.7988	0.655
	NDCG@5	0.8486	0.7989	<u>0.0997</u>
	NDCG@10	0.8832	0.8326	0.1251
W&D	AUC	0.9043	0.8105	0.6531
	NDCG@5	0.8538	0.8048	0.0959
	NDCG@10	0.8538	0.8381	0.1242
DeepFM	AUC	0.9049	0.8127	<u>0.655</u>
	NDCG@5	0.851	0.8088	0.0974
	NDCG@10	0.8848	0.84	0.1243
AutoInt	AUC	0.9034	0.813	0.6434
	NDCG@5	0.8619	0.8127	0.0924
	NDCG@10	0.8931	0.8472	0.1206
Fi-GNNs	AUC	<u>0.9063</u>	0.8136	0.6462
	NDCG@5	0.8705	0.8094	0.0986
	NDCG@10	0.9029	0.8522	0.1241
GMCF	AUC	0.8998	<u>0.8255</u>	**0.6566**
	NDCG@5	<u>0.9412</u>	<u>0.8843</u>	0.0995
	NDCG@10	<u>0.9413</u>	<u>0.8989</u>	<u>0.1347</u>
FMHA	AUC	**0.9093**	**0.8652**	0.6512
	NDCG@5	**0.9467**	**0.9127**	**0.102**
	NDCG@10	**0.9471**	**0.9243**	**0.1369**
Improve (%)	AUC	0.33	4.82	-0.44
	NDCG@5	0.59	3.21	2.37
	NDCG@10	0.62	2.82	1.65

4.1 Experimental Environments

Datasets. We conduct experiments on three public datasets: MovieLens 1M, Bookcrossing, and AliEC. MovieLens 1M is a datasets for movie recommendations, including user attributes, movie attributes, and user ratings for movies. Each sample in Bookcrossing consists of a user with attributes, a book with attributes, and the user's rating for the book. AliEC is a log datasets from Taobao, containing user attributes and ad attributes.

Baselines. We compare the proposed FMHA with other CTR models that take into account attribute information. All the baselines models are as follow: FM [5], NFM [6], W&D [21], DeepFM [7], AutoInt [8], Fi-GNNs [27], GMCF [9].

Experimental Settings. We randomly split the datasets into training, validation, and test sets with a ratio of 6:2:2. Our model is implemented in PyTorch. We set the size of ID embeddings to 64, batch size to 128, and learning rate (lr) to 10^{-3} with Adam optimizer. Testing is performed after each epoch, and the total number of epochs is 50 here. Through experiments comparing different combinations of the number of heads and layers, on the MovieLens and Bookcrossing datasets, the optimal configuration is found with 2 heads (H) and 4 layers (L). For the AliEC dataset, the best performance was achieved with 3 heads (H) and 1 layer (L).

4.2 Overall Performance and Comparisons

According to the experimental results presented in Table 1, we analyze the performance of our FMHA with its comparative models. The best results are highlighted in bold, and the second-best results are underlined. As depicted in Table 1, except a 0.44% decrease in AUC than that of CMCF model on ALiEC datasets, FMHA model consistently surpasses other baselines across other evaluation metrics, which average improvements on three datasets - a 1.57% increase in AUC, 2.06% increase in NDCG5, and 1.69% increase in NDCG10 - highlight its effectiveness in RS tasks. This success is largely attributed to its advanced filter-enhanced and MHA mechanisms, enhancing its capability in complex feature interactions.

4.3 Study on Number of Heads and Layers

During our experiments, we find that the results varies along with the optimal head number of MHA and layer number of GNNs on three benchmarks. As illustrated in Fig. 3, we compared the model AUC under different combinations of heads and layers. In particular, both head and layer number is varied from 1 to 4 step by step, respectively. The x-axis represents the number of heads, different-colored lines represent different numbers of layers, and the y-axis represents the experimental results under the three evaluation metrics. It can be observed that for MovieLens 1M and Book-Crossing datasets, with the increase in the number of attention heads and layers, there is a discernible upward trend in the AUC for the model, and for those two datasets, our model achieves the best results with 2 heads and 4 layers. In contrast, AliEC datasets has best performance at 1 layer, as the number of layers increases, the model's performance significantly decreases to indicate its over-fitting. Similarly, when the number of heads exceeds 3, a similar phenomenon occurs in AliEC datasets. The best results of AliEC datasets are achieved with 3 head and 1 layer. From above experimental results, we find that excessive smoothing problem is caused by the number of GNN

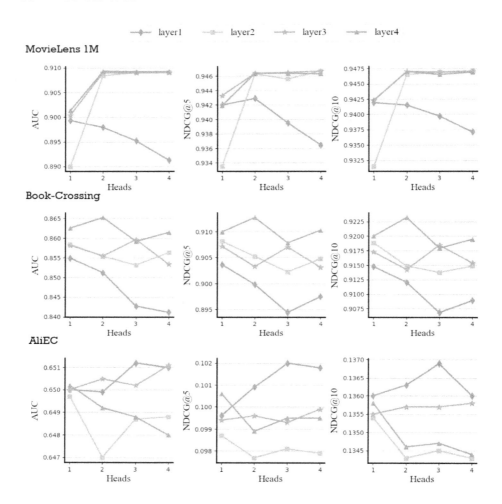

Fig. 3. Comparison of different numbers of heads and layers.

layers. This is because when the number of layers in GNN convolution operation is too large, each node can almost indirectly merge the information of all other nodes, resulting in the representation of all nodes tends to be the same. Our model achieves the best performance at 1 or 4 layers of GNNs. For the number of MHA heads, it can be seen that properly increasing the number of heads can improve the performance of the model, while excessive number of heads may lead to over-fitting or performance degradation. Therefore, appropriate parameters should be carried out according to different models.

4.4 Study on Ablation

We design a multi-head attention mechanism for feature interaction and a filter-enhanced component to filter noise in FHMA model. To assess the individual impacts of two principal components, we conduct an ablation study by independently disabling each component: Multi-Head Attention (MHA), Filter-Enhanced Algorithm(FEA). The findings from this study are presented in Table 2. The best results are highlighted in bold.

Excluding MHA from the model decreases averagely 1.61% in AUC, a 2.04% in NDCG@5, 1.61% in NDCG@10, identifies the strength of the correlations between features. The removal of FEA decrease averagely a 0.3% in NDCG@5, 0.46% in NDCG@10, indicating that the filter successfully attenuates the interference of signal noise on the model and enhances the representation of beneficial signals in the features. These results indicate that MHA and FEA component are useful for enhancing recommendation performance.

Table 2. ABLATION RESULTS

Models	Metrics	MovieLens 1M	Book-Crossing	AliEC
W/O MHA	AUC	0.8998	0.8255	0.6566
	NDCG@5	0.9412	0.8843	0.0995
	NDCG@10	0.9413	0.8989	0.1347
W/O FEA	AUC	0.909	0.8655	**0.6635**
	NDCG@5	0.9458	0.9114	0.1013
	NDCG@10	0.9461	0.9238	0.1353
FMHA	AUC	**0.9093**	**0.8656**	0.6512
	NDCG@5	**0.9467**	**0.9127**	**0.102**
	NDCG@10	**0.9471**	**0.9243**	**0.1369**

5 Conclusion

Improving the accuracy of CTR prediction models is of great significance for enhancing the practicality of recommendation systems. This paper aims to design an effective feature interaction mode that enhances useful feature interaction by introducing noise control. Our proposed FMHA model first identifies two types of attribute interactions, internal interaction and external interaction, and combines multi-head attention mechanism with interaction selection and filtering enhancement mechanism to provide guidance for extracting rich graph connections and feature interaction information. The empirical results indicate that our model performs better than existing methods on three datasets, and ablation experiments demonstrate the effectiveness of each part of the model. In future research, we plan to introduce higher-level feature interactions into the model structure, which may include considering edge learning and hypergraph matching.

References

1. Johnson, C.C., et al.: Logistic matrix factorization for implicit feedback data. Adv. Neural Inf. Process. Syst. **27**(78), 1–9 (2014)
2. McMahan, H.B., et al.: Ad click prediction: a view from the trenches. In Proceedings of the 19th ACM SIGKDD International Conference on Knowledge Discovery and Data Mining, pp. 1222–1230 (2013)
3. Yin, Y., Huang, C., Sun, J., Huang, F.: Multi-head self-attention recommendation model based on feature interaction enhancement. In: ICC 2022-IEEE International Conference on Communications, pp. 1740–1745. IEEE (2022)
4. Juan, Y., Zhuang, Y., Chin, W.-S., Lin, C.-J.: Field-aware factorization machines for CTR prediction. In: Proceedings of the 10th ACM Conference on Recommender Systems, pp. 43–50 (2016)
5. Rendle, S.: Factorization machines. In: 2010 IEEE International Conference on Data Mining, pp. 995–1000. IEEE (2010)
6. He, X., Chua, T.-S.: Neural factorization machines for sparse predictive analytics. In: Proceedings of the 40th International ACM SIGIR conference on Research and Development in Information Retrieval, pp. 355–364 (2017)
7. Guo, H., Tang, R., Ye, Y., Li, Z., He, X.: DeepFM: a factorization-machine based neural network for CTR prediction (2017). arXiv preprint arXiv:1703.04247
8. Song, W., et al.: AutoInt: automatic feature interaction learning via self-attentive neural networks. In: Proceedings of the 28th ACM International Conference on Information and Knowledge Management, pp. 1161–1170 (2019)
9. Su, Y., Zhang, R., Erfani, S.M., Gan, J.: Neural graph matching based collaborative filtering. In: Proceedings of the 44th International ACM SIGIR Conference on Research and Development in Information Retrieval, pp. 849–858 (2021)
10. Zhao, Z., Fang, Z., Li, Y., Peng, C., Bao, Y., Yan, W.: Dimension relation modeling for click-through rate prediction. In: Proceedings of the 29th ACM International Conference on Information & Knowledge Management, pp. 2333–2336 (2020)
11. Cheng, W., Shen, Y., Huang, L.: Adaptive factorization network: Learning adaptive-order feature interactions. In: Proceedings of the AAAI Conference on Artificial Intelligence, vol. 34, pp. 3609–3616 (2020)
12. Liu, B., et al. AutoFIS: automatic feature interaction selection in factorization models for click-through rate prediction. In: Proceedings of the 26th ACM SIGKDD International Conference on Knowledge Discovery & Data Mining, pp. 2636–2645 (2020)
13. Luo, Y., et al.: AutoCross: automatic feature crossing for tabular data in real-world applications. In: Proceedings of the 25th ACM SIGKDD International Conference on Knowledge Discovery & Data Mining, pp. 1936–1945 (2019)
14. Tao, Z., Wang, X., He, X., Huang, X., Chua, T.-S.: HoAFM: a high-order attentive factorization machine for CTR prediction. Inf. Process. Manag. **57**(6), 102076 (2020)
15. Heideman, M., Johnson, D., Burrus, C.: Gauss and the history of the fast fourier transform. IEEE ASSP Mag. **1**(4), 14–21 (1984)
16. Van Loan, C.: Computational frameworks for the fast Fourier transform. SIAM (1992)
17. Rabiner, L.R., Gold, B.: Theory and application of digital signal processing. Englewood Cliffs: Prentice-Hall (1975)
18. Soliman, S.S., Srinath, M.D.: Continuous and discrete signals and systems. Englewood Cliffs (1990)

19. Zhou, K., Yu, H., Zhao, W.X., Wen, J.-R.: Filter-enhanced MLP is all you need for sequential recommendation. In: Proceedings of the ACM Web Conference 2022, pp. 2388–2399 (2022)
20. Yu, T., Chen, J.: A novel sequential recommendation model based on the filter and model augmentation. In: 2023 International Joint Conference on Neural Networks (IJCNN), pp. 1–8. IEEE (2023)
21. Cheng, H.-T., et al.: Wide & deep learning for recommender systems. In: Proceedings of the 1st Workshop on Deep Learning for Recommender Systems, pp. 7–10 (2016)
22. Wang, R., Fu, B., Fu, G., Wang, M.: Deep & cross network for ad click predictions. In: Proceedings of the ADKDD'17, pp. 1–7 (2017)
23. Lian, J., Zhou, X., Zhang, F., Chen, Z., Xie, X., Sun, G.: xDeepFM: combining explicit and implicit feature interactions for recommender systems. In: Proceedings of the 24th ACM SIGKDD International Conference on Knowledge Discovery & Data Mining, pp. 1754–1763 (2018)
24. Xiao, J., Ye, H., He, X., Zhang, H., Wu, F., Chua, T.-S.: Attentional factorization machines: Learning the weight of feature interactions via attention networks (2017). arXiv preprint arXiv:1708.04617
25. Vaswani, A., et al.: Attention is all you need. In: Advances in Neural Information Processing Systems, vol. 30 (2017)
26. He, X., Deng, K., Wang, X., Li, Y., Zhang, Y., Wang, M.: LightGCN: simplifying and powering graph convolution network for recommendation. In: Proceedings of the 43rd International ACM SIGIR conference on research and development in Information Retrieval, pp. 639–648 (2020)
27. Li, Z., Cui, Z., Wu, S., Zhang, X., Wang, L.: Fi-GNN: modeling feature interactions via graph neural networks for CTR prediction. In: Proceedings of the 28th ACM International Conference on Information and Knowledge Management, pp. 539–548 (2019)
28. Guo, W., et al.: Dual graph enhanced embedding neural network for CTR prediction. In: Proceedings of the 27th ACM SIGKDD Conference on Knowledge Discovery & Data Mining, pp. 496–504 (2021)

Enhancing Sequential Recommendation via Aligning Interest Distributions

Yiyuan Zheng[1,2], Beibei Li[3], Beihong Jin[1,2(✉)], and Rui Zhao[1,2]

[1] Key Laboratory of System Software (Chinese Academy of Sciences) and State Key Laboratory of Computer Science, Institute of Software, Chinese Academy of Sciences, Beijing, China
Beihong@iscas.ac.cn
[2] University of Chinese Academy of Sciences, Beijing, China
[3] College of Computer Science, Chongqing University, Chongqing, China

Abstract. Contrastive learning improves the performance of sequential recommendation models by mining self-supervised information and mitigating the impact of data sparsity and noise interference. Existing contrastive sequential recommendation models pull the embeddings of positive sequence pairs close, and train sequence encoders to be invariant to data augmentations, e.g., reordering, which could destroy information beneficial for the recommendation task, e.g., the order of interactions. To alleviate the problem, we propose a contrastive sequential recommendation model IDARec, which adds projection heads between the sequence encoder and contrastive loss and builds the recommendation loss and contrastive loss in different hidden spaces. Specifically, IDARec introduces an interest distribution-based contrastive loss, which transforms sequence embeddings into multi-grained interest distributions and aligns the interest distributions of positive sequence pairs. Moreover, a clustering-classification approach is adopted to learn interest distributions, which learns interest prototypes by K-means first and then interest distributions by classification. We conduct extensive experiments on four public datasets, and the experimental results show that our model outperforms the state-of-the-art sequential recommendation models.

Keywords: Recommender Systems · Contrastive Learning · Interest Modeling

1 Introduction

Nowadays, to cope with the information explosion, recommender systems are widely applied in many scenarios such as e-commerce and video streaming applications. Sequential recommendation refers to achieving recommendation by predicting the next item of user historical interaction sequence, i.e., the target item. Deep sequential recommendation that incorporates deep learning for sequential recommendation has received a lot of attention. Generally speaking, these models leverage deep neural networks including recurrent neural networks (RNNs)

[7,10,25], graph neural networks (GNNs) [26] and attention networks (e.g. Transformer [11,24]) as the sequence encoder that encodes sequences into embeddings. Then, they predict the probability of the pairwise interactions between each user and each candidate item, and items with top-K interaction probabilities are selected as the final recommendation.

Since each user only interacts with a small percentage of items in the full item set and most of them are implicit feedback, sequential recommendation faces serious data sparsity and noise interference. To mitigate the impact of sparsity and noise on recommendation performance, contrastive learning is introduced into the sequential recommendation. So far, some contrastive sequential recommendation models have been developed [17,20,27], which perform contrastive learning as an auxiliary task to help the main task, i.e., the recommendation task, and improve the performance by mining self-supervised information. Moreover, these models construct contrastive losses related to sequence embeddings, that is, they obtain multiple positive sequence pairs, and then construct a contrastive loss to pull the embeddings of positive sequence pairs close, and push the embeddings of different sequences away.

Previous work [2] in computer vision points out contrastive loss in the representation space could remove information that may be useful for the downstream task, for example, it could optimize the encoder to neglect the orientation of objects that is useful for classification. Motivated by this, we argue that the sequence embedding-based contrastive loss in existing sequential recommendation models could result in an information loss problem, that is, it could destroy the information beneficial for the recommendation task, e.g., the order of interactions, since it requires the sequence embedding to be invariant to data augmentation including reordering.

A promising solution [2] to address the information loss problem is to project the sequence embeddings to another hidden space before calculating the contrastive loss. Along this line, in this paper, we design semantic projection heads that map sequence embeddings into interest distributions, so that the recommendation loss and contrastive learning loss can be built in different hidden spaces. Then, we construct a contrastive loss to align the interest distribution vectors implied by the two positive sequences. The insights behind this are that the user behaviors in the sequence are driven by multiple interests and the interest distribution is much more robust to data augmentation than sequence embeddings. Nevertheless, exploring the distribution of user interests is challenging, as they are unobserved variables hidden in user interactions. Considering the collaborative information between items, we utilize implicit item clusters to construct user interest distributions.

Our contributions are summarized as follows.

1. We propose a contrastive sequential recommendation model IDARec, which contains an interest distribution-based contrastive loss that transforms sequence embeddings to multi-grained interest distributions and aligns the interest distributions by positive sequence pairs via KL divergence.
2. We adopt a clustering-classification approach to learn the underlying interest distributions. Specifically, we cluster all the candidate items with K-means

and treat the cluster centers as interest prototypes, then classify the sequence embedding into each prototype and obtain the interest distributions.
3. We conduct extensive experiments on four public datasets. The experimental results show that our model outperforms the existing regular or contrastive sequential recommendation models in terms of HR and NDCG.

2 Related Work

2.1 General Sequential Recommendation

Sequential recommendation models learn user representations from their historical interaction sequences, assuming that historical behaviors reflect user preferences. Early sequential recommendation models adopt Markov Chains to capture transition relationships between items [21]. With the powerful modeling capability of deep learning, deep sequential recommendation models utilize deep neural networks to capture user interests more accurately. For example, GRU4Rec [10] adopts RNNs to learn the representation of user historical interaction sequences. Caser [23] treats the matrix formed by the embeddings of interacted items as an image and learns the sequence embedding with convolutional neural networks (CNNs). SR-GNN [26] and GC-SAN [28] take the items in each sequence as nodes, construct edges according to the transition relationship, and then learn the sequence representations with GNNs. Currently, attention-based sequential recommendation models achieve excellent performance [11,16,22]. For example, SASRec [11] follows the self-attention structures in Transformer to construct the sequence encoder and achieve promising recommendation performance and BERT4Rec [22] utilizes a bi-directional attention module to learn sequence representations by combining sequential information in both directions.

2.2 Contrastive Learning for Recommendation

Contrastive learning has achieved remarkable performance in computer vision and natural language processing [2–5,8,9]. It treats augmented instances of the same instance as positive pairs, and augmented instances of different instances as negative pairs, then constructs the contrastive loss to bring positive pairs closer and push the negative pairs away. Contrastive learning is also introduced into the recommendation to improve representation learning [12]. Combining contrastive learning into recommendation has attracted significant research attention in recent years and is widely applied in different scenarios [13,15,17–20,27,29,30]. For example, following contrastive learning in natural language processing, CL4SRec [27] proposes three data augmentation methods including random masking, cropping and reordering to construct positive sequence pairs, and build a contrastive sequential recommendation framework that jointly optimizes the main recommendation loss and the contrastive loss. CoSeRec [17] believes that item correlation-oriented augmentation schemes can be more informative, and design two augmentation schemes based on item correlation. Despite

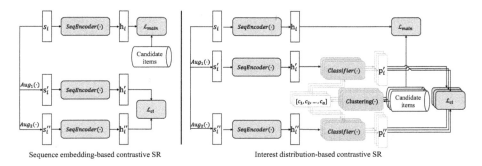

Fig. 1. *Sequence embedding-based contrastive sequential recommendation (SR) (left) vs. Interest distribution-based contrastive SR (right).* In the sequence embedding-based contrastive SR model, the embeddings from different augmentations of the same sequence are compared directly to each other. In interest distribution-based contrastive SR, we first obtain interest distributions by clustering and classifying, then align the two-view distributions.

self-supervised data-level augmentation, DuoRec [20] proposes a model-level data augmentation approach. PDMRec [29], a micro-video sequential recommendation model, proposes contrastive learning strategies that reduce the interference from micro-video positions in sequences. VA-GNN [12] introduces contrastive learning to align the item embedding across views. Besides, ICLRec [6] explores users latent intent and constructs an intent contrastive loss. However, existing contrastive sequential recommendation models construct embedding-oriented contrastive loss, which could remove information that is useful for the recommendation task.

3 Methodology

3.1 Overview

We denote the user set as $\mathcal{U}$, the item set as $\mathcal{V}$, and the embedding of an item $v \in \mathcal{V}$ as $\mathbf{v} \in \mathbb{R}^d$, where d is the dimension of embeddings. For a user $u_i \in \mathcal{U}$, we sort his/her interactions in ascending order by timestamp, obtaining an interaction sequence $\mathbf{s}_i = [v_{i1}, v_{i2}, \ldots, v_{i|s_i|}]$. Our goal is to build a model that predicts the next item that the user u_i is most likely to interact with at the $|s_i| + 1$ step among the item set $\mathcal{V}$ based on the sequence $\mathbf{s}_i$:

$$\text{argmax}_{v_j \in \mathcal{V}} P\left(v_{|s_i|+1} = v_j \mid s_i\right). \tag{1}$$

We select the items with top-K interaction probability as the final recommendation.

IDARec we propose is an interest distribution-based contrastive sequential recommendation model, as shown on the right in Fig. 1. It employs a Transformer-based sequence encoder the same as that in SASRec [11] and

CL4SRec [27], denoted as $SeqEncoder\,(\cdot)$, to obtain embedding $\mathbf{h}_i$ for sequence s_i, i.e., $\mathbf{h}_i = SeqEncoder(s_i)$. and collectively optimizes the recommendation task and the contrastive learning task. To achieve contrastive learning, it learns the user interest distributions reflected in each sequence by two steps, i.e., clustering and classifying, then constructs multi-granularity contrastive losses by aligning the interest distributions of two augmented sequences of the same sequence.

3.2 Multi-grained Interest Distribution Learning

To alleviate the information loss caused by sequence embedding-based contrastive loss, we design projection heads for mapping sequence embeddings into another latent space, i.e., the interest distribution space, and then construct a contrastive loss in the new space.

As we know, user interests are formed by user preferences for a series of correlated items, so learning user interest distributions requires discovering the implicit correlations between items. To achieve this target, motivated by that item embeddings reflect their collaborative information, we cluster all the items into several clusters according to their embedding similarity and treat the embeddings of the cluster centers as the user interest prototypes.

We apply K-means for clustering, where the item embeddings are normalized before clustering. Let the number of clusters be α, the average embedding of all the items in each cluster is regarded as the representation of the clustering centers. Thus, the user interest prototypes $[\mathbf{c}_1, \ldots, \mathbf{c}_\alpha]$ are calculated as:

$$[\mathbf{c}_1, \ldots, \mathbf{c}_\alpha] = \mathtt{Clustering}(\mathcal{V}, \alpha). \tag{2}$$

The probability of classifying sequence s_i to the j-th interest prototype is calculated as:

$$p_{ij} = \mathtt{Classifier}_\alpha(\mathbf{h}_i, \mathbf{c}_j) = \frac{e^{\mathrm{sim}(\mathbf{c}_j, \mathbf{h}_i)}}{\sum_{k=1}^{\alpha} e^{\mathrm{sim}(\mathbf{c}_k, \mathbf{h}_i)}}, \tag{3}$$

where $\mathrm{sim}\,(\mathbf{c}_j, \mathbf{h}_i) = \frac{\mathbf{c}_j^T \mathbf{h}_i}{\tau \|\mathbf{c}_j\|_2 \|\mathbf{h}_i\|_2}$, τ is the temperature parameter. Computing the classification probability of user u_i to each interest prototype successively, we obtain the interest distribution vector of the user u_i as $\mathbf{p}_i = [p_{i1}, \ldots, p_{i\alpha}]$.

By adjusting the number of clusters α, we can obtain multiple sets of interest prototypes with different granularity, and then derive multi-grained interest distributions.

3.3 Interest Distribution-Based Contrastive Loss

User interests implied by two augmented sequences of the same sequences are similar. Therefore, the interest distribution-based contrastive loss aims to keep the consistency of the interest distributions underlying the augmented sequences.

As for the two augmented sequences s_i' and s_i'' of the user u_i, we calculate their interest distributions as $\mathbf{p}_i'$ and $\mathbf{p}_i''$, respectively. Then, we utilize the KL divergence to build the contrastive loss, as shown in the Eq. 4.

$$\mathcal{L}_{cl}^{\alpha}(\boldsymbol{p}',\boldsymbol{p}'') = \frac{1}{2}\left(KL\left(\boldsymbol{p}_i'\|\boldsymbol{p}_i''\right) + KL\left(\boldsymbol{p}_i''\|\boldsymbol{p}_i'\right)\right)$$
$$= \frac{1}{2}\left(\sum_{k=1}^{\alpha} p_{ik}'\log\left(\frac{p_{ik}'}{p_{ik}''}\right) + \sum_{k=1}^{\alpha} p_{ik}''\log\left(\frac{p_{ik}''}{p_{ik}'}\right)\right). \quad (4)$$

Specifically, we build three contrastive losses of different granularity, where $\alpha_1 = 16, \alpha_2 = 64, \alpha_3 = 256$. The final contrastive loss is calculated as $\mathcal{L}_{cl} = \left(\mathcal{L}_{cl}^{\alpha_1} + \mathcal{L}_{cl}^{\alpha_2} + \mathcal{L}_{cl}^{\alpha_3}\right)/3$.

3.4 Loss Function

A recent work [1] reveals the essential of leveraging embedding normalization in the recommendation. Therefore, different from existing sequential recommendation models, such as SASRec and CL4SRec, IDARec adopts cosine similarity rather than dot product to measure user preferences, which leads to better performance in practice. For the user u_i, after obtaining the embedding of his/her interaction sequence $\boldsymbol{h}_i$, his/her preference score with each candidate item $v_j \in \mathcal{V}$ is predicted as:

$$\hat{y}_{ij} = \frac{\boldsymbol{h}_i^T \boldsymbol{v}_j / \epsilon}{\|\boldsymbol{h}_i\|_2 \|\boldsymbol{v}_j\|_2}, \quad (5)$$

where ϵ is the temperature parameter.

We treat sequential recommendation as a next-item prediction problem and regard the target item v_t as a positive sample and other candidate items as negative samples, then construct the cross-entropy loss as:

$$\mathcal{L}_{rec}(u_i) = -\ln \frac{\exp(\hat{y}_{it})}{\sum_{v_* \in \mathcal{V}} \exp(\hat{y}_{v_*})}. \quad (6)$$

The total loss is as Equation (7), where λ is the weight of the contrastive loss.

$$\mathcal{L} = \mathcal{L}_{rec} + \lambda \mathcal{L}_{cl}, \quad (7)$$

The procedure of model training is shown in Algorithm 1. Considering the time overhead of clustering, we perform item clustering once per epoch. For each mini-batch of user sequences, we calculate the recommendation loss and the contrastive loss, then update the network parameters via gradient descent.

3.5 Discussion

Connections with CL4SRec. Compared with CL4SRec [27], which builds upon the similarity of sequence embeddings, IDARec is equivalent to constructing the contrastive loss after projecting the sequence embeddings into another hidden space, i.e., the interest distribution space. Each set of interest prototypes can be

Algorithm 1: IDARec training.

Input: Hyper-parameters: λ, τ, ϵ; Numbers of clusters: $[\alpha_1, \alpha_2, \alpha_3]$; Batch size: N; Max training epochs: B; User interaction sequence set: $\mathcal{S}$.
Output: $SeqEncoder(\cdot)$.

1 Initialization the parameters in $SeqEncoder(\cdot)$;
2 **while** $epoch \leq B$ and not converged **do**
3   **for** $k \leftarrow 1$ to 3 **do**
4   Cluster candidate items into α_k clusters with K-means, and obtain the interest prototypes $[\boldsymbol{c}_1, \boldsymbol{c}_2, \ldots, \boldsymbol{c}_{\alpha_k}] = Clustering(\mathcal{V}, \alpha_k)$.
5   **end**
6   **for** sampled minibatch $\{s_i\}_{i=1}^N$ **do**
7   **for** $i \leftarrow 1$ to N **do**
8   Sample two augmentation operators:
                $Aug_1(\cdot), Aug_2(\cdot) \sim \{Mask, Crop, Reorder\}$;
9             $s'_i = Aug_1(s_i), s''_i = Aug_2(s_i)$;
10            $\boldsymbol{h}_i = SeqEncoder(s_i)$;
11            $\boldsymbol{h}'_i = SeqEncoder(s'_i), \boldsymbol{h}''_i = SeqEncoder(s''_i)$;
12   **for** $k \leftarrow 1$ to 3 **do**
13                 $\boldsymbol{p}'_i = Classifier_{\alpha_k}(\boldsymbol{h}'_i)$;
14                 $\boldsymbol{p}''_i = Classifier_{\alpha_k}(\boldsymbol{h}''_i)$;
15                 $\mathcal{L}_{cl}^{\alpha_k} = \frac{1}{N}\sum_{u=1}^{N} \mathcal{L}_{cl}^{\alpha_k}(\boldsymbol{h}'_i, \boldsymbol{h}''_i)$;
16   **end**
17   **end**
18         $\mathcal{L} = \frac{1}{N}\sum_{i=1}^{N} \mathcal{L}_{rec}(u_i) + \lambda(\mathcal{L}_{cl}^{\alpha_1} + \mathcal{L}_{cl}^{\alpha_2} + \mathcal{L}_{cl}^{\alpha_3})/3$;
19   Update network $SeqEncoder(\cdot)$ by minimizing $\mathcal{L}$;
20   **end**
21 **end**

regarded as a projection head $g(\cdot)$. Besides, IDARec employs a KL divergence-based contrastive loss rather than an InfoNCE loss. Last but not least, IDARec predicts interaction scores via cosine similarity rather than inner product, which contributes to the recommendation improvement a lot.

Time Complexity Analysis. In the training phase, the computation costs of our proposed method are mainly from two parts, i.e., K-means clustering and sequence encoding. For every epoch, the time complexity of K-means is $O(m|V|d\sum_{k=1}^{3}\alpha_k)$, where m is the maximum iteration number in clustering (m is set to 20 in this paper). And the time complexity of sequence encoding is 3 times of Transformer-based sequential recommendation with only next item prediction task, i.e., $O(\sum_{u_i \in \mathcal{U}} 3(|s_i|^2 d + |s_i| d^2))$, where the overall complexity is dominated by the term $O(\sum_{u_i \in \mathcal{U}} 3|s_i|^2 d)$. Thus, the time complexity for each epoch is $O(m|V|d\sum_{k=1}^{3}\alpha_k + \sum_{u_i \in \mathcal{U}} 3|s_i|^2 d)$. In the testing/predicting phase, the contrastive learning module in IDARec is no longer needed, and the time complexity is $O(\sum_{u_i \in \mathcal{U}} |s_i|^2 d)$.

Table 1. Statistics of the datasets after preprocessing

Dataset	#Users	#Items	#Interactions	Density
Beauty	22363	12101	198502	0.07%
Toy_and_Games	19412	11924	167597	0.07%
TikTok	30470	35713	684192	0.06%
WeChat_Channels	6631	8215	118984	0.22%

4 Experiments

4.1 Experimental Setting

Datasets. We select four public datasets to conduct experiments, including two e-commerce datasets and two micro-video datasets. Beauty and Toy_and_Games (Toy for short) are the sub-categories of the widely used Amazon datasets. WeChat_Channels (WeChat for short) and TikTok are two public datasets released by WeChat Big Data Challenge 2021[1] and the ICME Micro-Video Understanding Challenge 2019[2], respectively.

For the two Amazon datasets, we naturally treat user clicks as positive feedback. For the micro-video datasets, we take like as the positive feedback. We randomly select 10% of the users in the TikTok dataset to conduct our experiments. Users or items appearing less than 5 times are removed. The statistics of the processed datasets are shown in Table 1. For each positive interaction sequence $s_u = [v_1^u, v_2^u...v_{|s_u|}^u]$, we leave $v_{|s_u|-1}^u$ for validation and the last interaction $v_{|s_u|}^u$ for testing. The remaining interactions $[v_1^u, v_2^u..v_{|s_u|-2}^u]$ are used for training.

Evaluation Metrics. We follow the common evaluation metrics in the next item prediction task, i.e., HR@K and NDCG@K. In this paper, we report results when the value K is set to [20, 50].

Compared Methods. In this paper, we compare IDARec to four regular sequential recommendation models, i.e., NARM [14], GRU4Rec [10], SASRec [11] and BERT4Rec [22]. We also compare IDARec to three contrastive sequential recommendation models, i.e., CL4SRec [27], DuoRec [20] and ICLRec [6].

- **NARM** [14]: It captures the user's main purpose and models user preferences with an attention mechanism.
- **GRU4Rec** [10]: It is a sequential recommendation model that encodes sequences through a RNN composed of GRU units.
- **SASRec** [11]: One of the most popular sequential recommendation models. It adopts the multi-head self-attention mechanism to capture the dynamic user preferences, and achieves excellent performance on multiple public datasets.

---
[1] https://algo.weixin.qq.com/2021/problem-description.
[2] https://www.biendata.xyz/competition/icmechallenge2019.

- **BERT4Rec** [22]: It is a sequential recommendation model designed by BERT structure, which leverages a bidirectional self-attention mechanism to learn sequence representations.
- **CL4SRec** [27]: It introduces contrastive learning into sequential recommendation. Masking, cropping, and reordering are applied to construct positive sequence pairs and a sequence embedding-based contrastive loss is utilized to enhance sequence representations.
- **DuoRec** [20]: It improves the performance of the model by the self-supervised contrastive learning and the supervised contrastive learning, and alleviates the problem of representation degradation in contrastive sequential recommendation models.
- **ICLRec** [6]: It leverages the latent intents for the sequential recommendation and improves the robustness against data sparsity and noisy interaction via intent contrastive learning.

Implementation Details. We implement our model in PyTorch. The model is optimized by an Adam optimizer with a learning rate of 0.001. We set the dimension of embeddings to 64 and batch size to 512. The maximum sequence length is set to 100. For the self-attention module, we set the number of attention heads to 2, and the number of layers to 2. We initialize the model parameters with a normal distribution in the range $[-0.02, 0.02]$. We tune λ, τ and ϵ within $\{0.01, 0.1, 0.2, 1\}$, $\{0.01, 0.1, 0.2, 1\}$ and $\{0.01, 0.1, 0.2, 1\}$, respectively. λ is set as 0.2 for WeChat and 0.1 for the other datasets. τ and ϵ are set as 0.1 across all the datasets.

If a sequence is shorter than 100, we pad it with padding items, otherwise, we truncate a sub-sequence with the most recent 100 interactions. Besides, we expand the training set with prefix subsequences. Following the previous work [27], we apply masking, reordering and cropping to generate the augmented sequences. In the training process, for each user sequence, we randomly select two augmentation methods to obtain a positive sequence pair.

At the beginning of the training stage, we utilize the sequence embedding-based contrastive loss to warm up, which helps to build basic semantic information. The number of warm-up epochs is set to 2. In order to accelerate the training, Faiss[3], a library for efficient similarity search and clustering of dense vectors based quantization techniques, is used for K-means clustering. We train the model by an early stopping strategy, that is, we stop training the model if HR@20 does not improve on the validation set for 10 epochs consecutively and adopt model parameters achieving the best performance on the validation set for testing.

Among all the compared methods, NARM, GRU4Rec, SASRec, and BERT4Rec are implemented based on a popular open-source recommendation

---

[3] https://github.com/facebookresearch/faiss

Table 2. Recommendation performance on four datasets. We bold the best results and underline the second best results of each metric. The last column is the relative improvements compared with the best baseline results. Besides, H@K is short for HR@K, while N@K is short for NDCG@K.

Dataset	Metric	NARM	GRU4Rec	BERT4Rec	SASRec	CL4SRec	DuoRec	ICLRec	IDARec	Improv.
Beauty	H@20	0.0836	0.0917	0.0565	0.1120	0.1143	<u>0.1147</u>	0.0896	**0.1264**	10.20%
	H@50	0.1287	0.1459	0.1024	0.1684	0.1720	<u>0.1751</u>	0.1477	**0.1903**	8.68%
	N@20	0.0383	0.0409	0.0203	0.0465	0.0468	<u>0.0499</u>	0.0322	**0.0525**	5.21%
	N@50	0.0472	0.0516	0.0293	0.0577	0.0582	<u>0.0619</u>	0.0437	**0.0652**	5.33%
Toys	H@20	0.0710	0.0746	0.0592	0.1120	0.1143	<u>0.1203</u>	0.0930	**0.1309**	8.81%
	H@50	0.1147	0.1235	0.0993	0.1684	<u>0.1720</u>	0.1717	0.1483	**0.1918**	11.51%
	N@20	0.0322	0.0317	0.0216	0.0465	0.0468	<u>0.0520</u>	0.0344	**0.0539**	3.65%
	N@50	0.0408	0.0414	0.0294	0.0577	0.0582	<u>0.0622</u>	0.0453	**0.0642**	3.21%
TikTok	H@20	0.0491	0.0514	0.0618	0.0786	0.0753	<u>0.0808</u>	0.0445	**0.0881**	9.03%
	H@50	0.0932	0.0980	0.0922	0.1233	0.1213	<u>0.1254</u>	0.0580	**0.1399**	11.56%
	N@20	0.0192	0.0199	0.0397	0.0466	0.0397	<u>0.0467</u>	0.0341	**0.0480**	2.78%
	N@50	0.0279	0.0290	0.0457	0.0554	0.0488	<u>0.0555</u>	0.0368	**0.0582**	4.86%
Wechat	H@20	0.0704	0.0706	0.0460	0.0692	0.0733	<u>0.0787</u>	0.0600	**0.0946**	20.20%
	H@50	0.1351	0.1425	0.0992	0.1314	0.1428	<u>0.1484</u>	0.1240	**0.1810**	21.96%
	N@20	0.0261	0.0272	0.0164	0.0236	0.0260	<u>0.0275</u>	0.0221	**0.0364**	32.36%
	N@50	0.0388	0.0413	0.0268	0.0359	0.0397	<u>0.0413</u>	0.0347	**0.0534**	29.29%

framework Recbole[4]. CL4SRec, DuoRec[5] and ICLRec[6] are implemented with their open-source code. For fairness, The embedding dimension, batch size, the maximum sequence length and the hyperparameters in attention-modules are set to be consistent with our model. We search the weight of contrastive loss in {0.01,0.1,0.5,1} to achieve their best performance.

4.2 Performance Comparison

Table 2 reports the performance of different models on each dataset. Based on the results, we have the following observations.

Among all the regular sequential recommendation models, SASRec outperforms others in most datasets, which indicates that self-attentive blocks in Transformer can capture user preferences better than vanilla attention and RNNs, and verifies the effectiveness of SASRec. We observe that GRU4Rec achieves the best performance on the WeChat-Channels dataset compared to other regular sequential recommendation models, which suggests that GRU4Rec may be more suitable for denser datasets.

Compared to regular sequential recommendation models, contrastive sequential recommendation models that contain sequence embedding-based contrastive learning modules, i.e., CL4SRec and DuoRec, achieve better performance, which

---

[4] https://github.com/RUCAIBox/RecBole.
[5] https://github.com/RuihongQiu/DuoRec.
[6] https://github.com/salesforce/ICLRec.

Table 3. Ablation study of IDARec (HR@20).

Model	Datasets			
	Beauty	Toys	TikTok	WeChat
(A) w/o CL loss	0.1222 (↓ 0.42%)	0.1276 (↓ 0.33%)	0.0849 (↓ 0.32%)	0.0860 (↓ 0.86%)
(B) (A) + single-grained(16)	0.1248 (↓ 0.16%)	**0.1309** (—)	0.0870 (↓ 0.11%)	0.0881 (↓ 0.65%)
(C) (A) + single-grained(64)	0.1257 (↓ 0.07%)	0.1285 (↓ 0.24%)	**0.0893** (↑ 0.12%)	0.0938 (↓ 0.08%)
(D) (A) + single-grained(256)	0.1243 (↓ 0.21%)	0.1268 (↓ 0.41%)	0.0876 (↓ 0.05%)	0.0894 (↓ 0.52%)
(E) IDARec	**0.1264**	**0.1309**	0.0881	**0.0946**

illustrates the effectiveness of contrastive learning for sequential recommendation. Besides, DuoRec generally outperforms CL4SRec. The reason might be that DuoRec utilizes the sequences with the same target item to construct positive sequence pairs, and combines unsupervised and supervised contrastive learning to alleviate the representation degeneration problem. Though ICLRec declares that it enhances the robustness of the model by latent intents, its overall performance is not as good as DuoRec.

IDARec achieves the best performance in all metrics on the four datasets. For example, on WeChat_Channels, IDARec has an average improvement of 21.08% on HR. The excellent performance of IDARec can be attributed to the proposed interest distribution-based contrastive loss, which does not pull embeddings of positive sequence pairs close directly and alleviates the information loss problem.

4.3 Ablation Study

We conduct the ablation study on the multi-grained interest distribution-based contrastive loss. The results are shown in Table 3, where (A) is IDARec without contrastive loss. (B)-(D) are the results of (A) equipped with a single-grained interest distribution-based contrastive loss, and the number of clusters is in parentheses. (E) is the original model.

According to the results, the variant model (A) performs worse than IDARec on the four datasets, which illustrates the effectiveness of the strategy of aligning interest distributions. In most datasets, IDARec outperforms the variant models using single-grained contrastive loss. This suggests that mining multi-grained interest distributions may provide more information for the model.

4.4 Impact of Projection

IDARec maps sequence embeddings into interest distributions by the carefully designed projection heads, i.e., the clustering-classification approach. In order to analyze the impact of the semantic interest prototypes, we design a model variant $IDARec_{Random}$, which replaces interest prototypes with a group of random vectors that are not updated. Besides, we construct a model variant of CL4SRec, denoted as $CL4SRec_{cos}$, which is enhanced by calculating prediction score via cosine similarity instead of inner dot. $CL4SRec_{cos}$ does not conduct projection

Table 4. Impact of Projection.

	TikTok		Beauty	
	HR@20	NDCG@20	HR@20	NDCG@20
CL4SRec$_{cos}$	0.0847 (↓ 0.34%)	0.0417 (↓ 0.63%)	0.1235 ↓ (0.29%)	0.0509 ↓ (0.16%)
IDARec$_{random}$	0.0855 (↓ 0.26%)	0.0451 (↓ 0.29%)	0.1237 ↓ (0.27%)	0.0517 ↓ (0.08%)
IDARec	**0.0881**	**0.0480**	**0.1264**	**0.0525**

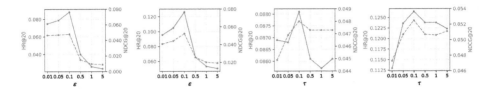

Fig. 2. Parameter sensitivity of the temperatures ϵ and τ on two datasets. (Left: TikTok, Right: Beauty)

before calculating the contrastive loss, thus can be regarded as a baseline model to analyze the impact of projection.

The experimental results are shown in Table 4. The results illustrate that replacing the semantic interest prototypes learned by K-means with random vectors leads to a performance decrease, which indicates that it is effective to design semantic projection heads related to the recommendation task. In addition, we find that CL4SRec$_{cos}$ performs worse than the other two model variants, even the IDARec$_{Random}$ whose projection head is based on a set of random and static vectors, which suggests that projecting embeddings into another hidden space does help the contrastive learning task achieve better compatibility with recommendation task, thus mitigate the information loss caused by contrastive learning.

4.5 Sensitivity of Temperature Parameters

In this section, the sensitivity of two temperature parameters ϵ and τ is analyzed. The two temperature parameters are used when calculating the interaction scores and the sequence-interest similarity, respectively. We tune the two parameters with [0.01,0.05,0.1,0.5,1,5]. The results are shown in Fig. 2.

From the results, we can find that the best value for ϵ and τ are both 0.1. As the temperature parameter increases, the performance rises first and then falls. Large ϵ makes performance drop rapidly. The reason may be that a large value of ϵ makes the model hard to distinguish positive and negative samples, which is harmful to the model convergence. On the contrary, the decrease of temperature τ has a greater impact on the performance than the increase. The reason may be that when the temperature is too small, the differences between interests may be over-amplified, so that it cannot tolerate the noise caused by augmentation, which makes aligning interest distributions extremely difficult so that the model cannot learn useful information.

5 Conclusion

In this paper, we propose a contrastive sequential recommendation model IDARec, which employs Transformer as the sequence encoder and includes a simple but effective contrastive loss based on interest distributions. Instead of pulling embeddings of positive sequence pairs closer directly, the proposed loss projects sequence embeddings to interest distributions and constrains the consistency of the interest distributions implied by augmented sequences of the same sequence. Extensive experiments show that IDARec outperforms existing sequential recommendation models. Besides, the proposed contrastive loss is generalized and can be applied to improve other sequential recommendation models.

Acknowledgment. This work was supported by the National Natural Science Foundation of China under Grant No. 62072450 and Meituan.

References

1. Chen, J., Wu, J., Wu, J., Cao, X., Zhou, S., He, X.: Adap-τ: adaptively modulating embedding magnitude for recommendation. In: WWW, pp. 1085–1096 (2023)
2. Chen, T., Kornblith, S., Norouzi, M., Hinton, G.: A simple framework for contrastive learning of visual representations. In: ICML, pp. 1597–1607 (2020)
3. Chen, T., Kornblith, S., Swersky, K., Norouzi, M., Hinton, G.E.: Big self-supervised models are strong semi-supervised learners. In: NeurIPS, pp. 22243–22255 (2020)
4. Chen, X., Fan, H., Girshick, R., He, K.: Improved baselines with momentum contrastive learning (2020). arXiv preprint arXiv:2003.04297
5. Chen, X., Xie, S., He, K.: An empirical study of training self-supervised vision transformers. In: CVPR, pp. 9640–9649 (2021)
6. Chen, Y., Liu, Z., Li, J., McAuley, J., Xiong, C.: Intent contrastive learning for sequential recommendation. In: WWW, pp. 2172–2182 (2022)
7. Chung, J., Gulcehre, C., Cho, K., Bengio, Y.: Empirical evaluation of gated recurrent neural networks on sequence modeling. In: NIPS 2014 Workshop on Deep Learning (2014)
8. Gao, T., Yao, X., Chen, D.: SimCSE: simple contrastive learning of sentence embeddings. In: EMNLP, pp. 6894–6910 (2021)
9. He, K., Fan, H., Wu, Y., Xie, S., Girshick, R.: Momentum contrast for unsupervised visual representation learning. In: CVPR, pp. 9729–9738 (2020)
10. Hidasi, B., Karatzoglou, A., Baltrunas, L., Tikk, D.: Session-based recommendations with recurrent neural networks. In: ICLR (2016)
11. Kang, W.C., McAuley, J.: Self-attentive sequential recommendation. In: ICDM, pp. 197–206 (2018)
12. Lai, W., Jin, B., Li, B., Zheng, Y., Zhao, R.: A vlogger-augmented graph neural network model for micro-video recommendation. In: De Francisci Morales, G., Perlich, C., Ruchansky, N., Kourtellis, N., Baralis, E., Bonchi, F. (eds.) Machine Learning and Knowledge Discovery in Databases: Applied Data Science and Demo Track. ECML PKDD 2023. LNCS(), vol. 14174. Springer, Cham (2023). https://doi.org/10.1007/978-3-031-43427-3_41

13. Li, B., Jin, B., Song, J., Yu, Y., Zheng, Y., Zhuo, W.: Improving micro-video recommendation via contrastive multiple interests. In: SIGIR, pp. 2377–2381 (2022)
14. Li, J., Ren, P., Chen, Z., Ren, Z., Lian, T., Ma, J.: Neural attentive session-based recommendation. In: CIKM, pp. 1419–1428 (2017)
15. Li, X., et al.: Multi-intention oriented contrastive learning for sequential recommendation. In: Proceedings of the Sixteenth ACM International Conference on Web Search and Data Mining, pp. 411–419 (2023)
16. Liu, Q., Zeng, Y., Mokhosi, R., Zhang, H.: STAMP: short-term attention/memory priority model for session-based recommendation. In: SIGKDD, pp. 1831–1839 (2018)
17. Liu, Z., Chen, Y., Li, J., Yu, P.S., McAuley, J., Xiong, C.: Contrastive self-supervised sequential recommendation with robust augmentation (2021). arXiv preprint arXiv:2108.06479
18. Ma, J., Zhou, C., Yang, H., Cui, P., Wang, X., Zhu, W.: Disentangled self-supervision in sequential recommenders. In: Proceedings of the 26th ACM SIGKDD International Conference on Knowledge Discovery & Data Mining, pp. 483–491 (2020)
19. Qin, X., Yuan, H., Zhao, P., Liu, G., Zhuang, F., Sheng, V.: Intent contrastive learning with cross subsequences for sequential recommendation (2023)
20. Qiu, R., Huang, Z., Yin, H., Wang, Z.: Contrastive learning for representation degeneration problem in sequential recommendation. In: WSDM, pp. 813–823 (2022)
21. Rendle, S., Freudenthaler, C., Schmidt-Thieme, L.: Factorizing personalized markov chains for next-basket recommendation. In: WWW, pp. 811–820 (2010)
22. Sun, F., et al.: BERT4Rec: sequential recommendation with bidirectional encoder representations from transformer. In: CIKM, pp. 1441–1450 (2019)
23. Tang, J., Wang, K.: Personalized top-n sequential recommendation via convolutional sequence embedding. In: WSDM, pp. 565–573 (2018)
24. Vaswani, A., et al.: Attention is all you need. In: NeurIPS (2017)
25. Wu, C.Y., Ahmed, A., Beutel, A., Smola, A.J., Jing, H.: Recurrent recommender networks. In: WSDM, pp. 495–503 (2017)
26. Wu, S., Tang, Y., Zhu, Y., Wang, L., Xie, X., Tan, T.: Session-based recommendation with graph neural networks. In: AAAI, pp. 346–353 (2019)
27. Xie, X., et al.: Contrastive learning for sequential recommendation. In: ICDE, pp. 1259–1273 (2022)
28. Xu, C., et al.: Graph contextualized self-attention network for session-based recommendation. In: IJCAI, pp. 3940–3946 (2019)
29. Yu, Y., Jin, B., Song, J., Li, B., Zheng, Y., Zhuo, W.: Improving micro-video recommendation by controlling position bias (2022)
30. Zhou, K., et al.: S3-Rec: self-supervised learning for sequential recommendation with mutual information maximization. In: CIKM, pp. 1893–1902 (2020)

LGCRS: LLM-Guided Representation-Enhancing for Conversational Recommender System

Ruobing Wang, Xin He, Hengrui Gu, and Xin Wang(✉)

School of Artificial Intelligence, Jilin University, Changchun 130012, China
{wangrb22,hexin20,guhr22}@mails.jlu.edu.cn, xinwang@jlu.edu.cn

Abstract. Conversational recommender systems primarily focus on acquiring user preferences through real-time interactions. However, their effectiveness in modeling user preferences is constrained by the amount of conversational information. Existing conversational recommendation methods typically leverage external knowledge graph to enhance recommendation performance, potentially overlooking valuable textual information related to users and items. With significant natural language understanding capabilities, large language models (LLMs) have the potential to improve the performance of conversational recommendation task by improving the quality of text representations. However, the incorporation of raw textual information may inadvertently introduce noise to recommendations. Additionally, the semantic gap between the textual information and the collaborative signals could potentially degrade the performance of the conversational recommender system. To tackle the aforementioned challenges, we propose a LLM-guided representation-enhancing method for conversational recommender system, which fuses collaborative signals and semantic information to improve recommendation performance and generate high-quality responses. Specifically, we leverage LLMs to refine item profiles while reducing noise in text information, construct fused item representations across multiple aspects, and align the LLMs-enhanced semantic representation with the CF-side rational representation through regularization terms in the training stage. Experiments on two publicly available conversational recommendation datasets show that our method exhibits better performance in both recommendation and conversation tasks.

Keywords: Conversational Recommender System · Large Language Models · Representation Learning

1 Introduction

In recent years, conversational recommender system (CRS) [6] has become an emerging research topic, which aims to obtain users' real-time interests through dialogue to provide high-quality recommendation results. CRSs focus more on extracting user preferences from users' natural language and explicit feedback during dialogue than traditional recommender systems.

Due to users' limited patience, CRSs frequently encounter brief conversations and insufficient contextual information [6]. To address this issue, existing research often incorporates external data sources to enrich contextual information: KBRD [1] and KGSF [23], incorporate knowledge graphs into the recommendation process to refine item representations; RevCore [12] integrates user reviews into conversations to enhance contextual information. However, previous approaches have primarily relied on external knowledge graphs to refine item representations, thereby overlooking the valuable textual information associated with items. The lack of the textual information may restrict the informativeness within the learned representations [14]. This motivated us to develop a method to enhance the understanding of items by integrating item-relevant text information into conversational recommender systems.

In recent years, prominent LLMs like GPT-4 [13] and LLaMA [17] have showcased remarkable capabilities in the domain of natural language processing. The phenomenon has sparked significant interest among researchers, who are actively exploring the capabilities of LLMs in handling textual information within recommender systems [5]. This motivates us to explore the utilization of LLMs for generating meaningful textual representations of items, aiming to enhance CRS performance. However, the incorporation of item-related textual information into conversational recommender systems presents two distinct challenges: (1) Unexpected noise in item-related text information can reduce the effectiveness of the learned item representation. (2) The semantic gap between the LLMs-enhanced semantic representation and the CF-side rational representation of item may decrease the performance of recommendation task.

To address the above challenges, we propose a LLM-Guided Representation-Enhancing method for Conversational Recommender System (LGCRS). Specifically, we first utilize LLMs to refine user reviews and generate textual item profiles. The core idea is to use manually designed prompts to guide LLMs in generating accuracy item profiles while reducing inherent noise in the raw textual information. In this way, we can introduce item semantic information into the CRS, thereby enhancing the system's understanding of the item. Subsequently, we employ distinct encoders to encode the conversational context, external item-oriented knowledge graph and LLMs-enhanced item profile, respectively. By utilizing regularization terms to reduce the semantic gap between the LLMs-enhanced semantic representations and the CF-side rational representations, along with employing a gate layer to fuse these two representations, we obtain more accurate item representations. Based on the fused item representations, we develop a recommendation module for making accurate recommendations and a conversation module for providing high-quality responses. Overall, our contributions can be summarized as follows:

- We introduce a conversational recommendation method that harnesses the textual item profiles refined by large language models to enhance item representations and improve recommendation performance.

- We introduced two regularization terms to reduce the semantic gap between the LLMs-enhanced semantic representation and the CF-side representation.
- Extensive experiments on two widely used conversational recommendation datasets demonstrate that our proposed model achieves concrete improvements in both recommendation accuracy and dialogue quality.

2 Related Work

2.1 Conversational Recommender System

In recent years, with the rapid advancement of dialogue systems, interactive conversations with users have become an attractive method for capturing dynamic user intents and preferences. CRSs typically consist of three main components: the conversation module, the policy module, and the recommendation module [6]. These components may have different emphases depending on the specific CRS type.

Attribute-based conversational recommender systems [9,10,16] primarily focus on constructing the policy module with the goal of making effective recommendations in the fewest dialogue turns possible. These models often employ reinforcement learning techniques to train the dialogue manager, aiming to maximize expected rewards over a longer time horizon. Consequently, they tend to have a simplified conversation module, such as using fixed templates with slots to populate recommendation results and generate the system's response text. These approach aim to streamline the recommendation process and provide users with relevant suggestions efficiently.

End-to-end conversational recommender systems [1,11,12,21,23,24] prioritize the delivery of smooth and coherent text responses to users. These systems seamlessly incorporate information about recommended items into their replies, thereby enhancing the interpretability and acceptability of the recommendations. In such models, the policy module is implicitly integrated within the conversation module. These approaches are designed to offer more natural and engaging conversational experiences while ensuring that the recommended items are presented in a user-friendly and comprehensible manner.

This paper extends the second category of research by incorporating item-related textual information into CRSs. We leverage large language models to reduce noise in textual information and build item semantic representations. By integrating the LLMs-enhanced semantic representations with the CF-side rational representations, we construct meaningful representations and improve the recommendation performance of the model.

3 Preliminaries

Formally, let u denotes a user from user set $\mathcal{U}$, i denotes an item from item set $\mathcal{I}$. A conversation C consists of a list of utterances, denoted by $C = \{d_t\}_{t=1}^n$, in which each utterance d_t is a conversation sentence at the t-th turn.

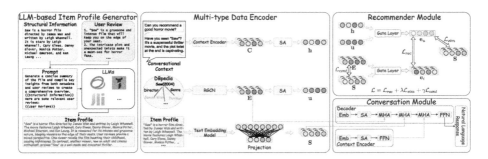

Fig. 1. The overview of our model LGCRS. First, we use LLM-based Item Profile Generator to generate accurate item profile. Then, Multi-type Data Encoder encodes conversational context, external item-oriented knowledge graph and LLMs-enhanced item profile, respectively. Finally, based on the learned representation, we develop a recommendation module for making accurate recommendations and a conversation module for providing high-quality responses. "SA", "MHA" denotes *self-attention* and *multi-head attention*, respectively.

To estimate the user's preferences based on their interactions, the CRS aggregates the utterances d_t in the conversation C and learns a vector $\mathbf{h}$ known as the conversation context. In the t-turn, the recommendation component employs estimated user preferences to generate a candidate item list $[i_1, i_2, \cdots, i_n]$ from the full range of available items $\mathcal{I}$. The conversation component then generates a response discourse that denotes the recommended items to the user.

In addition to the conversation history, previous work has introduced external data to enhance the performance of CRS. External data for CRS typically falls into two categories: structured external data, such as knowledge graphs, and unstructured external data, such as user-written reviews. The knowledge graph $\mathcal{G}$ is formed by a set of entities, denoted by $\mathcal{E}$, and a set of relations, denoted by $\mathcal{R}$. The entity set comprises all the items (*i.e.*, $\mathcal{I}$ is a subset of the entity set $\mathcal{E}$) and other item-related entities. Regarding each user, their interactive entity list $[e_1, e_2, \cdots, e_n]$ is denoted by all entities that have been mentioned in their conversation history C_u. Online reviews can be represented as a document consisting of sentences $R = \{r_1, r_2, \cdots\}$.

Based on the symbol definitions mentioned above, the goal of the CRS is to precisely furnish users with recommended items while generating intelligible responses, utilizing variety of contextual data such as conversational history C, knowledge graph $\mathcal{G}$, user reviews R, and other pertinent information.

4 Methodology

In this section, we introduce our proposed LLM-guided Representation-Enhancing method for Conversational Recommender System (LGCRS). Specifically, we first introduce our method for refining noisy user review information by leveraging a large language model generate item profiles. Secondly, we present the process of encoding the conversational context, external knowledge graph and LLMs-enhanced item profile. Finally, we discuss how to leverage the learned user representation and item representation for recommendation and dialogue generation (Fig. 1).

4.1 LLM-Based Item Profile Generator

In this section, we will introduce textual information such as user reviews to generate item profiles. However, employing raw item text attributes and user reviews as item profiles may introduce unexpected noise. We believe that by manually designed prompts, we can harness the powerful natural language processing capabilities of LLMs to generate accurate item profiles. Therefore, we propose an item profile generation paradigm based on large language models, which refines item-related textual information while reducing inherent noise.

Specifically, we integrate reasoning process into the LLM by incorporating carefully designed task prompt T to mitigate illusions and improve the quality of generated output. The task prompt clearly specifies its role in generating item profiles from user reviews by accurately specifying the input-output content and format. By combining manually designed task prompts T with raw item profile prompts Q_i, we can leverage LLM to generate accurate profiles. The specific process is outlined as:
$$\begin{aligned} \mathcal{P}_i &= \text{LLMs}(T, Q_i), \\ Q_i &= f_i(t_i, \mathcal{N}^i, R^i), \end{aligned} \quad (1)$$
where f_i represents the function that combines the item text attributes into a single string, t_i is the item title, $\mathcal{N}^i$ represents the one-hop neighborhood node set of the item in the knowledge graph, and R^i is the set of user reviews on the item. By combining structural item descriptions based on the $\mathcal{N}^i$ and user reviews R^i, our prompts Q_i provide precise information for LLMs, ensuring that the generated item profile accurately reflect users' general perspectives.

4.2 Multi-type Data Encoder

Context Encoder. The conversation history $C = \{d_t\}_{t=1}^n$ is a series of short sentences d_t communicated between the user and the system during a session, describing their informational interaction to achieve the recommendation goal. In order to obtain a representation of the conversation history, we use a standard Transformer to encode it:
$$\mathbf{C} = \text{Transformer}([d_1; d_2; \cdots ; d_t]), t \in [1, n], \quad (2)$$
where $[\cdot; \cdot]$ signifies the concatenation operation, and $\mathbf{C}$ represents the contextual word representation obtained from the top layer of Transformer. Subsequently, we employ a self-attention layer to consolidate the acquired contextual word representations into a representation for the conversation history:
$$\begin{aligned} \mathbf{h} &= \mathbf{C} \cdot \alpha_w, \\ \alpha_w &= \text{softmax}(\boldsymbol{b}_c^T \tanh(\mathbf{W}_c \mathbf{C})), \end{aligned} \quad (3)$$

where $\mathbf{W}_c$ and $\boldsymbol{b}_c$ denote the learnable parameter matrix and vector of the self-attention layer.

Knowledge Graph Encoder. To model the associations between items, we introduce an item-oriented knowledge graph denoted as $\mathcal{G}$, with the aim of enriching item-related information. The graph comprises a set of entities $\mathcal{E}$ and a set of relationships $\mathcal{R}$. In this study, we utilize DBpedia as our chosen item-oriented knowledge graph. Triplets in DBpedia can be represented by $\langle e, r, e' \rangle$, where $e, e' \in \mathcal{E}$ are entities and r is a relation from the relation set $\mathcal{R}$. Considering that the inter-entity relations of item-oriented knowledge graphs may have a considerable effect on understanding user preferences, we adopt RGCN [15] to encode knowledge graphs. The process of learning the representations of the next layer can be formalized as follows:

$$n_e^{(l+1)} = \sigma(\sum_{r \in \mathcal{R}} \sum_{e' \in \mathcal{E}_e^r} \frac{1}{Z_{e,r}} \mathbf{W}_r^{(l)} n_{e'}^{(l)} + \mathbf{W}_s^{(l)} n_e^{(l)}), \qquad (4)$$

where $n_e^{(l)}$ is the embedding of e at the l-th layer, $\mathcal{E}_e^r$ denotes the neighboring nodes for e with relation r, $\mathbf{W}_r$ is a learnable relation-specific matrix, $\mathbf{W}_s$ is a learnable matrx for transforming the node self-embedding and $Z_{e,r}$ is a normalization factor.

Item Profile Encoder. Utilizing the generated item profile $\mathcal{P}_i$ in Sect. 4.1, we can compute the semantic representation s_i of the item through the following procedure:

$$s_i = \mathcal{T}(\mathcal{P}_i), \qquad (5)$$

where $\mathcal{T}$ represents the text embedding model, a technology proven to effectively transform diverse textual inputs into fixed-length vectors that retain their inherent meaning and contextual information.

To integrate the semantic representation into the recommendation process, we employ a linear layer with dropout. This step serves to reduce the dimensionality of the large language model-enhanced semantic features and facilitates the mapping of semantic information into the collaboration space. The formal expression for this process is as follows:

$$\tilde{s}_i = \text{Linear}(s_i), \qquad (6)$$

where $s_i \in \mathrm{R}^{1 \times d_{\text{LLM}}}$ is the input feature and $\tilde{s}_i \in \mathrm{R}^{1 \times d}$ is the output feature after projection.

4.3 Recommendation Module

Item Representation Fusion. We introduce a gate layer to integrate the semantic representation $\tilde{s}_i$ and collaborative representation n_i of the item i:

$$\mathbf{e}_i = \beta_i \cdot \tilde{s}_i + (1 - \beta_i) \cdot n_i, \qquad (7)$$
$$\beta_i = \sigma(\mathbf{W}_{g_i}[\tilde{s}_i; n_i]), \qquad (8)$$

where $[\cdot;\cdot]$ denotes concatenation, and $\mathbf{W}_{g_i}$ is learnable parameters.

User Representation Fusion. We extract the following two parts from the user-system conversation history as user preferences $\mathbf{e}_u$: utterances in conversation $[d_1, d_2, \cdots, d_n]$ and mentioned entities $[e_1, e_2, \cdots, e_n]$. Similar to calculating conversation history $\mathbf{h}$ in Sect. 4.2, we set up a self-attention layer to aggregate entity representations mentioned in the conversation:

$$\begin{aligned} \mathbf{u} &= \mathbf{E} \cdot \beta, \\ \beta &= \mathrm{softmax}(\boldsymbol{b}_e^T \tanh(\mathbf{W}_e \, \mathbf{E})), \end{aligned} \quad (9)$$

where $\mathbf{E}$ denotes the representation matrix obtained by stacking the sequence of entity representations, $\mathbf{W}_c$ and $\boldsymbol{b}_c$ denote the learnable parameter matrix and vector of the self-attention layer.

To integrate the acquired user conversation history $\mathbf{h}$ and user interaction history $\mathbf{u}$, we establish a gate layer to derive the final user representation $\mathbf{e}_u$:

$$\mathbf{e}_u = \beta_u \cdot \mathbf{h} + (1 - \beta_u) \cdot \mathbf{u}, \quad (10)$$
$$\beta_u = \sigma(\mathbf{W}_{g_u}[\mathbf{h}; \mathbf{u}]), \quad (11)$$

where $\mathbf{W}_{g_u}$ is learnable parameters.

After learning the user's preference vector, we can calculate the user's potential rating for each item in the item set:

$$P_{rec}(u, i) = \mathrm{softmax}(\mathbf{e}_u^T \cdot \mathbf{e}_i). \quad (12)$$

We use the cross-entropy loss to calculate the loss of the item rating given by the recommendation module and the item mentioned in the actual response. The formal expression is as follows:

$$\mathcal{L}_{rec} = -\frac{1}{N} N \sum_{i=1}^{N} \log(\hat{p}_i), \quad (13)$$

where N is the total number of recommendations, and $\hat{p}_i$ represents the probability that the target item is recommended.

To ensure that the learned semantic representation is beneficial for recommendation and to reduce the semantic gap between the LLMs-enhanced semantic representation and the CF-side rational representation, we incorporated similarity loss and contrastive loss as regularization terms in the training process:

$$\mathcal{L}_{sim} = -\log(\mathrm{softmax}(e_u^T \cdot \tilde{s}_i)), \quad (14)$$
$$\mathcal{L}_{cont} = -\mathbb{E}\big(\log(\frac{\mathrm{sim}(\tilde{s}_i, n_i)}{\sum_{j \in \mathcal{I}} \mathrm{sim}(\tilde{s}_j, n_i)})\big). \quad (15)$$

The final objective function for the recommendation subtask is:

$$\mathcal{L} = \mathcal{L}_{rec} + \lambda \mathcal{L}_{sim} + \gamma \mathcal{L}_{cont}. \quad (16)$$

4.4 Conversation Module

In this section, we focus on how to create response utterances that are both easy to understand and flow naturally. We achieve this objective by employing a modified version of the Transformer model. While the encoder component adheres to the conventional Transformer architecture, we introduce modifications to the decoder component. More precisely, we stack the representations of mentioned entities $[e_1, e_2, \cdots, e_n]$ and the semantic representations of referenced items $[i_1, i_2, \cdots, i_n]$ in the conversation, forming matrices $\mathbf{E}$ and $\mathbf{S}$. After the self-attention layer, we introduce two attention layers that integrate both the structural information of KG and the semantic information of items into the context representation:

$$\mathbf{A}_0^n = \text{MHA}(\mathbf{R}^{n-1}, \mathbf{R}^{n-1}, \mathbf{R}^{n-1}), \tag{17}$$

$$\mathbf{A}_1^n = \text{MHA}(\mathbf{A}_0^n, \mathbf{E}, \mathbf{E}), \tag{18}$$

$$\mathbf{A}_2^n = \text{MHA}(\mathbf{A}_1^n, \mathbf{S}, \mathbf{S}), \tag{19}$$

$$\mathbf{A}_3^n = \text{MHA}(\mathbf{A}_2^n, \mathbf{C}, \mathbf{C}), \tag{20}$$

$$\mathbf{R}^n = \text{FFN}(\mathbf{A}_3^n), \tag{21}$$

where $\text{MHA}(\mathbf{Q}, \mathbf{K}, \mathbf{V})$ represents the multi-head attention function [18], which takes query, key, and value as input:

$$\text{MHA}(\mathbf{Q}, \mathbf{K}, \mathbf{V}) = [h_1; \cdots; h_n]\mathbf{W}_A, \tag{22}$$

$$h_i = \text{Attention}(\mathbf{Q}\mathbf{W}_i^\mathbf{Q}, \mathbf{K}\mathbf{W}_i^\mathbf{K}, \mathbf{V}\mathbf{W}_i^\mathbf{V}), \tag{23}$$

where $[\cdot; \cdot]$ denotes the concatenation operation, n is the number of heads, $\mathbf{W}_A$ is a learnable matrix. $\text{FFN}(x)$ is a two level fully-connected feed-forward network with ReLU activation function:

$$\text{FFN}(x) = \text{ReLU}(x\mathbf{W}_1 + \mathbf{b}_1)\mathbf{W}_2 + \mathbf{b}_2. \tag{24}$$

In the above equations, $\mathbf{R}^n$ is the embedding matrix from the decoder at n-th layer, $\mathbf{C}$ is the embedding matrix output by the conversation encoder and $\mathbf{A}_0^n, \mathbf{A}_1^n, \mathbf{A}_2^n, \mathbf{A}_3^n$ are the output after self-attention, cross-attention with item-oriented KG, cross-attention with items' semantic representations and cross-attention with conversation context, respectively. Additionally, we incorporated the copy mechanism [23] to generate more meaningful utterances.

The loss function for the conversation subtask is defined as follows:

$$\mathcal{L}_{gen} = -\frac{1}{|C|} \sum_{t=1}^{|C|} log(P(d_t|d_1, \cdots, d_{t-1})) \tag{25}$$

where $|C|$ represents the number of turns in a conversation C and we compute this loss for each utterance d_t within C.

5 Experiment

To verify the superiority of our proposed model, we conduct extensive experiments to answer the following research questions: **RQ1**: How does our proposed model perform on recommendation and conversation tasks compared with existing methods? **RQ2**: How much does each component of our proposed model contribute to its overall performance? **RQ3**: Does our proposed item profile generation paradigm reduce the impact of noisy text on recommendation performance compared to directly utilizing item-related textual information? **RQ4**: How does using different LLM to encode item-related textual information affect the recommendation performance? **RQ5**: How do different hyperparameter settings affect the performance of our model?

5.1 Experiment Setup

Dataset. We conduct our experiments on Redial [11] and TG-Redial [24] datasets. The Redial dataset is a widely used English movie conversational recommendation dataset that comprises 10,006 dialogues, involving 504 users and 51,699 movies. The TG-ReDial dataset is a Chinese conversational recommendation dataset that comprises 10,000 dialogues, involving 1482 users and 33,834 movies. To conduct our experiments, we randomly divide the dataset into a training set, a validation set and a test set, in 8:1:1 ratio.

Baselines. Since we mainly evaluate the performance of CRS models on recommendation and conversation tasks, we consider the following existing work as our baseline:

- Popularity [7]: This model generates a recommendation list for a user based solely on how frequently items appear in the corpus.
- BERT [3]: This model utilizes BERT, a pretraining [20] model, to encode the user's current utterances and provide recommendation results.
- TextCNN [8]: This model utilizes a convolutional neural network approach to process the input conversation data and extract user representation vectors.
- Transformer [18]: This model utilizes an encoder-decoder approach based on the Transformer architecture to generate conversation responses.
- ReDial [11]: This model comprises an autoencoder-based recommendation module alongside a conversation module based on the HRED architecture.
- KBRD [1]: This model incorporates the DBpedia into CRS to improve the representation of recommended items.
- KGSF [23]: This model introduces the DBpedia to enhance the item representation and utilizes the ConceptNet to enhance the word representation.
- RevCore [12]: This model improves the performance of the CRS by injecting user reviews with similar sentiments into the dialogue.
- VRICR [21]: This model achieves a more accurate user representation by introducing resampling conversation-specific subgraphs based on the variational Bayesian method.

Table 1. The recommendation performance.

Dataset	Redial									TG-Redial								
Metric	HR			MRR			NDCG			HR			MRR			NDCG		
Method	@1	@5	@10	@1	@5	@10	@1	@5	@10	@1	@5	@10	@1	@5	@10	@1	@5	@10
Popularity	0.0111	0.0359	0.0536	0.0111	0.0196	0.0220	0.0111	0.0236	0.0294	0.0004	0.0004	0.0031	0.0004	0.0004	0.0008	0.0004	0.0004	0.0013
TextCNN	0.0114	0.0359	0.0710	0.0114	0.0199	0.0243	0.0114	0.0239	0.0349	0.0024	0.0087	0.0131	0.0024	0.0046	0.0051	0.0024	0.0056	0.0070
Bert	0.0248	0.0900	0.1437	0.0248	0.0470	0.0540	0.0248	0.0576	0.0748	0.0002	0.0018	0.0034	0.0002	0.0007	0.0009	0.0002	0.0010	0.0015
Redial	0.0297	0.1035	0.1659	0.0297	0.0533	0.0589	0.0297	0.0637	0.0842	0.0003	0.0020	0.0037	0.0003	0.0008	0.0010	0.0003	0.0011	0.0016
KBRD	0.0338	0.1147	0.1764	0.0338	0.0610	0.0691	0.0338	0.0742	0.0941	0.0055	0.0169	0.0243	0.0055	0.0095	0.0104	0.0055	0.0113	0.0137
KGSF	0.0335	0.1164	0.1795	0.0335	0.0614	0.0698	0.0335	0.0750	0.0953	0.0054	0.0193	0.0295	0.0054	0.0100	0.0114	0.0054	0.0123	0.0156
RevCore	0.0398	0.1229	0.1886	0.0398	0.0692	0.0776	0.0398	0.0834	0.1039	0.0061	0.0191	0.0302	0.0061	0.0104	0.0119	0.0061	0.0126	0.0162
VRICR	0.0434	0.1249	0.1901	0.0434	0.0703	0.0797	0.0434	0.0847	0.1055	0.0053	0.0183	0.0290	0.0053	0.0098	0.0113	0.0053	0.0119	0.0154
LGCRS\S	0.0384	0.1184	0.1809	0.0384	0.0658	0.0741	0.0384	0.0788	0.0989	0.0055	0.0198	0.0322	0.0055	0.0101	0.0117	0.0055	0.0125	0.0164
LGCRS\C	0.0417	0.1217	0.1860	0.0417	0.0688	0.0773	0.0417	0.0818	0.1026	0.0060	0.0219	0.0342	0.0060	0.0110	0.0126	0.0060	0.0137	0.0177
LGCRS	**0.0445**	**0.1254**	**0.1909**	**0.0445**	**0.0720**	**0.0807**	**0.0445**	**0.0852**	**0.1063**	**0.0067**	**0.0227**	**0.0357**	**0.0067**	**0.0122**	**0.0138**	**0.0067**	**0.0148**	**0.0189**

Evaluation Metrics. We measure the performance of our recommendation and conversation modules separately. When evaluating the recommendation module, our primary focus lies in assessing the CRS's ability to accurately reflect user preferences within the Top-K recommendation list. To assess the performance of the recommendation module, we utilize metrics such as HR@k, MRR@k, and NDCG@k. Considering the limited length of response in dialogues, we use a small k value ($k \in [1, 5, 10]$) to evaluate the performance of the recommendation module, which is more suitable for practical application scenarios. When evaluating the conversation module, we assess both the accuracy and diversity of the generated response. Accuracy is measured by indicators such as BLEU-1,2,3,4. To evaluate diversity, we use Distinct n-gram ($n = 3, 4$) at the sentence level.

Implementation Details. We build upon the CRSLab [22] and implement our methodology. The hidden layer representations for the recommendation and conversation modules are both set to a dimension of 300. When aggregating information from the knowledge graph using GNN-based method, we set the depth to 1. The hyperparameters λ and γ in the recommendation module are set to 1.0 and 0.06, respectively. They are determined through grid search on the validation set. We choose vicuna-13b-v1.5-16k [2] as the LLM backend and utilize the Adam optimizer with default parameter settings.

5.2 Overall Performance (RQ1)

Recommendation. We perform several experiments to validate the effectiveness of our proposed model for the recommendation task. We present the results of various recommendation methods on distinct recommendation metrics in Table 1. According to the results, we can find that the conversational recommendation method (e.g., Redial, KBRD, KGSF) is significantly better than the traditional recommendation method (e.g., Popularity, TextCNN, Bert). This observation underscores the essential role of contextual information in the conversational recommendation scenario. Moreover, approaches incorporating external knowledge exhibit superior performance compared to those that do

Table 2. The conversation performance.

Dataset	Redial						TG-Redial					
Metric	BLEU@1	BLEU@2	BLEU@3	BLEU@4	DIST@3	DIST@4	BLEU@1	BLEU@2	BLEU@3	BLEU@4	DIST@3	DIST@4
Transformer	0.0983	0.0222	0.0114	0.0063	0.2258	0.3867	0.2667	0.0440	0.0145	0.0065	1.7392	2.7114
Redial	0.0409	0.0043	0.0007	0.0002	0.3064	0.7086	0.1177	0.0158	0.0021	0.0004	0.7564	1.1854
KBRD	0.1712	0.0314	0.0135	0.0085	0.3476	0.6231	0.2617	0.0446	0.0151	0.0075	2.3076	3.3234
KGSF	0.1666	0.0334	0.0176	0.0109	0.4458	1.0366	0.2632	0.0473	0.0134	0.0061	2.2022	3.4458
RevCore	0.1121	0.0327	0.0196	0.0113	0.4822	0.6064	0.2432	0.0490	0.0170	0.0090	1.7048	2.9774
VRICR	0.1036	0.0241	0.0099	0.0045	0.5129	0.7323	0.2274	0.0491	0.0177	0.0085	2.6649	4.6301
LGCRS	**0.1757**	**0.0418**	**0.0227**	**0.0139**	**0.5622**	**1.3600**	**0.2721**	**0.0510**	**0.0181**	**0.0092**	**3.7480**	**6.0010**

not use external knowledge, highlighting the effectiveness of incorporating item knowledge to enhance recommendation performance. Most notably, our model demonstrates superior performance when compared to all baseline methods in the recommendation task. This substantiates the effectiveness of our method in enhancing item representation through the utilization of an item profile generated by a large language model.

Conversation. We validate the performance of our proposed model on conversation task in terms of both accuracy and diversity, compare with baselines and conduct in-depth analysis. We show the performance of different conversation methods on conversation tasks in Table 2. Our model outperforms all the other models on both metrics, with the highest BLEU and DIST scores across all n-gram levels. The reason is that our model utilizes large language models to enhance item representation to reduce irrelevant information, thereby producing responses closer to ground truth. Overall, our model is capable of generating diverse responses while preserving similarity to ground truth responses.

5.3 Ablation Study (RQ2)

In this section, we conduct an ablation study to assess the contribution of each component to recommendation performance. The results are presented in Table 1. We consider two variants of our method: (1) LGCRS\S removes the semantic information contained in the item representation. (2) LGCRS\C removes the alignment constraints of the semantic representation and the collaborative filtering representation.

It is apparent that each component plays a substantial role in recommendation task, as evidenced by the notable decline in recommendation performance observed in both variants. Particularly, when compared with LGCRS\C, LGCRS\S exhibits a greater performance decrease, suggesting that the item profile generated by integrating LLMs effectively enhances the understanding of items in recommender systems. Moreover, LGCRS\C leads to performance degradation, indicating that LGCRS effectively reduce the semantic gap between the LLMs-enhanced semantic representation and the CF-side rational representation.

Table 3. Text Enhance Method Analysis.

Dataset	Redial									TG-Redial								
Method	H@1	H@5	H@10	M@1	M@5	M@10	N@1	N@5	N@10	H@1	H@5	H@10	M@1	M@5	M@10	N@1	N@5	N@10
sep.	0.0415	0.1174	0.1856	0.0415	0.0673	0.0763	0.0415	0.0796	0.1016	0.0058	0.0192	0.0299	0.0058	0.0101	0.0114	0.0058	0.0123	0.0156
agg.	0.0417	0.1197	0.1809	0.0417	0.0689	0.0770	0.0417	0.0814	0.1012	0.0058	0.0223	0.0352	0.0058	0.0112	0.0129	0.0058	0.0139	0.0181
enhance	0.0445	0.1254	0.1909	0.0445	0.0720	0.0807	0.0445	0.0852	0.1063	0.0067	0.0227	0.0357	0.0067	0.0122	0.0138	0.0067	0.0148	0.0189

5.4 Text Enhance Method Analysis (RQ3)

In this section, we investigate the influence of employing various methods to introduce item-related text information on recommendation performance. We primarily focus on the following three methods:

- sep.: This method employs LLMs to generate a representation for each review associated with the item. Subsequently, the generated representations are fed into the self-attention layer to derive the semantic representation of the item.
- agg.: This method directly utilizes LLMs to encode concatenated item reviews, thereby obtaining item semantic representations.
- enhance: This method initially employs LLMs to refine item profiles based on descriptions generated from reviews, utilizing manually designed prompts. Subsequently, a text encoding model is employed to obtain semantic representations of the items.

In Table 3, we showcase the performance on recommendation task using vicuna-13b-v1.5-16k as the backend for text encoding on two datasets. The results reveal that the performance of the 'enhance' method surpasses that of 'agg.' and 'sep.'. This can be attributed to the following two factors: 1) Integration of Information: Compared with 'sep.' that considers each item review separately, 'agg.' leverages the excellent natural language understanding capability of the large language model. It integrates the information contained in reviews more effectively than the simple self-attention layer of 'sep.'. 2) Logical Reasoning: With manually designed prompts, the 'enhance' method utilizes the logical reasoning ability of the large language model to reduce noise in the review data, thereby enhancing the quality of the item profile. In summary, our proposed item profile generation paradigm effectively reduces noise and improves recommendation performance compared to directly utilizing item-related textual information.

5.5 LLM Backend Analysis (RQ4)

In this section, we investigate the impact of employing different large language model for processing text information on recommendation performance. We consider the utilization of the following five open-source large language models: chatglm3-6b [4], baichuan2-7b-chat [19], baichuan2-13b-chat, vicuna-7b-v1.5-16k [2] and vicuna-13b-v1.5-16k. In the Table 4, we present the results of employing the 'enhance' method to process item-related text information for the recommendation task on the Redial dataset.

Table 4. LLM Backend Analysis.

LLM Backend	H@1	H@5	H@10	M@1	M@5	M@10	N@1	N@5	N@10
chatglm3-6b	0.0410	0.1206	0.1844	0.0410	0.0683	0.0767	0.0410	0.0812	0.1017
baichuan2-7b	0.0420	0.1184	0.1886	0.0420	0.0685	0.0777	0.0420	0.0809	0.1034
baichuan2-13b	0.0426	0.1213	0.1832	0.0426	0.0695	0.0777	0.0426	0.0823	0.1022
vicuna-7b	0.0422	0.1234	0.1846	0.0422	0.0699	0.0781	0.0422	0.0830	0.1029
vicuna-13b	0.0445	0.1254	0.1909	0.0445	0.0720	0.0807	0.0445	0.0852	0.1063

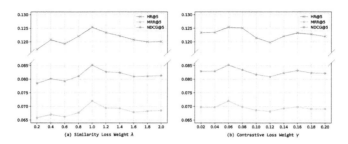

Fig. 2. Hyper-parameters sensitive analysis on the Redial dataset.

From the results, it is evident that whether using baichuan2 or vicuna-v1.5, the recommendation outcomes corresponding to the 13b model are marginally superior to those of the 7b model. The recommendation performance using vicuna-v1.5 as the backend surpasses that of chatglm3 and baichuan2. This difference in performance may be attributed to the fact that the latter two models have undergone fine-tuning on Chinese conversation data, resulting in a partial loss of English understanding and processing capabilities.

5.6 Parameter Sensitive Analysis (RQ5)

In this section, we examine the sensitivity of two hyperparameters: the contrast loss weight λ and the similarity loss weight γ. For each hyperparameter, we explore its values within the empirical space and report ten representative values in Fig. 2. In general, as the weight increases, the recommendation performance initially rises and subsequently descends. In a word, employing regularization terms to align the semantic representation space and collaborative signal space of items proves beneficial in enhancing the recommendation performance of conversational recommender systems, and we should tune them in fine-grained.

6 Conclusion

In this paper, we introduce a novel method LGCRS that utilizes item semantic representations to improve recommendation performance and generate more natural responses. Through the use of a large language model to refine user reviews, we have obtained a high-quality item profile for generating semantic representations. LGCRS enhances item understanding and improves recommendation effectiveness by constructing a multi-aspect item representation that integrates semantic representations with CF-side rational representations. Experiments on publicly available conversational recommendation datasets show that our method performs better in both recommendation and conversation tasks.

References

1. Chen, Q., et al.: Towards knowledge-based recommender dialog system. In: Proceedings of the 2019 Conference on Empirical Methods in Natural Language Processing and the 9th International Joint Conference on Natural Language Processing, Hong Kong, China, pp. 1803–1813. Association for Computational Linguistics (2019)
2. Chiang, W.L., Li, Z., Lin, Z., Sheng, Y., Wu, Z., Zhang, H., et al.: Vicuna: an opensource chatbot impressing GPT-4 with 90%* ChatGPT quality (2023). https://lmsys.org/blog/2023-03-30-vicuna/
3. Devlin, J., Chang, M., Lee, K., Toutanova, K.: BERT: pre-training of deep bidirectional transformers for language understanding. In: Proceedings of the 2019 Conference of the North American Chapter of the Association for Computational Linguistics: Human Language Technologies, Minneapolis, MN, USA, pp. 4171–4186. Association for Computational Linguistics (2019)
4. Du, Z., et al.: GLM: general language model pretraining with autoregressive blank infilling. In: Proceedings of the 60th Annual Meeting of the Association for Computational Linguistics, Dublin, Ireland, pp. 320–335. Association for Computational Linguistics (2022)
5. Fan, W., et al.: Recommender systems in the era of large language models (LLMS) (2023). CoRR **abs/2307.02046**, https://doi.org/10.48550/ARXIV.2307.02046
6. Gao, C., Lei, W., He, X., de Rijke, M., Chua, T.: Advances and challenges in conversational recommender systems: a survey. AI Open **2**, 100–126 (2021)
7. Ji, Y., Sun, A., Zhang, J., Li, C.: A re-visit of the popularity baseline in recommender systems. In: Proceedings of the 43rd International ACM SIGIR Conference on Research and Development in Information Retrieval, Virtual Event, China, pp. 1749–1752. ACM (2020)
8. Kim, Y.: Convolutional neural networks for sentence classification. In: Proceedings of the 2014 Conference on Empirical Methods in Natural Language Processing, Doha, Qatar, pp. 1746–1751. ACL (2014)
9. Lei, W., He, X., Miao, Y., Wu, Q., Hong, R., Kan, M., Chua, T.: Estimation-action-reflection: towards deep interaction between conversational and recommender systems. In: The Thirteenth ACM International Conference on Web Search and Data Mining, Houston, TX, USA, pp. 304–312. ACM (2020)
10. Lei, W., et al.: Interactive path reasoning on graph for conversational recommendation. In: The 26th ACM SIGKDD Conference on Knowledge Discovery and Data Mining, CA, USA, pp. 2073–2083. ACM (2020)
11. Li, R., Kahou, S.E., Schulz, H., Michalski, V., Charlin, L., Pal, C.: Towards deep conversational recommendations. In: Annual Conference on Neural Information Processing Systems, Montréal, Canada, pp. 9748–9758 (2018)

12. Lu, Y., et al.: Revcore: review-augmented conversational recommendation. In: Findings of the Association for Computational Linguistics. vol. ACL/IJCNLP 2021, pp. 1161–1173. Association for Computational Linguistics (2021)
13. OpenAI: GPT-4 technical report (2023). CoRR **abs/2303.08774**, https://doi.org/10.48550/ARXIV.2303.08774
14. Ren, X., e al.: Representation learning with large language models for recommendation (2023). CoRR **abs/2310.15950**, https://doi.org/10.48550/ARXIV.2310.15950
15. Schlichtkrull, M., Kipf, T.N., Bloem, P., van den Berg, R., Titov, I., Welling, M.: Modeling relational data with graph convolutional networks. In: Gangemi, A., et al. (eds.) ESWC 2018. LNCS, vol. 10843, pp. 593–607. Springer, Cham (2018). https://doi.org/10.1007/978-3-319-93417-4_38
16. Sun, Y., Zhang, Y.: Conversational recommender system. In: The 41st International ACM SIGIR Conference on Research & Development in Information Retrieval, Ann Arbor, MI, USA, pp. 235–244. ACM (2018)
17. Touvron, H., Lavril, T., Izacard, G., Martinet, X., Lachaux, M., Lacroix, T., et al.: LLaMA: Open and efficient foundation language models (2023). CoRR **abs/2302.13971**, https://doi.org/10.48550/ARXIV.2302.13971
18. Vaswani, A., et al.: Attention is all you need. In: Annual Conference on Neural Information Processing Systems, Long Beach, CA, USA, pp. 5998–6008 (2017)
19. Yang, A., Xiao, B., Wang, B., Zhang, B., Bian, C., Yin, C., et al.: Baichuan 2: Open large-scale language models (2023). CoRR abs/2309.10305
20. Zeng, Z., et al.: Knowledge transfer via pre-training for recommendation: a review and prospect. Front. Big Data **4**, 602071 (2021)
21. Zhang, X., et al.: Variational reasoning over incomplete knowledge graphs for conversational recommendation. In: Proceedings of the Sixteenth ACM International Conference on Web Search and Data Mining, Singapore, pp. 231–239. ACM (2023)
22. Zhou, K., et al.: CRSLab: an open-source toolkit for building conversational recommender system. In: Proceedings of the Joint Conference of the 59th Annual Meeting of the Association for Computational Linguistics and the 11th International Joint Conference on Natural Language Processing, Online, pp. 185–193. Association for Computational Linguistics (2021)
23. Zhou, K., Zhao, W.X., Bian, S., Zhou, Y., Wen, J., Yu, J.: Improving conversational recommender systems via knowledge graph based semantic fusion. In: The 26th ACM SIGKDD Conference on Knowledge Discovery and Data Mining, CA, USA, pp. 1006–1014. ACM (2020)
24. Zhou, K., Zhou, Y., Zhao, W.X., Wang, X., Wen, J.: Towards topic-guided conversational recommender system. In: Proceedings of the 28th International Conference on Computational Linguistics, Barcelona, Spain (Online), pp. 4128–4139. International Committee on Computational Linguistics (2020)

Multi-intent Aware Contrastive Learning for Sequential Recommendation

Junshu Huang[1], Zi Long[2]($^{\boxtimes}$), Xianghua Fu[2], and Yin Chen[2]

[1] Shenzhen University, Shenzhen, China
[2] Shenzhen Technology University, Shenzhen, China
longzi@sztu.edu.cn

Abstract. Intent is a significant latent factor influencing user-item interaction sequences. Prevalent sequence recommendation models that utilize contrastive learning predominantly rely on single-intent representations to direct the training process. However, this paradigm oversimplifies real-world recommendation scenarios, attempting to encapsulate the diversity of intents within the single-intent level representation. SR models considering multi-intent information in their framework are more likely to reflect real-life recommendation scenarios accurately. To this end, we propose a **M**ulti-intent Aware **C**ontrastive **L**earning for Sequential **Rec**ommendation (MCLRec). It integrates an intent-aware user representation learning method to enable multi-intent recognition within interaction sequences through the spatial relationships between user and intent representations. We further propose a multi-intent aware contrastive learning strategy to mitigate the impact of pair-wise representations with high similarity. Experimental results on widely used four datasets demonstrate the effectiveness of our method for sequential recommendation.

Keywords: Sequential Recommendation · Contrastive learning · Multi-intent aware

1 Introduction

Recommendation systems assist users in capturing helpful information and deliver personalized recommendations to diverse users from extensive collections of items in reality. Sequential recommendation (SR) models [16,34], which can effectively capture similar patterns of user behavior across different user-item interaction sequences, have become the state-of-the-art recommendation systems [5,26,32,49]. SR models encode sequences into user representations by deep neural networks and finally make accurate next-item predictions that users would be interested in. Importantly, these predictions are consistent with the underlying logic of real-world recommendation systems.

Traditional SR-based approaches [21,22,25] focus on learning from chronological sequences. This approach enables them to capture the sequential dynamics of user-item interactions. However, they exhibit limitations in identifying

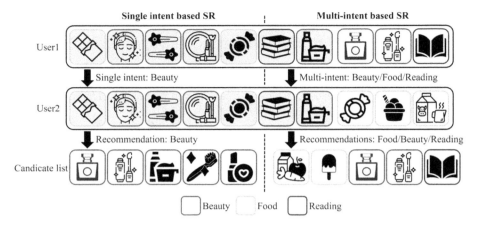

Fig. 1. The figure demonstrates the variation in candidate item propensity when the training of SR models is guided by single-intent or multi-intent information. Items in the sequence of User2 that are identical to those of user1 have been highlighted with a blue background. (Color figure online)

inter-sequence correlations, constraining their capacity to understand intricate user behavior patterns and preferences. Sequences that exhibit similar purchase intentions in real-world shopping contexts provide valuable reference points for enhancing recommendation accuracy when predicting the next item. While reliable and precise labeled data is lacking, recent works have demonstrated that leveraging the intent similarity across diverse users to guide contrastive self-supervised learning (SSL) tasks can enhance the performance of SR models. Among those methods, ICLRec [5] employs an expectation-maximization (EM) framework to maximize the agreement between a view of a sequence and its corresponding single intent, whose distributions are learned from all interaction sequences. It attempts to encapsulate the complex and diverse intents in real-world interaction sequences with a single-intent representation.

However, representing user-item interaction sequences with a single intent inevitably leads to losing multi-intent information. The essence of Sequence Recommendation (SR) models lies in learning sequential patterns from the training dataset to predict the next item in the testing dataset's sequences. Each user-item interaction sequence contains distinct intentions, yet models considering single-intent capture only the primary intent of each user, resulting in a loss of multi-intent information. Consider the example illustrated in Fig. 1. User1 and User2 have a portion of the same interaction pattern. For instance, both have engaged with face masks and hair clips in the 'Beauty' category and chocolates and candies in the 'Food' category. However, there are also some distinct interactions, such as the presence of perfume in User1's sequence, which is absent in User2's. The SR model on the left infers the 'Beauty' intent by extracting single-intent information from User1, tending to recommendations of items from the 'Beauty' category to a similar User2. Conversely, a multi-intent aware SR model considers a range of intents such as 'Beauty', 'Food' and 'Reading', as learned from User1's behavior. The model then leverages its comprehensive understanding to offer diverse recommendations to User2, thus potentially enhancing the effectiveness of the recommendations.

SR models with multi-intent modeling are still underexplored. IOCRec [20] features a global module designed to capture user preferences by disentangling the intent dimensions, thus separating global and local representations of a sequence. The sum of these two representations is utilized as the intent representation for contrastive learning (CL). This approach disentangles the multiple intents within a sequence, yet there is no information crossover in the intent dimension; in other words, a single-intent level representation is employed in the CL process. SR models incorporating multi-intent considerations into their structural design can offer a more accurate representation in actual recommendation scenarios.

To address the issues mentioned above, we propose a novel approach, **M**ulti-**I**ntent Aware **C**ontrastive **L**earning for Sequential **Rec**ommendation (MCLRec), which utilizes multi-intent level information for model construction. Specifically, we apply an intent-aware user representation learning approach to infer a variety of intents within sequences and leverage the spatial relationship between user and intent representations in the latent space. To reduce the impact of irrelevant data, the model filters out a given number of main intents to enhance the quality of the intent-aware user representation learning. Then, we propose a multi-intent aware contrastive learning strategy, which aims to mitigate the influence of the pairwise representations with similar multiple intents on learning, thereby improving the model's performance. We summarize the contributions of this work below:

- MCLRec learns intent-aware user representations in the latent space from a multi-intent perspective for user-item interaction sequences.
- We propose a multi-intent aware contrastive learning strategy to mitigate the impact of the pair-wise representations with high similarity in their representations.
- Experimental results on four datasets verify the effectiveness of our proposed method.

2 Related Work

2.1 Sequential Recommendation

Sequential recommendation (SR) aims to disentangle users' interest according to historical interactions, which has been widely researched [2,8,13,33,46]. Early works on SR usually extract sequential patterns based on the Markov Chain (MC) assumption. FPMC [30] fuses sequential patterns and users' general interests, combining a first-order MC and Matrix Factorization. Fossil [10] fuses similarity-based models with a high-order MC to tackle data sparsity issues for clarity. Recent models have begun to integrate deep neural networks into SR, such as Recurrent Neural Networks (RNN)-based [7,12,38,45] and Convolutional

Neural Networks (CNN)-based [34,47,48] models. GRU4Rec [12] first introduces RNN in session-based recommendation trained with a ranking loss function. Caser [34] embeds interaction sequences into images and extracts sequential patterns with CNN. BERT4Rec [32] leverages a deep bidirectional self-attention network to model interaction sequences, utilizing the Cloze task to capture the sequential dependency effectively. However, the methods mentioned above for SR often struggle to address issues of data sparsity and noise effectively.

2.2 Contrastive SSL for SR

SSL has emerged as a significant trend in CV [3,9,14], NLP [6,19,23] and recommendation [40,42–44]. Contrastive SSL aims to extract correlation within vast amounts of unlabeled data to enhance the capability to discern negative samples simultaneously. S^3-Rec [49] proposes four self-supervised optimization objectives to capture the interrelations between items, attributes, sequences and subsequences. SGL [39] adopts a multi-task framework with SSL and maximizes the agreement between different augmented views of the same node to improve node representation learning. CL4SRec [41] maximizes the agreement between differently augmented views of the same sequence in the latent space and utilizes a multi-task framework to encode the user representation. CoSeRec [24] advances CL4SRec by proposing two additional data augmentation techniques to exploit item correlations. However, the above mentioned methods do not account for users' latent intent in applying contrastive SSL. This can limit the model's capacity to discern the nuanced motivations driving user behavior, which is critical for tailoring recommendations that align with underlying user preferences.

2.3 Latent Intent for Recommendation

Many recent works have focused on learning intent representation to enhance the performance and robustness of models [1,4,28,35,36]. ASLI [35] leverages a temporal convolutional network alongside user side information to decode latent user intents, incorporating an attention mechanism to address the complexities of long-term and short-term item dependencies. DSSRec [26] introduces a SSL task in the latent space and designs a sequence encoder to infer and disentangle the latent intents under interaction sequences. ICLRec [5] introduces a latent intent variable to maximize the agreement between user representations through an EM framework. The intent representations are used to supervise user representations clustered by K-means. IOCRec [20] suggests a novel sequence encoder integrating global and local representations to select the primary intents. Distinct from these works, our approach incorporates multi-intent level information within one interaction sequence when learning another sequence. This enables our model to learn multi-intent aware user representations and amplify the efficacy of contrastive learning tasks.

3 Preliminaries

3.1 Problem Definition

Sequential Recommendation (SR) predicts the next item users would be interested in based on their interaction sequences. We denote a set of users and items as $\mathcal{U}$ and $\mathcal{V}$, respectively. Given a user $u \in \mathcal{U}$, the user sequence is a sequence of user-item interactions $\mathcal{S}^u = [s_1^u, s_2^u, ..., s_{|\mathcal{S}^u|}^u]$, where $|\mathcal{S}^u|$ is the total number of interactions and $s_t^u \in \mathcal{V}$ denotes the item that user u interacts with at time step $t \in [1, |\mathcal{S}^u|]$. The sequence $\mathcal{S}^u$ is usually truncated by maximum length T. If $|\mathcal{S}^u| \geqslant T$, the latest T item interactions are considered, expressed as $\mathcal{S}^u = [s_{|\mathcal{S}^u|-T+1}^u, s_{|\mathcal{S}^u|-T+2}^u, ..., s_{|\mathcal{S}^u|}^u]$; otherwise, zero items are padded before $\mathcal{S}^u$ until $|\mathcal{S}^u| = T$. For convenience, $\mathcal{S}^u$ is denoted as $[s_1^u, s_2^u, ..., s_T^u]$. The goal of SR is to predict the next item s_{T+1}^u with the highest probability of interaction, which is formulated as follows:

$$\arg\max_{v \in \mathcal{V}} P(s_{T+1}^u = v | \mathcal{S}^u). \tag{1}$$

3.2 Next Item Prediction

The main objective of the next item prediction is to develop an encoder $f_\theta(\cdot)$ that takes interactions $\mathcal{S} = \{\mathcal{S}^u\}_{u=1}^N$ as input and generates user representations $\mathcal{H} = \{\mathcal{H}^u\}_{u=1}^N$ as output in a batch with N users. Here, $\mathcal{H}^u = [\mathbf{h}_1^u, \mathbf{h}_2^u, ..., \mathbf{h}_T^u]$ and $\mathbf{h}_t^u$ represents interacting item of user u at step t. According to Eq. (1), parameters θ can be optimized by maximizing the log-likelihood of the next items of N sequences, as expressed by the formula:

$$\arg\max_{\theta} \sum_{u=1}^{N} \sum_{t=2}^{T} \ln P_\theta(\mathbf{s}_t^u), \tag{2}$$

where $\mathbf{s}_t^u$ represents the embedding of target item s_t^u. To achieve this, the adapted binary cross entropy loss can be equivalently minimized, defined as:

$$\mathcal{L}_{Rec} = -\sum_{u=1}^{N} \sum_{t=2}^{T} \left[\log \sigma \left(\mathbf{h}_{t-1}^u \cdot \mathbf{s}_t^u \right) + \sum_{neg} \log \left(1 - \sigma \left(\mathbf{h}_{t-1}^u \cdot \mathbf{s}_{neg}^u \right) \right) \right], \tag{3}$$

where $\mathbf{s}_{neg}^u$ is the embedding of item never interacted with by user u and σ represents the sigmoid function. A sampled softmax technique is adopted to reduce computational complexity, following the approach in S^3-Rec [49], where a negative item is randomly sampled for each time step in each sequence.

3.3 Contrastive Learning in SR

By adopting the mutual information maximization principle, the contrastive learning (CL) paradigm for SR leverages correlations among different views of the same sequence, maximizing user representations' agreement and enhancing the learning process. InfoNCE, as one of the CL approaches, aims at optimizing a lower bound of mutual information [3,9]. Given a user-item interaction sequence $\mathcal{S}_u$, sequence augmentation is operated to create two positive views α and β denoted as $\tilde{\mathcal{S}}_\alpha^u$ and $\tilde{\mathcal{S}}_\beta^u$, that can be formulated as follows:

$$\tilde{\mathcal{S}}_\alpha^u = g_\alpha(\mathcal{S}^u), \tilde{\mathcal{S}}_\beta^u = g_\beta(\mathcal{S}^u), \tag{4}$$

where g_α and g_β are randomly chosen as augmentation approaches from 'crop', 'mask', or 'reorder' like BERT4Rec [32] and CL4SRec [41]. We usually treat two views created from the same sequence as positive pairs and, conversely, created from different sequences as negative pairs. These views are encoded to two-dimensional $\tilde{\mathcal{X}}_\alpha^u, \tilde{\mathcal{X}}_\beta^u \in \mathbb{R}^{T \times d}$ with the sequence encoder $f_\theta(\cdot)$, expressed as $\tilde{\mathcal{X}}_\alpha^u = f_\theta(\tilde{\mathcal{S}}_\alpha^u)$ and $\tilde{\mathcal{X}}_\beta^u = f_\theta(\tilde{\mathcal{S}}_\beta^u)$, and then are concatenated into one-dimensional vectors as $\tilde{\mathbf{x}}_\alpha^u, \tilde{\mathbf{x}}_\beta^u \in \mathbb{R}^{Td}$, where d is the embed size of the encoder $f_\theta(\cdot)$. Finally, parameters θ can be optimized by the InfoNCE loss function:

$$\mathcal{L}_{CL} = \mathcal{L}_{CL}\left(\tilde{\mathbf{x}}_\alpha^u, \tilde{\mathbf{x}}_\beta^u\right) + \mathcal{L}_{CL}\left(\tilde{\mathbf{x}}_\beta^u, \tilde{\mathbf{x}}_\alpha^u\right), \tag{5}$$

and

$$\mathcal{L}_{CL}\left(\tilde{\mathbf{x}}_\alpha^u, \tilde{\mathbf{x}}_\beta^u\right) = -\log \frac{exp(\tilde{\mathbf{x}}_\alpha^u \cdot \tilde{\mathbf{x}}_\beta^u)}{\sum_{neg} exp(\tilde{\mathbf{x}}_\alpha^u \cdot \tilde{\mathbf{x}}_{neg}^u)}, \tag{6}$$

where $\mathbf{x}_{neg}^u$ is a negative view representation of $\mathbf{x}_\alpha^u$.

4 The Proposed Method

Figure 2 shows the framework of the proposed MCLRec. It first estimates intent representations from all interaction sequences. Subsequently, it computes similarity metrics by comparing intent representations with user representations and further obtains the intent-aware user representations after masking low correlation metrics. We propose a CL method to enhance the learning quality of intent-aware user representations. Instead of distinguishing between positive and negative examples in the traditional sense, we employ multi-intent aware weights, which are continuous, to quantify the similarity between samples and the learning objective by weight decay approach. The framework is designed to leverage multi-intent level information throughout the training process.

4.1 Intent-Aware User Representation

The intent-aware user representations are estimated through clustering by $\mathcal{H}^{all} = \{\mathbf{h}^u\}_{u \in \mathcal{U}}$, which denotes the set of all user representations. We utilize $\mathcal{H}^{all}$ to calculate a set of K cluster centroids $\mathcal{C} = \{\mathbf{c}^i\}_{i=1}^K$ as latent intent representations by a K-means clustering $\mathcal{K}(\cdot)$, expressed as $\mathcal{C} = \mathcal{K}(\mathcal{H}^{all})$. Each centroid $\mathbf{c}_i$ represents a cluster corresponding to a specific latent intent inferred from users' interaction patterns in $\mathcal{U}$.

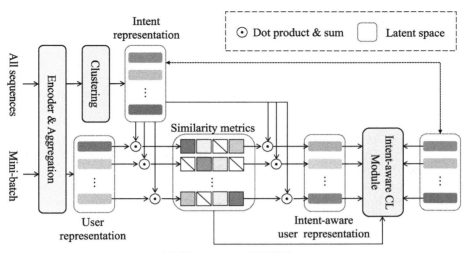

(a) The proposed MCLRec.

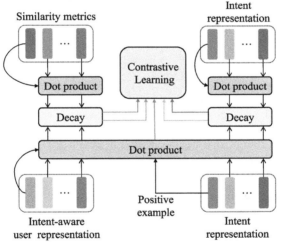

(b) The process of the multi-intent aware contrastive learning task.

Fig. 2. Overall framework.

As shown in Fig. 2(a), we can aggregate the user representations $\mathcal{X}_\alpha^u = [\mathbf{x}_{\alpha,1}^u, \mathbf{x}_{\alpha,2}^u, ..., \mathbf{x}_{\alpha,T}^u] \in \mathbb{R}^{T \times d}$ mentioned in Sect. 3.3 into $\bar{\mathbf{x}}_\alpha^u = \frac{1}{T}\sum_{t=1}^{T} \mathbf{x}_{\alpha,t}^u$ ($\bar{\mathbf{x}}_\alpha^u \in \mathbb{R}^d$) to reduce computational complexity through mean pooling. Similarly, we can aggregate $\mathcal{X}_\beta^u$ into $\bar{\mathbf{x}}_\beta^u$. For the sake of convenience in description, we denote $\bar{\mathbf{x}}_\alpha^u$ and $\bar{\mathbf{x}}_\beta^u$ collectively as $\bar{\mathbf{x}}^u$ to indicate the application of the same operation to both, that is, $\bar{\mathbf{x}}^u = \{\bar{\mathbf{x}}_\alpha^u, \bar{\mathbf{x}}_\beta^u\}$.

Given N user representations $\{\bar{\mathbf{x}}^u\}_{u=1}^N$ in a mini-batch, the correlation of intent i to user u can be calculated from $\mathcal{C}$ as follows:

$$w^{u,i} = \frac{1}{|\bar{\mathbf{x}}^u - \mathbf{c}^i|}, \tag{7}$$

where $w^{u,i}$ means the reciprocal of Euclidean distance between user representation $\bar{\mathbf{x}}^u$ and intent representation $\mathbf{c}^i$ in the same latent space. u, i are in range of $[1, ..., N], [1, ..., K]$, respectively.

For user u, we assume that there are only R relevant intents $\hat{\mathcal{C}}^u = \{\mathbf{c}^{u,k}\}_{k=1}^R$ ($\mathbf{c}^{u,k} \in \mathcal{C}$) that mainly influence user's decisions within interaction sequence $\mathcal{S}^u$, where $R \in (0, K)$ is a hyper-parameter and $\mathbf{c}^{u,k} \in \mathcal{C}$ denotes the intent representation corresponding to the k-th largest weight for user u. Since the remaining $K - R$ intents have little impact on user decision-making, we uniformly filter out these lesser weights and set them to constant zero[1]. After normalization and softmax, we have found an approach to describe the weight of diverse intents to user u as follows:

$$\hat{w}^{u,i} = \frac{\exp \omega^{u,i}}{\sum_{j=1}^K \exp \omega^{u,j}}, \tag{8}$$

and

$$\omega^{u,i} = \begin{cases} 0, & \mathbf{c}^i \notin \hat{\mathcal{C}}^u \\ w^{u,i}, & \mathbf{c}^i \in \hat{\mathcal{C}}^u \end{cases}, \tag{9}$$

where u, i are in range of $[1, ..., N], [1, ..., K]$, respectively.

The metrics $\bar{\mathbf{w}}^u = [\bar{\mathbf{w}}^{u,1}, \bar{\mathbf{w}}^{u,2}, ..., \bar{\mathbf{w}}^{u,K}]$ can customize an intent-aware user representation $\bar{\mathbf{c}}^u$ based on multiple intents for sequence $\mathcal{S}^u$:

$$\bar{\mathbf{c}}^u = \sum_{i=1}^K \hat{w}^{u,i} \cdot \mathbf{c}^{u,i}. \tag{10}$$

4.2 Multi-intent Aware Contrastive Learning

We have estimated intent-aware user representation set $\bar{\mathbf{c}}^u$ for user u in a mini-batch. However, directly contrastive learning is not effective enough since pair-wise representations with highly divergent multi-intent are far more valuable than those with minimal differences. We suggest assessing the relationship between representations by employing spatial distance in the latent space as shown in Fig. 2(b).

In a mini-batch, we construct a merged representation set $\mathcal{B} = \{\bar{\mathbf{x}}^u, \bar{\mathbf{c}}^u\}_{u=1}^N$ consisting of $2N$ representations for CL. For the user representation $\bar{\mathbf{x}}^u$, we treat $\bar{\mathbf{c}}^u$ as a learning target, and the remaining $2N - 1$ representations denoted by the set $\mathcal{B}^-$. The loss function about $\bar{\mathbf{x}}^u$ is optimized according to the following formulation:

$$\mathcal{L}_{MCL}(\bar{\mathbf{x}}^u, \bar{\mathbf{c}}^u) = -\log \frac{\exp(sim(\bar{\mathbf{x}}^u, \bar{\mathbf{c}}^u))}{\sum_{\mathbf{b} \in \mathcal{B}^-} \exp(sim(\bar{\mathbf{x}}^u, \mathbf{b}))}. \tag{11}$$

---

[1] We set the values to zero before normalization and softmax operations for smoothing the weights.

Analogous to Eq. (11), the loss function with respect to $\bar{\mathbf{c}}^u$ is optimized as follows:

$$\mathcal{L}_{MCL}(\bar{\mathbf{c}}^u, \bar{\mathbf{x}}^u) = -\log \frac{\exp(sim(\bar{\mathbf{c}}^u, \bar{\mathbf{x}}^u))}{\sum_{\mathbf{b} \in \mathcal{B}^-} \exp(sim(\bar{\mathbf{c}}^u, \mathbf{b}))}. \tag{12}$$

Overall, the loss function for multi-intent aware contrastive learning tasks can be articulated as follows:

$$\mathcal{L}_{MCL} = \mathcal{L}_{MCL}(\bar{\mathbf{x}}^u, \bar{\mathbf{c}}^u) + \mathcal{L}_{MCL}(\bar{\mathbf{c}}^u, \bar{\mathbf{x}}^u). \tag{13}$$

In Eq. (11) and Eq. (12), the function $sim(\cdot)$ acts as a similarity metric that quantifies the agreement between two representations $\mathbf{b}^p, \mathbf{b}^q \in \mathcal{B}$, as defined below:

$$sim(\mathbf{b}^p, \mathbf{b}^q) = \mathbf{b}^p \cdot \mathbf{b}^q - \mathcal{D}(\mathbf{b}^p, \mathbf{b}^q), \tag{14}$$

and

$$\mathcal{D}(\mathbf{b}^p, \mathbf{b}^q) = \begin{cases} +\infty, & \mathbf{b}^p = \mathbf{b}^q \\ \log_2 \dfrac{2}{1 - sim_{cos}(\bar{\mathbf{w}}^p, \bar{\mathbf{w}}^q)}, & \mathbf{b}^p \neq \mathbf{b}^q \text{ and } \mathbf{b}^p, \mathbf{b}^q \notin \{\bar{\mathbf{c}}^u\}_{u=1}^N \\ \log_2 \dfrac{2}{1 - sim_{cos}(\bar{\mathbf{c}}^p, \bar{\mathbf{c}}^q)}, & otherwise \end{cases}, \tag{15}$$

where $sim_{cos}(\cdot)$ denotes the cosine similarity function. We introduce a decay function $\mathcal{D}(\cdot)$ to control the impact of representations $\mathbf{b}^p$ and $\mathbf{b}^q$ in CL according to their similarity. When considering solely the similarity between user representations, the decay is determined by the parameter $\bar{\mathbf{w}}^u$. Otherwise, it is governed by the intent-aware user representation $\bar{\mathbf{c}}^u$. Two proximate representations in the latent space exhibit a high degree of similarity from a multi-intent perspective, necessitating a more pronounced decay. Conversely, greater distance warrants less decay. This ensures that the similarity function respects the underlying multi-intent aware framework in the latent space by mitigating the impact of pair-wise representations with high similarity.

4.3 Multi-task Learning

We employ a multi-task learning framework that simultaneously optimizes the main task of sequential prediction alongside three auxiliary learning objectives. In the framework, Eq. (3) is to optimize the main next item prediction task, Eq. (5) is to optimize the sequential contrastive learning task, and Eq. (13) is to optimize the multi-intent aware contrastive learning task. Following ICLRec [5], we can optimize the intent contrastive learning task as $\mathcal{L}_{ICL}$. Mathematically, we jointly train the model as follows:

$$\mathcal{L} = \mathcal{L}_{Rec} + \beta \cdot \mathcal{L}_{CL} + \lambda \cdot \mathcal{L}_{ICL} + \gamma \cdot \mathcal{L}_{MCL}, \tag{16}$$

where β, λ and γ control the strength of CL, ICL and multi-intent aware contrastive learning tasks, respectively, to be tuned.

Table 1. Statistics of four experimented datasets.

Dataset	Beauty	Sports	Toys	Yelp
# Users	22,363	35,598	19,412	30,431
# Items	12,101	18,357	11,924	20,033
# Actions	198,502	296,337	167,597	316354
# Avg.length	8.9	8.3	8.6	10.4
Sparsity	99.93%	99.95%	99.93%	99.95%

5 Experiments

5.1 Experimental Setting

Datasets. We conduct experiments on Amazon [11,27], a public dataset of product reviews, and Yelp[2], a dataset for business recommendation. In this work, we select three experimental subcategories of Amazon: 'Beauty', 'Toys' and 'Sports'. Following SASRec [16], we retain only the 5-core datasets, where users and items with fewer than five interactions have been removed. Interactions are grouped by user and arranged in ascending chronological order. For each user, the last item in the interaction sequence is used for testing, the second-to-last item is reserved for validation, and the remaining items are utilized for training. The statistics of three subcategories are displayed in Table 1.

Evaluation Metrics. Evaluation involves ranking predictions over the complete set of items without employing negative sampling [18,37]. We use two widely-used evaluation metrics to evaluate the model, including Hit Ratio@k (HR@k) and Normalized Discounted Cumulative Gain@k (NDCG@k) where $k \in \{5, 10\}$.

Baselines Models. We compare our model with baseline methods categorized into four groups. The first group comprises a non-sequential model, BPR-MF [31]. The second group includes traditional sequential recommendation models such as GRU4Rec [12], Caser [34] and SASRec [16]. The third group encompasses models that integrate self-supervised learning (SSL) within a sequential framework, including BERT4Rec [32], S^3-Rec [49] and CL4SRec [41]. The fourth group consists of intent-based sequential models, namely DSSRec [26], ICLRec [5] and IOCRec [20].

Implementation Details. The implementations of Caser, BERT4Rec, S^3-Rec, ICLRec and IOCRec are provided by the authors. BPR-MF, GRU4Rec, DSSRec, SASRec and CL4SRec are implemented based on public resources. The parameters for these methods are set as described in their respective papers,

---
[2] https://www.yelp.com/dataset.

Table 2. Performance Comparison on HIT and NDCG Metrics. Best baseline scores are underlined; scores where our model outperforms the baseline are in bold. The last row indicates the performance increase of our model over the best baseline as a percentage.

(a) HIT

Metric	HIT@5				HIT@10			
DataSet	Beauty	Sports	Toys	Yelp	Beauty	Sports	Toys	Yelp
BPR-MF	0.0178	0.0123	0.0122	0.0131	0.0296	0.0215	0.0197	0.0246
GRU4Rec	0.0180	0.0162	0.0121	0.0154	0.0284	0.0258	0.0184	0.0265
Caser	0.0251	0.0154	0.0205	0.0164	0.0342	0.0261	0.0333	0.0274
SASRec	0.0377	0.0214	0.0429	0.0161	0.0624	0.0333	0.0652	0.0265
BERT4Rec	0.0360	0.0217	0.0371	0.0186	0.0601	0.0359	0.0524	0.0291
S^3-Rec	0.0189	0.0121	0.0456	0.0175	0.0307	0.0205	0.0689	0.0283
CL4SRec	0.0401	0.0231	0.0503	0.0218	0.0642	0.0369	0.0736	0.0354
DSSRec	0.0408	0.0209	0.0447	0.0171	0.0616	0.0328	0.0671	0.0297
ICLRec	0.0500	0.0290	<u>0.0597</u>	<u>0.0240</u>	0.0744	0.0437	<u>0.0834</u>	<u>0.0409</u>
IOCRec	<u>0.0511</u>	<u>0.0293</u>	0.0542	0.0222	<u>0.0774</u>	<u>0.0452</u>	0.0804	0.0394
MCLRec	**0.0566**	**0.0308**	**0.0635**	**0.0255**	**0.0811**	**0.0465**	**0.0896**	**0.0421**
improv.	10.76%	5.19%	6.31%	6.38%	4.78%	2.92%	7.41%	3.01%

(b) NDCG

Metric	NDCG@5				NDCG@10			
DataSet	Beauty	Sports	Toys	Yelp	Beauty	Sports	Toys	Yelp
BPR-MF	0.0109	0.0076	0.0076	0.0760	0.0147	0.0105	0.0100	0.0119
GRU4Rec	0.0116	0.0103	0.0077	0.0104	0.0150	0.0142	0.0097	0.0137
Caser	0.0145	0.0114	0.0125	0.0096	0.0226	0.0135	0.0168	0.0129
SASRec	0.0241	0.0144	0.0245	0.0102	0.0342	0.0177	0.0320	0.0134
BERT4Rec	0.0216	0.0143	0.0259	0.0118	0.0300	0.0190	0.0309	0.0171
S^3-Rec	0.0115	0.0084	0.0314	0.0115	0.0153	0.0111	0.0388	0.0162
CL4SRec	0.0268	0.0146	0.0264	0.0131	0.0345	0.0191	0.0339	0.0188
DSSRec	0.0263	0.0139	0.0297	0.0112	0.0329	0.0178	0.0369	0.0152
ICLRec	<u>0.0326</u>	<u>0.0191</u>	<u>0.0404</u>	<u>0.0153</u>	<u>0.0403</u>	<u>0.0238</u>	<u>0.0480</u>	<u>0.0207</u>
IOCRec	0.0311	0.0169	0.0297	0.0137	0.0396	0.0220	0.0381	0.0192
MCLRec	**0.0377**	**0.0201**	**0.0433**	**0.0166**	**0.0455**	**0.0252**	**0.0519**	**0.0220**
improv.	15.64%	5.34%	7.55%	8.63%	12.90%	5.71%	8.06%	6.04%

and the best settings are selected according to the performance of models on the validation dataset. We implement our method in PyTorch and use Faiss [15] for K-means clustering to speed up training. For the encoder part, we set self-attention blocks and attention heads as 2 and embedding dimension as 64. The model is optimized by Adam optimizer [17] where learning rate, β_1 and β_2 are set to 0.001, 0.9 and 0.999, respectively. The maximum sequence length is set to 50. For hyper-parameters of MCLRec, we tune λ, β and γ all within the set $\{0.1, 0.2, ..., 0.8\}$. K is tuned in the range of $\{32, 64, 128, 256, 512, 1024, 2048\}$. The ratio of R to K is set to specific values, with R/K taking on the following proportions: $\{0.125, 0.25, 0.375, 0.5, 0.625, 0.75, 0.875\}$. The model's training incorporates an early stopping mechanism guided by its performance metrics on validation data. All experiments are run on a single NVIDIA A100 GPU.

5.2 Performance Comparison

Table 2 shows the performance comparison of different methods on four datasets. We have the following observations. BPR underperforms compared to conventional sequential models, underscoring the significance of extracting sequential patterns from interaction sequences. In sequential models, those employing the attention mechanism, such as SASRec and BERT4Rec, outperform models like Caser and GRU4Rec, which do not apply the attention mechanism, demonstrating the attention mechanism's effectiveness in modeling interaction sequences. Sequential models integrating the attention mechanism, such as SASRec and BERT4Rec, outperform models like Caser and GRU4Rec, which lack this feature. This highlights the effectiveness of the attention mechanism in modeling interaction sequences. Besides, SSL-based models like BERT4Rec and S^3-Rec, which utilize Masked Item Prediction (MIP) tasks to learn user representation, significantly underperform compared to other CL-based models including CL4SRec, ICLRec, IOCRec. The reason might be that MIP tasks for SSL require sufficient context information.

Intent-based sequential models, including ICLRec, IOCRec and the proposed MCLRec, perform better than SSL-based models, including BERT4Rec and S^3-Rec, which indicates the importance of learning intent representations under user-item interaction sequences. However, ICLRec is not as effective as MCLRec, probably because they do not consider the multi-intent level information within one interaction sequence. The performance of MCLRec is also superior to that of IOCRec, which may be attributed to the lack of intersecting multi-intent level information in IOCRec.

Leveraging multi-intent aware contrastive learning, MCLRec demonstrates notably enhanced performance compared to alternative methods across various metrics within three subcategories. The superiority of the best outcome over the top baseline ranges from 2.92% to 15.64% in HR and NDCG. Low improvement in Sports could be attributed to the increasing of candidate item number and relatively insufficient interaction sequences, which likely introduces greater complexity to contrastive learning tasks in the latent space.

5.3 Hyper-Parameter Sensitivity

Given space constraints, we only report the effect of hyper-parameters R/K and K on model performance. We conduct four experiments in the Beauty dataset to

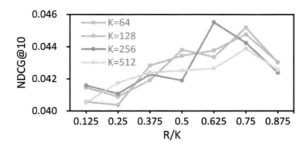

Fig. 3. Performance comparison w.r.t. hyper-parameters K and R/K.

investigate the impact of hyper-parameters, including the ratio of relevant intent R/K and the number of intent K. We keep other hyper-parameters unchanged for all models in the following experiments and consider NDCG@10. We show the results of experiments in Fig. 3 and our observations as follows.

Our method surpasses the baseline across all values of K and R/K (The best NDCG@10 of ICLRec is 0.0403), which indicates the effectiveness of MCLRec. The model's performance peaks at a ratio of $R/K = 0.625$ given $K = 256$ and $R/K = 0.75$ given K values of $64, 128, 512$. Subsequently, performance begins to decline as R/K increases further. This decline in performance is likely attributable to introducing an excessive number of representations with low weights $\hat{w}^{u,i}$ into the CL process. This can lead to reduced learning effectiveness and a diminished robustness to noisy data. Additionally, the model's optimal performance improves with increasing values of K up to 256, beyond which a decline is observed. This suggests that introducing an excessive number of intent categories does not further facilitate multi-intent learning, likely due to the diminished average number of samples available per intent category.

6 Conclusion

In this paper, we propose a framework called MCLRec, which can filter relevant intents of interactions by similarity of representations in the latent space. We implement a novel multi-intent aware contrastive learning approach to mitigate the impact of the pair-wise representations with high similarity during contrastive learning. Experiments conducted on four public datasets further validate the effectiveness of the proposed model. Several existing studies [29] have discovered that adopting the perspective of utilizing high-quality positive examples can further enhance the performance and robustness of the SR model, which we leave for future studies.

Acknowledgement. This research is supported by the Research Promotion Project of Key Construction Discipline in Guangdong Province (2022ZDJS112) and Shenzhen Technology University Research Start-up Fund (GDRC202133).

References

1. Cen, Y., et al.: Controllable multi-interest framework for recommendation. In: Proceedings of the 26th ACM SIGKDD International Conference on Knowledge Discovery & Data Mining, pp. 2942–2951 (2020)
2. Chang, J., et al.: Sequential recommendation with graph neural networks. In: Proceedings of the 44th International ACM SIGIR Conference on Research and Development in Information Retrieval, pp. 378–387 (2021)
3. Chen, T., et al.: A simple framework for contrastive learning of visual representations. In: International Conference on Machine Learning, pp. 1597–1607. PMLR (2020)
4. Chen, W., et al.: Improving end-to-end sequential recommendations with intent-aware diversification. In: Proceedings of the 29th ACM International Conference on Information & Knowledge Management, pp. 175–184 (2020)
5. Chen, Y., et al.: Intent contrastive learning for sequential recommendation. In: Proceedings of the ACM Web Conference 2022, pp. 2172–2182 (2022)
6. Devlin, J., et al.: BERT: Pre-training of deep bidirectional transformers for language understanding (2018). arXiv preprint arXiv:1810.04805
7. Donkers, T., et al.: Sequential user-based recurrent neural network recommendations. In: Proceedings of the Eleventh ACM Conference on Recommender Systems, pp. 152–160 (2017)
8. Fang, H., et al.: Deep learning for sequential recommendation: algorithms, influential factors, and evaluations. ACM Trans. Inf. Syst. (TOIS) **39**(1), 1–42 (2020)
9. He, K., et al.: Momentum contrast for unsupervised visual representation learning. In: Proceedings of the IEEE/CVF Conference on Computer Vision and Pattern Recognition, pp. 9729–9738 (2020)
10. He, R., et al.: Fusing similarity models with markov chains for sparse sequential recommendation. In: 2016 IEEE 16th International Conference on Data Mining (ICDM), pp. 191–200. IEEE (2016)
11. He, R., et al.: Ups and downs: modeling the visual evolution of fashion trends with one-class collaborative filtering. In: Proceedings of the 25th International Conference on World Wide Web, pp. 507–517 (2016)
12. Hidasi, B., et al.: Session-based recommendations with recurrent neural networks (2015). arXiv preprint arXiv:1511.06939
13. Ji, W., et al.: Sequential recommender via time-aware attentive memory network. In: Proceedings of the 29th ACM International Conference on Information & Knowledge Management, pp. 565–574 (2020)
14. Jing, L., et al.: Self-supervised visual feature learning with deep neural networks: a survey. IEEE Trans. Pattern Anal. Mach. Intell. **43**(11), 4037–4058 (2020)
15. Johnson, J., et al.: Billion-scale similarity search with GPUs. IEEE Trans. Big Data **7**(3), 535–547 (2019)
16. Kang, W.C., et al.: Self-attentive sequential recommendation. In: 2018 IEEE International Conference on Data Mining (ICDM), pp. 197–206. IEEE (2018)
17. Kingma, D.P., et al.: Adam: A method for stochastic optimization (2014). arXiv preprint arXiv:1412.6980
18. Krichene, W., et al.: On sampled metrics for item recommendation. In: Proceedings of the 26th ACM SIGKDD International Conference on Knowledge Discovery & Data Mining, pp. 1748–1757 (2020)
19. Lan, Z., et al.: ALBERT: A lite BERT for self-supervised learning of language representations (2019). arXiv preprint arXiv:1909.11942

20. Li, X., et al.: Multi-intention oriented contrastive learning for sequential recommendation. In: Proceedings of the Sixteenth ACM International Conference on Web Search and Data Mining, pp. 411–419 (2023)
21. Lian, D., et al.: Geography-aware sequential location recommendation. In: Proceedings of the 26th ACM SIGKDD International Conference on Knowledge Discovery & Data Mining, pp. 2009–2019 (2020)
22. Liu, Q., et al.: Context-aware sequential recommendation. In: 2016 IEEE 16th International Conference on Data Mining (ICDM), pp. 1053–1058. IEEE (2016)
23. Liu, Y., et al.: RoBERTa: A robustly optimized BERT pretraining approach (2019). arXiv preprint arXiv:1907.11692
24. Liu, Z., et al.: Contrastive self-supervised sequential recommendation with robust augmentation (2021). arXiv preprint arXiv:2108.06479
25. Ma, C., et al.: Hierarchical gating networks for sequential recommendation. In: Proceedings of the 25th ACM SIGKDD International Conference on Knowledge Discovery & Data Mining, pp. 825–833 (2019)
26. Ma, J., et al.: Disentangled self-supervision in sequential recommenders. In: Proceedings of the 26th ACM SIGKDD International Conference on Knowledge Discovery & Data Mining, pp. 483–491 (2020)
27. McAuley, J., et al.: Image-based recommendations on styles and substitutes. In: Proceedings of the 38th international ACM SIGIR Conference on Research and Development in Information Retrieval, pp. 43–52 (2015)
28. Pan, Z., et al.: An intent-guided collaborative machine for session-based recommendation. In: Proceedings of the 43rd International ACM SIGIR Conference on Research and Development in Information Retrieval, pp. 1833–1836 (2020)
29. Qin, X., et al.: Intent contrastive learning with cross subsequences for sequential recommendation (2023). arXiv preprint arXiv:2310.14318
30. Rendle, S., et al.: Factorizing personalized markov chains for next-basket recommendation. In: Proceedings of the 19th International Conference on World Wide Web, pp. 811–820 (2010)
31. Rendle, S., et al.: BPR: Bayesian personalized ranking from implicit feedback (2012). arXiv preprint arXiv:1205.2618
32. Sun, F., et al.: BERT4Rec: sequential recommendation with bidirectional encoder representations from transformer. In: Proceedings of the 28th ACM International Conference on Information and Knowledge Management, pp. 1441–1450 (2019)
33. Tan, Q., et al.: Sparse-interest network for sequential recommendation. In: Proceedings of the 14th ACM International Conference on Web Search and Data Mining, pp. 598–606 (2021)
34. Tang, J., et al.: Personalized top-n sequential recommendation via convolutional sequence embedding. In: Proceedings of the eleventh ACM International Conference on Web Search and Data Mining, pp. 565–573 (2018)
35. Tanjim, M.M., et al.: Attentive sequential models of latent intent for next item recommendation. In: Proceedings of The Web Conference 2020, pp. 2528–2534 (2020)
36. Wang, S., et al.: Modeling multi-purpose sessions for next-item recommendations via mixture-channel purpose routing networks. In: International Joint Conference on Artificial Intelligence. International Joint Conferences on Artificial Intelligence (2019)
37. Wang, X., et al.: Neural graph collaborative filtering. In: Proceedings of the 42nd International ACM SIGIR Conference on Research and Development in Information Retrieval, pp. 165–174 (2019)

38. Wu, C.Y., et al.: Recurrent recommender networks. In: Proceedings of the tenth ACM International Conference on Web Search and Data Mining, pp. 495–503 (2017)
39. Wu, J., et al.: Self-supervised graph learning for recommendation. In: Proceedings of the 44th International ACM SIGIR Conference on Research and Development in Information Retrieval, pp. 726–735 (2021)
40. Xia, X., et al.: Self-supervised hypergraph convolutional networks for session-based recommendation. In: Proceedings of the AAAI Conference on Artificial Intelligence. vol. 35, pp. 4503–4511 (2021)
41. Xie, X., et al.: Contrastive learning for sequential recommendation. In: 2022 IEEE 38th International Conference on Data Engineering (ICDE), pp. 1259–1273. IEEE (2022)
42. Xin, X., et al.: Self-supervised reinforcement learning for recommender systems. In: Proceedings of the 43rd International ACM SIGIR Conference on Research and Development in Information Retrieval, pp. 931–940 (2020)
43. Yao, T., et al.: Self-supervised learning for deep models in recommendations (2020). arXiv preprint arXiv:2007.12865
44. Yao, T., et al.: Self-supervised learning for large-scale item recommendations. In: Proceedings of the 30th ACM International Conference on Information & Knowledge Management, pp. 4321–4330 (2021)
45. Yu, F., et al.: A dynamic recurrent model for next basket recommendation. In: Proceedings of the 39th International ACM SIGIR Conference on Research and Development in Information Retrieval, pp. 729–732 (2016)
46. Yu, L., et al.: Multi-order attentive ranking model for sequential recommendation. In: Proceedings of the AAAI Conference on Artificial Intelligence. vol. 33, pp. 5709–5716 (2019)
47. Yuan, F., et al.: A simple convolutional generative network for next item recommendation. In: Proceedings of the Twelfth ACM International Conference on Web Search and Data Mining, pp. 582–590 (2019)
48. Zheng, L., et al.: Joint deep modeling of users and items using reviews for recommendation. In: Proceedings of the Tenth ACM International Conference on Web Search and Data Mining, pp. 425–434 (2017)
49. Zhou, K., et al.: S3-Rec: self-supervised learning for sequential recommendation with mutual information maximization. In: Proceedings of the 29th ACM International Conference on Information & Knowledge Management, pp. 1893–1902 (2020)

Subgraph Collaborative Graph Contrastive Learning for Recommendation

Jie Ma[1,2], Jiwei Qin[1,2(✉)], Peichen Ji[1,2], Zhibin Yang[1,2], Donghao Zhang[1,2], and Chaoqun Liu[3]

[1] School of Computer Science and Technology, Xinjiang University, Xinjiang Urumqi 830046, Xinjiang Uygur Autonomous Region, China
{107552201394,jipeichen,107552203961,107552203965}@stu.xju.edu.cn,
jwqin@xju.edu.cn
[2] Key Laboratory of Signal Detection and Processing, Xinjiang Uygur Autonomous Region, Xinjiang University, Urumqi 830046, China
[3] School of Computer Science and Technology, Soochow University, SuZhou, China

Abstract. Graph Collaborative Filtering discovers potential connections between users and items using graph neural networks. However, graph neural networks may aggregate the noise in the user-item interaction data when updating node features. In addition, increasing the number of network layers may lead to the over-smoothing problem, whereby the representations of distinct nodes become excessively similar.

We propose a new <u>S</u>ubgraph Collaborative <u>G</u>raph <u>C</u>ontrastive <u>L</u>earning framework (SGCL) to mitigate these issues. The SGCL model mainly consists of a self-alignment module and a subgraph module. The self-alignment module uses two encoders to create contrasting views, optimizing alignment and uniformity to improve user and item representations. The subgraph module consists of three parts: Initially, we select neighbors that most accurately represent each node for subgraph sampling, aiming to lessen the impact of high neighbors. This method helps alleviate the effect of over-smoothing. Furthermore, we filter out noisy nodes in the frequency domain to achieve the subgraph denoising. Lastly, we apply subgraph collaborative representation to enhance the node representations. The SGCL mitigates noisy nodes and over-smoothing issues. Our experiments across three public datasets show the SGCL performance improvement of about 7.8% .

Keywords: Graph Collaborative Filtering · Graph Contrastive Learning · Subgraph · Recommendation

1 Introduction

In application scenarios such as e-commerce and music platforms, recommender systems can effectively provide personalized services to users. The accuracy of recommendation results depends on recommendation algorithms, among which

Collaborative Filtering (CF) is an effective recommendation algorithm. CF analyzes a user's historical interaction information to predict items that might capture the user's interest.

Researchers have combined graph neural networks and collaborative filtering into the Graph Collaborative Filtering (GCF) method to explore more potential relationships between users and items. GCF method models user-item relationships with graph data, leveraging the principle that similar users have similar preferences. LightGCN [2] stands as a classic model in recommender system, which simplifies graph neural network structures. However, as the number of layers increases, each node repeatedly summarizes information from higher-order neighboring nodes. This phenomenon leads to the over-smoothing problem, in which the feature representations of nodes become increasingly similar, reducing the model's ability to distinguish between different nodes.

Recommender systems often face data sparsity issues. Many GCF methods mitigate this by using contrastive learning, which creates new data samples through data augmentation to alleviate data sparsity. These enhances user and item embeddings by employing a contrastive loss function, which reduces the distance between similar samples while increasing the distance between dissimilar ones. Combining these two approaches can significantly improve the recommendation efficiency, e.g., SGL [8], SimGCL [13], SimRec [10], DCCF [5], LightGCL [1] and AutoCF [9]. However, the data augmentation methods used by the SGL model, including nodes or edges dropout and random walking, may lose semantic information.

User-item interactions may contain noise. As shown in the Fig. 1. For user u_1, item i_1 is a noise node. Similarly, for user u_2, item i_5 is a noise node. In the graph convolution process, nodes are represented by aggregating information from neighboring nodes, which will affect the node's feature representation accuracy if they contain noisy nodes. Currently, most denoising research focuses on the spatial domain. For example, RGCF [6] and GSocRec [4] . The denoising module in RGCF consists of hard and soft denoising to minimize the effects of the impact of noise interactions on the learning of GNN representations. GSocRec removes the noise using the CDMSR module, which considers high-trust relations and deletes low-trust relations.

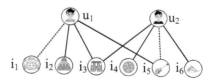

Fig. 1. This figure shows a simplified user-item bipartite graph. The graph has two user nodes: user u_1 and user u_2. user u_1 is connected to four item nodes (i_1, i_2, i_3, i_5) and user u_2 is connected to four item nodes (i_3, i_4, i_5, i_6). The solid line indicates the user's preferences and the dashed line indicates the noise nodes. User u_1 mistakenly touches item i_1. similarly, user u_2 mistakenly touches item i_5.

Given this, we construct the SGCL model. The contribution of this paper is as follows:

1. Our SGCL model comprises a self-alignment module and a subgraph module. The self-alignment module uses two encoders to create contrastive views, alleviate data sparsity from the feature augment perspective, and prevent semantic information loss. Users and items optimize alignment and uniformity within the two views to improve their representations.
2. To mitigate the impact of over-smoothing and noisy nodes. We construct the subgraph module to enhance subgraph representation by creating user, positive item, and negative item subgraphs. This method aims to lessen the impact of high neighbors, thus alleviating the over-smoothing effect. Subsequently, we implement denoising techniques for subgraphs from the frequency domain perspective, aiming to diminish the effects of noisy nodes.
3. We validate the superior performance of the SGCL model compared to several state-of-the-art (SOTA) baseline models by experimenting with three public datasets.

2 Problem Formulation

We define a set of users as $\mathcal{U} = \{u_1, u_2, \ldots, u_i, \ldots, u_{|\mathcal{U}|}\}$ and a set of items as $\mathcal{V} = \{v_1, v_2, \ldots, v_j, \ldots, v_{|\mathcal{V}|}\}$, where $|\mathcal{U}| = \mathcal{I}$ and $|\mathcal{V}| = \mathcal{J}$ denote the total number of users and items, respectively. The interactions between users and items are denoted by the interaction matrix $\mathcal{A}$, with elements $\mathcal{A}_{ij}$ set to 1 if user u_i has purchased item v_j, and 0 otherwise. We have an embedding vector for each user u_i as $e_i^u \in \mathbb{R}^d$, and an embedding vector for each item v_j as $e_j^v \in \mathbb{R}^d$.

3 Method

This section provides an in-depth description of the framework of the SGCL model, as illustrated in Fig 2. The SGCL model includes self-alignment and subgraph modules. The self-alignment module consists of two encoders that perform self-alignment for the generated views. The subgraph module includes subgraph sampling, denoising, and collaboration representation submodules.

3.1 The Self-alignment Module

We propose the self-alignment module to alleviate data sparsity and preserve information integrity during data augmentation.

Dual Encoder. The self-alignment module uses two different encoders to capture local collaborative relationships between users and items. One encoder uses LightGCN [2] as its backbone network. The other encoder uses LightGCN as the backbone network and adds gaussian noise when performing graph convolution.

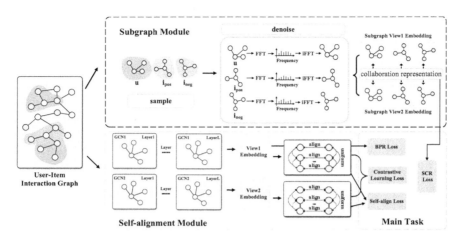

Fig. 2. The general framework of the SGCL model that we propose consists of the self-alignment module and the subgraph module. SCR representation subgraph collaboration representation submodule. In these three subgraphs, each color signifies a different subgraph type. Selecting negative item subgraphs aims to represent negative item samples more accurately.

Each encoder processes the user-item bipartite graph to generate a dual-view representation of each node. This data augmentation method alleviates data sparsity and avoids causing semantic loss of samples.

All user embeddings are $E^{(u)} \in \mathbb{R}^{I \times d}$, and all item embeddings are $E^{(v)} \in \mathbb{R}^{J \times d}$, where $\mathcal{I}$ and $\mathcal{J}$ are user counts and item counts, respectively. LightGCN is the backbone for message passing, aggregating neighborhood information from each node. From layer l-1 up to layer l, we define the messaging process. Eq 1 displays the embeddings of users and items from the last layer.

$$e_i^u = \sum_{l=0}^{L} e_{i,l}^u \quad ; \quad e_j^v = \sum_{l=0}^{L} e_{j,l}^v \quad . \tag{1}$$

The user's preference is the inner product between the final embedding of the user e_i^u and the item e_j^v. As in the Eq 2.

$$y_{u,v} = e_i^u \cdot e_j^v \tag{2}$$

The second encoder is to add fine-grained gaussian noise when performing graph convolution as shown in Eq 3. Adding gaussian noise avoids losing semantic information and extends the feature space of users and items. η is a parameter.

$$e_i^{u'} = e_i^u + \eta \cdot \Delta e_i^u \quad ; \quad \Delta e_i^u \sim \mathcal{N}\left(0, \sigma^2\right). \tag{3}$$

We can denote these as the representations of user $e_i^{u'}$ and item $e_j^{v'}$, respectively. The last layer's embeddings of users and items, with added perturbation, are shown as in Eq 4.

$$e_i^{u'} = \sum_{l=0}^{L} e_{i,l}^{u'} \quad ; \quad e_j^{v'} = \sum_{l=0}^{L} e_{j,l}^{v'} \quad . \tag{4}$$

Self-alignment. We use a self-alignment module for the views generated by the two encoders. Specifically, for the user-item bipartite graph, we aim to increase the proximity between users and their relevant items while creating a greater distance between them and unrelated items. This module allows for a more even spatial distribution of features across all users and items. It helps reduce the preference for popular items, which implicitly mitigates the popularity bias problem.

Optimize two attribute information, alignment and uniformity, within the view. Considering that the data distributions of users and items are inherently different, we optimize the two attribute information of user representation and item representation separately.

Optimizing alignment between users and items, as shown in Fig 2, uses alignment to represent users and items better. Aligning users and items brings users nearer to their related items and pulls them farther away from their unrelated items. As described in Eq 5.

$$\mathcal{L}_{align} = \|f(e^u) - f(e^v)\|_2^2 + \|f(e^{u'}) - f(e^{v'})\|_2^2 \tag{5}$$

where $(u,v) \sim p_{pos}$, $(u',v') \sim p_{pos}$, $f(\cdot)$ represents the vector representation normalized by L2. u, v denotes the representation of the user and the item. We optimize the uniformity for users and items within the views generated by the two encoders (view1, view2) to achieve an even distribution of their representations across the hypersphere space. As shown in Eq 6.

$$\mathcal{L}_{uniform1}(e^u, e^v) = \log\left(\frac{1}{N}\sum_{a \neq b} \exp\left(\|e_a^u - e_b^u\|_2^2\right)\right) \\ + \log\left(\frac{1}{N}\sum_{a \neq b} \exp\left(\|e_a^v - e_b^v\|_2^2\right)\right) \tag{6}$$

where $\mathcal{L}_{uniform1}(e^u, e^v)$ is for the view1. Similarly, the term $\mathcal{L}_{uniform2}(e^{u'}, e^{v'})$ can be expressed in the same manner. $\mathcal{L}_{uniform} = \mathcal{L}_{uniform1} + \mathcal{L}_{uniform2}$. The loss function for the self-alignment module is as follows:

$$\mathcal{L}_{self-align} = \mathcal{L}_{align} + \mathcal{L}_{uniform} \tag{7}$$

3.2 The Subgraph Module

To mitigate the impact of over-smoothing and noisy nodes. The subgraph module comprises subgraph sampling, denoising, and subgraph collaboration representation modules.

Subgraph Sample. We aim to improve node embedding representations by creating subgraphs for users, positive items, and negative items, selecting those that best represent each node.

We use the user-item bipartite graph to construct user-related items in the user subgraph to achieve this goal. Suppose the number of selected items exceeds those related to the user. In that case, we randomly select k items. Suppose the number of items in the selection may be less than that in the user's relationship. In that case, the user will select items related to the item. And so on, we can finally get the representation user subgraph, positive and negative item subgraph. $G_u \in \{1, \ldots, u_k\}$ denotes a user subgraph. $G_{ipos} \in \{1, \ldots, i_k\}$ denotes a positive item subgraph. $G_{ineg} \in \{1, \ldots, i_k\}$ denotes a negative item subgraph.

Subgraph Denoise. Noise is inevitable when users interact with the item. We have drawn inspiration from filtering algorithms used in signal processing, which can reduce noise in the frequency domain [14]. We construct a subgraph denoising module to study it from the frequency domain perspective. We perform the Fast Fourier Transform (FFT) on the input subgraph $G_{(\bullet)}$ to obtain its frequency domain representation X_G.

$$X_{G_u} = \mathcal{F}(G_u)$$
$$X_{G_{ipos}} = \mathcal{F}(G_{ipos}) \quad (8)$$
$$X_{G_{ineg}} = \mathcal{F}(G_{ineg})$$

Since noise in the frequency domain manifests itself as a high frequency portion, i.e., a rapidly changing portion, we remove noise in the frequency domain for the subgraphs G_u, G_{ipos}, and G_{ineg}, which requires preserving low-frequency information and deleting high-frequency information. Filtering each feature dimension in the frequency domain can be seen as removing item nodes with noise for the operation of G_u.

Based on the specified ratio rate γ, we randomly mask elements to generate a mask matrix M_γ with the same shape as the subgraph. Then, utilizing the number of masks, we create a dominant frequency mask matrix M_{mask}. This matrix serves to preserve lowfrequency information while setting highfrequency information to zero.

Subsequently, we combine the random mask matrix M_γ with the dominant frequency mask matrix M_{mask} through a bitwise-and operation to preserve the random mask at low-frequency locations.

$$M_\gamma = M_\gamma \& M_{mask} \quad (9)$$

The mask is then used to zero out the elements in the natural and imaginary parts of the frequency domain representation $X_{G_{(\bullet)}}$ that correspond to the positions where the mask is True, i.e., to zero out the high-frequency information.

$$f_{real} = Re(X_{G_{(\bullet)}}) \odot (M_\gamma \cdot 0)$$
$$f_{imag} = Im(X_{G_{(\bullet)}}) \odot (M_\gamma \cdot 0) \quad (10)$$

$$X_{G_{(\bullet)}} = f_{real} + i \cdot f_{imag} \tag{11}$$

In order to restore the data from the frequency domain to the spatial domain, we apply an inverse Fast Fourier Transform. As shown in Eq 12.

$$\tilde{G_{(\bullet)}} = \mathcal{F}^{-1}(X_{G_{(\bullet)}}) \tag{12}$$

The noise in the historical behavioral data can be well removed by denoising the subgraphs in the frequency domain.

Subgraph Collaboration Representations Module. Encoding the denoised subgraph results in two different subgraph views. One view, $G'_{(\bullet)}$, is generated by the perturbed GCN encoder, while the other view, $G_{(\bullet)}$, is generated by the unperturbed GCN encoder. By learning from these two subgraph views, the model can obtain more effective feature representations and increase the diversity of node features, thus improving the robustness of the model.

$$\begin{aligned}\mathcal{L}_{SCR}((G'_{(\bullet)}),(G_{(\bullet)})) &= \mathcal{L}_{KL}(G'_u, G_u) \\ &+ \mathcal{L}_{KL}(G'_{ipos}, G_{ipos}) \\ &+ \mathcal{L}_{KL}(G'_{ineg}, G_{ineg}) \end{aligned} \tag{13}$$

3.3 Model Training

The model uses the Bayesian Personalized Rank (BPR) loss as the primary loss function and introduces auxiliary loss functions: contrastive learning loss, self-alignment loss, and subgraph collaboration representation loss .

Contrastive Learning. We consider views of the same node as positive pairs and different views as negative pairs. We define a contrastive loss function and construct training samples using positive and negative sample pairs. Positive sample pairs are views of the same node $(\mathbf{e}'_i, \mathbf{e}''_i)$. Negative sample pairs in different views $(\mathbf{e}'_i, \mathbf{e}''_j)$.

$$\mathcal{L}_{cl}^{user} = \sum_{u_i \in \mathcal{U}} -\log \frac{\exp\left(s\left(\mathbf{e}'_i, \mathbf{e}''_i\right)/\tau\right)}{\sum_{u_j \in \mathcal{U}} \exp\left(s\left(\mathbf{e}'_i, \mathbf{e}''_j/\tau\right)\right)} \tag{14}$$

where $s(\cdot, \cdot)$ calculates their cosine similarity. Similarly, the loss function of the item node is $\mathcal{L}_{cl}^{item}$, and the total contrastive learning loss function is $\mathcal{L}_{cl}$.

$$\mathcal{L}_{cl} = \mathcal{L}_{cl}^{user} + \mathcal{L}_{cl}^{item} \tag{15}$$

Contrastive learning aims to close the distance between positive sample pairs and far away from negative pairs in the feature space. By increasing the similarity between positive pairs and reducing it between negative pairs, the model learns to distinguish positive from negative samples, resulting in more distinctive feature representations.

BPR Loss. Recommender systems use the BPR loss function to actively learn personalized user preferences.

$$\mathcal{L}_{bpr} = \sum_{(u,v^+,v^-) \in S} -\log \sigma \left(\hat{y}_{uv^+} - \hat{y}_{uv^-} \right) \tag{16}$$

In the training data, we use a set $S = (u, v^+, v^-) \mid (u,i) \in S^+, (u, v^-) \in S^-$ that consists of a triad of samples of the form (u, v^+, v^-). Here, (u, v^+) denotes positive samples, and (u, v^-) denotes negative samples. In particular, let S^+ denote the set of all positive samples, and let S^- denote the set of all negative samples.

We combine these losses with traditional recommendation losses to create the following multi-task learning objective:

$$\mathcal{L} = \mathcal{L}_{bpr} + \lambda_1 \mathcal{L}_{cl} + \lambda_2 \mathcal{L}_{self-align} + \lambda_3 \mathcal{L}_{SCR}(G'_{(\bullet)}, G_{(\bullet)}) \tag{17}$$

4 Experiment

This section describes our experimental environment and answers four specific research questions (RQs). We performed all experiments using a NVIDIA GeForce RTX 3090 Ti with 24GB.

- **RQ1:** How does our SGCL model compare to SOTA's contrastive learning-based graph collaborative filtering recommendation model?
- **RQ2:** How does each module in our SGCL model affect its overall performance on different datasets?
- **RQ3:** How do variations in hyperparameters impact the performance outcomes of the SGCL model?
- **RQ4:** Does our SGCL model mitigate the common over-smoothing problem?

4.1 Experimental Settings

Datasets: Our model uses three public datasets; table 1 shows the details of these three public datasets.

- **LastFM**: The LastFM dataset is a widely used public dataset containing accurate interaction data between users and music.

Table 1. Basic information about the three datasets.

Dataset	Users	Items	Interactions	Desity
LastFM	1,892	17,632	92,834	2.8×10^{-3}
Yelp	42,712	26,822	182,357	1.6×10^{-4}
BeerAdvocate	10,456	13,845	1,381,094	9.5×10^{-3}

- **Yelp**: It contains reviews, ratings, and other relevant information about merchants from users of the Yelp platform.
- **BeerAdvocate**: It comes from the BeerAdvocate website and contains information about user ratings of beer.

Evaluation Protocols. Each dataset is randomly divided into the train, validation, and test sets, with proportions of 70%, 20%, and 10%, respectively. Recall and NDCG are used as evaluation metrics to assess the effectiveness of top-k recommendations. Recall@k measures the ability of a recommender system to capture the user's interest by evaluating how many of the top 20 or 40 recommendations match the user's preferences. NDCG@k evaluates a recommendation list's ranking quality by considering the items' relevance and their positions within the list. We set k to 20 by default.

Compared Baseline Methods. We assess the SGCL model by comparing various baseline models. The following section details these baseline models.

- **LightGCN** [2]: This widely used backbone network simplifies GCN structure by eliminating nonlinear activation functions and linear eigenvalue transformation.
- **HCCF** [11]: The model extends the Hypergraph Comparative Learning framework for collaborative filtering modeling approaches to graph neural networks.
- **SLRec** [12]: The framework utilizes comparative learning between node features for data enhancement. It mitigates the issue of data sparsity in recommender systems.
- **SGL** [8]: The framework proposes three data augmentation methods to change the graph's structure in different dimensions. These include node dropout, edge dropout, and random walk.
- **NCL** [3]: The model enhances neural graph collaborative filtering by integrating structural and semantic neighbors. It employs the EM algorithm to optimize the proposed prototype comparison objective.
- **DirectAU** [7]: The model argues that alignment and uniformity positively impact the quality of representations based on collaborative filtering (CF) models and is validated by theoretical analysis and empirical studies.
- **SimGCL** [13]: The model performs feature enhancement by adding noise to the representations of users and items.
- **DCCF** [5]: DCCF utilizes the learned global contextual de-entanglement representation to extract finer potential factors to mitigate noise.

Implementation Details: Our SGCL model was implemented in PyTorch and trained using the Adam optimizer. The learning rate sets $1e^{-3}$. The embedding size was 32, the batch size was 4096, and the model utilized convolutional layers with filter sizes of 2 and 3. η takes the value of 0.001 in the Yelp dataset, 0.08 in the LastFM dataset, and 0.01 in the BeerAdvocate dataset.

4.2 Performance Comparison (RQ1)

We have drawn the following conclusions from the summary of our experimental results (see Table 2) :

1. Outperforming SOTA contrastive learning methods. Evaluation results show that SGCL outperforms other SOTA models in the top 20 and 40 settings, achieving the best performance. Our SGCL model outperforms existing methods. Other contrastive learning models are susceptible to noise and need to better generate effective user and item representations.
2. Compared to other model. SGCL has two main advantages:

- The model effectively learns useful embeddings of users and items by employing the self-alignment method. These embeddings capture the relevant features and relationships between users and items, which helps to improve the performance of the recommendation task and the quality of the learned embeddings.
- We utilize the Fast Fourier Transform to transform the spatial domain into frequency domain information and use the frequency domain for denoising operations.

Table 2. Our model SGCL performs better than baseline methods on three public datasets. % column refers to the improvement relative to the best competitor.

Dataset	Metric	LightGCN	SLRec	SGL	HCCF	SimGCL	DirectAU	DCCF	Ours	%
LastFM	Recall@20	0.2349	0.1957	0.2427	0.2410	0.2417	0.2422	0.2299	**0.2616**	**7.8**
	NDCG@20	0.1704	0.1442	0.1761	0.1773	0.1788	0.1727	0.1715	**0.1911**	**6.9**
	Recall@40	0.3220	0.2792	0.3405	0.3232	0.3381	0.3356	0.3228	**0.3614**	**6.1**
	NDCG@40	0.2022	0.1737	0.2104	0.2051	0.2111	0.2042	0.2031	**0.2242**	**6.2**
Yelp	Recall@20	0.0761	0.0665	0.0803	0.0789	0.0814	0.0818	0.0849	**0.0892**	**5.1**
	NDCG@20	0.0373	0.0327	0.0398	0.0391	0.0406	0.0424	0.0431	**0.0452**	**4.9**
	Recall@40	0.1175	0.1032	0.1226	0.1210	0.1223	0.1226	0.1301	**0.1337**	**2.8**
	NDCG@40	0.0474	0.0418	0.0502	0.0492	0.0524	0.0524	0.0541	**0.0559**	**3.3**
BeerAdvocate	Recall@20	0.1102	0.1048	0.1138	0.1156	0.1159	0.1182	0.1184	**0.1244**	**5.1**
	NDCG@20	0.0943	0.0881	0.0959	0.0990	0.0972	0.0981	0.1001	**0.1057**	**5.6**
	Recall@40	0.1757	0.1723	0.1776	0.1847	0.1833	0.1797	0.1867	**0.1952**	**4.6**
	NDCG@40	0.1113	0.1068	0.1122	0.1176	0.1151	0.1139	0.1182	**0.1241**	**5.0**

4.3 Ablation Study (RQ2)

We performed an ablation study on two public datasets (LastFM and BeerAdvocate) to analyze the impact of each component (contrastive learning, self-alignment module, and subgraph module). The results, summarized in Table 3, clearly demonstrate that removing any of the components leads to a decline in performance. We draw the following important conclusions.

- **Contrastive learning (-CL).** Contrastive learning allows similar users and similar items to map into a shared embedding space, thereby enhancing the representation of their interrelations. This approach can effectively capture the otential connections and interactions between users and items. The performance gap between SGCL and -CL shows the contribution of contrastive learning to the overall performance.
- **The self-alignment module (-self-alignment).** The performance gap between SGCL and -self-alignment shows the contribution of the self-alignment module to overall performance. Alignment makes the positive samples closer and more similar in the embedding space. Uniformity aims to make the positive samples closer and more similar in the embedding space. The self-alignment module improves the recommender system's performance.
- **The subgraph module (-subgraph).** The ablation experiment proves the usefulness of the subgraph module. The subgraph denoising submodule can effectively remove the noise in the subgraphs of user items. The subgraph collaboration representation submodule can prevent the subgraphs from not causing the deviation of semantic information. This module can improve the model's robustness and the recommendation's performance.

Table 3. Analysis of SGCL components through ablation experiments.

Category	Data Variants	LastFM Recall	NDCG	BeerAdvocate Recall	NDCG
w/o CL		0.2422	0.1811	0.1180	0.1010
w/o self-alignment		0.2561	0.1872	0.1200	0.1014
w/o subgraph		0.2525	0.1875	0.1233	0.1040
Ours		**0.2616**	**0.1911**	**0.1244**	**0.1057**

4.4 Effect of the Hyperparameters (RQ3)

This section details the effects of hyperparameters within the model, and the effects of each hyperparameter are summarized and explained below. Hyperparameters including $\lambda_1, \lambda_2, \lambda_3$, mask num (mask), ratio rate (γ), and subgraph num (k) . When adjusting one parameter, we set the others to their optimal values.

Hyperparameter λ_1 : When λ_1 is low, the model's ability to distinguish between samples decreases due to the reduced influence of positive example similarity in the loss function. As λ_1 increases, the model better recognizes and distinguishes positive sample. Therefore, when adjusting λ_1 , it's crucial to balance the influence of positive and negative samples.

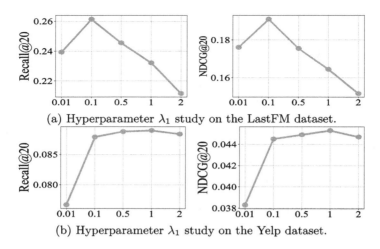

Fig. 3. Hyperparameter λ_1 study for the proposed SGCL regarding R@ 20 and N@ 20 changes on the Yelp and LastFM datasets.

Hyperparameter λ_2 : Constructing the two attributes of alignment and uniformity loss functions helps to obtain more effective user and item representations. When the value of λ_2 is too large, it leads to tight lustering of users and positive items in the model. Conversely, when λ_2 is too small, it leads to insufficient embedding of users and items to distinguish, and the model struggles to distinguish between positive and negative items.

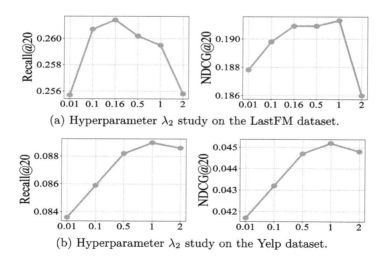

Fig. 4. Hyperparameter λ_2 study for the proposed SGCL regarding R@ 20 and N@ 20 changes on the Yelp and LastFM datasets.

Hyperparameter λ_3 : We observed that the model achieves optimal performance on the LastFM dataset when we set the value of λ_3 to 18. This result indicates that a significant percentage of our subgraph representation module plays a more significant role in enhancing performance. As shown in Fig 5.

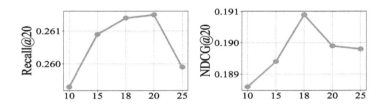

Fig. 5. Hyperparameter λ_3 study for the proposed SGCL regarding R@ 20 and N@ 20 changes on the LastFM datasets.

Hyperparameter Mask : As shown in Fig 6, the model performance is optimized when the number of masks is 3.

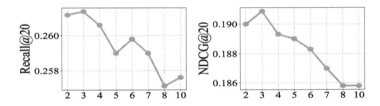

Fig. 6. Hyperparameter mask study for the proposed SGCL regarding R@ 20 and N@ 20 changes on the LastFM datasets.

Hyperparameter γ : As shown in Fig. 7, when γ is 0.4, the model is the optimal result among the two evaluation metrics. When γ is 0.8, the result is the worst. Since we are denoising the frequency domain through subgraphs, we remove the high-frequency and low-frequency information in the subgraphs when the value is too large. The larger the value, the more critical information is lost, which affects the model's performance. On the contrary, if the selected value is too small, little high-frequency information is filtered. Therefore, this experiment proves that the model can perform better in removing noise when γ is 0.4.

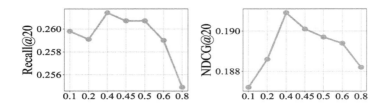

Fig. 7. Hyperparameter γ study for the proposed SGCL regarding R@ 20 and N@ 20 changes on the LastFM datasets.

Hyperparameter K : We select k nodes to form a subgraph as shown in Fig8. Since LastFM is a sparse dataset, this dataset provides better proof of the validity of choosing the count of subgraphs. Since the user-item interactions amount is too sparse when there are fewer than k neighbor nodes when constructing a subgraph for each node, it selects the second-order neighbors of the node for each node, and so on. This approach increases the sample size and makes the selected subgraph more representative of the nodes.

We can see from the figure that the more nodes in the selected subgraph, the better the model performance, which proves that our module is practical. The effect becomes worse when the selected nodes are 20. This method may be because there are too few first-order neighbors, and in turn, it selects higher-order neighbors, which may not be related to the central node.

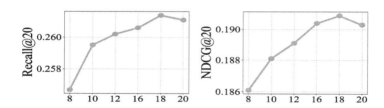

Fig. 8. Hyperparameter k study for the proposed SGCL regarding R@ 20 and N@ 20 changes on the LastFM datasets.

4.5 The Effect of GCN Layer Numbers. (RQ4)

Too many layers in graph collaborative filtering models can lead to over-smoothing problems. In this section, we verify whether our SGCL model was over-smoothing the problem, as shown in Fig. 9 . As layers increase, SGCL continues to provide stable performance. The SGCL model mitigates the over-smoothing problem.

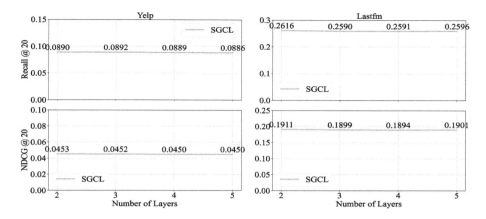

Fig. 9. Verification that our SGCL model has been mitigated from the over-smoothing problem.

5 Conclusion

This paper proposes a subgraph collaborative contrastive learning model for recommendation, including a self-alignment module and a subgraph module. The self-alignment module gets more effective user and item representations. It utilizes the two most crucial attributes of information (uniformity and alignment) in contrastive learning. The subgraph module comprises subgraph sampling, denoising, and collaborative representation submodules .

The process denoises the subgraphs from a frequency domain perspective. We have substantiated the practicality of our model through extensive experimentation. Our focus for future work is primarily on negative sampling and denoising samples.

Acknowledgments. This work was supported by the Science Fund for Outstanding Youth of Xinjiang Uygur Autonomous Region under Grant No. 2021D01E14.

References

1. Cai, X., Huang, C., Xia, L., Ren, X.: LightGCL: simple yet effective graph contrastive learning for recommendation. arXiv preprint arXiv:2302.08191 (2023)
2. He, X., Deng, K., Wang, X., Li, Y., Zhang, Y., Wang, M.: LightGCN: simplifying and powering graph convolution network for recommendation. In: Proceedings of the 43rd International ACM SIGIR Conference on Research and Development in Information Retrieval, pp. 639–648 (2020)
3. Lin, Z., Tian, C., Hou, Y., Zhao, W.X.: Improving graph collaborative filtering with neighborhood-enriched contrastive learning. In: Proceedings of the ACM Web Conference 2022, pp. 2320–2329 (2022)
4. Quan, Y., Ding, J., Gao, C., Yi, L., Jin, D., Li, Y.: Robust preference-guided denoising for graph based social recommendation. In: Proceedings of the ACM Web Conference 2023, pp. 1097–1108 (2023)

5. Ren, X., Xia, L., Zhao, J., Yin, D., Huang, C.: Disentangled contrastive collaborative filtering. arXiv preprint arXiv:2305.02759 (2023)
6. Tian, C., Xie, Y., Li, Y., Yang, N., Zhao, W.X.: Learning to denoise unreliable interactions for graph collaborative filtering. In: Proceedings of the 45th International ACM SIGIR Conference on Research and Development in Information Retrieval, pp. 122–132 (2022)
7. Wang, C., Yu, Y., Ma, W., Zhang, M., Chen, C., Liu, Y., Ma, S.: Towards representation alignment and uniformity in collaborative filtering. In: Proceedings of the 28th ACM SIGKDD Conference on Knowledge Discovery and Data Mining, pp. 1816–1825 (2022)
8. Wu, J., et al.: Self-supervised graph learning for recommendation. In: Proceedings of the 44th International ACM SIGIR Conference on Research and Development in Information Retrieval, pp. 726–735 (2021)
9. Xia, L., Huang, C., Huang, C., Lin, K., Yu, T., Kao, B.: Automated self-supervised learning for recommendation. In: Proceedings of the ACM Web Conference 2023, pp. 992–1002 (2023)
10. Xia, L., Huang, C., Shi, J., Xu, Y.: Graph-less collaborative filtering. In: Proceedings of the ACM Web Conference 2023, pp. 17–27 (2023)
11. Xia, L., Huang, C., Xu, Y., Zhao, J., Yin, D., Huang, J.: Hypergraph contrastive collaborative filtering. In: Proceedings of the 45th International ACM SIGIR Conference on Research and Development in Information Retrieval, pp. 70–79 (2022)
12. Yao, T., et al.: Self-supervised learning for large-scale item recommendations. In: Proceedings of the 30th ACM International Conference on Information & Knowledge Management, pp. 4321–4330 (2021)
13. Yu, J., Yin, H., Xia, X., Chen, T., Cui, L., Nguyen, Q.V.H.: Are graph augmentations necessary? simple graph contrastive learning for recommendation. In: Proceedings of the 45th International ACM SIGIR Conference on Research and Development in Information Retrieval, pp. 1294–1303 (2022)
14. Zhou, K., Yu, H., Zhao, W.X., Wen, J.R.: Filter-enhanced MLP is all you need for sequential recommendation. In: Proceedings of the ACM Web Conference 2022, pp. 2388–2399 (2022)

Time-Aware Squeeze-Excitation Transformer for Sequential Recommendation

Hongwei Chen[1], Luanxuan Liu[1(✉)], Zexi Chen[2], and Xia Li[3]

[1] School of Computer Science, Hubei University of Technology, Wuhan, China
chw2001@sina.com, 102211110@hbut.edu.cn
[2] Xiaomi Technology (Wuhan) Co., Ltd., Wuhan, China
[3] Wuhan Donghu University, School of Computer Science, Wuhan 430212, China

Abstract. Sequential recommendation models user preferences based on the sequence of user interactions. A key challenge in this context is the variability of user behavior. While Transformer-based models demonstrate unique superiority, existing models still struggle with modeling temporal information and adequately capturing users' long-term and short-term preferences. This paper proposes a Time-Aware Squeeze-Excitation Transformer for sequential recommendation (TASESRec). The model has two salient features: (1) TASESRec preserves the continuity dependency within timestamps from both duration and spectrum perspectives using a time window function, and integrates timestamps and user interactions through a multi-layer encoder-decoder structure. (2) Given the non-uniqueness of users' latent purchasing behavior, characterized by multiple potential purchasing behaviors, the model utilizes Squeeze-Excitation Attention (sigmoid activation) to comprehensively capture relevant items, thus enhancing prediction accuracy. Extensive experiments validate the superiority of the proposed model over various state-of-the-art models under several widely used evaluation metrics.

Keywords: Sequential recommendation · Self-attention · Transformer

1 Introduction

Recommendation systems play a crucial role in helping users navigate through redundant and complex information to find content of interest, widely applied on online platforms such as Google [1] and Facebook [2], thus shaping the landscape of the digital information era. This has led to the development of various methods and ideas in recommendation systems, from early Markov chain models [3] to the continuous advancement in deep learning, achieving significant success in modeling sequence dependencies and representations. For instance, models based on RNNs [4,5], incorporating CNN methods [6], and integrating multi-node information in graph neural networks (GNNs) [7]have all demonstrated considerable effectiveness. Additionally, recent variations of Transformer models have shown promising performance [8,9], primarily modeling long-term and short-term preferences through self-attention mechanisms and benefiting from their parallel computation capability, resulting in faster processing speeds compared to the aforementioned methods.

For sequences with inherent relative order, incorporating temporal information can significantly enhance model performance. However, we observe that the majority of current models do not utilize temporal information [10,11], often relying on positional embeddings to determine the relative positions of elements. Some models that incorporate temporal information struggle to capture the dependency relationship between context and timestamps, as well as the simultaneous consideration of multiple timestamps and interaction events [12]. In this study, we propose a Time-Aware Squeeze-Excitation Transformer (TASE-TRec) for sequential recommendation. Specifically, we first capture temporal dependencies through the design of a novel window function, followed by the separate modeling of time and interaction information. Additionally, we employ Squeeze-Excitation Attention to handle time and preference weights, addressing the constraint in attention mechanisms where weights sum up to 1. This allows for the simultaneous consideration of multiple potential temporal and purchasing behaviors, thereby enhancing the model's expressive power and modeling accuracy. We outperform state-of-the-art methods on several commonly used evaluation metrics. The main contributions of this paper are as follows:

- We introduce a novel Transformer architecture where the encoder part transforms discrete timestamps into continuous dependency time embeddings, while the decoder part first captures preferences of interaction items and then combines them with the temporal information from the encoder.
- We model time information from two perspectives duration, and spectrum, utilizing a novel time window function that better maintains the continuity of timestamps. Subsequent visualization experiments validate this operation.
- Squeeze-Excitation Attention is employed to handle time and preference weights, allowing multiple potential temporal and interaction information to be simultaneously considered, thus improving the model's expressive power.

2 Related Work

2.1 Sequential Recommendation

Traditional recommendation systems, such as content-based [13] and collaborative filtering [14] systems, capture the relationships between users and items in a static manner. However, in numerous real-world scenarios, the interests of users and the popularity of items undergo temporal fluctuations. Users may develop interests in new movies, books, and music, as previous preferences may slowly diminish over time. Certain items may also experience sudden popularity due to time-sensitive events or marketing campaigns [15]. Moreover, interactions between users and items often exhibit sequential dependencies. Sequential recommendation leverages the dynamic dependencies between sequences to infer users' latent preferences, which is why it has garnered significant attention in recent years.

Early research primarily utilized Markov chain models [3], which could only capture dependencies with small spans and simple dimensions. With the rise of neural networks, Recurrent Neural Networks (RNNs) [4,5] enhanced performance by leveraging sequential dependencies and memory mechanisms. However, they also faced challenges in capturing global dependencies. Convolutional Neural Networks (CNNs) [6] emphasized capturing local features in sequences, with models like GRU4Rec+ emerging to address complex dependency relationships [4].

Additionally, Graph Neural Networks capture sequence topology by introducing graph structures with node and edge attributes [16]. Attention modules are widely applied in the sequential recommendation, with self-attention mechanisms allowing direct feature interactions between any two points in a project sequence, thus addressing global dependency issues [10–12]. Unlike RNNs and LSTMs that require iterative computations, self-attention mechanisms offer parallel computation capabilities, enhancing computational efficiency. Due to these advantages, many models based on self-attention mechanisms have emerged, achieving state-of-the-art performance in sequential recommendation tasks.

2.2 Attention Mechanism

The attention mechanism represents a pivotal technique within the realm of deep learning [17], drawing inspiration from the selective focus inherent in human cognitive processes. It assigns varying weights to different parts of input sequences to extract the most relevant information and finds extensive application in various tasks such as video captioning [18], machine translation [19], and visual recognition. Specifically, the key to the Self-Attention (SA) model lies in its ability to effectively capture long-range dependencies within sequences, with multi-head attention being an integral component allowing the model to focus on different segments of the input sequence simultaneously. Although most existing SA-based models have demonstrated commendable performance, many existing models tend to disregard timestamp information. Only TiSASRec integrates timestamp data by considering time intervals as relative positional embeddings [12], surpassing the performance of SASRec [10].However, TiSASRec overlooks the temporal sequence and dependency of timestamps. In contrast, this study maintains the continuity dependency within timestamps by employing a time window function from both duration and spectrum perspectives. Additionally, it combines timestamps and user interactions through a multi-layer encoder-decoder structure.

3 Methodology

3.1 Problem Definition

Similar to other approaches [12], Each user's interaction sequence $C^u = [(S_1^u, t_1^u), (S_2^u, t_2^u), \ldots, (S_{(|s^u|-1)}^u, t_{(|s^u|-1)}^u)]$ is initially transformed into a fixed-length item sequence $S = (s_1, s_2, \ldots, s_L)$ and timestamp sequence $T = (t_1, t_2, \ldots, t_L)$, where L represents the maximum length of interaction sequences accepted by the model. If the length of the user interaction sequence exceeds L, only the most recent L interactions are considered. If the length of the interaction sequence is less than L, zero padding is applied to the left side of the item sequence until its length reaches L.

We create a learnable embedding matrix $M \in \mathbb{R}^{(N \times d)}$ for item information, where d is the latent dimension [11,12]. Embedding vectors provide flexibility and expressive power to the model for representing item features. Thus, we obtain the embedding matrix $E_I \in \mathbb{R}^{(L \times d)}$, where $E_{(l_i)} = M_{(s_i)}$. Additionally, to incorporate the item sequence, a trainable positional embedding $P \in \mathbb{R}^{(L \times d)}$ is added to the item embedding matrix. Using the function $f(\cdot)$ to concatenate the item embeddings and positional embeddings, the final item embedding matrix is given by the following equation:

$$\hat{E}_I = f(M_{(s_i)}, P_i), \; i \in \{1, 2, \ldots, l\} \tag{1}$$

3.2 Personalized Timestamp Embedding

Specifically, for a given user u, with a fixed-length time sequence $T = (t_1, t_2, \ldots, t_L)$, where the time interval between two items i and j is $|t_i - t_j|$, and R^U denotes the set of all users' relative time intervals. To optimize modeling the memory cost of time information, we introduce an innovative approach here: categorizing time intervals into different classes, representing the maximum relative time interval as $T_{\max} = \max(R^U)$, and introducing a scaling factor ϕ to control the maximum time interval. Subsequently, all relative time intervals are restricted and scaled to a fixed range of $[0, k]$, where k denotes the number of time windows, generating the final interval sequence T_s as follows:

$$T_s = (t_1^s, t_2^s, \ldots, t_L^s) = \left(\frac{k \times \min(T_r, T_{\max}) \times \phi}{\max(R^U)} \right) \tag{2}$$

We utilize a learnable embedding matrix $M_T = e_i \in \mathbb{R}^{k \times d}$, where d represents the dimension size. To mitigate the impact of time intervals with large spans but lacking significant temporal features on the time embedding matrix, we employ a window-based embedding approach to transform the scaled relative time interval sequence T_s into a time embedding matrix. The time weight for each row is given by the following formula:

$$e_{(t_i)'} = \sum_{j \in (0,k)} w_{(t_i, j)} e_j \tag{3}$$

The final time embedding matrix $E_T = e_{(t_l)} \in \mathbb{R}^{L \times d}$, where $l \in \{1, 2, \ldots, L\}$, and the weights $w_{(t_i, j)}$ are calculated using a window function:

$$w_{(t_i, j)} = F_{\text{win}}(t_i - j) \tag{4}$$

$$F_{\text{win}}(x) = \left(\alpha - \beta \cdot \cos\left(\frac{2\pi x}{w}\right)\right) \times \frac{1}{2} \times \left(sign\left(x + \frac{w}{2}\right) - sign\left(x - \frac{w}{2}\right)\right) \tag{5}$$

In the above formulas, the function $sign(x) = \begin{cases} 1, & x > 0 \\ 0, & x = 0 \\ -1, & x < 0 \end{cases}$, the parameters $\alpha = 0.51$ and $\beta = 0.46$ serve as weights for the window function, and w is the size of the window function. It can be observed that this window function adopts a composite function, combining a rectangular window. Smooth transitions are employed in the transition region from the window boundary to gradually change values. This approach allows for more precise calculation of relative time interval sequences, thus more accurately revealing potential implicit information in past and present behaviors.

3.3 Time-Aware Squeeze-Excitation Transformer

The overall framework of our Time-Aware Squeeze-Excitation Transformer (TASETRec) is illustrated in the Fig. 1. TASETRec comprises three primary components: an encoding layer that incorporates temporal information, a global attention module (Transformer) enhanced with Squeeze-Excitation Attention, and a decoding layer that merges temporal weight information with item features. Finally, there is a task prediction layer for predicting the score of the next item. Although the encoding and decoding layers handle different types of information, they both utilize the pivotal global attention module. Therefore, we first introduce the global attention module.

3.4 Global Attention Module

As illustrated in the Fig. 1, the global attention layer consists of two parts: the multi-head attention module and the Squeeze-Excitation Attention module. The multi-head attention layer first passes through a fully connected layer, followed by layer normalization to adjust the latent dimensions and enhance generalization. It then employs multi-head attention to capture information from different dimensional spaces. Finally, residual connections and layer normalization are applied to address network degradation caused by neural networks and to maintain stable gradient descent. Subsequently, the Squeeze-Excitation Attention layer performs weighted calculations on multiple items in the sequence.

$$\delta_{\text{Global}}^{\text{LN}}(H_{l-1}) = \text{LN}(\text{Dpt}(\text{Linear}(H_{l-1}))), \tag{6}$$

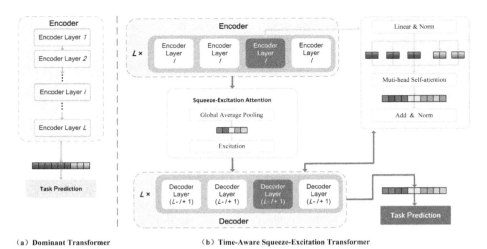

Fig. 1. Comparison between existing mainstream Transformers and the proposed Time-Aware Squeeze-Excitation Transformer.

$$\text{Att}(Q, K, V) = softmax\left(\frac{QK^T}{\sqrt{d_{\text{model}}/h}}\right)V, \quad (7)$$

$$\text{head}_i = \text{Att}\left(\underbrace{\delta^{\text{LN}}_{\text{Global}}(H_{l-1})W_i^Q}_{Q}, \underbrace{\delta^{\text{LN}}_{\text{Global}}(H_{l-1})W_i^K}_{K}, \underbrace{\delta^{\text{LN}}_{\text{Global}}(H_{l-1})W_i^v}_{V}\right), \quad (8)$$

$$\text{MHSA}(\delta^{\text{LN}}_{\text{Global}}(H_{l-1})) = \text{LN}(\text{Dpt}(\text{Concat}(\text{head}_1, \ldots, \text{head}_h)W^o)), \quad (9)$$

Here, H_{l-1} represents the output of the global attention layer at the $l-1$th layer, LN, and Dpt denote layer normalization and dropout operations, respectively, h represents the number of attention heads. Matrices $W_i^Q \in \mathbb{R}^{d \times d/h}$, $W_i^K \in \mathbb{R}^{d \times d/h}$, $W_i^v \in \mathbb{R}^{d \times d/h}$, and $W^o \in \mathbb{R}^{d \times d/h}$ are considered as learnable parameters used to learn queries Q, keys K, values V, and weights of the multi-head attention layer. The scaling factor $\sqrt{d_{\text{model}}/h}$ is introduced to avoid gradient vanishing or exploding [9,10].

3.5 Squeeze-Excitation Attention Module

The majority of contemporary approaches neglect to consider the attention span across multiple highly correlated interactions, leading to suboptimal predictive accuracy. In response to this issue [20]. drawing inspiration from [21], we introduce Squeeze-Excitation Attention (SEAtt), specifically designed for sequence recommendation tasks. The computational process is as follows:

$$O_{\text{sq}} = F_{\text{sq}}(\text{Out}(H_{l-1})) = \frac{1}{d_{\text{model}}} \sum_{j=1}^{d_{\text{model}}} \text{Out}(H_{l-1})_{[:,j]}, \quad (10)$$

$$F_{\text{ex}}(O_{\text{sq}}) = \sigma(g(O_{\text{sq}})) = \sigma(W_2\delta(W_1 O_{\text{sq}})), \tag{11}$$

$$\text{SEAtt}(\text{Out}(H_{l-1})) = F_{\text{ex}}(O_{\text{sq}}), \tag{12}$$

Here, $SEAtt(\text{Out}(H_{l-1})) \in \mathbb{R}^{L \times 1}$ represents the Squeeze-Excitation scores, $\text{Out}(H_{l-1}) \in \mathbb{R}^{L \times d}$ denotes the output of the multi-head attention layer, $F_{\text{sq}}(\text{Out}(H_{l-1}))$ denotes the global average pooling operation applied to $\text{Out}(H_{l-1})$, $W_1 \in \mathbb{R}^{(L/r) \times L}$, $W_2 \in \mathbb{R}^{(L/r) \times L}$ represent the Squeeze-Excitation parameters, respectively, where r is a scaling parameter (default set to 2), and δ and σ represent the ReLU and Sigmoid activation functions, respectively.

The utilization of Squeeze-Excitation Attention confers two benefits: (1) SEAtt dynamically assigns weights to different channels, intensifying focus on pivotal channels, thereby enhancing the expressiveness of the model. This enhancement is further validated through ablation experiments conducted to assess the model's performance. (2) In this context, the sigmoid activation function is employed. While the commonly used softmax activation function confines the sum of all item scores to 1, resulting in mutually exclusive relationships among multiple historical interactions, the sigmoid function permits simultaneous activation of multiple similar items.

Due to the disparate inputs in the multi-head attention of the encoder and decoder, for the sake of simplicity and accuracy in describing the entire process, we define the output of the global attention as follows:

$$O = G_{\text{Tr}}(Q, K, V) \tag{13}$$

3.6 Encoder Module and Decoder Module

The encoder block receives temporal information from the personalized time embedding module. Here, in the global attention, the same object $H_{l-1}^{(T)}$ is utilized as query Q, key K, and value V, allowing it to distinguish useful temporal information across different contexts.

$$H_{l-1}^{(T')} = G_{Tr}(H_{l-1}^{(T)}, H_{l-1}^{(T)}, H_{l-1}^{(T)}) \tag{14}$$

Although the utilization of global attention facilitates adaptive weighting of all-time embedding vectors, it inherently represents a linear transformation. Consequently, two learned linear transformations with ReLU activation are subsequently applied following attention to introduce nonlinearity into the model. Lastly, residual connections and layer normalization are incorporated to mitigate network degradation and ensure stability throughout network training.

$$FFN(H_{l-1}^{(T')}) = ReLU(H_{l-1}^{(T')}W_1 + b_1)W_2 + b_2 \tag{15}$$

$$O_{l-1}^E = H_{l-1}^{(T')} + \text{Dpt}(FFN(LN(H_{l-1}^{(T')}))) \tag{16}$$

Here, $W_1, W_2 \in \mathbb{R}^{d \times d}$, $b_1, b_2 \in \mathbb{R}^d$, and $O_{l-1}^E \in \mathbb{R}^{L \times d}$ represents the output of the $l - 1$th layer of the encoder.

In the decoder, item information is initially received from the item embedding module. Similar to the encoder, the same object H_{l-1}^I is used as query Q, key K, and value V for global attention:

$$H_{l-1}^{(I')} = G_{Tr}(H_{l-1}^I, H_{l-1}^I, H_{l-1}^I) \tag{17}$$

To integrate both temporal and item information, $Q = H_{l-1}^I$, $K = O_E$, and $V = O_E$ are set here. By incorporating temporal information into K and V, the model can more easily distinguish and integrate relevant information from different time points. H_{l-1}^{IE} is used to denote the combination of $l - 1$th layer item and temporal information.

$$H_{l-1}^{IE} = G_{Tr}(H_{l-1}^{(I')}, O_E, O_E), \quad H_{l-1}^{IE} \in \mathbb{R}^{L \times d} \tag{18}$$

To capture key features within the sequence and enhance the model's performance and expressive capacity, a method similar to Eq. 14 is employed here. O_{l-1}^D represents the output of the $l - 1$th layer of the decoder.

$$O_{l-1}^D = H_{l-1}^{IE} + \text{Dpt}(\text{FFN}(\text{LN}(H_{l-1}^{IE}))) \tag{19}$$

3.7 Output Layer

After aggregating all hierarchical information for layer l, the calculation of user preferences for each item is as follows:

$$R_{i,t} = O_l^D M_i^T \tag{20}$$

where $M_i^T \in \mathbb{R}^{L \times d}$ denotes the embedding matrix for the items. To prevent overfitting and enhance model performance, the same matrix M_i^T is used in the item embedding layer [18]. $R_{i,t}$ represents the preference score given the history of the first t items $(s_1, s_2, s_3, \ldots, s_t)$ and their timestamps $(t_1, t_2, t_3, \ldots, t_t)$. A higher preference score $R_{i,t}$ implies a higher likelihood of interaction with item i. Given an input interaction sequence $[(s_1, t_1), (s_2, t_2), \ldots, (s_{n-1}, t_{n-1})]$, the model's expected output is the prediction for the next interaction in the input sequence $(s_2, s_3, s_4, \ldots, s_n)$, utilizing the last row of O_l^D for prediction.

Model Training. The binary cross-entropy loss is employed as the objective function:

$$L = -\sum_{S^u \in S} \sum_{l \in [1,2,\ldots,L]} \left[\log(\sigma(r_{o_l,l})) + \log(1 - \sigma(r_{o'_i,l})) \right] \tag{21}$$

The sigmoid function is defined as $\sigma(x) = \frac{1}{1+e^{-x}}$. It is vital to emphasize the masking of losses for items. The proposed model optimizes using the Adam optimizer, and for enhanced training efficiency, it applies the t-fixup, a Transformer weight initialization approach.

4 EXPERIMENTS

4.1 Experimental Settings

Datasets. The two datasets used for experimental evaluation are from the real world and widely employed in relevant research [10–12].

Steam [10]: This benchmark dataset is derived from the prominent online video game distribution platform, Steam. The time span roughly covers the period from 2010 to 2017.

Userbehavior [22]: Provided by Alibaba, this dataset includes user behavior from the online shopping platform Taobao. The included user data comprises clicks, purchases, items added to the shopping cart, and product preferences.

For both datasets, actions such as reviews, clicks, or adding items to the cart are considered implicit feedback and are sorted by timestamp. We follow a similar preprocessing procedure for Steam, discarding items with less than 5 interactions that lack meaningful features. For Userbehavior, we sort users by their user IDs, extract the interaction sequences of the top 100,000 users, and users with more than 300 interactions or fewer than 20 interactions were excluded.

Table 1. Basic Dataset Statistics.

Dataset	Steam	Userbehavior
Users	334,700	100,000
Items	13,047	677,456
Actions per user	10.59	76.43
Actions per item	4.2M	7.8M
Time span	7 years	7 days

Evaluation Metrics. Following evaluation standards [10–12], we employed five evaluation metrics: NDCG@5, NDCG@10, Hit Rate@5, Hit Rate@10, and MRR. HR focuses on the accuracy of the model, NDCG emphasizes the position of the user's desired items in the model's recommended list, favoring higher positions. MRR highlights the position of the user's desired items in the model's recommended list, favoring higher positions.

To tackle the computational time challenge of ranking items based on preference scores in large datasets, we randomly selected 100 users who had not previously interacted with the items and ranked these newly sampled items alongside the actual items [11,12]. Subsequently, the evaluation metrics were computed based on these 101 items.

Baseline Models. To demonstrate the effectiveness of our proposed TASE-TRec, we compare it with various categories of recommendation models, including classical general recommendation models (PopRec) that do not consider sequential patterns, matrix factorization (MC) based models (TransRec [23], Caser [6]), recurrent neural network (RNN) based models (GRU4Rec [24], GRU4Rec+ [4]), and self-attention (SA) based models (SASRec [10], BERT4Rec [9], FDSA [11], TiSASRec [12]).Due to space constraints, we will not provide an introduction to each method here.

Implementation Details. We implemented TASESRec using PyTorch, with a maximum sequence length (L) set to 100 for all datasets. The batch size is 128, the learning rate is 0.001, and the dropout rate for Userbehavior is 0.4, while for Steam it is 0.2. The experiment was terminated if there was no improvement after 40 epochs. The default window function (w) size is set to 20. We adjusted the bin number (k) in the range $\{256, 512, 1024, 2048\}$ and the scale factor in the range $\{2, 4, 6, 8, 10\}$. The optimizer used is the Adam optimizer with momentum decay rates $\beta_1 = 0.9$ and $\beta_2 = 0.98$. All experiments were conducted on a server equipped with 18 vCPU AMD EPYC 9754 and Nvidia GTX 3090 GPU.

4.2 Overall Performance

Table 2 presents the recommendation performance of our model and baseline methods across two datasets. Specifically, the matrix factorization-based TransRec method demonstrates superior performance compared to UserBehavior on the relatively sparse Steam dataset. Models utilizing neural networks, such as Caser, GRU4Rec, and GRU4Rec+, exhibit significantly better performance than TransRec, attributed to their enhanced capability in capturing complex sequential behaviors. Notably, models leveraging self-attention mechanisms, including SASRec, BERT4Rec, FDSA, and TiSARec, outperform those relying on MC and RNN across both datasets. This superiority is attributed to the effectiveness of self-attention mechanisms in capturing both short-term and long-term relationships.

We observe a significant lead of our proposed TASESRec over current state-of-the-art methods on both datasets. The remarkable improvement in the Userbehavior dataset can be attributed to the denser nature of the data, where window-based methods provide time embeddings with continuous dependencies. By incorporating innovative Squeeze-Excitation Attention mechanisms, TASESRec enhances prediction accuracy by capturing multiple potential targets. For datasets characterized by higher sparsity and larger time spans, predictive difficulty increases linearly. Nevertheless, TASESRec achieves substantial improvement even on the sparser Steam dataset, highlighting the model's superiority in complex environments.

Subsequently, Fig. 2 illustrates the impact of a key hyperparameter - dimension size. It is evident that, as dimension size varies, TASESRec consistently outperforms other models on both datasets. This highlights the robustness and high performance of the TASESRec model. Particularly in the Userbehavior interaction dataset, the influence of dimension size on SA/RNN-based models is

Table 2. Performance comparison of various models on Userbehavior and Steam datasets. The optimal outcome in each row is highlighted in bold, and the second-best outcome is underlined.

Model	Userbehavior					Steam				
	NDCG@5	Hit@5	NDCG@10	Hit@10	MRR	NDCG@5	Hit@5	NDCG@10	Hit@10	MRR
PopRec	0.2679	0.3658	0.3182	0.4856	0.2887	0.4235	0.5775	0.4727	0.7493	0.4061
TransRec	0.3298	0.4342	0.3673	0.5314	0.3276	0.4742	0.6414	0.5222	0.7891	0.4495
Caser	0.2856	0.3413	0.3071	0.4763	0.2610	0.4883	0.6517	0.5347	0.7945	0.4637
GRU4Rec	0.5168	0.6173	0.5415	0.6629	0.3582	0.2287	0.3169	0.2703	0.4464	0.2393
GRU4Rec+	0.6189	0.7210	0.6306	0.7768	0.6018	0.4533	0.6327	0.5488	0.7984	0.4692
SASRec	0.6176	0.7195	0.6286	0.7543	0.6050	0.6081	<u>0.7617</u>	<u>0.6427</u>	0.8680	0.5783
FDSA	0.6197	0.7186	0.6260	0.7672	0.5958	<u>0.6102</u>	0.7611	0.6422	<u>0.8704</u>	<u>0.5794</u>
BERT4Rec	0.6295	0.7220	0.6340	0.7723	0.6094	0.6052	0.7550	0.6350	0.8523	0.5730
TiSASRec	<u>0.6403</u>	<u>0.7285</u>	<u>0.6472</u>	<u>0.7834</u>	<u>0.6208</u>	0.6069	0.7599	0.6392	0.8503	0.5767
TASESRec	**0.6981**	**0.7713**	**0.7083**	**0.8071**	**0.6841**	**0.6274**	**0.7803**	**0.6598**	**0.8827**	**0.5965**
Improve.	9.03%	5.86%	9.44%	3.03%	10.20%	2.82%	2.44%	2.66%	1.41%	2.96%

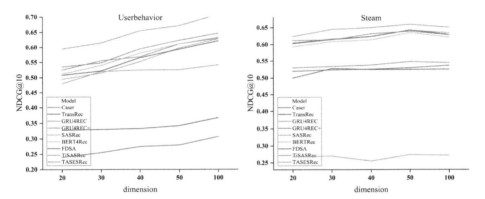

Fig. 2. Models' performance (NDCG@10) under different dimension size.

more pronounced, indicating that neural network models are more sensitive to changes in dimensionality in dense user behavior data.

To assess the effectiveness and importance of different modules in our model, we conducted thorough ablation studies. In Fig. 2, we analyzed the impact of each component, revealing a substantial decrease in performance upon their removal from TASESRec. This emphasizes the pivotal role played by each module. Specifically, in the densely populated Userbehavior dataset, the exclusion of the time window function led to a significant drop in performance. This resulted from the inability to transform similar timestamps into comparable embedding vectors, thereby compromising the accuracy of time embedding vector computations. Similarly, in the sparser Steam dataset, the absence of the merging time module resulted in a notable decrease in performance. This underscores the significance of timestamp information, which encompasses critical user interaction

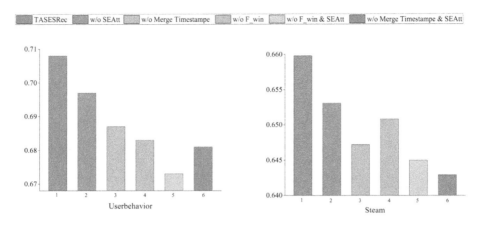

Fig. 3. Performance comparison of the proposed components (NDCG@10) including Squeeze-Excitation Attention, integration of timestamps, and window function, with (w) or without (w/o).

features, facilitating more precise user modeling. Furthermore, the elimination of the Squeeze-Excitation mechanism led to performance decrements across both datasets. This mechanism plays a crucial role in concurrently capturing multiple latent behaviors, thereby enhancing prediction accuracy.

4.3 Hyperparameter Study

This subsection illustrates the impact of the number of bins for timestamps and dropout rates on two datasets. A larger number of bins suggests that the time embedding layer can employ a more intricate trainable embedding matrix $M_T \in \mathbb{R}^{k \times d}$ to capture the temporal information inherent in timestamps. As depicted in Table 3, the number of bins exhibits minimal impact on the performance of TASESRec on the Steam dataset, with consistent strong performance. However, in the Userbehavior dataset, a smaller number of bins results in decreased performance. This trend can be attributed to the higher frequency of item interactions in the Userbehavior dataset, necessitating a larger embedding matrix M_T to effectively capture intricate user behavior patterns. Moreover, Table 3 illustrates a more pronounced variation in model performance with changes in the dropout rate for the Userbehavior dataset. This sensitivity could stem from the heightened susceptibility of more intensive user behavior to dropout effects, contrasting with the Steam dataset.

4.4 Visualization

To assess the effectiveness of the personalized timestamp embedding layer and visualize the effects of embedding methods with and without the window function, we selected five timestamps: 1 min, 15 min, 30 min, 8 h, and 72 h. We then

Table 3. The impact of bins and dropout rates on recommendation performance. Optimal outcomes are highlighted in bold, while the second-best outcomes are underscored.

Configuration	Userbehavior		Steam	
	NDCG@5	Hit@5	NDCG@5	Hit@5
256 bins	0.6824	0.7634	0.6192	0.7703
512 bins	0.6901	0.7627	**0.6274**	0.7740
1024 bins	**0.6981**	**0.7713**	0.6199	0.7803
2048 bins	0.6910	0.7663	0.6213	**0.7814**
Dropout Rate: 0.2	0.6724	0.7454	**0.6274**	0.7803
Dropout Rate: 0.3	0.6865	0.7627	0.6156	0.7764
Dropout Rate: 0.4	**0.6981**	**0.7713**	0.6180	0.7720

conducted experiments using these two different embedding modules, transforming the timestamps into time embedding vectors. Subsequently, we projected these vectors into a two-dimensional space using T-SNE. Figure 4 illustrates the distribution of the five timestamp vectors under different embedding methods. It is evident that without the window function, these timestamps exhibit irregular distributions. However, when the window function is applied, timestamps with similar time information (1 min, 15 min, and 30 min) display clustered distributions, indicating their temporal similarity. In contrast, timestamps with larger time span (8 h and 72 h) are positioned farther apart, suggesting distinct time characteristics.

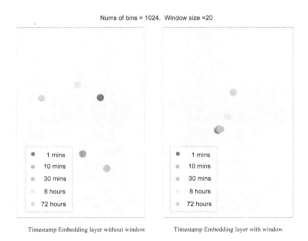

Fig. 4. Comparison of personalized timestamp embedding layer without and with window by visualization.

5 Conclusion and Future Work

This paper addresses the limitations of existing sequential recommendation models based on self-attention mechanisms (SA) in leveraging temporal information. It proposes a novel Time-Aware Squeeze-Excitation Transformer (TASESRec), which creates a trainable timestamp matrix from the perspectives of temporal continuity and temporal spectrum. The model employs an encoder-decoder architecture to separately model continuous-time information and project IDs. Additionally, it introduces Squeeze-Excitation Attention to enhance the learning of sequence dependencies across multiple related items. Experimental results on two real-world datasets demonstrate the effectiveness of the TASESRec approach. This opens up potential applications in practical scenarios such as e-commerce (Taobao, Amazon), short video platforms (TikTok), and music recommendations.

For future endeavors, several pathways warrant exploration: proposing lighter-weight methodologies to enhance computational efficiency; integrating supplementary personal data, including user age, gender, and location, to augment recommendation personalization; and refining specialized strategies for forecasting interactions spanning broader time intervals and involving sparser data interactions, thereby attaining heightened precision and accuracy in results.

References

1. Joglekar, M.R., et al.: Neural input search for large scale recommendation models. In: Proceedings of the 26th ACM SIGKDD International Conference on Knowledge Discovery & Data Mining, pp. 2387–2397 (2020)
2. Huang, J.T., et al.: Embedding-based retrieval in facebook search. In: Proceedings of the 26th ACM SIGKDD International Conference on Knowledge Discovery & Data Mining, pp. 2553–2561 (2020)
3. He, R., McAuley, J.: Fusing similarity models with markov chains for sparse sequential recommendation. In: 2016 IEEE 16th International Conference on Data Mining (ICDM), pp. 191–200 (2016)
4. Hidasi, B., Karatzoglou, A.: Recurrent neural networks with top-k gains for session-based recommendations. In: Proceedings of the 27th Acm International Conference on Information and Knowledge Management, pp. 843–852 (2018)
5. Liu, Q., Zeng, Y., Mokhosi, R., Zhang, H.: STAMP: short-term attention/memory priority model for session-based recommendation. In: Proceedings of the 24th ACM SIGKDD International Conference on Knowledge Discovery & Data Mining, pp. 1831–1839 (2018)
6. Tang, J., Wang, K.: Personalized top-n sequential recommendation via convolutional sequence embedding. In: Proceedings of the Eleventh ACM International Conference on Web Search and Data Mining, pp. 565–573 (2018)
7. Wang, C., Zhang, M., Ma, W., Liu, Y., Ma, S.: Make it a chorus: knowledge- and time-aware item modeling for sequential recommendation. In: Proceedings of the 43rd International ACM SIGIR Conference on Research and Development in Information Retrieval, pp. 109–118 (2020)

8. Yu, Z., Lian, J., Mahmoody, A., Liu, G., Xie, X.: Adaptive user modeling with long and short-term preferences for personalized recommendation. In : IJCAI, pp. 4213–4219 (2019)
9. Sun, F., et al.: BERT4Rec: Sequential recommendation with bidirectional encoder representations from transformer. In: Proceedings of the 28th ACM International Conference on Information and Knowledge Management, pp. 1441–1450 (2019)
10. Kang, W.C., McAuley, J.: Self-attentive sequential recommendation. In: 2018 IEEE International Conference on Data Mining (ICDM), pp. 197–206. IEEE (2018)
11. Zhang, T., et al.: Feature-level deeper self-attention network for sequential recommendation. In: IJCAI, pp. 4320–4326 (2019)
12. Li, J., Wang, Y., McAuley, J.: Time interval aware self-attention for sequential recommendation. In: Proceedings of the 13th International Conference on Web Search and Data Mining, pp. 322–330 (2020)
13. Lin, Y., Lin, F., Yang, L., Zeng, W., Liu, Y., Pengcheng, W.: Context-aware reinforcement learning for course recommendation. Appl. Soft Comput. **125**, 109189 (2022)
14. Yang, B., Chen, J., Kang, Z., Li, D.: Memory-aware gated factorization machine for top-n recommendation. Knowl.-Based Syst. **201**, 106048 (2020)
15. Zhang, P., Kim, S.: A survey on incremental update for neural recommender systems. arXiv preprint arXiv:2303.02851 (2023)
16. Shu, W., Tang, Y., Zhu, Y., Wang, L., Xie, X., Tan, T.: Session-based recommendation with graph neural networks. In: Proceedings of the AAAI Conference on Artificial Intelligence, vol. 33, pp. 346–353 (2019)
17. Vaswani, A., et al.: Attention is all you need. Adv. Neural Inf. Proc. Syst. **30** (2017)
18. Chen, L., et al.: SCA-CNN: spatial and channel-wise attention in convolutional networks for image captioning. In: Proceedings of the IEEE Conference on Computer Vision and Pattern Recognition, pp. 5659–5667 (2017)
19. Chen, K., Wang, R., Utiyama, M., Sumita, E., Zhao, T.: Syntax-directed attention for neural machine translation. In: Proceedings of the AAAI Conference on Artificial Intelligence, vol. 32 (2018)
20. Ji, M., Joo, W., Song, K., Kim, Y.-Y., Moon, I.-C.: Sequential recommendation with relation-aware kernelized self-attention. In: Proceedings of the AAAI Conference on Artificial Intelligence, vol. 34, pp. 4304–4311 (2020)
21. Hu, J., Shen, L., Sun, G.: Squeeze-and-excitation networks. In: Proceedings of the IEEE Conference on Computer Vision and Pattern Recognition, pp. 7132–7141 (2018)
22. Zhu, H., et al.: Learning tree-based deep model for recommender systems. In: Proceedings of the 24th ACM SIGKDD International Conference on Knowledge Discovery & Data Mining, pp. 1079–1088 (2018)
23. He, R., Kang, W.C., McAuley, J.: Translation-based recommendation. In: Proceedings of the Eleventh ACM Conference on Recommender Systems, pp. 161–169 (2017)
24. Hidasi, B., Karatzoglou, A., Baltrunas, L., Tikk, D.: Session-based recommendations with recurrent neural networks. arXiv preprint arXiv:1511.06939 (2015)

Environment and Climate

Carbon Price Forecasting with LLM-Based Refinement and Transfer-Learning

Haiqi Jiang, Ying Ding, Rui Chen, and Chenyou Fan(✉)

South China Normal University, Guangdong, China
fanchenyou@scnu.edu.cn

Abstract. We propose a unified forecasting framework for accurately predicting carbon markets of EU Emission Trading Scheme (EU ETS) and Chinese Emission Allowance (CEA). Our framework utilizes a Time-Series Model (TSM) for initial prediction followed by applying a Large Language Model (LLM) to refine the forecasts. We prompt the LLM to refine the TSM forecasts by demonstrating an example pair of past TSM predictions and their corresponding true future prices to the LLM as a chain-of-thought. The in-context learning capacity of the LLM allows the LLM to rectify inaccurate predictions to reflect on TSM predictions and refine the forecasts. To further reduce the prompting delays and expenses involving LLMs, we innovate a post-finetuning approach to train a Gated Linear Unit (GLU) model to condense the LLM's in-context learning capability. This enables direct fine-tuning of TSM outputs without the need for explicit prompting LLM during inference. Experimental results show that our method can refine the TSM prediction by 10% to 40% in various zones, as well as enhance transfer learning by 10% to 21% through the inclusion of market context of the source zone when predicting the target zone. Remarkably, our GLU model achieves comparable, and in some cases superior, performance compared to LLM prompting. It effectively combines the short-term forecasting capability of classical Time Series Models with the long-term trend prediction ability typically associated with the LLMs.

Keywords: Carbon Future Market · Price Forecasts · Large Language Models · Transfer Learning · Time-Series Prediction · Gated Linear Unit · Post-Finetuning

1 Introduction

The recent proposal of carbon neutrality aims to eliminate net carbon emissions in the next 20 to 30 years. To regulate economic activities towards this goal, countries like China and the European Union have established carbon markets where emission allowances can be traded. Industrial manufacturers can either purchase more allowances or reduce their own emissions, promoting cleaner energy and encouraging innovation.

Accurately predicting carbon prices can help manufacturing companies minimize costs through effective planning, while also offering valuable insights to governments for regulating domestic industrial sectors. The challenge in predicting the carbon market

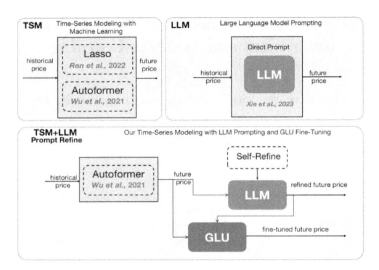

Fig. 1. Motivation of using both TSM (Time-Series Model) and LLM (Large Language Model) for carbon market forecasting. The top-left box shows studies of using TSM methods such as Lasso and Autoformer for time-series prediction. The top-right box shows recent works of prompting LLM with historical prices for future predictions. The bottom box shows our methods of using Autoformer for raw prediction followed by prompting the LLM to refine the model outputs. We also design fine-tuning the TSM output with a GLU (Gated Linear Unit) to replace LLM for efficient inference.

arises from the lack of sufficient data in emerging novel markets, as well as the presence of non-linearity and volatility, rendering traditional prediction methods less effective. To tackle this challenge, we propose enhancing carbon price forecasting by integrating machine-learning-based Time-Series Models with recent advancements in Pre-trained Large Language Models in the field of AI.

The recent emerging Large Language Models (LLMs) have been proven to possess robust Few-Shot learning capability [2, 10]. Through pre-training on extensive human knowledge, LLMs acquire a rich understanding not only of human languages, but also in mathematical deduction and reasoning [19], as well as financial marketing [7]. Our paper tries to answer the following important research questions: Do LLMs understand the price trends of the EU Emission Trading Scheme and the Chinese carbon emission market? Can LLMs have abilities to refine the prediction results of the machine learning models? We address these questions by constructing a Time-Series Model and Large Language Model based two-stage time-series forecasting framework.

Firstly, we gather data of EU ETS carbon prices from 2009–2020, as well as Chinese Emission Allowance (CEA) prices from 2015–2022. Following [14], we also collect macro-economic influencing factors as predictors such as commodity prices of oil and coal, as well as stock indices regarding cleaner energy for both EU and China. We pre-process the data with missing data filling and feature selection using a traditional Lasso-based method [11].

Next, we train a state-of-the-art deep-learning-based Time-Series Model (TSM) to fit temporal patterns of historical carbon prices in a standard supervised way using historical data. To this end, we employ the state-of-the-art attention-based Autoformer [21] as our TSM backbone Deep Neural Network and perform end-to-end training. Subsequently, we employ the well-trained TSM to generate raw future predictions based on the learned market context patterns condensed in the model parameters.

Secondly, we leverage the capabilities of the Large Language Models to incorporate market context, world knowledge, and associations of regional markets into predictions. To accomplish this, we incorporate advanced prompting techniques such as Chain-of-Thought (CoT) [19] and Self-Refine [9] into the prompting LLM procedures. The CoT technique enables successful predictions based on observable historical patterns, such as preceding time periods, thereby providing the LLM with relevant market context. The Self-Refine method encourages the LLM to actively reflect on past predictive inaccuracies and refine its current predictions accordingly.

Moreover, we consider a realistic case of predicting carbon price at a novel market even without supervised training on its historical data. We innovate an approach of demonstrating the price trend of a mature regional market in a global context followed by predicting the future of the novel market.

Finally, we aim to minimize communication and prompting costs associated with involving LLMs. We introduce a post-finetuning approach, condensing the LLM's in-context learning capability into a gated linear unit model. This enables direct fine-tuning of TSM outputs without the need for explicit LLM prompting during inference. We will demonstrate that this TSM post-finetuning approach can effectively match the capacity of the LLM refinement process and achieve comparable or superior prediction accuracy.

Overall, we conduct a comprehensive series of empirical analyses to show that the LLM could significantly refine the TSM prediction by 9.6% to 31% by merely one demonstration, and improve transfer learning by 14% to 20% without any historical training data of a novel market. Moreover, our post-finetuning process with designed GLU model achieves comparable, and in some cases superior, performance compared to LLM prompting.

In summary, the main contributions of our work include:

1. We are among the first to study the critical topic of predicting EU Emission Trading Scheme and Chinese Emission Allowances for social good. We leverage advanced AI models to incorporate diverse economic influencers as market contexts effectively into our forecasting methodology.
2. We propose a novel two-stage framework that utilizes a capable Time-Series Model (TSM) for initial prediction, then prompting the state-of-the-art LLMs to enhance the forecasts with demonstrated market contexts by learning from past deviations.
3. We also show to utilize the LLM to transfer future predictions generated by TSM from one mature region to an emerging market, thereby improving prediction accuracy even in the absence of historical data.
4. We further condense the in-context learning capability of the LLM into a GLU model, allowing for direct fine-tuning of TSM outputs without prompting the LLM during inference.
5. Our methods improve the prediction accuracy of carbon market prices by 2%–57% in movement trend classification and 9.6%–40% decrease in regression MSE. In

transfer learning, our method shows 3–22% and 10–21% improvement in trend classification and regression, respectively. Our GLU model also achieves comparable performance in both ETS and CEA markets without need of prompting the LLMs.

2 Related Work

Our research is based on the following aspects and tries to explain whether the LLMs can understand and accurately predict the price of EU-ETS and China CEA markets.

2.1 Time-Series Modeling (TSM) and Carbon Market Prediction

As the problem of carbon emissions has become increasingly prominent, the literature begins to study how to predict carbon prices more accurately through advanced artificial intelligence algorithms. Previous studies focus on using time-series models to predict the price and they find that advanced machine learning methods can improve the prediction accuracy. [26] developed a general carbon price prediction framework based on decomposition-synthesis. [25] proposed to decompose multi-dimensional data to capture both long-term trends and short-term fluctuations. [14] recently proposed to use Quantile Group Lasso for feature selection and carbon futures price prediction in EU ETS market. [11] further proposed to use of adaptive sparse Quantile Group Lasso for more robust price predictions. However, the research on the price prediction of China's carbon emission market is not comprehensive, and some of the studies are only aimed at the earliest carbon emission markets in Beijing and Guangdong. Our study uses data from carbon markets in four regions of China and studies the potential correlation between different markets.

2.2 AI and Large Language Models (LLMs) for Finance

Our study is related to applying advanced AI techniques to interpret finance market such as making predictions [4] and analyzing the market trends [3]. Recently, LLMs such as the GPT family [2,10] and LLaMA [17] have shown great advantages in modeling language tasks, such as arithmetic reasoning and question answering. Several studies propose to simulate the human thinking process by LLMs, such as thinking step-by-step with Chain-of-Thought [19], reflecting on past experience [16] and making further refinement over past decisions such as Self-refine [9]. Some works also studied utilizing LLMs for financial tasks such as finance-related content generation and question answering [7,22], extracting information from corporate policy [6], as well as mining trading signals or factors [18]. Due to the lack of specific domain knowledge, the LLMs could under-perform on specific queries such as medical QA and time-series forecasting. Researchers have adopted methods such as low-parameter fine-tuning [5,8], external knowledge retrieval augmentation [12,15], post-pretraining [20] and prompt-based in-context learning [1,19,24] to improve the output of LLMs in vertical domains and make them more professional and precise.

Table 1. CEA Feature List

Panel A: Numerical features with descriptions.	
p_{close}	The closing price of CEA.
vol/amt	The quantity and total amount of traded CEAs.
f_{ind}	The features of carbon indices including price, vol. & amt.
VXFXI	The China ETF Volatility Index as global market sentiment.
p_{oil}	China Daqing crude oil spot price.
p_{coal}	China Qinhuangdao coal spot price.

Panel B: Selected Carbon Indices of various CEA Zones	
GD	Carbon Tech 30 & 60, Mainland L-C Index
HB	Carbon Tech 60, Mainland L-C Index
SZ	Carbon Tech 60, Mainland L-C Index
SH	Carbon Tech 30, Carbon Tech 60

3 Data and Methodology

We define the carbon price forecasting task as follows. By observing the past T_h steps $T_h = \{1, 2, ..., T_h\}$, we predict the subsequent T_f future steps $T_f = \{T_h + 1, ..., T_h + T_f\}$. For the market of EU-ETS, we obverse the past 20 monthly prices and predict the future 12 monthly prices. For China CEA markets, we observe the past 48 d and predict the next 36 d. Next, we describe the data features.

3.1 Carbon Price Data and Feature Selection with Lasso-Based Method

We adopt the EU carbon future price data from previous works [11, 14], which also includes 13 selected factors including crude oil and natural gas production, imports and exports of European countries, as well as economic indices such as FTSE100 index, M2 values, inflation rate and interest rates, etc. The factors have been extensively discussed in previous work [11].

Similarly, we self-collect the Chinese Carbon Emission Allowance (CEA) data from four CEA zones in China, including Hubei (HB), Shenzhen (SZ), Shanghai (SH), Guangdong (GD). We collect from each CEA market the closing price, trade volume and amount at each day. In addition, we also collect auxiliary economic data, including the China ETF volatility index (VXFXI), China Daqing crude oil spot price, and China Qinhuangdao coal spot price. Detailed descriptions are presented in Table 1 Panel A. We also collect from China's stock market 25 carbon economy stock indices with values of closing price, volume, and amount. With this large of stock indices, we select the most relevant stock indices for predicting a specific CEA zone.

Following previous work [14], we utilize the classical Lasso method to perform stock index selection for each CEA zone. We select the top 2 or 3 most relevant features

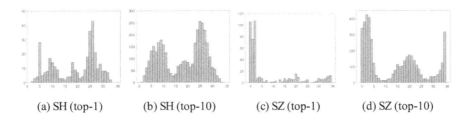

(a) SH (top-1)   (b) SH (top-10)   (c) SZ (top-1)   (d) SZ (top-10)

Fig. 2. Statistics of learned lags for SH and SZ.

if the coefficient score is above a threshold of 0.1. In Table 1 Panel B, we show the selected carbon indices for different CEA zones. We provide the details as follows.

Mainland L-C Index is the CSI mainland low-carbon economy index, which is composed of 50 stocks in China's A-share market involving clean energy power generation, energy conversion and storage, cleaner production and consumption, and waste disposal. Carbon Tech 60 is the CNI CIKD Carbon Neutral Technology Power Index, which selects 60 stocks from the Shenzhen Stock Exchange in the A-share market as sample stocks. Carbon Tech 30 is the SZSE ChiNext Carbon Neutral Technology Power Index, which selects 30 stocks from the Growth Enterprise Market of the Shenzhen Stock Exchange. The index compilation is based on the classification of the carbon-neutral technology service industry of the listed companies.

3.2 Time-Series Modeling (TSM) with Autoformer (AF)

We utilize Autoformer [21] as the state-of-the-art backbone Deep Neural Network for TSM to provide initial coarse carbon price forecasts. The Autoformer has two main components. The **Auto-Correlation** is a mechanism that can capture the period-based dependencies of historical price and factors X_h by computing the correlation of sub-series with different time delays and aggregating to a new series denoted by $\mathcal{X}$.

The **Series-Decomp**($\mathcal{X}$) denotes the historical series decomposition which can separate the series into trend-cyclical $\mathcal{T}$ and seasonal parts $\mathcal{S}$ from the input series. Thus the two parts add to the future prediction $\hat{Y} \leftarrow \mathcal{S} + \mathcal{T}$.

We fit the Autoformer to the CEA data as TSM and denote this process as

$$\hat{Y}^{AF} \leftarrow AF(X_h) . \tag{1}$$

We can visualize the cyclic trends of the CEA price. A lag τ reflects the time-delay similarity between $\mathcal{X}_t$ and its τ lag series $\mathcal{X}_{t-\tau}$. We iterate $\tau \in [1, 2, ..., L]$ and show the top-1 and top-10 most correlated lags in Shanghai and Shenzhen CEA markets, respectively, in Fig. 2. The X-axis is τ, while the Y-axis is the count of history periods that match that specific lag. In Fig. 2 (a), we observe two prominent peaks in the top-1 lags across all test sequences, namely the 5-th day (weekly) and the 26-th day (monthly). We further explore this pattern in (b) by considering the top-10 lags and finding consistent results. Analyzing (c) and (d), we observe that the top-1 lags in Shenzhen are concentrated within the first few days, while the top-10 lags exhibit either short-term or long-term (three weeks) trends. This suggests the presence of volatility and non-linearity in the CEA markets.

4 Apply Large Language Model for Forecast Refining

In this study, we utilize GPT-3.5 as the underlying LLM for refining the forecasts of CEA markets based on in-context learning capacities.

4.1 DP-Direct Prompting LLM Methodology

We directly utilize the LLM to predict the carbon price for future steps in the future without using TSM. To this end, we prompt the LLM with "Give you historical carbon price: [...], please predict the price for next 48 d." We extract the predicted sequence from the LLM's response and denote this methodology by *DP*.

This method is also known as "zero-shot learning" which relies entirely on the world-knowledge of the LLM to predict the future [2]. We will take this as a basic baseline. We will show that our observation is consistent with previous work [23] that current LLMs are not capable of precisely predicting the stock or market price without giving enough market contexts.

4.2 CoT-RF-Joint Time-Series and Large Language Modeling

Fig. 3. Flowchart of refining Autoformer predictions with LLM by CoT-Refine. Step (a): we apply trained AF to predict the future prices. Step (b): we apply LLM to enhance AF and obtain refined future predictions.

In contrast to direct prompting methods, our approach involves utilizing TSM to generate initial predictions with one selected period of past steps, along with the corresponding true prices as references. We then employ the LLM to refine these predictions for future time steps. This methodology is similar to Chain-of-Thought (CoT) prompting, as described by Wei et al. [19]. We utilize CoT prompting to leverage the contextual learning capabilities of the LLM to extrapolate from limited examples (past predictions) and generalize to new scenarios (future predictions). We give a concrete example below.

We design the CoT prompting template as "We give you some historical carbon prices [...] and AF's predicted prices. Then I need you to improve AF's predictions and give you the real prices and let you reflect on your predictions. Then we also give you AF's predictions for next 48 d [...]. Please improve AF's next 48 d prediction."

We call this method as *CoT-RF* since we feed the step-wise future raw prediction of Autoformer $\hat{Y}^{TSM}$ to the LLM to refine. This is denoted as:

$$\hat{Y}^{RF} \leftarrow \text{CoT-RF}(\hat{Y}^{AF}) . \tag{2}$$

We demonstrate this pipeline in Fig. 3. Step (a) corresponds to Eq. (1) which applies trained *AF* to predict the future prices. Step (b) corresponds to Eq. (2) which applies LLM to enhance AF and obtain refined future predictions.

4.3 FT-Efficient Low-Data Fine-Tuning with Gated Linear Unit

The CoT-RF described above requires feeding the Autoformer predictions to the LLM for refinement, thus causing potential high communication delays and high computational costs. Additionally, utilizing commercial LLMs like ChatGPT can be prohibitively expensive due to the large number of tokens required for each prompt. Furthermore, there are considerable risks and legal issues associated with uploading sensitive private data and proprietary features to LLM providers.

We introduce an innovative post-finetuning approach to mitigate the computational expenses and privacy issues while upholding the efficacy of the LLM. We train a supplementary GLU (Gated Linear Unit) model to fine-tune the Autoformer outputs with GPT refinement results efficiently. Consequently, during inference, the GLU incurs minimal computational overhead and entirely removes the necessity for frequent communications with a cloud-deployed LLM like ChatGPT.

Inspired by the recent embedding fine-tuning of LLMs, we design a two-layer GLU with Input-Linear-Swish-Linear-Output architecture. The Swish activation function [13] is defined as:

$$\text{Swish}_{\beta_1,\beta_2}(\boldsymbol{x}) = (\beta_1^\top \boldsymbol{x}) \cdot \sigma(\beta_2^\top \boldsymbol{x}) . \tag{3}$$

Thus our designed GLU model with trainable parameters β_1, β_2, W_1 and W_2 (biases omitted) can be formulated as:

$$GLU(\boldsymbol{x}) = W_2 \cdot \text{Swish}_{\beta_1,\beta_2}(W_1 \boldsymbol{x}) \tag{4}$$

We fine-tune with GLU unit by learning to transform the AF raw predictions to LLM refined predictions. This post-finetuning process is as follows:

$$\begin{aligned} \hat{Y}^{AF} &\leftarrow AF(X_h) \\ \hat{Y}^{FT} &\leftarrow GLU(\hat{Y}^{AF}) \end{aligned} \tag{5}$$

Given $\hat{Y}^{RF}$ is from LLM refinement of Eq. (2), the loss function is to minimize the fine-tuning output with LLM responses, such that:

$$\mathcal{L} \leftarrow ||\hat{Y}^{FT} - \hat{Y}^{RF}||_2^2 . \tag{6}$$

We call this process as post-finetuning as the GLU unit refines the Autoformer without altering its heavy parameters. The compact GLU unit has only 27952 parameters, amounting to only 0.3% compared with 10587153 parameters of the Autoformer. We will demonstrate in experiments that our designed post-finetuning can quickly adapt the GLU to the LLM capacity with just a few hundred training examples.

5 Empirical Results

We compare the following methods of predicting the future CEA in three different regions, Hubei (HB), Shenzhen (SZ), and Guangdong (GD). There is a total number of 960 test steps with timestamps from 2021–03 to 2022–02.

Lasso [14] uses Lasso regression to fit historical data. **AF** (Sect. 3.2) trains the Autoformer to make future predictions with supervised learning. **DP** (Sect. 4.1) relies on the LLM to directly prompt the future price given the history. We take AF and DP as two baselines. **CoT-RF** (Sect. 4.2) is based on LLM Refinement which prompts the LLM to refine AF predictions by demonstrating its predictions and true price sequences over past steps. **GLU** (Sect. 4.3) is our proposed post-finetuning process which trains a compact GLU to simulate LLM refinement process of enhancing the AF predictions.

5.1 Evaluation Tasks and Metrics

We evaluate all methods with the future regression task and the trend classification task.

MSE. The regression task is measured with Mean Squared Error (MSE, lower is better) averaged over all future 30 predicted steps.

Accuracy. The 3-way future trend classification task evaluates the predicted price on Day-10, 20, and 30 as up, neutral or down, indicating the relative position at a future step compared with the mean observed price p over the historical 18 steps. We define the neutrality class as a price range within $[(1 - \tau)p, (1 + \tau)p]$. The range of upward and downward classes are $((1 + \tau)p, \infty)$ and $(-\infty, (1 - \tau)p)$, respectively. We choose τ to be 2%. For example, if the price on Day-10 is 36.0 and the mean historical price over the past window of 18 steps is 35.0, the increase is approximately 2.8%, indicating an upward trend.

5.2 EU Emission Trading Scheme (EU ETS) Forecasting Result Analysis

We show the EU-ETS results in Table 2. Since the ETS carbon future data consists only monthly prices from 2009–03 to 2020–12, there is a total number of 182 data samples. Following [11], we split the data split to training/validation samples from 2009- 03 to 2019–12 and test on 12 monthly prices from 2020–01 to 2020–12. Due to the lack of sufficient data, we failed to train a proper deep-learning based Autoformer model. Thus we did not provide AF results.

Instead, we take the **Lasso** as the baseline and report relative increases or decreases in MSE in Table 2 of LLM-based methods. The trend classification labels are converted from the regressed price and compared with actual trend labels (up, neutral or down).

Table 2. EU-ETS carbon future price forecasting results.

Method	MSE ↓	Accuracy ↑
Lasso ([14])	7.46 (0%)	58%
DP ([19])	10.16 (36% ↑)	33%
CoT-RF (ours)	6.50 (13% ↓)	**67%**
GLU (ours)	**6.41** (14% ↓)	**67%**

This table presents the Mean Squared Error (MSE) for predicting the subsequent 12 months of European Union Emission Trading Scheme (EU-ETS) data for year 2022, along with the mean Accuracy of 3-way trend classification. The benchmark model is **Lasso** [14], which uses Lasso as the baseline which regresses on historical data. **DP** relies on the LLM to directly prompt the future price given the history. **CoT-RF** uses LLM to learn patterns from a complete Lasso prediction example on past steps as chain-of-thought then improves the AF predictions over new observed history. **GLU** performs post-finetuning over Lasso prediction by using LLM refined results.

CoT-RF refines Lasso predictions with LLM in-context learning ability. **GLU** performs post-finetuning over Lasso prediction by using LLM refined results. This is achieved by minimizing the disparity between GLU predictions and CoT-RF predictions across training timesteps. Subsequently, we employ the trained GLU model to make inferences on test timesteps.

We have observed the following trends in our analysis: The GLU model exhibits the lowest MSE, with CoT-RF following closely in second place. Both GLU and CoT-RF demonstrate a highest 67% accuracy rate in predicting trend classes. Notably, GLU, which is post-finetuned from CoT-RF results across training steps, exhibits very similar performance to CoT-RF. In fact, GLU even surpasses CoT-RF in predicting future steps, indicating its superior generalizability to unobserved future data. On the other hand, DP achieves the lowest performance among the models considered.

5.3 The Chinese Emission Allowance (CEA) Forecasting Result Analysis

We show the CEA price predictions over 3 zones in Table 3. We take the AF as the baseline and report relative increases or decreases in MSE of other methods. We observe trends as follows.

Table 3. CEA forecasting results in regions of HB, SZ and GD.

Hubei (HB)

Method	MSE	Accuracy		
		Day-10	Day-20	Day-30
Lasso ([14])	13.83 (664% ↑)	25%	30%	25%
AF ([21])	7.68 (0%)	**64%**	53%	50%
DP ([19])	10.12 (32% ↑)	42%	42%	40%
CoT-RF (ours)	6.94 (9.6% ↓)	63%	**65%**	52%
GLU (ours)	**6.88 (10% ↓)**	61%	61%	**55%**

Shenzhen (SZ)

Method	MSE	Accuracy		
		Day-10	Day-20	Day-30
Lasso ([14])	82.35 (183% ↑)	40%	35%	40%
AF ([21])	40.72 (0%)	35%	31%	28%
DP ([19])	71.73 (43% ↑)	21%	22%	21%
CoT-RF (ours)	**31.72 (22% ↓)**	**74%**	**78%**	**85%**
GLU (ours)	32.41 (20% ↓)	68%	71%	82%

Guangdong (GD)

Method	MSE	Accuracy		
		Day-10	Day-20	Day-30
Lasso ([14])	14.03 (81% ↑)	45%	35%	40%
AF ([21])	7.80 (0%)	38%	27%	26%
DP ([19])	9.44 (21% ↑)	31%	22%	25%
CoT-RF (ours)	5.37 (31% ↓)	34%	**65%**	**67%**
GLU (ours)	**4.71 (40% ↓)**	**49%**	62%	**67%**

This table presents the Mean Squared Error (MSE) for each method over the subsequent 30 trading days, along with the Accuracy of 3-way trend classification at Day-10, Day-20, and Day-30. The benchmark model is **AF**, which trains the Autoformer to make future predictions with supervised learning. **Lasso** uses Lasso regression to fit historical data. **DP** relies on the LLM to directly prompt the future price given the history. **CoT-RF** leverages the LLM to learn patterns from a comprehensive AF prediction instance over past steps. It then refines AF predictions based on newly observed historical data. **GLU** performs post-finetuning over AF prediction by using CoT-RF refined results.

Firstly, the **Lasso** method exhibits significantly larger MSE when compared to other methods, revealing its struggle in accurately learning highly non-linear CEA prices. Secondly, the advanced machine-learning method **AF** outperforms **DP** with $21\% - 43\%$ decrease in MSE. This shows that AF is more effective in learning temporal patterns from long-term sequences than a general LLM, due to AF's time-series modeling architecture and multi-period learning [21].

Thirdly, the **CoT-RF** applies the LLM to refine the AF predictions, significantly outperforming the AF. Compared to AF, CoT-RF gives 9.6% (6.94 vs. 7.68) MSE reduction in HB, 22% (31.72 vs. 40.72) in SZ, and 31% (5.37 vs. 7.80) in GD. Compared to AF, GLU gives 10% (6.88 vs. 7.68) MSE reduction in HB, 20% (32.41 vs. 40.72) in SZ, and 40% (4.71 vs. 7.80) in GD. That is, CoT-RF consistently achieves state-of-the-art performance. These results show that LLMs have the capable in-context learning ability to learn from past AF mistakes and refine AF future predictions.

Finally, our **GLU** method performs closely to CoT-RF in HB and SZ in MSE, achieving 0.4% lower MSE in HB (6.88 vs. 6.94) and 2% higher MSE in SZ (32.41 vs. 31.72). However, GLU significantly outperforms in GD, with 9% less in MSE of CoT-RF (4.71 vs. 5.37). By checking the Day-10 and Day-30 accuracy metrics, the GLU demonstrates superior performance in both short-term prediction, akin to AF, and long-term prediction, akin to CoT-RF. This observation suggests that post-finetuning effectively inherits the predictive capabilities of both AF and CoT-RF, encompassing both short-term and long-term forecasting capacities.

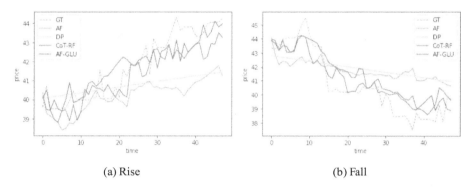

(a) Rise   (b) Fall

Fig. 4. Four methods for CEA price forecasting (Color figure online)

We observe the prediction results in Fig. 4. We visualize two randomly selected testing sequences in which (a) exhibits a rising trend and (b) exhibits a declining trend. The CoT-RF (red) and GLU (green) fit the ground-truth (dashed blue) best. The AF (orange) has large deviations in long-term future, while DP (yellow) can only provide a straight line as an estimation of the trend. Therefore, the GLU inherits both the short-term fitting capacity of AF but also the long-term trend prediction capacity of CoT-RF.

5.4 Transfer Learning Result Analysis

We show the transfer-learning results of three paired zones in Table 4. For example, the SZ-GD panel shows results of transferring predictions from SZ (source zone) to GD (target zone) market. For **AF**, we directly use the Autoformer trained on source zone to infer the target zone's future prices, which takes as the baseline. For **DP**, we directly ask the LLM to predict on target zone's future prices, without using information on source zone. **CoT-RF** demonstrates the LLM with AF predictions on source zone and true prices on both source and target zone over aligned past timesteps. This gives contexts of the market differences of source and target zones. Then it asks the LLM to transfer

Carbon Price Forecasting with LLM-Based Refinement and Transfer-Learning   151

and refine AF predictions on source zone of future timesteps to target zone. **GLU** post-finetunes AF predictions on source zone with transferred predictions of CoT-RF on target zone. This integration enables GLU to simultaneously refine AF predictions and transfer market knowledge.

We observe the following trends. Firstly, **CoT-RF** still consistently performs the best among all methods. In SZ-GD case, where we transfer the market context from Shenzhen to Guangdong, CoT-RF outperforms DP by a 23% (17.63 vs. 27.87) decrease

Table 4. Transfer-learning of paired CEA regional markets.

Shenzhen → Guangdong (SZ-GD)

Method	MSE	Accuracy		
		Day-10	Day-20	Day-30
Lasso ([14])	94.1 (992% ↑)	10%	15%	15%
AF ([21])	20.41 (0%)	**46%**	40%	24%
DP ([19])	27.87 (37% ↑)	23%	26%	21%
CoT-RF (ours)	**17.63 (14% ↓)**	35%	40%	**40%**
GLU (ours)	18.30 (10% ↓)	33%	**41%**	38%

Shanghai → Hubei (SH-HB)

Method	MSE	Accuracy		
		Day-10	Day-20	Day-30
Lasso ([14])	35.3 (697% ↑)	20%	15%	30%
AF ([21])	9.43 (0%)	57%	57%	61%
DP ([19])	14.53 (54% ↑)	21%	20%	17%
CoT-RF (ours)	7.56 (20% ↓)	64%	64%	64%
GLU (ours)	**7.49 (21% ↓)**	**65%**	**65%**	**67%**

Hubei → Shenzhen (HB-SZ)

Method	MSE	Accuracy		
		Day-10	Day-20	Day-30
Lasso ([14])	181.1 (20% ↑)	10%	5%	20%
AF ([21])	84.73 (0%)	44%	58%	51%
DP ([19])	131.42 (55% ↑)	12%	23%	30%
CoT-RF (ours)	**69.29 (18% ↓)**	**66%**	65%	**59%**
GLU (ours)	69.97 (17% ↓)	65%	**66%**	55%

This table presents the Mean Squared Error (MSE) for each transferring market pair over the subsequent 30 trading days, along with the Accuracy of 3-way trend classification at Day-10, Day-20, and Day-30. The benchmark model is **AF**, which trains the Autoformer on source zone while making future predictions on target zone. **Lasso** fits historical data of source zone while predicting on target zone. **DP** uses the LLM to directly prompt the future price given the history. **CoT-RF** demonstrates the LLM with AF predictions on source zone and true prices on both source and target zones. The LLM learns from the market contexts and refines AF predictions on source zone of future timesteps to target zone. **GLU** post-finetunes AF predictions on source zone with transferred predictions of CoT-RF on target zone.

in MSE and 10% higher in movement trend classification accuracy. In SH-HB, we observe a 34% (7.56 vs. 14.53) decrease in MSE, and 43 − 47% increase in Accuracy. A similar trend has also appeared in HB-SZ case. Secondly, CoT-RF also significantly outperforms the baseline AF. Compared to AF, CoT-RF gives 14%(17.63 vs. 20.41) MSE decrease for SZ-GD, 20% (7.56 vs. 9.43) decrease for SH-HB, and 18% (69.29 vs. 84.73) for HB-SZ.

Furthermore, our **GLU** can also achieve reasonably good results. GLU underperforms the best CoT-RF in 4% in SZ-GD case (18.30 vs. 17.63), while comes very close ±1% in SH-HB and HB-SZ in MSE. The above observation shows that for SZ-GD and SH-HB cases, the source zone actually provides a general trend of the market which can be effectively utilized as extra information to boost target zone prediction.

While for the HB-SZ case, the MSE is as large as over 69, indicating a substantial market difference between these two zones. Indeed, the SZ market exhibits significant daily price fluctuations around 3–10 CNY, whereas the HB market remains notably more stable, with daily fluctuations around just 1 CNY. Despite these disparities, our transferring technique effectively captures the overarching trend, resulting in a respectable trend accuracy of over 60%.

Lastly, the performance of Lasso is highly abnormal with an exceptionally high MSE, which strongly indicates that using Lasso for fitting on one zone and predicting on another zone is inadequate.

6 Conclusion

We introduce a framework that integrates supervised Time-series Modeling (TSM) with LLMs prompting to enhance the accuracy of future forecasts for CEA markets. By presenting TSM predictions alongside relevant market contexts to the LLM through textual prompts, we effectively improve forecasting outcomes by leveraging the LLM's inherent few-shot learning capabilities. Additionally, we exhibit the LLM's capacity to learn from past errors and refine its predictive abilities through self-reflection. Furthermore, we confirm the LLM's capability to extrapolate global market information from a source market to predict future trends in another target market, even in the absence of TSM data. Finally, we innovate a post-finetuning process which distills the refinement capacity of the LLM into a compact GLU model. As a result, we can bypass the LLM prompting phase and effectively mitigate the expenses associated with utilizing commercial LLMs.

Acknowledgment. This work is supported by the National Natural Science Foundation of China (Project 62106156), and the South China Normal University, China. We also thank Tianqi Pang for providing the implementations of Lasso methods on EU ETS forecasting.

References

1. Besta, M., et al.: Graph of thoughts: solving elaborate problems with large language models. arXiv preprint arXiv:2308.09687 (2023)
2. Brown, T., et al.: Language models are few-shot learners. Adv. Neural. Inf. Process. Syst. **33**, 1877–1901 (2020)
3. Chen, R., Ren, J.: Do AI-powered mutual funds perform better? Financ. Res. Lett. **47**, 102616 (2022)
4. Fan, C., Pang, T., Huang, A.: Pre-trained financial model for price movement forecasting. In: International Conference on Neural Information Processing, pp. 216–229. Springer (2023). https://doi.org/10.1007/978-981-99-8184-7_17
5. Hu, E.J., et al.: Lora: low-rank adaptation of large language models. arXiv preprint arXiv:2106.09685 (2021)
6. Jha, M., Qian, J., Weber, M., Yang, B.: ChatGPT and corporate policies. Tech. rep, National Bureau of Economic Research (2024)
7. Li, Y., Wang, S., Ding, H., Chen, H.: Large language models in finance: a survey. In: Proceedings of the Fourth ACM International Conference on AI in Finance, pp. 374–382 (2023)
8. Liu, X., Ji, K., Fu, Y., Tam, W.L., Du, Z., Yang, Z., Tang, J.: P-tuning v2: Prompt tuning can be comparable to fine-tuning universally across scales and tasks. arXiv preprint arXiv:2110.07602 (2021)
9. Madaan, A., Tandon, N., et al.: Self-refine: iterative refinement with self-feedback. arXiv preprint arXiv:2303.17651 (2023)
10. Ouyang, L., et al.: Training language models to follow instructions with human feedback. arXiv preprint arXiv:2203.02155 (2022)
11. Pang, T., Tan, K., Fan, C.: Carbon price forecasting with quantile regression and feature selection. In: 2023 7th International Symposium on Computer Science and Intelligent Control (ISCSIC) (2023)
12. Pang, T., Tan, K., Yao, Y., Liu, X., Meng, F., Fan, C., Zhang, X.: REMED: retrieval-augmented medical document query responding with embedding fine-tuning. IJCNN (2024)
13. Ramachandran, P., Zoph, B., Le, Q.V.: Searching for activation functions. arXiv preprint arXiv:1710.05941 (2017)
14. Ren, X., Duan, K., Tao, L., Shi, Y., Yan, C.: Carbon prices forecasting in quantiles. Energy Econ. **108**, 105862 (2022)
15. Shi, W., Min, S., Yasunaga, M., Seo, M., James, R., Lewis, M., Zettlemoyer, L., Yih, W.t.: Replug: Retrieval-augmented black-box language models. arXiv preprint arXiv:2301.12652 (2023)
16. Shinn, N., Labash, B., Gopinath, A.: Reflexion: an autonomous agent with dynamic memory and self-reflection. arXiv preprint arXiv:2303.11366 (2023)
17. Touvron, H., Lavril, T., et al.: LLaMA: open and efficient foundation language models. arXiv preprint arXiv:2302.13971 (2023)
18. Wang, S., Yuan, H., Zhou, L., Ni, L.M., Shum, H.Y., Guo, J.: Alpha-GPT: human-AI interactive alpha mining for quantitative investment. arXiv preprint arXiv:2308.00016 (2023)
19. Wei, J., et al.: Chain of thought prompting elicits reasoning in large language models. arXiv preprint arXiv:2201.11903 (2022)
20. Wu, C., et al.: LLaMA pro: progressive llama with block expansion. arXiv preprint arXiv:2401.02415 (2024)
21. Wu, H., Xu, J., Wang, J., Long, M.: Autoformer: decomposition transformers with autocorrelation for long-term series forecasting. Adv. Neural. Inf. Process. Syst. **34**, 22419–22430 (2021)

22. Wu, S., et al.: BloombergGPt: a large language model for finance. arXiv preprint arXiv:2303.17564 (2023)
23. Xie, Q., Han, W., Lai, Y., Peng, M., Huang, J.: The wall street neophyte: a zero-shot analysis of chatGPT over multimodal stock movement prediction challenges. arXiv preprint arXiv:2304.05351 (2023)
24. Yao, S., et al.: Tree of thoughts: deliberate problem solving with large language models. arXiv preprint arXiv:2305.10601 (2023)
25. Zhou, F., Huang, Z., Zhang, C.: Carbon price forecasting based on CEEMDAN and LSTM. Appl. Energy **311**, 118601 (2022)
26. Zhu, B.: A novel multiscale ensemble carbon price prediction model integrating empirical mode decomposition, genetic algorithm and artificial neural network. Energies **5**(2), 355–370 (2012)

Challenges, Methods, Data–A Survey of Machine Learning in Water Distribution Networks

Valerie Vaquet[(✉)], Fabian Hinder[(✉)], André Artelt, Inaam Ashraf[(✉)], Janine Strotherm, Jonas Vaquet, Johannes Brinkrolf[(✉)], and Barbara Hammer

Bielefeld University, Bielefeld, Germany
{vvaquet,fhinder,jbrinkro}@techfak.uni-bielefeld.de

Abstract. Research on methods for planning and controlling water distribution networks gains increasing relevance as the availability of drinking water will decrease as a consequence of climate change. So far, the majority of approaches is based on hydraulics and engineering expertise. However, with the increasing availability of sensors, machine learning techniques constitute a promising tool. This work presents the main tasks in water distribution networks, discusses how they relate to machine learning and analyses how the particularities of the domain pose challenges to and can be leveraged by machine learning approaches. Besides, it provides a technical toolkit by presenting evaluation benchmarks and a structured survey of the exemplary task of leakage detection and localization.

Keywords: Water Distribution Networks · Survey · Concept Drift

1 Introduction

High levels of threat in water security concern almost 80% of the world's population [47]. Recent studies show that this effect will aggravate as water resources become more scarce due to climate change [33]. Having well-working water distribution networks (WDNs) in place plays a crucial role in using the limited resources most efficiently and ensuring the quality of the available drinking water. As reliable and clean drinking water is essential to the health and well-being of the population, similar to electrical grids and transportation systems, WDNs are considered to be part of the critical infrastructure [13]. As a consequence, approaches using artificial intelligence (AI) need to obey specific regulations on safety, robustness, and human agency as specified in the European AI-ACT [1].

Since planning and controlling these systems is a non-trivial task, WDNs are an active research area for experts from hydro-informatics and control theory [13,18,27] with multiple scientific challenges being hosted, e.g. [28,42,49]. While systems are traditionally planned, maintained, and controlled using engineering expertise – e.g. by modeling and control based on knowledge about the hydraulic and chemical processes in the system – with increasing availability of (mobile) sensor devices, machine learning (ML) techniques gain relevance as they can have a profound impact in this area [19]. Facing upcoming challenges that will expand in the future [33] contributions by ML experts have the potential to accelerate research efforts in this domain. Besides, as the domain has some specific properties causing challenges but also chances for ML approaches, considering WDNs as a benchmark scenario to evaluate algorithmic contributions is promising.

The goal of this work is to provide a formalization and structured survey of challenges in the WDN domain which can be targeted as ML tasks, and references to possible benchmark data. To the best of our knowledge so far no such survey exists. In contrast to [9], which performs a bibliometric analysis on the body of the literature, as our main contributions, we outline and structure the characteristics of the domain with a focus on ML applications, propose how to connect the main tasks in WDNs to classical ML tasks, and specify the main challenges and chances induced by the characteristics of the domain. Besides, focusing on the exemplary task of leakage detection and localization as an exemplary assignment we structure the current state-of-the-art methods from the perspective of *drift detection and analysis* and examine the strengths and limitations of different methodological families. Additionally, we provide an overview of benchmarks and suitable evaluation strategies for this task. This survey equips ML researchers with a toolkit to work in the domain of WDNs, with our more detailed analysis of leakage detection and localization serving as a blueprint for the other presented tasks.

This paper is structured as follows: Sect. 2 discussed the characteristics of the water domain while Sect. 3 provides an overview of the main tasks in the domain, alongside the connected challenges and opportunities. Then, focusing on leakage detection and localization, we present evaluation benchmarks (Sect. 4), existing solution schemes, and categorize the latter (Sect. 5).

2 Particularities of the Domain

When designing ML approaches for the domain of WDNs, one needs to account for several particularities, which can pose challenges and offer potential for possible solutions. As WDNs are part of the critical infrastructure, next to technical particularities, one needs to account for environmental and human factors. These give rise to data-level aspects which need to be accounted for as well. We briefly discuss these key points which are summarized in the left part of Fig. 1.

2.1 Technical Aspects

The *complexity of problems* in WDNs is a key challenge. WDNs are very complex systems with many different components (e.g. valves, pumps, and reservoirs) which can change the direction of the water flow and the system's dynamics. This yields mixed integer/real-valued problems which are paired with *complex data structures* such as complex dynamic graphs [27]. While the hydraulic dynamics pose challenges, they also offer opportunities: As the dynamics in the networks underly the *laws of physics and chemistry*, models might be improved by considering physics-informed ML and incorporating domain-specific engineering

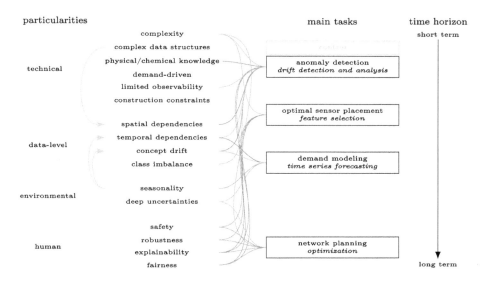

Fig. 1. Overview of the particularities of and key tasks in WDNs (in the boxes) and their dependencies. Red connections mark key challenges, green potentials, and blue additional constraints. (Color figure online)

expertise. However, WDNs are generally *demand-driven systems*. Thus, a system is highly influenced by the behavior of customers resulting in daily and weekly patterns that inflict themselves on the system. Another obstacle is the *limited observability* of water networks. Due to the associated cost and installation efforts, WDNs are only starting to be scarcely equipped with sensor technology [11].

2.2 Environmental Aspects

Next to daily and weekly patterns, the water demand is also influenced by environmental effects, e.g. *seasonalities*, and short-term factors, e.g. the weather conditions or vacation times [10]. On a longer time horizon, one needs to account for *deep uncertainties* [50]: As we cannot describe environmental and societal developments in the future, we often have to assume a range of possible future scenarios when working on long-term solutions.

2.3 Data-Level Aspects

The discussed technical and environmental aspects yield some data-level aspects. While the complex graph structure results in *spatial dependencies*, seasonal effects and the influence of demands result in *temporal dependencies*. Besides, seasonalities and rare events can result in distributional changes, also referred to as *concept drift* [15]. Finally, *class imbalance* in the measurements has to be expected, as usually much fewer measurements of anomalous behavior or rare events such as leakages or extreme weather are collected. This can pose a problem for ML-based approaches.

2.4 Human Factor

In addition to the more technical aspects, some additional constraints are imposed on WDNs as there are additional societal requirements on critical infrastructure. Since physical destruction or poisoning pose high risks as contaminated water can spread in large areas and may threaten the health of many people [18] and WDNs are vulnerable to attacks due to their size, paying attention to safety when developing methods is crucial. Thus, ensuring the *safety* and *robustness* of the developed approaches is a key criterium.

Besides, according to the European AI-ACT [1], *explainability* [30] and *fairness* [7] need to be ensured. Providing explainable AI models not only increases the acceptance of the general public but also facilitates the possibility of blending domain knowledge provided by human experts with the models. Further, explainability enables verification in case models with unclear black-box generalization ability are used. As it is crucial to provide reliable services to different locations with varying demographics throughout a city, considering the fairness of ML models is essential when developing new methods.

3 Tasks in Water Distribution Networks

The main tasks in WDNs can be categorized according to their time horizon ranging from *real-time control* to the *long-term planning* of the network layout as visualized in the right part of Fig. 1. In this section, we will introduce them, propose how they can be connected to classical ML assignments, and discuss which particularities need to be accounted for when designing approaches. Note that we will not cover the task of control due to space constraints.

3.1 Anomaly Detection

Next to the actual operation of WDNs, *real-time control* systems aim at supervising the network and resolving anomalous behavior. Anomalies generally can be categorized into hydraulic anomalies, e.g. leakages, and water quality anomalies, e.g. contamination [12]. Both might occur by accident or be caused intentionally. Usually, the implementation of monitoring systems is split into the following subtasks [13]: First, anomalies need to be detected, i.e. an alarm should be raised whenever there is atypical behavior in the system. In the second step, more information about the anomalous event is collected. The goal is to identify the type of anomaly (e.g. sensor fault, leakage, pipe burst, fire hydrant usage), localize its position in the network, and determine the exact starting time of the anomaly. Finally, some kind of accommodation is required. This can range from physical actions, e.g. repairing a leaking pipe, to software solutions, e.g. imputing missing data in case of failing sensors [31].

ML techniques are increasingly used for anomaly detection as they require less knowledge than traditional engineering approaches. In contrast, localizing the anomalies in the network frequently relies mostly on hydraulic approaches.

From a ML perspective, modeling these tasks as *drift detection and analysis* is suitable [44]. While this modeling exploits the fact that anomalies inflict themselves as concept drift, solutions need to account for the complexity, and limited observability of WDNs. Besides, temporal dependencies and class imbalances concerning anomalies pose challenges. In converse, it might be beneficial to leverage domain knowledge and spatial dependencies when designing methodologies. Fairness and explainability requirements should be met. We discuss leakage detection and localization as an exemplary assignment in Sect. 5.

Imputation for missing or faulty sensor measurements has mostly been done by standard imputation techniques [31]. Recently using virtual sensors has been proposed, e.g. by simple linear models [45] or graph convolutional networks [4].

3.2 Optimal Sensor Placement

To implement the discussed tasks of anomaly detection and control, the installation of adequate sensor technology is required. Optimal sensor placement focuses on how to find an appropriate placement of sensors across the network.

Finding an optimal sensor placement is a difficult task due to the complexity of the network. The majority of work in this field focuses on heuristics and optimization methods [18]. Many approaches rely on having a simulation of the network available to evaluate the suitablility of different sensor placements for downstream tasks like anomaly detection. Besides, most works assume that the sensors have perfect measurement capabilities, e.g. that quality sensors pick up even the smallest concentration of a given substance [18]. Recently, some initial works incorporated ML methods into the optimization process [5].

Assuming a simulation is available, from an ML perspective, this task can be formalized as a *feature selection* problem, where all possible sensor types and locations are features to select from. However, when performing feature selection in WDNs, this should ideally be done with multiple downstream tasks in mind as finding one suitable sensor placement reduces cost and construction efforts. For example, it would be desirable to find a placement putting hydraulic and quality sensors in the same spots as each installation requires construction work to access the pipes. Additionally, due to the nature of the domain of WDNs, this sensor placement needs to enhance safety and be robust, explainable, and fair.

3.3 Demand Modeling

Modeling demands on short time horizons is important to inform real-time operations while on long time horizons, it is key to planning WDNs. Modeling this task, we can benefit from the property of temporal dependencies in the data: We can assume time series data, making methods from *time-series forecasting* available. However, one needs to account for seasonality and concept drift. Considering long time horizons deep uncertainties pose additional challenges.

There are some approaches relying on ML with the majority being neural network models for time series prediction or regression [2]. ML-based methods have the advantage of making fewer assumptions than conventional models [10]. Focusing on short-term modeling, applying adaptive methods is promising as less data is required initially and trends can be captured more successfully. Here, ML techniques are not widely explored yet [32].

3.4 Long-Term Planning

Long-term planning of WDNs aims at finding an optimal network topology given multiple constraints while accounting for deep uncertainties, e.g. the expansion of cities, and external factors, e.g. climate change, are not clearly predictable. Planning is a very complex *optimization* problem. In addition to the underlying complexity, requirements regarding safety and robustness, and deep uncertainties, one needs to consider additional constraints on fairness and explainability.

Considerable research has been conducted on planning under these uncertainties by so-called staged and flexible design [27,43]. While intelligent optimization technologies such as evolutionary strategies have extensively been used to target the task of planning as an optimization problem [27], ML methods have only occasionally been integrated mostly for medium or short-term subtasks such as prioritizing pipelines for rehabilitation [14].

In the remainder of this paper, we will exemplarily discuss the tasks of leakage detection and localization as there exist already several ML approaches due to available sensor data and benchmark scenarios for evaluation. The specific modeling for dealing with leakages can serve as a blueprint for other challenges, where training samples for ML are not as readily available yet.

4 Data and Evaluation for Anomaly Detection

As mentioned before evaluation benchmarks for the tasks of leakage detection and localization are available. We will present available data benchmarks and commonly used evaluation schemes. There are three potential options for evaluating leakage detection and localization algorithms concerning data sets. (i) *Real-world measurements* are difficult to collect and frequently not publicly available. Besides, usually, it is not possible to obtain reliable ground truth about the underlying anomalies, e.g. it is hard to figure out the precise starting point of a small leak. Further, data-related particularities as discussed in Sect. 2 like limited observability and class imbalance pose problems. For this reason, research projects in this domain frequently rely on (ii) simulated data. While there exists (ii-a) *a collection of ready-to-use simulated scenarios* which are available as part of LeakDB [48], it might be favorable to (ii-b) *generate custom scenarios* as one can control the exact type, timing, and location of the anomalies and generate multiple similar scenarios to evaluate novel methodologies more thoroughly.

We will briefly describe how WDNs are modeled, as this is a necessary prerequisite for the data simulation process. Besides, we present both standard network layouts and the role of demand patterns when simulating data and cover evaluation strategies.

Modeling WDNs. Water supply networks can be modeled as graphs consisting of nodes representing junctions and undirected edges representing pipes as the flow direction of the water is not pre-defined and can change over time due to different inputs and demands in the network. As usually real-world water supply systems are extended over time, it is possible that multiple pipes lie parallel to each other [27]. Hence, in this domain, we deal with multigraphs that might change over time. Frequently, water systems also contain some sensor technology that can be installed at the connections of the pipes or in basins (e.g. level or pressure sensors) or in the pipes (e.g. flow sensors). While it is also possible to measure the demand of the end-users by so-called smart meters, usually this is not done exhaustively due to the associated cost and privacy concerns [6].

WDNs and Simulation Tools. There are few popular openly available water networks. They range from very small systems like the Hanoi with 32 nodes and 34 pipes or the Anytown network with 20 pipes and 34 nodes to larger, more realistic ones like L-Town containing 785 nodes and 909 pipes. Some scenarios are available as part of LeakDB [48].

As these scenarios are limited, simulating custom ones with a flexible choice of anomalies, e.g. leakages or sensor faults is a valuable option when one wants to evaluate a potential solution. The standard simulation tool is EPANET [36] which is commonly used through the python package WNTR [22]. To simplify the process of generating multiple scenarios with different types of anomalies in easy-to-use workflows, we provide the Automation Toolbox for ML in water Networks (ATMN)[1]. Next to the possibility of easily defining various scenarios containing multiple leakages and sensor faults, the toolbox provides parallelized, resource-aware automated generation with efficient storage of the simulation results. Besides, there is an easy-to-use data-loading API that is compatible with LeakDB, and ATMN contains a visualization tool for WDNs including sensors, leakages, and sensor faults. Note that there also exist simulation tools for water quality simulations [23].

Demands. Next to the size and structure of the network, the demands influence how difficult the problem of leakage detection is in a given scenario. As stated in Sect. 3 demand modeling itself is a difficult problem and thus an active research area. Many contributions evaluate their methodologies on artificial demands made up of simple sinusoidal curves or averages over a few collected exemplary demands. One prominent example providing a large network and realistic demands is the BattLeDIM challenge [49].

Evaluation. Evaluating detection algorithms is usually done by considering the trade-off between the true positive rate and the false positive rate, which can be further analyzed by considering the ROC-AUC score. Since the goal is to keep

---

[1] https://pypi.org/project/atmn/.

potential water loss low, reporting detection delays [19] and water losses [49] is valuable.

Table 1. Summary of approaches for leakage detection and localization alongside their data requirements and a summary of how the stages of drift detection are realized.

Method		Data requirements			Detection		Localization	Data
ref	strategy	leak-free historic data	precise topology	realtime demand	predict (h)/process	detect		
[25]	supervised		✗	✗	hydraulic model	STL decomposition k-means on trend	inverse problem: simulation (Euclidean and cosine dist)	L-Town
[39]	supervised		✗	✗	hydraulic model	dual model with artificial reservoirs CUSUM	Pearson correlation on residuals in virtual leak flows	L-Town
[8]	supervised		✗	✗	linear regression	thresholding on CUSUM	inverse problem: minimization by LP	L-Town
[51]	supervised		✗	✗	hydraulic model time series analysis	change point detection rolling mean	inverse problem	L-Town
[29]	supervised		✗	✗	hydraulic model	visual inspection	inverse problem	L-Town
[35]	supervised		✗	✗	demand model	thresholding	graph-based interpolation geometric comparison	L-Town
[24]	supervised				evol. polynomial regression	thresholding	–	UK
[21]	supervised			✗	group demands Kallman filters	statistics (CUSUM hotelling T^2)	–	Austin
[20]	unsupervised	✗			Z-scoring	statistics (WEC CUSUM hotelling T^2 EWMA)	–	Austin, Rhine
[26]	unsupervised	✗			grouping by daytime	outlier detection by median based statistics	–	5 custumn
[52]	unsupervised	✗			data denoising by wavelets	outlier detection by iforest	–	Anytown, Z-City
[46]	unsupervised				standard methods suitable window selection	standard unsupervised drift detection methods	feature-wise KS test-based drift detection	L-Town
[44]	unsupervised				–	–	model-based drift explanations	L-Town
[34]	comb	✗			pressure forcast neural network	thresholding on residuals from prediction and historical data	–	UK

Regarding the localization of leakages, the goal is to pinpoint the location as precisely as possible. Measuring and comparing the topological distances is suitable [19]. Assuming that only scarce sensor data is available, analyzing whether the leakage lies in the area spanned by several sensors is a reasonable choice [44] as then this area can be further investigated by traditional methodologies, e.g. sound-based localization techniques [37]

5 Approaches for Leakage Detection and Localization

Considerable research has been conducted on leakage detection and localization [19]. In general, these approaches are frequently tested in very small benchmark networks and often rely on strongly simplified artificial demands. In this work, we will provide a categorization of approaches (i) that perform leakage detection and/or localization based on pressure, flow, and demand measurements and (ii) that are evaluated in larger networks with realistic demands or real-world measurements used as we assume better generalization and scalability in these cases. We summarize them in Table 1. Due to space constraints, we only include the six best-performing contributions from the BattLeDIM challenge [49].

As discussed before, leakages or anomalies more generally can be modeled as *concept drift*, i.e. as distributional changes in the data generating distribution [15]. While some works attempt leakage detection as end-to-end ML, many other approaches implement some kind of drift detection scheme.

In end-to-end ML, e.g. [41,53], observations x_t are directly mapped to the output "leak/no leak" given training samples. However, this approach is suffering from a strong class imbalance. In our literature review, we only found solutions that were evaluated on very small networks and thus excluded them from further analysis as this indicates that end-to-end classification on more realistic and complex networks is too challenging.

Drift detection schemes aim at deciding whether distributional changes occur given a stream of data [17]. There are two options: one can apply *supervised drift detection* [15], where one analyses the model loss of some inference model as a proxy, or *unsupervised drift detection* [17], where one analyses the data distribution directly. Summarizing the body of work on leakage detection, we found that all approaches can be categorized as drift detection schemes. In the following, we will provide a survey on this task structured according to these options. Afterward, we focus on the assignment of leakage localization. A summary of all methods is provided in Table 1.

5.1 Prediction-Residual-Based Approaches

(a) Prediction-residual-based approaches   (b) Observation-residual-based

Fig. 2. Schematic of approaches used in the literature

As visualized in Fig. 2a supervised drift detection-based approaches, commonly also referred to as residual-based, fit a predictive model h based on the normal operational state of the network and use it in the first step to compute the expected state of the network $\hat{x}_{t+1}$. In the second step, they analyze the residual of $\hat{x}_{t+1}$ and the measured – possibly disturbed – data x_{t+1}. Assuming h fits the data this discrepancy is small. Under the assumption that h does not generalize well to out-of-distribution samples, the model no longer fits the observed data if a leakage or another anomaly occurs, and thus the residual rises. From a theoretical viewpoint, these strategies are not generally suitable for monitoring tasks like anomaly detection since for ML models, the connection between model loss and drift is rather loose [16], i.e. a model would need to perfectly replicate the reality to not raise false alarms or miss leakages. This claim was empirically supported for leakage detection by [46].

Prediction Phase. For the choice of the predicting model h there are two main options: hydraulics-based approaches and ML-based models[2] [19]. *Hydraulics-based* approaches realize h as a hydraulic simulation model replicating the WDN [25,29,39,51]. This choice requires knowledge of the exact network topology, including pipe diameters and elevation levels. Further characteristics, e.g. pipe roughness coefficients, are usually calibrated using demand data. While building a precise hydraulic simulation model provides an accurate model of the network at hand, the requirements are very limiting: Usually, the precise network topology is not known, and real-time demands are not available. While these models can be easily adapted in case a new leakage has been detected and further analyzed, applying this strategy to new or evolving WDNs requires repeating the creation of the hydraulic simulation, a costly and time-intensive task. Thus, the generalizability of this approach is limited.

ML models constitute an alternative realization of h and generally do not require knowledge of the precise network topology and demand measurements. However, usually, these methods assume the availability of leakage-free historical measurements of pressure (and flow) measurements. [8] rely on linear regression to predict the pressure at each node. Similarly, [24] apply evolutionary polynomial regression to keep an updated model of the network pressures and flows. [34] use a neural-network-based approach to predict the network state. Besides, [46] apply some standard regression models to investigate the suitability of supervised drift detection. While the hydraulic-based methods take the demands as an input, this family of models generalizes over different possible network states which are caused by the demand patterns as inputs.

In contrast, there are some approaches leveraging the temporal dependencies in the data which are induced by the seasonality: [51] propose to simulate future pressures and demands using time series analysis. They assume that the water consumption is composed of different components, i.e. trends, periodicities, and random factors, and use empirical mode decomposition (EMD) to decompose the data and consequently model normal leak-free operation. Similar to the hydraulic model-based approaches this method shares the weakness that it relies on demands for the EMD. [21] are predicting group demands by applying

---

[2] Note that in the water community, hydraulics-based approaches are called model-based and ML-based approaches are called data-driven.

Kalman filters. Those group demands can be used as input for the detection step. [35] rely on a model predicting future demands.

Given knowledge of the topology and a less scarce sensor availability, one could explore ML models leveraging the spatial dependencies, for example, as done by [4] for missing value imputation.

Detection Phase. The proposed methods for detecting the leakages range from visual inspection by a domain expert [29], simple thresholding [24,34,35], over standard statistical approaches like CUSUM and Hotelling scores [8,21,51] to more complex approaches. [39] perform the CUSUM method on a dual model with artificial reservoirs at each node which captures the leaking water. An example of a more sophisticated approach using ML is the contribution by [25]. They perform a signal decomposition and apply k-means clustering on the trend component to find time periods containing leakages.

5.2 Oberservation-Residual-Based Approaches

As shown in Fig. 2b, observation-residual-based approaches that can be categorized as unsupervised drift detection analyze the differences of observation statistics [17]: the comparison is based on a compact description of the data characteristics observed in a reference and a current window. Anomalies are related to a significant change in these statistical compressions over time based on observed values.

Processing Phase. In the first step, approaches aim to summarize and clean the data. [52] applies a wavelet transformation to obtain data smoothing. [20] eliminate demand patterns by applying z-scoring. [26] discard large consumers and summarize the network flow data to estimate the total network consumption. [46] leverages unsupervised drift detection schemes. To account for temporal dependencies in the data, they propose to consider two weekly windows to eliminate daily, weekly, and long-term patterns.

Detection Phase. The detection phase based on the residuals between the considered windows is mostly done by simple statistics: [26] rely on robust statistics, e.g. the median absolute deviation. [34] rely on standard statistical methods and combine their findings with the prediction-based approach. [20] apply methods from statistical process control, e.g. Western Electric Company rules, CUSUM, exponentially weighted moving average control charts and Hotelling T^2 control charts. [52] use an outlier detection by an isolation forest on incoming data and historical data samples which were collected at the same time of day (as similar patterns are to be expected). [46] evaluates a range of standard unsupervised drift detection schemes ranging from statistical tests, over virtual classifiers to block-based strategies. This contribution has the advantage of not requiring historical leakage-free data.

5.3 Leakage Localization

Tackling the task of leakage localization there are mainly hydraulic-based approaches relying on the hydraulic simulators obtained in the first step of leakage detection as described in Sect. 5.1. Here, the idea is to localize leakages by inverting the problem, i.e. by simulating leakages in different locations and minimizing the error over the possible leakage locations [8,25,29,35]. [25] first narrow the possible leakage locations by a heuristic simulation relying on leakage flows. [8] uses linear programming to find the most realistic leakage location. Additionally, [25] rely on localizing one leakage at a time and then updating the hydraulic model with the leakage information.

However, there are also a few approaches considering this task through the lens of concept drift and analyzing their intermediate results closer. Some rely on an analysis of the residuals: In the areas with sufficient sensor information, [35] rely on graph-based interpolation and a geometric comparison of the measured and historical leak-free data to localize the leakage. Similarly, [38] rely on Kringing interpolation and Bayesian reasoning to increase the localization accuracy when compared to historical leakage-free data. [39] compute the residuals of the virtual flows described in the prediction section. Leakages are associated with a high Pearson correlation. [46] analyze the results of feature-wise drift detection with a statistical test to find the sensors most affected by the leakage. Besides, [44] further analyze the concept drift by employing model-based explanations.

5.4 The Role of Machine Learning in Leakage Detection and Localization

So far ML plays a limited role in solving leakage detection. For the briefly discussed end-to-end strategy, approaches suffer from too strong class imbalances. Looking at the more popular supervised approaches a mixed picture emerges. Considerable research has been conducted on using ML models for predicting the network state under normal operation instead of relying on hydraulic modeling. As discussed before and can be seen in Table 1, these methods are advantageous as they do not require real-time demands and the precise network topology. However, they rely on historical leakage-free data, and their detection capabilities are limited as shown theoretically in [16] and experimentally in [46]. While still limited directly relying on unsupervised schemes seems very promising as recently shown by [46].

Regarding the localization, one heavily relies on hydraulic simulations which has the advantage that the entire network state is simulated, and thus, the precise leakage location can be determined. However, these strategies are not easily adaptable to changing and new networks and require real-time demands which is a strong limitation for their applicability in real-world applications. Relying on hydraulic measurements only, ML-based localization techniques seem a promising alternative assuming an increasing sensor availability.

We mainly focused on the key challenges raised in Sect. 2. However, there are additional first works focusing on incorporating the additional constraints we discussed. For instance, there are first investigations on how to guarantee fairness in anomaly detection systems [40]. Besides, [3] investigate explainability in such systems. More research on these initial contributions needs to be conducted to develop holistic solutions incorporating all requirements. Additionally, more research on leveraging domain knowledge, e.g. by developing physical-informed ML approaches seems promising.

6 Conclusion

In this work, we presented open challenges in the domain of WDNs as an interesting and impactful endeavor for ML research. In addition to presenting the main tasks and arguing why ML can play an important role in the future, we discussed the particularities of the domain and proposed how to link them to the main tasks in WDNs. As a practical guide aiding first research efforts in the domain of WDNs, we presented suitable evaluation methods and provided an easy-to-use data generation tool. Besides, we considered leakage detection and localization as an exemplary task for which we provided a survey of the current methods.

Acknowledgments. We gratefully acknowledge funding from the European Research Council (ERC) under the ERC Synergy Grant Water-Futures (Grant agreement No. 951424). This research was supported by the Ministry of Culture and Science NRW (Germany) as part of the Lamarr Fellow Network. This publication reflects the views of the authors only.

References

1. European Commission and Directorate-General for Communications Networks and Content and Technology: Proposal for a Regulation laying down harmonised rules on Artificial Intelligence (Artificial Intelligence Act) and amending certain Union legislative acts (2021)
2. Alvisi, S., Franchini, M.: Assessment of predictive uncertainty within the framework of water demand forecasting using the model conditional processor (MCP). urban water J **14**(1), 1–10 (2017)
3. Artelt, A., Vrachimis, S., Eliades, D., Polycarpou, M., Hammer, B.: One explanation to rule them all – ensemble consistent explanations (2022)
4. Ashraf, I., Hermes, L., Artelt, A., Hammer, B.: Spatial graph convolution neural networks for water distribution systems. In: IDA 2023 (2023)
5. Candelieri, A., Ponti, A., Giordani, I., Archetti, F.: Lost in optimization of water distribution systems: better call bayes. Water **14**(5), 800 (2022)
6. Cardell-Oliver, R., Carter-Turner, H.: Activity-aware privacy protection for smart water meters. pp. 31–40. BuildSys '21, Association for Computing Machinery (2021)
7. Castelnovo, A., et al.: A clarification of the nuances in the fairness metrics landscape. Sci. Rep. **12**(1), 1–21 (2022)
8. Daniel, I., et al: A sequential pressure-based algorithm for data-driven leakage identification and model-based localization in water distribution networks. J. Water R. Pl **148**(6), 04022025 (2022)

9. Denakpo, H., Houngue, P., Dagba, T., Degila, J.: Machine learning applied to water distribution networks issues: a bibliometric review. EAI Endorsed Trans. Energy Web **11** (2024). https://doi.org/10.4108/ew.5567, https://publications.eai.eu/index.php/ew/article/view/5567
10. Donkor, E.A., Mazzuchi, T.A., Soyer, R., Alan Roberson, J.: Urban water demand forecasting: review of methods and models. J. Water Res. Pl. **140**(2), 146–159 (2014)
11. Eggimann, S., et al.: The potential of knowing more: a review of data-driven urban water management. Environ. Sci. Technol. **51**(5), 2538–2553 (2017)
12. Eliades, D.G., et al.: Contamination event diagnosis in drinking water networks: a review. Annu. Rev. Control. **55**, 420–441 (2023)
13. Eliades, D.G., Polycarpou, M.M.: A fault diagnosis and security framework for water systems. IEEE Trans. Control Syst. Technol. **18**(6), 1254–1265 (2010)
14. Elshaboury, N., Marzouk, M.: Prioritizing water distribution pipelines rehabilitation using machine learning algorithms. Soft. Comput. **26**(11), 5179–5193 (2022)
15. Gama, J., Žliobaitė, I., Bifet, A., Pechenizkiy, M., Bouchachia, A.: A survey on concept drift adaptation. ACM comput. surv. (CSUR) **46**(4), 1–37 (2014)
16. Hinder, F., Vaquet, V., Brinkrolf, J., Hammer, B.: On the hardness and necessity of supervised concept drift detection (2023)
17. Hinder, F., Vaquet, V., Hammer, B.: One or two things we know about concept drift - a survey on monitoring evolving environments. ArXiv (2023)
18. Hu, C., Li, M., Zeng, D., Guo, S.: A survey on sensor placement for contamination detection in water distribution systems. Wireless Netw. **24**(2), 647–661 (2018)
19. Hu, Z., et al.: Review of model-based and data-driven approaches for leak detection and location in water distribution systems. Water Supply **21**(7), 3282–3306 (2021)
20. Jung, D., Kang, D., Liu, J., Lansey, K.: Improving the rapidity of responses to pipe burst in water distribution systems: a comparison of statistical process control methods. J. Hydroinform **17**(2), 307–328 (2015)
21. Jung, D., Lansey, K.: Water distribution system burst detection using a nonlinear kalman filter. J. Water Res. Plann. Manage. **141**(5), 04014070 (2015)
22. Klise, K.A., Bynum, M., Moriarty, D., Murray, R.: A software framework for assessing the resilience of drinking water systems to disasters with an example earthquake case study. Environ. Model. Software **95**, 420–431 (2017)
23. Kyriakou, M.S., et al: EPyT: An EPANET-python toolkit for smart water network simulations. J. Open Source Softw. **8**(92), 5947 (2023)
24. Laucelli, D., Romano, M., Savić, D., Giustolisi, O.: Detecting anomalies in water distribution networks using EPR modelling paradigm. J. Hydroinform **18**(3), 409–427 (2016)
25. Li, Z., et al: Fast detection and localization of multiple leaks in water distribution network jointly driven by simulation and machine learning. J. Water Res. Plann. Manage. **148**(9), 05022005 (2022)
26. Loureiro, D., et al: Water distribution systems flow monitoring and anomalous event detection: a practical approach. urban water J. **13**(3), 242–252 (2016)
27. Mala-Jetmarova, H., Sultanova, N., Savic, D.: Lost in optimisation of water distribution systems? A literature review of system design. Water **10**(3), 307 (2018)
28. Marchi, A., et al: Battle of the water networks II. J. Water Res. Plann. Manage. **140**(7), 04014009 (2014)
29. Marzola, I., Mazzoni, F., Alvisi, S., Franchini, M.: Leakage detection and localization in a water distribution network through comparison of observed and simulated pressure data. J Water. Res. Plann. Manage. **148**(1), 04021096 (2022)

30. Molnar, C.: Interpretable Machine Learning (2020)
31. Osman, M.S., Abu-Mahfouz, A.M., Page, P.R.: A survey on data imputation techniques: water distribution system as a use case. IEEE Access **6**, 63279–63291 (2018)
32. Pacchin, E., Gagliardi, F., Alvisi, S., Franchini, M.: A comparison of short-term water demand forecasting models. Water Res. Manage. **33**(4), 1481–1497 (2019)
33. Rodell, M., et al.: Emerging trends in global freshwater availability. Nature **557**(7707), 651–659 (2018)
34. Romano, M., Kapelan, Z., Savić, D.A.: Automated detection of pipe bursts and other events in water distribution systems. J. Water Res. Plann. Manage. **140**(4), 457–467 (2014)
35. Romero-Ben, L., et al: Leak localization in water distribution networks using data-driven and model-based approaches. J. Water. Res. Plann. Manage. **148**(5), 04022016 (2022)
36. Rossman, L.A.: EPANET 2: Users Manual. US Environmental Protection Agency, Office of Research and Development (2000)
37. Sitaropoulos, K., Salamone, S., Sela, L.: Frequency-based leak signature investigation using acoustic sensors in urban water distribution networks. Adv. Eng. Inf. **55**, 101905 (2023)
38. Soldevila, A., et al: Data-driven approach for leak localization in water distribution networks using pressure sensors and spatial interpolation. Water **11**(7), 1500 (2019)
39. Steffelbauer, D.B., et al: Pressure-leak duality for leak detection and localization in water distribution systems. J. Water Res. Plann. Manage. **148**(3), 04021106 (2022)
40. Strotherm, J., Hammer, B.: Fairness-enhancing ensemble classification in water distribution networks. In: LNCS, vol. 14134, pp. 119–133 (2023)
41. Sun, C., Parellada, B., Puig, V., Cembrano, G.: Leak localization in water distribution networks using pressure and data-driven classifier approach. Water **12**(1), 54 (2019)
42. Taormina, R., et al: Battle of the attack detection algorithms: disclosing cyber attacks on water distribution networks. J. Water. Res. Plann. Manage. **144**(8), 04018048 (2018)
43. Tsiami, L., Makropoulos, C., Savic, D.: A review on staged design of water distribution networks. In: WDSA CCWIs (2022)
44. Vaquet, V., et al: Localization of small leakages in water distribution networks using concept drift explanation methods (2023)
45. Vaquet, V., Artelt, A., Brinkrolf, J., Hammer, B.: Taking care of our drinking water: dealing with sensor faults in water distribution networks. In: ICANN 2022, LNCS, vol. 13530, pp. 682–693. Springer (2022). https://doi.org/10.1007/978-3-031-15931-2_56
46. Vaquet, V., Hinder, F., Hammer, B.: Investigating the suitability of concept drift detection for detecting leakages in water distribution networks. In: Proceedings of the 13th International Conference on Pattern Recognition Applications and Methods - ICPRAM, pp. 296–303 (2024). https://doi.org/10.5220/0012361200003654
47. Vörösmarty, C.J., et al: Global threats to human water security and river biodiversity. nature **467**(7315), 555–561 (2010)
48. Vrachimis, S.G.: LeakDB: a Benchmark Dataset for Leakage Diagnosis in Water Distribution Networks: (146). In: WDSA CCWI vol. 1 (2018)
49. Vrachimis, S.G., et al: Battle of the leakage detection and isolation methods. J Water. Res. Plann. Manage. **148**(12), 04022068 (2022)
50. Walker, W.E., Lempert, R.J., Kwakkel, J.H.: Deep uncertainty. In: Encyclopedia of Operations Research and Management Science, pp. 395–402. Springer US (2013)

51. Wang, X., et al: Multiple leakage detection and isolation in district metering areas using a multistage approach. J Water. Res. Plann. Manage. **148**(6), 04022021 (2022)
52. Xu, W., et al: Disturbance extraction for burst detection in water distribution networks using pressure measurements. Water Res. Res. **56**(5), e2019WR025526 (2020)
53. Zhou, M., et al: An integration method using kernel principal component analysis and cascade support vector data description for pipeline leak detection with multiple operating modes. Processes **7**(10), 648 (2019)

Day-Ahead Scenario Analysis of Wind Power Based on ICGAN and IDTW-Kmedoids

Yun Wu[1], Wenhan Zhao[1]([✉]), Yongbin Zhao[2], Jieming Yang[1], Diwen Liu[3], Ning An[4], and Yifan Huang[1]

[1] Department of Computer Science, Northeast Electric Power University, Jilin, China
[2] State Grid Baicheng Power Supply Company, Jilin, China
yemachuan0117@163.com
[3] Shanghai University, School of Computer Engineering and Science, Shanghai, China
[4] Northeast Branch of State Grid Corporation of China, Liaoning, China

Abstract. Aiming at the problem that current scenario analysis methods fail to fully capture complex time series correlations during scenario generation and do not consider time series similarities during scenario reduction, a wind power day-ahead scenario analysis method based on ICGAN and IDTW-Kmedoids is proposed. First, introducing a multi-time scale convolution layer into the CGAN scenario generation model(ICGAN) comprehensively extracts wind power time series correlation information, thereby improving scenario set generation quality. Secondly, the Kmedoids clustering algorithm (IDTW-Kmedoids) is used for scenario reduction. This algorithm uses an improved DTW algorithm to calculate the distance between clusters, which can better calculate the similarity of time series data and improve the effect of scenario reduction. The calculation results show that compared with traditional scenario analysis methods, this method can better capture the correlation and similarity of complex time series and can derive more representative typical scenarios.

Keywords: Scenario analysis · Scenario generation · Multiple time scales · CGAN · Scenario reduction · DTW · Kmedoids

1 Introduction

As the penetration rate of renewable energy power generation continues to increase, the intermittency and randomness of its output seriously affect the stable operation of the power system. How to describe the uncertainty of renewable energy output is a key issue in overcoming these challenges [1]. Scenario analysis technology is a common method for characterizing renewable energy output, including scenario generation and scenario reduction. Scenario generation obtains a large number of scenarios with random characteristics through sampling based on the probability distribution function or statistical characteristics of the research object; scenario reduction reduces the number of similar scenarios and reduces computational complexity through data analysis [2]. In the study of scenario generation, traditional methods are usually based on mathematical probability models. Zhong et al. [3] first proposed a method to calculate the joint probability distribution of wind energy and solar energy output based on Copula, and then used a clustering algorithm to conduct model error analysis. This model takes into account the influence and changing relationship between the two under different environments, greatly improving the prediction accuracy of wind and solar output power; Duan et al. [4] analyze the shortcomings of the traditional Copula method and propose a new dynamic Copula function model. The model calculates dynamic correlations for eight different groups of instances, validating the accuracy and precision of the proposed model. Scenario renewable energy output is a set of random variables with time series characteristics, and there is an inherent correlation between wind and solar output at different times, scenario generation methods based on mathematical probability models have great limitations. With the development of artificial intelligence, generative models are widely used in scenario generation [5,6], which can better capture complex nonlinear relationships, and can perform end-to-end learning from large amounts of data to improve the accuracy of scenario generation. Wang et al. [7] proposed a renewable energy output scenario generation model based on conditional VAE. This model can efficiently and unsupervisedly extract the spatiotemporal characteristics and fluctuation characteristics of renewable energy output, and has a strong generalization ability; Chen et a. [8] first proposed using the GAN network to learn the spatiotemporal correlation of renewable energy output. This approach generates realistic, high-quality spatiotemporal behavior scenarios of renewable energy sources without any explicit model building; Chen et al. [9] aiming at the problem of mismatch between generated scenarios and actual scenarios, renewable energy data are divided according to power thresholds, and the CGAN network is used to generate renewable energy output scenarios under different weather conditions. However, these current scenario generation models do not consider the correlation of wind power series at different time scales, resulting in the generated scenario set being difficult to fully capture the correlation of real scenarios time series.

In the research on scenario reduction, the cluster analysis method is mostly used to reduce similar scenarios. This type of method first classifies the initial scenario set and then selects the cluster center of each category as a typical scenario. Bai et al. [10] used the improved Kmedoids algorithm for scenario reduction, which effectively improved the computing speed of converting stochastic optimization problems into deterministic optimization problems and increased the upper limit of the scale of the problem. Zeng et al. [11] combined self-organizing feature map neural network and ppaper swarm optimization to improve the Kmeans algorithm and reduced the initial scenario set based on the improved Kmeans algorithm. However, current research on scenario reduction models lacks consideration of the similarity of time series, and it is easy to lose valuable scenarios and reduce the accuracy of scenario reduction.

In summary, this paper proposes a wind power day-ahead scenario analysis method based on ICGAN and IDTW-Kmedoids. First, an ICGAN scenario generation model that introduces multi-time scale convolution layers into CGAN is proposed to improve the quality of the generated scenario set; then a DTW algorithm that improves the distance calculation method(IDTW) is proposed to improve computational efficiency and improve outliers. The IDWT-Kmedoids scenario reduction method uses IDTW to replace the Kmedoids clustering distance calculation method, which achieves effective reduction considering the similarity of the scenario set and improves the scenario reduction accuracy.

2 Relevant Principles and Methods

2.1 Conditional Generative Adversarial Network (CGAN)

GAN is an unsupervised learning model proposed by Goodfellow et al. Year 2014. It contains two structurally independent deep learning networks, a generator and a discriminator. The authenticity of the samples generated by the generator is improved through mutual game learning between the generator and the discriminator. The conditional generative adversarial network (CGAN) is an improved generative network based on generative adversarial networks [12], which combines supervised learning and unsupervised learning techniques. It has a better generalization effect on specified types of data samples. The difference from GAN is that CGAN adds conditional value input to the input of both the generator and the discriminator, which enables the generator in CGAN to learn the sample probability distribution mapping relationship that satisfies the corresponding conditions. The structure of CGAN is shown in Figure 1. At the input of the CGAN generator, the noise z is combined with the condition c as the input quantity of the generator G, and the sample $x' = G(z|c)$ is generated through the generator G output.

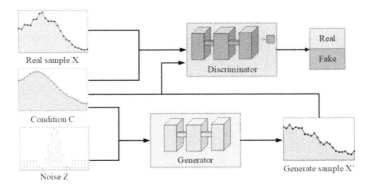

Fig. 1. Basic structure of CGAN .

The discriminator D of CGAN not only needs to judge the similarity between the generated sample distribution $P(x')$ and the real sample distribution $P(x)$, but also needs to judge whether the generated sample x' satisfies the condition c. Therefore, the loss functions of the generator and discriminator in CGAN are as shown in Eq. 1 and Eq. 2.

$$Loss_G = -E_{x' \sim p(x')}[D(x'|c)] \tag{1}$$

$$Loss_D = -E_{x \sim p(x)}[D(x \mid c)] + E_{x' \sim p(x')}[D(x' \mid c)] \tag{2}$$

where E represents the expected value of the corresponding distribution; $D(\sim)$ represents the discriminator function. The generator hopes to increase the discriminator output value of the generated sample x', while the discriminator hopes to decrease the discriminator output value of the generated sample x' and increase the discriminator output value of the real sample x. Therefore, the training goal of CGAN is a conditional minimax game, as shown in Eq. 3 [13].

$$\min_G \max_D V(D,G) = E_{x \sim p(x)}[D(x \mid c)] - E_{x' \sim p(x')}[D(x' \mid c)] \tag{3}$$

Wasserstein distance can well describe the distance between real samples and generated samples. However, in GAN, Wasserstein distance is difficult to calculate directly. Therefore, the Kantorovich-Rubinstein dual form is used to describe the distance between generated samples and real samples. When it is applied to CGAN as shown in Eq. 4:

$$W(p(x), p(x')) = \sup_{\|f_D\|_L \leq 1} E_{p(x)}[D(x \mid c)] - E_{p(x')}[D(x' \mid c)] \tag{4}$$

where $\|f_D\|_L \leq 1$ represents that the black box function composed of the discriminator D needs to satisfy 1-Lipschitz continuity, and the upper bound of the absolute value of its derivative is 1. Introducing the function D into the gradient penalty function in the definition domain of equation (3) can ensure that the discriminator function approximately satisfies 1-Lipschitz continuity, so that the discriminator loss function can effectively describe the distance between real samples and generated samples. Therefore, the objective function of CGAN is converted to :

$$\min_G \max_D V(D,G) = E_{x-p(x)}[D(x \mid c)] - E_{x'-p(x)}[D(x' \mid c)] - \lambda E[\|\nabla D(\sim)\| - 1]^2 \tag{5}$$

2.2 DTW Algorithm

Dynamic Time Warping (DTW) is a technique used to measure the similarity between two sequences. It is particularly suitable for processing time series data and can capture the time offset and deformation between sequences. The DTW algorithm measures the similarity between two sequences by finding the best matching path between them to minimize the total distance between them. The calculation formula is shown in Eq. 6 [14].

$$DTW(A,B) = \min\left\{\sqrt{\sum_{i=1}^{n}(a_i - b_i)^2}\right\} \quad (6)$$

2.3 Kmedoids Clustering Algorithm

Kmedoids is a method of grouping data points into clusters with similar characteristics. Data points can be divided into different clusters by assigning them to the closest representative points. At the same time, Kmedoids is a variation of the Kmeans clustering algorithm [15]. Taking Kmeans as an example, the cluster center is defined through Eq. 7. The cluster center is the mean value of all objects in the cluster in each dimension:

$$C_l = \frac{\sum_{x_i \in S_i} X_i}{|S_l|} \quad (7)$$

where C_l represents the center of the first cluster, $1 \leq l \leq k$. $|S_l|$ represent the number of objects in the first cluster, X_i represents the i-th object in the first cluster, $1 \leq i \leq |S_l|$.

Calculate the Euclidean distance from each point to the central cluster through Eq. 8:

$$dis(X_i, C_j) = \sqrt{\sum_{i=1}^{m}(x_i - c_{ji})} \quad (8)$$

where X_i represents the i-th object, $1 \leq i \leq n$, C_j represents the j-th cluster center, x_{it} represents the t-th attribute of the i-th object, $1 \leq t \leq m$. c_{jt} represents the t-th attribute of the j-th cluster center.

In cluster analysis, the ultimate goal is to obtain multiple sample groups surrounding the centers of different clusters, so as to achieve the clustering effect. Kmeans mutation algorithm Kmedoids no longer calculates the cluster center, but sets a certain sample as a cluster center, and the Euclidean distance can also be changed to calculate the difference value.

3 Scenario Analysis Based on ICGAN and IDTW-Kmedoids

3.1 Wind Power Power Day-Ahead Scenario Generation Model Based on ICGAN

Current scenario generation methods make it difficult to fully capture the correlation information of wind power time series. Therefore, this paper proposes the ICGAN scenario generation model, introduces multi-time scale convolution in CGAN, fully considers the correlation information of time series, designs appropriate generator and discriminator network structures, and generates high-quality initial wind power generation scenarios.

3.1.1 Input Information Structure

To enable the convolution kernel to extract the effective part of the input information, this paper studies the splicing of noise input and condition value input. Considering that the renewable energy output at each moment can be regarded as within a probability distribution, a noise is matched to the predicted value at each moment. The noise input length is consistent with the condition input length. The noise input z and the condition input c are vertically spliced into a matrix and then input into the generator, so that the noise and the prediction information form an upper and lower correspondence at each moment, which is convenient for the convolution kernel to perform correlation analysis on adjacent noise and prediction conditions. The input information is spliced and used as input to the generator network, and its input structure is shown in Fig. 2.

Noise input Z	0	Z_1	Z_2	Z_3	Z_4	...
Conditional input C	0	C_1	C_2	C_3	C_4	...

Fig. 2. Input Information Structure.

3.1.2 Multi-time Scale Convolution Design

Wind power output has strong time-series autocorrelation, which means that the wind and solar output at a certain moment is related to the wind and solar output at adjacent moments. Considering that the output at a certain moment is related to the previous moment, the first two moments, the next moment, and the last two moments. This paper designs a multi-scale convolution layer to extract and fuse time series feature information at different scales. To achieve

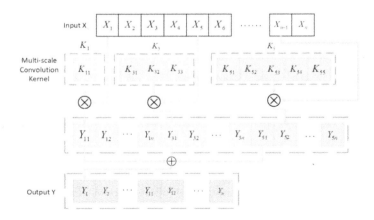

Fig. 3. Demonstration Diagram of Multi-scale Convolution Layers.

the purpose of fully extracting the correlation information of time series. The structure is shown in Fig. 3, and the mathematical operation is shown in Eq. 9:

$$Y = K_1 \otimes X + K_3 \otimes X + K_5 \otimes X \tag{9}$$

where Y is the output of the multi-scale convolution layer, X is the input of the multi-scale convolution layer, K_1 is the convolution kernel size of 2×1, K_3 is the convolution kernel size of 2×3, K_5 is the convolution kernel size of 2×5, $\otimes$ is the convolution operations and $+$ is splicing operation.

3.1.3 Generator Structure Design

In generative networks, a regularization layer is introduced after multi-scale convolution of the input layer and one-dimensional convolution of the hidden layer. The purpose is to set the data to a standard normal distribution. It can prevent it from entering the saturation interval or dead zone of the activation function and enhance the training ability of the network. However, if a regularization layer is introduced in the output layer, the output value will be biased towards the standard normal distribution, which will affect the distribution of the generator output scenario. Therefore, there is no need to introduce a regularization layer in the output layer.

Since there is no value less than 0 for renewable energy output, data less than 0 may appear during the training process. To prevent data from entering the ReLU dead zone prematurely and losing training ability, we choose to use the LeakyReLU activation function in the hidden layer of the network so that data less than 0 still has a certain training ability in the network. LeakyReLU has a non-zero slope on negative inputs, which makes it responsive to inputs less than zero and avoids the dead zone problem of ReLU. The output layer uses the ReLU activation function to ensure that the generated output value is always positive.

To sum up, the input layer and hidden layer in the generator network include a multi-scale convolution layer, one-dimensional convolution layer, regularization layer, and activation function. The output layer includes the convolution layer and activation function.

The network structure of the generator is shown in Fig. 4:

Fig. 4. Generator Structure.

3.1.4 Discriminator Structure Design

The structure of the discriminator is symmetrical to that of the generator network, and the input layer uses multi-scale convolution networks for comprehensive feature extraction and differentiation of real/generated samples. The input of the discriminator is similar to the generator, and real samples, generated samples, and condition values are vertically spliced. Then feature extraction is performed through a one-dimensional convolution layer and a regularization layer, and finally a discriminant value is output through the fully connected layer. Do not introduce a regularization layer or activation function in the output layer to avoid the output data being biased towards the same probability distribution, maintain the difference between the discriminator output value distribution of real samples and generated samples, and help the discriminator describe the difference between sample distributions.

To sum up, the input layer and hidden layer in the discriminator network include multi-scale convolution layers, one-dimensional convolution layers, regularization layers, and activation functions. The output layer only contains fully connected layers.

The network structure of the discriminator is shown in Fig. 5:

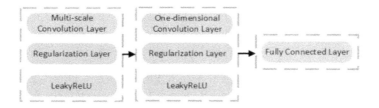

Fig. 5. Discriminator Structure.

3.2 Scenario Reduction Based on IDTW-Kmedoids

Existing scenario reduction methods lack consideration of time series similarity. Therefore, this paper proposes the IDWT-Kmedoids scenario reduction method that replaces the Kmedoids cluster distance calculation method with IDTW to improve the effect of scenario reduction.

3.2.1 IDTW Algorithm

DTW uses Euclidean Distance as the distance metric, and its computational complexity $0(n)$ will increase significantly as the length of the time series increases. At the same time, it is very sensitive to noise outliers when processing data, which will affect its accuracy. Therefore, this paper proposes an IDTW algorithm that improves the distance calculation method, which can better capture the shape similarity between time series and improve the robustness against outliers. Its calculation is Eq. 10:

$$D(A,B) = \sum_{i=1}^{n}(\mid a_i - b_i \mid + \min((a_{i-1}, b_i), (a_i, b_{i-1}), (a_{i-1}, b_{i-1}))) \qquad (10)$$

3.2.2 IDWT-Kmedoid Algorithm

This paper proposes the IDWT-Kmedoid algorithm, which replaces the cluster distance calculation method in Kmedoid with IDTW. This algorithm uses IDTW to reduce the large amount of scenario information generated by ICGAN and select the most typical scenario in the scenario set. The specific steps are as follows:

1) Initialization: randomly select the initial cluster center from the wind power scenario set;
2) Iterative clustering: for each sample, calculate the IDTW distance from the current cluster center and assign it to the nearest cluster; for each cluster, select a new center sample to minimize the IDTW distance between samples within the cluster;
3) Determine convergence: if the new cluster center is the same as the old cluster center or reaches the maximum number of iterations, the algorithm converges;
4) Select representative scenario: select the cluster center with the smallest sum of distances from the final cluster centers as the final representative scenario;
5) Output results: return the selected representative scenario.

4 Experimental Results and Analysis

4.1 Experimental Environment

The details of the experimental environment of this paper are shown in Table 1.

Table 1. Experimental environment

Configuration item	Version
Python	3.8
Deep learning framework	Pytorch 1.8
Operating system	Windows 11
Memory size	16 GB
CPU	Intel(R)Core(TM)i5-12600KF
GPU	NVIDIA GeForce RT4060ti

4.2 Experimental Dataset

The calculation example uses the day-ahead forecast value and actual value of wind power in NREL as the sample set. The time interval is 1h, and there are 52608 sets of data in total, which is 2192 d of data. Each set of data consists of two values: actual measured output and day-ahead prediction. The calculation example uses 24 sets of continuous data in one day as one day's samples and divides the training set and test set at 8:1. The calculation example first verifies the effectiveness of the wind power day-ahead scenario analysis method proposed in this paper, then compares the ICGAN method with the traditional CGAN scenario generation method, and finally compares the IDWT-Kmedoid scenario reduction method with common scenario reduction clustering algorithms.

4.3 Evaluation Indicators

Autocorrelation Coefficient: the wind power output curve has strong time autocorrelation. Therefore, in addition to having similar probability distribution characteristics to the actual scenario, the generated scenario should also take into account its time correlation. This paper uses the commonly used autocorrelation coefficient to describe the time correlation of wind power scenarios. The autocorrelation coefficient reflects the correlation between the original time series and the time series lagged by a fixed time interval. Its calculation is Eq. 11 [16]:

$$R(\tau) = \frac{E[(S_t - \mu)(S_{t+\tau} - \mu)]}{E(S_t - \mu)^2} \tag{11}$$

where S_t is the output value of the actual sample or generated sample at time t, μ is the expected value of the output sequence, τ is the time interval.

Continuous Ranked Probability Score (CRPS) : CRPS is evaluated from the perspective of the probability distribution of generated data and actual data. It is regarded as an extension of Mean Square Error (MSE) in the field of probability. It is one of the most widely used indicators in probability forecasting. At the same time, the smaller CRPS value proves that the prediction effect is better. Its calculation is Eq. 12 [17].

$$CPRS(F, y_t) = \frac{1}{N} \sum_{t=1}^{N} \int_{-\infty}^{\infty} (F(\bar{y}_t) - \varepsilon(\bar{y}_t - y_t))^2 d\bar{y}_t \tag{12}$$

where y_t is the actual data at time t, $\bar{y}_t$ is the generated data at t time, and $F(\bar{y}_t)$ represents the cumulative distribution function (CDF) of the data generated at time t.

Wasserstein Distance: Wasserstein distance measures the distance between the probability distributions of two samples of actual data and generated data. The smaller the Wasserstein distance mean value, the higher the accuracy of fitting the generated data to the actual data probability distribution. Its calculation is Eq. 13 [18].

$$W(p(x), p(x')) = \inf_{\pi(x,x')} \int d(x, x') \pi(dx, dx') \tag{13}$$

where $\pi(x,x')$ is the joint probability density distribution that satisfies the marginal distribution of $p(x)$ and $p(x')$; $d(x,x')$ is the distance measure between scenarios.

Coverage Rate (CR): The coverage rate is to evaluate whether the scenario reduction results can represent the actual data. Its calculation is Eq. 14 and Eq. 15 [18].

$$N_{t,\alpha}^h \begin{cases} 1, if\ y_t \times (1-\alpha) \le \bar{y}_t \le y_t \times (1+\alpha) \\ 0, else \end{cases} \tag{14}$$

$$CR_\alpha^h = \sum_{i=1}^{N} \frac{N_{t,\alpha}^h}{N} \tag{15}$$

where y_t is the actual data at time t, $\bar{y}_t$ is the generated data at time t, α is the percentage, N represents the total number of generated scenarios, $N_{t,\alpha}^h$ represents the number of scenarios falling within the interval of the percentage α of the actual data at time t, CR_α^h represents the percentage of coverage in the actual data for the α band.

4.4 Scenario Generation Effectiveness Analysis

This sample is taken from an untrained test set. The generator input is the wind power prediction value and 200 sets of noise. The noise and prediction values are spliced and then input into the generator to generate a wind power day-ahead scenario set based on the prediction value. A set of wind power day-ahead

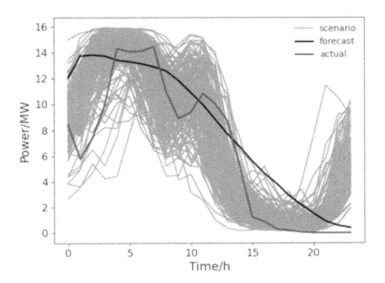

Fig. 6. Wind Power Day-ahead Scenario Set.

scenarios based on the predicted values was generated to verify the generalization of the model, as shown in Fig. 6.

The gray curve in Fig. 6 shows the 200 day-ahead scenarios generated by the generator. It can be seen that the wind power scenarios generated based on ICGAN have no obvious disordered fluctuations, the output trend is consistent with the predicted value, and the measured values can be better included in the scenario set interval. This paper randomly selected ten pieces of data from the wind power scenario set in Fig. 6, and drew the corresponding autocorrelation coefficient diagram, as shown in Fig. 7. It can be seen from this figure that the generated scenario set has a certain degree of autocorrelation within a day. As the time interval becomes longer, the autocorrelation coefficient gradually decreases and then increases. The autocorrelation coefficient diagram of the generated wind power day-ahead scenario is very close to the autocorrelation diagram of the actual data on that day, which means that the generated wind power day-ahead scenario can capture the correct time correlation. This further proves the effectiveness of the model proposed in this paper.

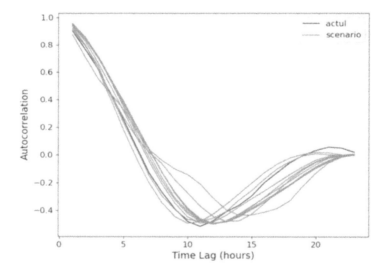

Fig. 7. Wind Power Autocorrelation Coefficient Graph.

4.5 Scenario Generation Result Analysis and Ablation Experiment

To verify the effectiveness of the multi-scale convolution layer of the model in this paper, IGCAN was compared with the CGAN before modification. Since the generated data is a scenario set, this evaluation is done by calculating the average CRPS value and Wasserstein value between all scenarios in the scenario set and the real scenario. The experimental results here are shown in Table 2:

Table 2. Performance Analysis of Scenario Generation

Evaluation index	CGAN	ICGAN
CPRS	1.23	1.03
Wasserstein	1.75	1.52

As can be seen from Table 2, compared to the unmodified CGAN model, the CPRS value and Wasserstein of the ICGAN proposed in this paper were reduced by 16.26% and 13.14% respectively. It shows that the proposed ICGAN model can better extract relevant information from time series and better capture the distribution of actual wind power output.

4.6 Scenario Reduction Result Analysis and Comparative Experiment

Based on the scenario set obtained by the ICGAN day-ahead scenario generation model, this paper uses the IDWT-Kmedoids scenario reduction model to reduce the scenario set. The CRPS value, CR value, and Wasserstein value of the traditional scenario reduction clustering method and the IDWT-Kmedoids method proposed in this paper are calculated below. Table 3 shows the analysis of CRPS value and Wasserstein value results; Fig. 8 shows the result analysis of CR value under different percentages. As can be seen from Table 3, compared with other comparison methods, the results of the model proposed have the smallest Wasserstein and CRPS values. There is almost no difference in the results between Kmeans and PCA-Kmeans. Kmedoids are better than PCA-Kmeans in the CPRS index, but weaker than PCA-Kmeans in the Wasserstein index. The IDWT-Kmedoid method proposed in this paper reduces the CPRS value by 10.45% compared to Kmedoid and reduces the Wasserstein index by 5.2% compared to PCA-Kmeans. These results indicate that the proposed model can better extract scenarios close to the actual data distribution. This paper further calculates the CR value between the reduced scenario and the real scenario, and draws the CR comparison chart based on the actual value of the reduced scenario at different percentages of α, as shown in Fig. 8. Overall, the CR value of the scenario reduction method proposed in this paper is greater than the CR value of other comparison methods, and the CR value becomes larger as the percentage increases. It can be seen from the figure that among the listed percentages, the CR value increased by 1.67% on average and the maximum increase was 8.4%. It shows that the scenario reduction model proposed in this paper can cover more points within a certain interval. The scenario after scenario reduction can better take into account the actual data information.

Table 3. Scenario Reduction Performance Analysis

	CPRS	Wasserstein
Kmeans	11.93	0.97
PCA-Kmeans	11.93	0.96
Kmedoids	10.04	1.19
IDWT-Kmedoids	8.99	0.91

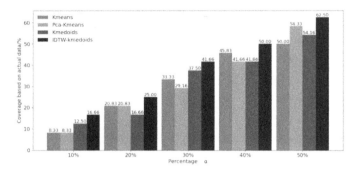

Fig. 8. Comparison Chart of CR_α^k-values Based on Actual Data at Different Percentages of α.

5 Conclusions

This paper proposes a wind power day-ahead scenario analysis method based on ICGAN and IDTW-Kmedoids. First, the ICGAN scenario generation model is used to better extract the correlation of wind power generation power series at different time scales. Secondly, the IDWT-Kmedoids scenario reduction method is proposed to replace the Kmedoids cluster distance calculation method with IDTW, which fully considers the similarity of time series during the scenario reduction process and improves the effect of scenario reduction. Experimental results show that the scenario analysis method proposed in this paper has a stronger ability to capture complex time series correlations and similarities, and can derive more representative typical operating scenarios. In future work, we will further consider the joint correlation of photovoltaic and wind power output to extract the joint distribution characteristics of wind power and photovoltaic to improve the quality of scenario generation. We will also try to combine the clustering algorithm with the autoencoder to improve scenario reduction.

Acknowledgements. This work is supported by the Science and Technology Projects of Northeast Branch of State Grid Corporation of China(No:52992623001P).

References

1. Xue, Y.S., Lei, X., Xue, F.: A review on impacts of wind power uncertainties on power systems. Proc. CSEE **34**(29), 5029–5040 (2014)
2. Ai, X., Zhou, S.P., Zhao, Y.Q.: Research on optimal dispatch model considering interruptible loads based on scenario analysis. Proc. CSEE **34**(S1), 25–31 (2014)
3. Zhong, J.Q., Li, M.J., Jiang, J.: Method of wind/solar output forecast error analysis based on Copula theory. Adv. Technol. Electr. Eng. Energy **36**(06), 39–46 (2017)
4. Duan, S.M., Miao, S.H., Huo, X.S.: Modeling and dynamic correlation analysis of wind/solar power joint output based on dynamic Copula. Power Syst. Prot. Control **47**(05), 35–42 (2019)
5. Huang, T.N., Wang, W.T., Cai, G.W.: The joint scenario generation of multi source-load by modular denoising variational autoencoder considering the complex coupling characteristics of meteorology. Proc. CSEE **39**(10), 2924–2934 (2019)
6. Kingma, D. P., Welling, M.: Auto-Encoding Variational Bayes. In: International Conference on Learning Representations. Ithaca, NY arXiv.org (2013)
7. Wang, S.X., Chen, H.W., Li, X.: Conditional variational automatic encoder method for stochastic scenario generation of wind power and photovoltaic system. Power Syst. Technol. **42**(6), 1860–1867 (2018)
8. Chen, Y., Wang, X., Zhang, B.: An Unsupervised Deep Learning Approach for Scenario Forecasts, pp. 1–7 (2018)
9. Chen, Y., Wang, Y., Kirschen, D.: Model- free renewable scenario generation using generative adversarial networks. IEEE **33**, 3265–3275 (2019)
10. Bai, B., Han, M.L., Lin, J.: Scenario reduction method of renewable energy including wind power and photovoltaic. Power Syst. Prot. Control **49**(15), 141–149 (2021)
11. Zeng, Y., Li, C., Wang, H.: Scenario-set-based economic dispatch of power system with wind power and energy storage system. IEEE Access **8**(99), 109105–109119 (2020)
12. Mirza, M., Osindero, S.: Conditional Generative Adversarial Nets. Comput. Sci. 2672–2680 (2014)
13. Dong, X.Z., Sun, Y.Y., Pu, T.J.: Day-ahead scenario generation of renewable energy based on conditional GAN. Proc. CSEE **40**(17), 5527–5536 (2020)
14. Sun, Z.Y., Ren, R., Wei, X.Z.: Research on stock classification and forecast based on DTW-TCN. Comput. Modernization (08), 31–37 (2023)
15. Xu, L., Ma, J.X., Li, G.:Post epidemic traffic vitality and economic recovery based on DTW and K-medoids clustering method. Comput. Digit. Eng. **51**(10), 2287-2292+2317 (2023)
16. Chen, F., Chen, L.M., Wang, M.: Controllable scenario generation method for wind power and photovoltaic output based on improved InfoGAN. Power System Technol., 1–14 (2024)
17. Arjovsky, M., Chintala, S., Bottou, L.: Wasserstein GAN (2017)
18. Sheng, Y.W.: Day-ahead typical scenario generation method of wind and solar power generation and its application power system unit commitment. North China Electric Power University (Beijing) (2023)

Enhancing Weather Predictions: Super-Resolution via Deep Diffusion Models

Jan-Matyáš Martinů[✉] and Petr Šimánek[✉]

Faculty of Information Technology, Czech Technical University in Prague,
Prague 16000, Czech Republic
jelliklp@seznam.cz, petr.simanek@fit.cvut.cz

Abstract. This study investigates the application of deep-learning diffusion models for the super-resolution of weather data, a novel approach aimed at enhancing the spatial resolution and detail of meteorological variables. Leveraging the capabilities of diffusion models, specifically the SR3 and ResDiff architectures, we present a methodology for transforming low-resolution weather data into high-resolution outputs. Our experiments, conducted using the WeatherBench dataset, focus on the super-resolution of the two-meter temperature variable, demonstrating the models' ability to generate detailed and accurate weather maps. The results indicate that the ResDiff model, further improved by incorporating physics-based modifications, significantly outperforms traditional SR3 methods in terms of Mean Squared Error (MSE), Structural Similarity Index (SSIM), and Peak Signal-to-Noise Ratio (PSNR). This research highlights the potential of diffusion models in meteorological applications, offering insights into their effectiveness, challenges, and prospects for future advancements in weather prediction and climate analysis.

Keywords: Weather modelling · Super-resolution · Denoising diffusion probabilistic models

1 Introduction

The application of deep learning models, particularly diffusion models, to enhance the resolution of weather data represents a significant advancement in the field of artificial intelligence and meteorology. This research area is driven by the critical need to improve the accuracy and detail of weather predictions, which are essential for a wide range of applications, from agriculture and aviation to disaster management and climate research. Super-resolution (SR) techniques, traditionally applied to image processing, are being adapted to meteorological data to generate high-resolution (HR) outputs from low-resolution (LR) inputs, thereby providing more detailed and accurate weather information.

Diffusion models, a class of generative models that have shown remarkable success in generating high-quality images, offer a promising approach to super-resolution. These models work by gradually denoising a signal, starting from a random distribution and moving towards the data distribution, effectively refining the details of images in a controlled manner. In the context of weather data, this technique can be used to enhance the resolution of various meteorological variables, such as temperature, precipitation, and wind speed maps, which are crucial for accurate weather forecasting and climate analysis.

The motivation behind using deep-learning diffusion models for the super-resolution of weather data lies in their ability to capture complex atmospheric patterns and details that are often missed or smoothed out in lower-resolution datasets. By generating higher-resolution weather data, we can achieve more precise local weather predictions, improve the understanding of microclimates, and enhance climate models. This not only benefits scientific research but also has practical implications for agriculture, urban planning, and emergency response strategies, where detailed weather information can lead to better decision-making and outcomes.

The application of such advanced super-resolution techniques to weather data is still an emerging field, with significant research potential. This paper aims to contribute to this field by exploring the use of deep-learning diffusion models for the super-resolution of weather data, focusing on their effectiveness, challenges, and potential improvements. By doing so, we seek to open new pathways for enhancing the quality and utility of meteorological data, contributing to more accurate and detailed weather forecasts and climate models.

2 Related Literature

The literature on the super-resolution of weather data using deep learning, particularly diffusion models, intersects with several research domains: image super-resolution, generative models, and meteorological data enhancement.

Image Super-Resolution with Deep Learning: The foundation of using deep learning for super-resolution is well-established, with Convolutional Neural Networks (CNNs) being the most common approach. Techniques like SRCNN [1] and ESRGAN [13] have shown significant improvements in generating high-resolution images from low-resolution counterparts. These methods have set the stage for applying deep learning to various SR tasks, including weather data.

Diffusion Models in Generative Tasks: Diffusion models, such as those proposed by [4] and further developed by [10], represent a newer class of generative models that have demonstrated remarkable capabilities in generating high-quality images. These models operate by reversing a diffusion process, gradually denoising an image from a purely noisy state to a detailed high-resolution image. Their application to image super-resolution, termed SR3 by [10], has shown promising results in enhancing image details while maintaining natural textures and patterns.

Super-Resolution of Meteorological Data: While the application of super-resolution techniques to meteorological data is less explored, there is a growing interest in this area. Studies like those by [6,12] have applied machine learning to downscale climate and weather models, improving the spatial resolution of precipitation forecasts and cloud structures. These approaches highlight the potential of machine learning in enhancing the resolution and accuracy of weather predictions but also point to the need for specialized models that can handle the unique challenges of meteorological data, such as its spatiotemporal dynamics and non-linear relationships.

Challenges and Opportunities: The adaptation of diffusion models to weather data super-resolution presents both challenges and opportunities. One challenge is the need to model the complex and dynamic nature of weather patterns accurately. However, the inherent flexibility and capability of diffusion models to capture intricate details offer a unique opportunity to significantly improve the resolution and quality of weather data.

In conclusion, the literature indicates a promising intersection of deep learning, diffusion models, and meteorological data enhancement. This paper builds on these foundations, aiming to advance the field by specifically focusing on the application of deep-learning diffusion models for the super-resolution of weather data, addressing both the challenges and potential of this innovative approach.

3 Methods

3.1 SR3

SR3, as first proposed by [10], is a model developed for image super-resolution (SR) that incorporates diffusion models [4]. It enhances low-resolution images by iteratively refining their details and quality, aiming for high-resolution outputs.

Diffusion models are characterized by two primary processes described through Markov chains: the forward process and the reverse process. The forward process progressively adds Gaussian noise to an initial image y_0, following a predefined noise schedule $(\beta_1, \beta_2, ..., \beta_T)$. For each step t, let $\alpha_t = 1 - \beta_t$ and $\bar{\alpha}_t = \prod_{i=1}^{t} \alpha_i$. The distribution of the noised image y_t given the original image y_0 at step t can then be expressed as:

$$q(y_t \mid y_0) = \mathcal{N}\left(y_t; \sqrt{\bar{\alpha}_t} y_0, (1 - \bar{\alpha}_t)I\right)$$

where

In the reverse process, the goal is to reconstruct the original image y_0 from the noise, where the distribution of noise is $\mathcal{N}(\mathbf{0}, \mathbf{I})$. Direct computation of $q(x_t|x_{t-1})$ is impossible; instead, reconstruction is approximated by a learned model $p_\theta(x_{t-1}|x_t)$ with parameters θ where x is low resolution source image:

$$p_\theta(y_{t-1} \mid y_t, x) = \mathcal{N}\left(y_{t-1}; \mu_\theta(x, y_t, \bar{\alpha}_t), \sigma_t^2 I\right)$$

Training involves simulating both the forward and reverse processes on training images, and optimizing the model to maximize the likelihood of reconstructing the high-resolution (HR) target image. This optimization targets minimizing the negative log-likelihood, which, after applying Jensen's Inequality and a series of derivations including reparametrization, leads to the following objective function:

$$\mathbb{E}_{t,y_0,x,\mathbf{e}_t}\left[\left\|\mathbf{e}_t - f_\theta\left(x, \sqrt{\bar{\alpha}_t} y_0 + \sqrt{1 - \bar{\alpha}_t}\mathbf{e}_t, t\right)\right\|^2\right]$$

Here, f_θ is the model predicting the noise added at step t.

Inference with SR3 begins by sampling a noisy image $y_t \sim \mathcal{N}(\mathbf{0}, \mathbf{I})$ and iteratively denoising it towards a high-resolution output. At each step t, noise $z \sim \mathcal{N}(\mathbf{0}, \mathbf{I})$ is sampled and used to update the image according to:

$$y_{t-1} = \frac{1}{\sqrt{\alpha_t}}\left(y_t - \frac{1-\alpha_t}{\sqrt{1-\bar{\alpha}_t}} f_\theta(x, y_t, \bar{\alpha}_t)\right) + \sqrt{1-\alpha_t}\, z$$

This iterative process is carried out for steps $T, ..., 1$, ultimately yielding the denoised, high-resolution image.

The denoising function's f_θ architecture, employs a U-net structure. The input to this function consists of an interpolated image concatenated with a noisy image. The U-net incorporates ResNet blocks [2] with self-attention and skip connections. A more comprehensive description of the architecture can be found in [10] and in Fig. 1, where an enhanced version of the SR3 architecture, resdiff [11], is presented. All the other details regarding SR3 are provided in [10].

3.2 ResDiff

The Residual-structure-based diffusion model [11] is an architecture based on SR3 that combines Convolutional Neural Networks (CNNs) [8] and modified Diffusion Probabilistic Models.

ResDiff substitutes the initial bicubic interpolation prediction with a pre-trained CNN, specialized in capturing major low-frequency components and partial high-frequency components. However, we do not employ this pre-trained CNN because it performs less effectively with climatic data compared to classical bicubic interpolation.

Furthermore, ResDiff introduces High-Frequency Guided Diffusion that modifies the diffusion model to focus on and enhance high-frequency image components. This method ensures that the details lost in the SR3 diffusion process, are meticulously reconstructed. As shown in Fig. 1, HF Guided Diffusion adds the FD Info Splitter and HF-guided Cross-Attention into the architecture.

FD Info Splitter. The Frequency-Domain Information Splitter (FD Info Splitter) [11] is a novel component of ResDiff that segregates image data into high and low-frequency bands. This segregation is fundamental for focusing the model's attention on preserving or enhancing the details critical for high-quality image super-resolution. The process begins with the application of a 2D Fast Fourier Transform (FFT) to both the interpolated and the noised images. This transformation allows the model to analyze and manipulate the frequency components of the image data directly.

Another innovation within the FD Info Splitter is the use of a Residual Squeeze-and-Excitation (ResSE) block, proposed and described in [5]. The ResSE block processes the FFT-transformed feature maps, dynamically adjusting the importance of different channels in the output feature map. Following

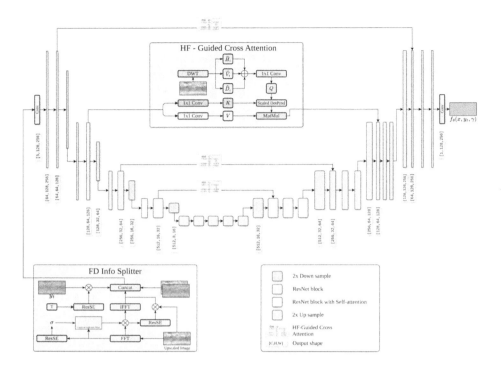

Fig. 1. Architecture of Resdiff model used for Climate Variable downscaling.

this, the computation of the standard deviation, σ, is carried out with formula:

$$\sigma = \min\left(|\operatorname{ResSE}(M)| + \frac{l}{2}, l\right)$$

where M represents the feature map obtained from the FFT, and l denotes a predetermined upper limit for the filter's standard deviation. This calculation ensures that the emphasis on high-frequency details is adaptively scaled, enhancing the model's focus on areas of the image requiring finer detail reconstruction. The high-pass filtering, represented by $H(u,v)$, is then applied to isolate and enhance these high-frequency components:

$$H(u,v) = 1 - e^{-\frac{D^2(u,v)}{2\sigma^2}}$$

where $D(u,v)$ measures the distance of a frequency component from the origin in the frequency domain. This filtering process results in a feature map L, enriched with high-frequency details [11].

Following the high-pass filtering, an inverse FFT is applied to L, facilitating the generation of a low-frequency focused image representation, x_{LF}. Concurrently, the ResSE block refines L to extract attention weights that are specific to the high-frequency domain. These weights are then applied to the interpolated image to isolate its high-frequency components, resulting in x_{HF}.

The culmination of this process produces a set of five distinct feature maps: $[x, y, x_{HF}, x_{LF}, x'_t]$, where x and y represent the source and the noise-added images, respectively, x_{HF} and x_{LF} denote the high and low-frequency components extracted, and x'_t is the target high-resolution output at iteration t. These feature maps are integral to the ResDiff model's ability to reconstruct images with enhanced detail and clarity, particularly in the high-frequency bands that are crucial for perceptual quality [11].

HF-Guided Cross-Attention. The HF-guided Cross-Attention [11] mechanism further refines the model's capacity to focus on and enhance high-frequency details. By leveraging the high-frequency component obtained by Discrete wavelet transformation [7], this mechanism directs the model's attention specifically towards areas of the image that benefit most from detail enhancement. This targeted approach ensures that the diffusion process is finely tuned to the nuances of image textures and edges, significantly improving the reconstruction of details lost in the initial SR3 diffusion stages.

In summary, ResDiff's architecture, with its emphasis on high-frequency detail preservation through innovative mechanisms like the FD Info Splitter and HF-guided Cross-Attention, represents a significant advancement in the field of image super-resolution. It addresses the limitations of previous models by providing a more focused and effective method for reconstructing images with high fidelity, especially in applications sensitive to the loss of high-frequency information.

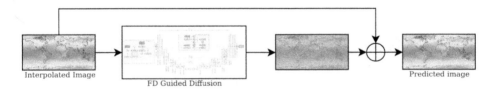

Fig. 2. Resdiff image sampling architecture.

It's significant to mention that, the ResDiff architecture diverges from SR3 by adding the image outputted by the U-net to the bicubic-interpolated image, as depicted in Fig. 2.

3.3 ResDiff Enhanced with Physics-Inspired Convolutional Filters

In an effort to further refine the ResDiff architecture for applications requiring nuanced understanding of dynamic systems, such as weather prediction, we introduce a novel modification termed "ResDiff + Physics". This adaptation integrates physics-inspired convolutional filters that mimic finite difference schemes used for computing derivatives. This approach allows the model to explicitly focus on derivative features that are fundamental to the Navier-Stokes equations, which govern fluid dynamics and are critical in meteorological modeling.

Motivation. The integration of physics-based principles into deep learning models offers a promising direction for improving the accuracy of predictions in fields heavily reliant on physical laws. By embedding convolutional filters that approximate spatial derivatives, "ResDiff + Physics" aims to capture the underlying physical processes that drive weather patterns. This method not only enhances the model's capacity for detail and pattern recognition but also aligns its internal feature extraction mechanisms more closely with the real-world phenomena it seeks to emulate.

Convolutional Derivative Filters. To achieve this, we replace the Frequency-Domain Information Splitter with a set of three specialized convolutional filters designed to approximate first and second-order spatial derivatives, key components in differential equations like the Navier-Stokes. These filters are applied to the interpolated image, capturing the gradient and curvature information relevant to fluid motion and atmospheric dynamics. The filters are defined as follows:

$$\partial_x = \begin{bmatrix} 0 & 0 & 0 \\ 0 & -1 & 1 \\ 0 & 0 & 0 \end{bmatrix} \qquad \partial_y = \begin{bmatrix} 0 & 0 & 0 \\ 0 & -1 & 0 \\ 0 & 1 & 0 \end{bmatrix} \qquad \nabla^2 = \begin{bmatrix} 0 & 1 & 0 \\ 1 & -4 & 1 \\ 0 & 1 & 0 \end{bmatrix}$$

Each filter is applied to the interpolated image using a reflect padding of 1 to ensure boundary consistency. The resulting derivative feature maps are then concatenated, forming a comprehensive tensor of shape $[B, 3, 128, 256]$ that encapsulates spatial variation information.

Integration into ResDiff. Following the convolutional operation, the derivative feature maps are concatenated with the noise-added image and the original interpolated image. This enriched tensor serves as the input to the U-net architecture, conditioning the model on physically meaningful derivatives rather than solely on frequency-domain information. This modification aims to ground the model's learning process in the physical reality of atmospheric dynamics, providing a more informed basis for generating high-resolution weather predictions.

Furthermore, we have refined the HF-guided cross-attention mechanism by incorporating a 1×1 convolution directly applied to the high-frequency components derived from the Discrete Wavelet Transform (DWT). This adjustment further enhances the model's ability to focus on and reconstruct high-frequency details, now informed by derivative-based features that reflect underlying physical processes.

"ResDiff + Physics" represents the first advancement in the application of deep learning to weather prediction and other fields governed by complex physical laws. By integrating physics-inspired convolutional filters, the model is well-equipped to interpret and reconstruct images in a manner that is both high-fidelity and physically coherent, promising improvements in the accuracy and reliability of predictions for dynamic systems.

4 Experiments

4.1 Data

For model training and evaluation, we used the WeatherBench dataset [9], a benchmark dataset designed to evaluate and compare machine learning models on weather forecasting tasks. It includes a range of meteorological variables derived from the ERA5 [3] reanalysis dataset, which is a comprehensive dataset combining model data with observations from across the world to provide a consistent, gridded view of the weather at various points in the past.

In this work, our focus is on the downscaling of the T2M (two-meter temperature) variable. This variable represents temperature readings in Kelvin, mapped across a latitude-longitude grid situated two meters above the ground, with data recorded hourly. Consequently, each day is represented by 24 distinct temperature measurements, Since our work is a super-resolution task, we used data pairs with different grid spacings of 5.625° and 1.40525°. As a result, the low-resolution image consists of 32 × 64 pixels, while the high-resolution image contains 128 × 256 pixels. These images consist of only one channel as we are working with a single variable. For model training, we used data pairs collected between January 1, 1979, and February 1, 2015. The validation set covers the period from January 1, 2016, to February 1, 2016. Due to limited computational resources and the slow validation of our model when utilizing 1000 timesteps for sampling, we downsized the dataset by using only the values measured in January for both the validation and training sets. These data are then standardized separately for each resolution.

4.2 Description of the Experiments

All experiments were conducted on a single NVIDIA A100 graphics card. We trained models for 200,000 iterations, using various batch sizes. Implementation of the code was done in PyTorch. Specifically, SR3 models utilized a batch size of 16, while ResDiff required a batch size of 4 due to its higher memory demands during experimentation. Validation was performed every 10,000 iterations across the entire validation set. Despite the computational intensity of diffusion models during inference, as they must utilize the entire U-net for each timestep T, we kept T=1000 for both training and validation to have accurate validation error estimates. As a result, the total training time for a single run was approximately 50 h. We used a linear noise schedule from 1e−6 to 1e−2 and a dropout rate of 0.2. Additionally, similar to the authors in the original paper [4], we applied Exponential Moving Average to model parameters with a decay rate of 0.9999. For optimization, we utilized Adam with a learning rate of 1e−4. Overall, we aimed to maintain the hyperparameters close to those of the original ResDiff [11] model.

As validation metrics, we used MSE and MAE for error analysis, along with SSIM, which assesses image quality by structural similarity, and PSNR, measuring the signal-to-noise ratio to gauge image reconstruction precision [14].

5 Results

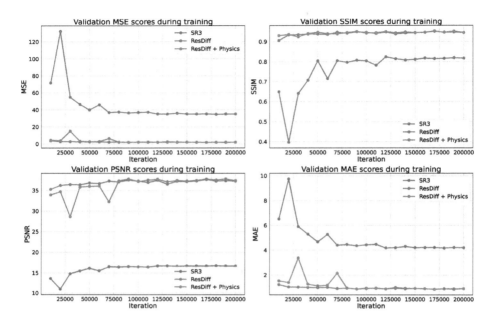

Fig. 3. Validation scores during training across models: SR3, ResDiff, ResDiff + Physics.

Using SR3 as a benchmark we can observe from Fig. 3, that both ResDiff and ResDiff+Physics significantly outperform the SR3 method throughout the entire 200,000 iterations. The validation accuracy of SR3 is not stable and exhibits considerable fluctuation during training. In contrast, ResDiff-based methods maintain an SSIM above 0.9 consistently. SR3 achieves its best validation SSIM at 120,000 iterations, while ResDiff+Physics and ResDiff reach their peak validation SSIM at 190,000 and 170,000 iterations, respectively.

Overall, as shown in Table 1, the ResDiff model achieves the best results, except for PSNR, where it is outperformed by the ResDiff+Physics architecture. In contrast, the SR3 model shows a much higher MSE.

The outputs from SR3 are worse than interpolated outputs. On the other hand, ResDiff and ResDiff+Physics capture low-frequency information much more effectively, yielding results that are difficult to distinguish from high-resolution reference images, as illustrated in Fig. 4.

6 Discussion

The findings from this study underscore the potential of deep-learning diffusion models, particularly SR3 and its enhancements via the ResDiff and ResDiff+Physics architectures, in the super-resolution of weather data. These models

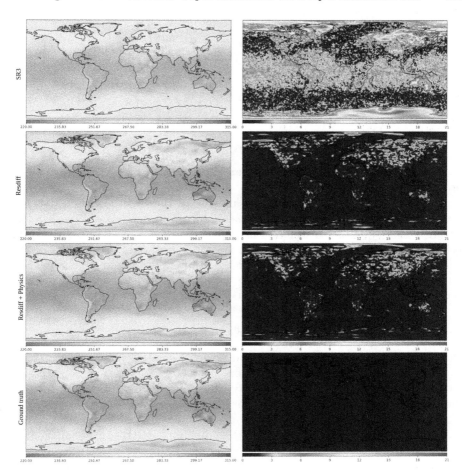

Fig. 4. In the left column, the HR reference image and images generated by the models are displayed, annotated with their corresponding temperatures in Kelvin. The right column shows the absolute error between HR reference and super-resolution images.

have demonstrated the ability to significantly improve the resolution and detail of meteorological variables, offering promising avenues for enhancing weather prediction accuracy and climate model fidelity.

The comparison of SR3 with ResDiff and ResDiff+Physics highlights the importance of model architecture and training strategy in achieving high-quality super-resolution. The superior performance of ResDiff and ResDiff+Physics over SR3, as evidenced by the validation metrics, can be attributed to their tailored design for capturing and enhancing high-frequency details, which are crucial in representing the fine-scale features of weather patterns. This suggests that the integration of domain-specific knowledge, such as the physics of weather

Table 1. Best validation scores across all models.

Method	MSE ↓	SSIM ↑	PSNR ↑	MAE ↓
Ground Truth	0	1	∞	0
SR3	35.15	0.824	16.71	4.477
ResDiff	**1.768**	**0.952**	37.65	**0.813**
ResDiff + Physics	1.789	0.951	**37.78**	0.821

phenomena, into the diffusion model architecture can further improve super-resolution outcomes.

Moreover, the application of these models to the WeatherBench dataset - a benchmark dataset for evaluating machine learning models on weather forecasting tasks - provides a concrete example of how advanced super-resolution techniques can be applied to real-world meteorological data. The ability of ResDiff and ResDiff+Physics to outperform traditional SR methods, including SR3, in terms of metrics such as MSE, SSIM, and PSNR, demonstrates their potential to contribute meaningfully to the fields of meteorology and climate science.

6.1 Implications and Future Work

The implications of this research extend beyond the academic domain, offering valuable insights for practical applications in weather forecasting, climate modeling, and environmental monitoring. By improving the resolution and accuracy of weather data, these models can aid in more precise local weather predictions, enhance the understanding of climate change impacts, and support decision-making in sectors such as agriculture, urban planning, and disaster management.

Future work should focus on further refining diffusion model architectures and training strategies to enhance their efficiency and effectiveness in super-resolving weather data. This includes exploring the integration of additional physical constraints and variables into the models to ensure that the super-resolved data adheres closely to real-world atmospheric dynamics. Additionally, expanding the application of these models to a broader range of meteorological variables and datasets will be crucial in assessing their generalizability and utility in diverse climatic and geographical contexts.

7 Conclusion

This study has demonstrated the effectiveness of deep-learning diffusion models, specifically SR3, ResDiff, and ResDiff+Physics, in the super-resolution of weather data. By leveraging the capabilities of these models to capture and enhance fine-scale details in meteorological variables, we have shown that it is possible to significantly improve the resolution and quality of weather data, thereby contributing to more accurate and detailed weather forecasts and climate models. The success of these models underscores the value of interdisciplinary research, blending advances in artificial intelligence with meteorological science, to address some of the most pressing challenges in weather prediction and climate analysis. As we continue to refine these models and explore their applications, we move closer to realizing the full potential of deep learning in enhancing our understanding and prediction of weather and climate phenomena.

References

1. Dong, C., Loy, C.C., He, K., Tang, X.: Image super-resolution using deep convolutional networks (2015)
2. He, K., Zhang, X., Ren, S., Sun, J.: Deep residual learning for image recognition (2015)
3. Hersbach, H., et al.: The era5 global reanalysis. Q. J. Royal Meteorol. Soc. **146**(730), 1999–2049 (2020). https://doi.org/10.1002/qj.3803, https://rmets.onlinelibrary.wiley.com/doi/abs/10.1002/qj.3803
4. Ho, J., Jain, A., Abbeel, P.: Denoising diffusion probabilistic models (2020)
5. Hu, J., Shen, L., Albanie, S., Sun, G., Wu, E.: Squeeze-and-excitation networks (2019)
6. Leinonen, J., Guillaume, A., Yuan, T.: Reconstruction of cloud vertical structure with a generative adversarial network. Geophys. Res. Lett. **46**(12), 7035–7044 (2019). https://doi.org/10.1029/2019GL082532, https://agupubs.onlinelibrary.wiley.com/doi/abs/10.1029/2019GL082532
7. Mallat, S., Hwang, W.: Singularity detection and processing with wavelets. IEEE Trans. Inf. Theory **38**(2), 617–643 (1992). https://doi.org/10.1109/18.119727
8. O'Shea, K., Nash, R.: An introduction to convolutional neural networks (2015)
9. Rasp, S., Dueben, P.D., Scher, S., Weyn, J.A., Mouatadid, S., Thuerey, N.: WeatherBench: a benchmark data set for data-driven weather forecasting. J. Adv. Model. Earth Syst. **12**(11) (2020). https://doi.org/10.1029/2020ms002203
10. Saharia, C., Ho, J., Chan, W., Salimans, T., Fleet, D.J., Norouzi, M.: Image super-resolution via iterative refinement (2021)
11. Shang, S., Shan, Z., Liu, G., Zhang, J.: ResDiff: combining CNN and diffusion model for image super-resolution (2023)
12. Vandal, T., Kodra, E., Ganguly, S., Michaelis, A., Nemani, R., Ganguly, A.R.: DeepSD: generating high resolution climate change projections through single image super-resolution (2017)
13. Wang, X., Yu, K., Wu, S., Gu, J., Liu, Y., Dong, C., Qiao, Yu., Loy, C.C.: ESRGAN: enhanced super-resolution generative adversarial networks. In: Leal-Taixé, L., Roth, S. (eds.) ECCV 2018. LNCS, vol. 11133, pp. 63–79. Springer, Cham (2019). https://doi.org/10.1007/978-3-030-11021-5_5
14. Wang, Z., Bovik, A., Sheikh, H., Simoncelli, E.: Image quality assessment: from error visibility to structural similarity. IEEE Trans. Image Process. **13**(4), 600–612 (2004). https://doi.org/10.1109/TIP.2003.819861

Hybrid CNN-MLP for Wastewater Quality Estimation

Marco Cardia[1](✉), Stefano Chessa[1], Alessio Micheli[1], Antonella Giuliana Luminare[2], and Francesca Gambineri[2]

[1] University of Pisa, 56127 Pisa, PI, Italy
marco.cardia@phd.unipi.it, {stefano.chessa,alessio.micheli}@unipi.it
[2] ARCHA S.R.L., 56121 Pisa, PI, Italy
{giuliana.luminare,francesca.gambineri}@archa.it

Abstract. This study explores the application of Machine Learning (ML) models in conjunction with Ultraviolet-Visible spectroscopy for real-time prediction of chemical oxygen demand in industrial wastewater. To this purpose we propose a novel soft sensor architecture that makes use of a supervised regressor based on ensemble of Convolutional Neural Network (CNN) for absorbance spectra and Multi-Layer Perceptron (MLP) for extracted and non-optical features. In our evaluation based on a real dataset built on purpose, beyond models already experimented on these tasks in the literature, we also compare against Recurrent Neural Networks and Echo State Networks that are tailored to process series, such as absorbance spectra. The experimental results show an improvement given by the proposed Hybrid CNN-MLP ensemble model with respect to other models used in the literature.

Keywords: Deep Learning · Machine Learning · Soft Sensing · Convolutional Neural Network · Echo State Network · Ensemble Model

1 Introduction

The management of risks due to highly pollutant industrial processes requires continuous monitoring of the environmental footprint associated with these processes. This necessitates the development of reliable and efficient methods for real-time assessment of environmental burdens. A fundamental parameter in wastewater management is the Chemical Oxygen Demand (COD), that indicates the amount of oxygen required to degrade organic matter present [13,17]. Traditional COD determination methods in laboratory, such as the titrimetric analysis, are time-consuming, labor-intensive, and involve hazardous chemical, e.g. the dichromate potassium [12,23]. This necessitates exploring alternative methods for faster, more efficient, and environmentally friendly COD estimation based on soft sensing [12].

Soft sensing is the combination of non-specific sensors (such as spectrometers) and advanced signal analysis (usually based on Machine Learning). Soft sensing techniques offer the advantage of real-time monitoring and reduced reliance on labor-intensive laboratory procedures [6,11]. Central to the concept of soft sensing are optical techniques, which leverage the interaction of light with matter to extract valuable information about chemical composition. Among these techniques, Ultraviolet-Visible (UV-Vis) spectroscopy stands out as a powerful tool for water quality assessment [6]. UV-Vis spectroscopy involves measuring the absorbance of light across a range of ultraviolet and visible wavelengths, providing a fingerprint of the chemical composition of a sample. In particular, the correlation between the absorbance of light at a specific wavelength and the concentration of a particular substance in a solution is stated by the Beer-Lambert law [20]. By analyzing the UV-Vis absorbance spectrum (hereafter called absorbance), it is then possible to gain insights on various parameters of a solution, including the presence of organic compounds, contaminants, and other constituents.

Although wastewater quality assessment based on Machine Learning and spectroscopy has been successfully used in several applications with low-polluted waters [1,3,5,7,14,18,25–27], with highly polluted waters they show unsatisfactory performances. This is most likely due to the large number of polluting compounds of highly polluted waters and, according to our preliminary experiments, also to the limited set of features they consider. In order to capture the complexity of the relationships between absorbance profile and quality in highly polluted wastewater, we focus on soft sensing based on UV-Vis spectroscopy and on the COD indicator, which is the most widely used parameter. The absorbance produced by the spectrometer is a vector of reals, and in the case of highly polluted waters it presents a typical structure. It is possible to interpret the absorbance both as a time series signal, or as a uni-dimensional image. Hence the straightforward approach in the analysis of this signal is either by means of Recurrent Neural Networks (RNNs) if it is interpreted as a time series, or by means of a Convolutional Neural Networks (CNNs) if it is interpreted as an image. On the other hand, the human-based interpretation of the absorbance by specialists makes use of features like the presence and height of peaks, the value of absorbance at specific wavelengths. Moreover, the wastewater is automatically diluted with distilled water before the measurement with the UV-Vis spectrometer so to reduce the saturation of the signal. The measurement of absorbance is complemented with measurements from simple sensors (pH and conductivity). The analysis of the signal is both on the absorbance and on features extracted by the absorbance combined with sensor data and dilution factor. On the base of these considerations, in this work we address the problem by proposing a more complex architecture of the soft sensor. The analysis of the signal is implemented with an ensemble Machine Learning model named Hybrid CNN-MLP, that combines a CNN for the analysis of the absorbance and a Multi-Layer Perceptron (MLP) that deals with sensor data of pH and conductivity, the dilution factor and the absorbance features. Concerning the features, we consider common ones plus a specific set of features that capture signal characteristics that are commonly observed in human-based interpretation. We make a general assessment of the impact of these features on the COD estimation by using a

Random Forest model, and we are currently extending this analysis to assess their impact in the proposed model.

We evaluate our solution against several other approaches by using a self-produced dataset obtained by analyzing wastewater of tanneries in an industrial leather area in Tuscany. We compare our solution against models already experimented in the literature such as MLP and CNN based on the only absorbance as input data. However, differently than the literature approaches, we also experiment solutions based on RNNs and Echo State Networks (ESNs). The results show that our solution slightly improves all the other considered approaches.

In the following, Sect. 2 reports the state of the art, Sect. 3 presents the construction of the dataset and its preprocessing. In Sect. 4, we detail the configuration of the models, and in Sect. 5 we present the full ensemble model. Section 6 describes the experimental setting and Sect. 7 discusses the results. Section 8 draws the conclusions.

2 Related Work

Traditionally, physical, chemical, and biological analyses conducted in laboratories played a crucial role in assessing wastewater quality. These methods provide accurate measurements for various parameters, including Biochemical Oxygen Demand (BOD), Chemical Oxygen Demand (COD), Total Suspended Solids (TSS), and specific contaminants. However, these methods are hampered by several limitations, such as the time-consuming nature, labor intensiveness, and the environmental concerns, since specific analyses, like the titrimetric method for COD determination, involve hazardous chemicals (e.g. potassium dichromate) posing safety risks and environmental concerns. This dependence on harmful chemicals underscores the need for developing safer and more sustainable approaches. These limitations emphasize the critical need for new strategies that address these shortcomings and pave the way for faster, more efficient, and environmentally friendly wastewater analysis [21].

Recognizing the limitations of traditional methods, researchers have explored the potential of optical sensors as an alternative approach for specific applications in water quality monitoring. One such example involves exploiting the spectral absorption coefficient at 254 nm for specific water quality assessments [16]. While these approaches offered a quicker and more straightforward alternative compared to traditional methods, their reliance on linear regression models using single-wavelength data limited their scope and overall performance [22].

This highlights the need for exploring more sophisticated approaches that can overcome the limitations of single-wavelength models and leverage the rich information contained within full spectral data. This pursuit has led to the integration of full-spectrum Ultraviolet-Visible (UV-Vis) spectroscopy with Machine Learning models, marking a significant development in wastewater quality prediction. This approach leverages the comprehensive information captured by the full UV-Vis spectrum, enabling ML models like Support Vector Machine (SVM), Random Forest (RF), and Neural Networks to perform detailed analyses of water quality [1–5,14,18,27]. Ye et al. employed a CNN to predict COD using UV-Vis spectroscopy in a context of mildly polluted wastewater ($COD \leq 500 mg/L$) [26]. Similarly, Guan et al. utilized a CNN for COD prediction with UV-Vis spectroscopy, incorporating spectrum preprocessing algorithms to mitigate noise sensitivity [7]. Xu et al. investigated the combined use of UV-Vis and near-infrared spectra for predicting COD, ammonia nitrogen (AN), and total nitrogen (TN). The study demonstrated improved prediction performance compared to single-spectroscopic models, particularly in settings with lower pollution levels ($COD < 82 mg/L$) [25]. These advancements facilitate real-time and on-site water quality monitoring, offering a more efficient and environmentally friendly alternative to traditional chemical methods. On the other hand, our work makes a step ahead considering also RNNs and ensemble approaches. Specifically, highly noisy signals had been dealt with RNN and ESN in rather different contexts such as e-health [19,24], and they well suit the nature of the absorbance produced by the UV-Vis spectroscopy.

Our study employs ensemble methods in a context similar to few-shot learning scenarios where the dataset is limited. They prove efficiency in solving few-shot learning for image classification, where they are often configured using two modules: one for the feature extraction and the other one for computing the non-linear distance, both exploiting the CNN but in different configurations [28]. An ensemble of MLP and of CNN has been used by Hartpence and Kwasinski for packet classification and by Moon et al. to detect the development stage of fruits in images [8,15]. In our work we reconsider RNN, ESN and ensemble models over one-dimensional absorbance signals, and we demonstrate the effectiveness of our approach in a challenging and realistic scenario often encountered in industrial settings, like the highly polluted wastewater of tanneries.

3 Data Acquisition and Preprocessing

Wastewater samples are obtained from different tanneries located in Tuscany (Italy). To capture different wastewater compositions, samples are collected from different treatment processes. The collected samples are analyzed by an UV-Vis optical probe that captures absorbance spectra within the wavelength range of 200 nm to 727.5 nm, with a resolution of 2.5 nm, and an optical path length of 1 mm. The corresponding COD concentration of each sample is determined using the small-scale sealed-tube method [9]. Absorbance of samples is analyzed both without dilution and using different dilution factors (from 2 up to 10) to overcome the possibility of signal saturation. An example of absorbance sample is shown in Fig. 1.

The industrial nature of our dataset is characterized by significant variability and complexity, and thus it requires a meticulous approach to data preprocessing, as straightforward training of Machine Learning models for the prediction of water quality indicators proved non-optimal.

The target variable, i.e. COD, displays a pronounced right skewness. To overcome this, and achieve a more symmetrical distribution, we apply a logarithmic

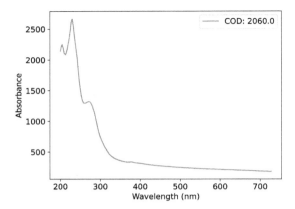

Fig. 1. Absorbance of the sample with the median value of COD.

transformation to the target variable. To ensure comparability across features, we also apply min-max scaling to reduce the absorption spectra in the range $[0, 1]$.

Furthermore, the dataset contains outliers that negatively affect both the model's learning process and the accuracy of the assessments. To mitigate this effect, we remove outliers from the dataset with two methods. The first one consists in identifying outliers in the COD values of the samples as data points whose logarithm exceeded 1.5 times the interquartile range (IQR) above the upper quartile or below the lower quartile. Overall, this method results in the removal of two outliers. In the second method, we remove samples with a skewness of the absorbance below 0.5. This threshold is computed by analyzing the distribution of skewness in the spectra. Samples falling below this skew threshold are considered outliers, because signals with very small or negative skewness exhibit flat absorbance spectra, indicating saturation or excessive absorption, typically observed in cases of overly concentrated samples where the spectrometer's measurement capacity is exceeded. According to our experiments, the removal of these samples is crucial to maintain the integrity and reliability of our predictive models.

Following the removal of outliers, the total number of records in the dataset is reduced to 244.

4 Machine Learning Models

In our experiments, we use several models, namely the MLP, RNN, ESN and CNN. We briefly summarize, one by one, their configuration.

In the following, let x_{abs} be the vector of reals representing the absorbance, and x_t be any single element of index t of this vector, that corresponds to the absorbance at wavelength t.

We considered a MLP architecture with $l_{mlp} \in [3,5]$ fully connected layers (1 input layer, $[1,3]$ hidden layers, 1 output layer). The hyperparameters for the model selection of the MLP are sampled from the range values reported in Table 1.

The RNN architecture includes an input layer, from 1 to 3 recurrent layers (depending on the specific configuration), and an output layer. At processing cycle t the RNN receives as input x_t and computes the following function:

$$h_t = tanh(W^{(ih)} x_t + b_{ih} + W^{(hh)} h_{t-1} + b_{hh}) \tag{1}$$

where h_t is the hidden state at cycle t and h_{t-1} is either the hidden state of the previous layer at cycle $t-1$ or the initial hidden state h_0, $W^{(ih)}$ is the input-hidden weights matrix, b_{ih} is the input-hidden bias, $W^{(hh)}$ is the hidden-hidden weights matrix, and b_{hh} is the hidden-hidden bias. Table 1 reports the hyperparameters and the range of values for the model selection of the RNN.

For the ESN, we adopt the Leaky Integrator Echo State Network (LI-ESN) [10] architecture that consists of an input layer, a reservoir layer with hidden, and an output layer. The leaky rate controls the leakiness of the reservoir nodes, influencing the rate at which information decays over time. The LI-ESN is configured using a set of parameters and hyperparameters to define its internal structure and behavior.

The configuration includes the following components: the matrix of input weights $W^{(in)}$, the matrix of the recurrent weights W, and the bias vector W_{bias} that adds a constant term to the reservoir's internal dynamics. Given input x_t at processing cycle t, the equation that rules the LI-ESN is as follows:

$$h_t = (1-a)h_{t-1} + a \cdot tanh(W^{(in)} x_t + W h_{t-1} + W_{bias}) \tag{2}$$

where a is the reservoir neuron's leaky rate and h_t is the reservoir activation state at processing cycle t. The output layer is defined by the equation:

$$y_t = W^{(out)}[h_t; x_t] \tag{3}$$

where $W^{(out)}$ is the matrix of output weights, and $[h_t; x_t]$ denotes the concatenation of state at the cycle t.

The hyperparameters are selected from the ranges reported in Table 1. The spectral radius determines the time constant of the ESN (a larger spectral radius allows information to persist and propagate for a longer duration and longer-range interactions). The input scaling is related to the degree of nonlinearity of the reservoir dynamics.

Our CNN architecture consists of N_b convolutional blocks to capture a hierarchical representation across the signal. Each block comprises a 1D convolutional layer, a rectified linear unit (ReLU) activation function, a max pooling layer, and a batch normalization (BN). Formally, given an input signal $A^{(0)} = x_{abs}$, a convolutional layer at depth $m \in [1, N_b]$ receives input $A^{(m-1)}$ and produces output $A^{(m)}$ (also named feature map), computed as:

$$A^{(m)} = BN(ReLU(W^{(m)} \odot A^{(m-1)} + b^{(m)})) \tag{4}$$

here, $W^{(m)}$ and $b^{(m)}$ are the weight matrix and the bias at layer m, respectively, BN is the batch normalization operator, and $\odot$ is the 1D convolutional operator. Table 1 contains the range values of hyperparameters used for model selection for the CNN.

Table 1. Range of Machine Learning models hyperparameters values for model selection. Values are uniformly sampled from the reported ranges.

MLP		RNN		LI-ESN		CNN	
Hyperparameter	Range	Hyperparameter	Range	Hyperparameter	Range	Hyperparameter	Range
Learning Rate	[0.0001, 0.0005]	Learning Rate	[0.0001, 0.0005]	Spectral radius	[0.7, 1]	Learning Rate	[0.0001, 0.0005]
Weight Decay	[0.001, 0.0015]	Weight Decay	[0.001, 0.0015]	Leaky rate a	[0.2, 1.1]	Weight Decay	[0.001, 0.0015]
Hidden Size	[300, 900]	Hidden Size	[300, 800]	Input scaling	[0.5, 1.0]	Hidden Size	[300, 900]
Hidden Layers	[1,3]	Hidden Layers	[1,3]	Hidden dimension (Reservoir size)	[20, 50]	Convolutional blocks	[2,5]
				Connectivity	[0.10, 0.8]	Kernel Size	[2, 12]
				Bias scaling	[0.01, 0.3]	Stride Size	[2, 10]

5 The Soft Sensor Architecture

The soft sensor architecture integrates data from absorbance alongside extracted features and other sensor data, including pH, conductivity, and dilution factor. Figure 2 presents the dataflow diagram of the soft sensor. The soft sensor takes in input the absorbance and three scalars that are the dilution factor and the values of pH and conductivity measured by the sensors used in combination with the spectrometer. The absorbance is input both to the CNN of the Hybrid CNN-MLP ensemble model and to the feature extractor. The extracted features, along with the three scalars, are instead input to the MLP of the Hybrid CNN-MLP ensemble model. The fusion component aggregates the outputs of the CNN and of the MLP by using another MLP model.

5.1 Features Extraction

Starting from the absorbance, we extract a set of informative features aiming at enhancing the predictive performance of the Hybrid CNN-MLP. We include statistic-based features of data, that provides insights of the distribution and variability of the absorbance. Among these we consider: skewness, kurtosis, and standard deviation. We also consider other features taking inspiration from the human-based interpretation of samples. We group them into three disjoint categories:

1. Peak-based features. We distinguish peaks in the absorbance from noise by considering those with a prominence greater than 5 (the vertical distance between the peak and the lowest contour connecting neighboring valleys) and a minimum width of 7.5 nm. Extracted features in this category include: the total numbers of peaks within the spectrum, the first three wavelengths (in nm) at which the i-th peak was identified, the absorbance value at the i-th

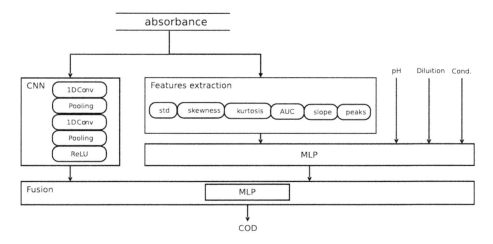

Fig. 2. Dataflow diagram of the soft sensor.

peak, the prominence of the i-th peak, representing its relative height compared to neighboring valleys, and the full width at half maximum (FWHM) of the i-th peak, indicating its breadth.
2. Slope-based features. For an interval $[t_0, t_1]$, the slope is defined as $m = \frac{x_{t_1} - x_{t_0}}{t_1 - t_0}$ where x_t is the absorbance at wavelength t. Slopes of the absorbance curve is calculated between various wavelength ranges to capture different aspects of the spectral behavior. This includes the overall trend in the UV region (between 230 nm and 400 nm), insights into the high-wavelength behavior (between the wavelength of the maximum peak and 727.5 nm), and information about the slope around the peak (between the wavelength of the maximum peak and 400 nm).
3. Area-Based Features. Area Under the Curve (AUC) measurements are computed across different wavelength ranges to quantify the total absorbance. This includes the AUC of the absorbance spectrum between 200 nm and 730 nm (total absorbance across the measured range), the AUC between 200 nm and 400 nm (focused on the UV region), and the AUC between 400 nm and 730 nm (focused on the visible region).

By integrating these diverse features, our models is equipped to capture a wide range of spectral characteristics and contextual factors.

5.2 The Hybrid CNN-MLP Ensemble Model

The Hybrid CNN-MLP ensemble model comprises three components of the soft sensor shown in Fig. 2: the CNN, the MLP, and the fusion component, which is also based on MLP. The CNN processes the vector representing the absorbance through convolutional and max pooling layers. The MLP handles tabular data, specifically the extracted features along with the dilution factor, pH, and conductivity, through fully connected layers.

The fusion component is represented by the following equation:

$$z = ReLU(W_f[A^{(N_b)}; H^{(l_{mlp})}] + b_f) \qquad (5)$$

where $[A^{(N_b)}; H^{(l_{mlp})}]$ denotes the concatenation of the outputs of the CNN and MLP, W_f represents the weight matrix of the fusion component, and b_f denotes the bias vector. The fusion component uses backpropagation with shared gradients. Thus, the fusion component aggregates complementary information from the two CNN and MLP sources. The output layer of the MLP in the fusion component is the linear combination:

$$COD_{pred} = W^{(out)T} z + b_{out} \qquad (6)$$

where W^{out} is the vector of weights of the output layer of the MLP in the fusion component, b_{out} is the corresponding bias, and COD_{pred} is the Hybrid CNN-MLP COD prediction. We set the hyperparameters of the Hybrid CNN-MLP by uniform sampling in the ranges reported in Table 2.

Table 2. Range of Hybrid CNN-MLP hyperparameters values for model selection. Values are uniformly sampled from the reported ranges.

Hyperparameter	Range
Learning Rate	[0.0001, 0.0005]
Number of units MLP	[200, 700]
Hidden Layers MLP	[1,3]
Weight Decay	[0.001, 0.0014]
Activation function	[ReLU, Sigmoid]
Hidden Size CNN	[300, 900]
Convolutional blocks CNN	[2,5]
Kernel Size	[2, 12]
Stride Size	[2, 10]

6 Experimental Design

Our experiments focus on two tasks: the prediction of the COD using the absorbance and the analysis of the relevance of the extracted features. This dual focus allow us to assess the significance of individual features extracted from the absorbance in improving COD estimation accuracy.

Concerning COD prediction, we make a first test on its prediction directly on the absorbance. To this purpose we experiment with MLP, CNN, and LI-ESN models. In a second experiment we compare different ensemble models that take in input both absorbance, extracted features and pH, dilution factor and conductivity. To this purpose we compare our Hybrid CNN-MLP ensemble model against similar ensembles such as Hybrid MLP-MLP and averaging ensembles such as CNN-MLP, MLP-MLP, MLP-Random Forest (RF), and LI-ESN-RF. In the averaging ensemble, each model provides its own prediction for a given input. Then, the final prediction is obtained by averaging the predictions of all the models, combining multiple models to improve predictive performance by leveraging the diversity of their individual predictions.

For model selection and assessment, we employ double k-fold cross-validation, comprising 4 inner folds for model selection and 5 outer folds for model assessment. Note that this methodology not only validates the robustness of our chosen model but also mimics real-world conditions, offering insights into its effectiveness across possible scenarios that can be encountered in the field. To avoid an unbiased assessment of the models, we make sure that the test set does not contain samples of the training set even if they have a different dilutions.

For the performance evaluation, we select four evaluation metrics: Root Mean Squared Error (RMSE), logarithmic Root Mean Squared Error (logRMSE), Coefficient of Determination (R^2), and Mean Absolute Percentage Error (MAPE). Model selection is conducted using random search with 128 different combinations of hyperparameters, and the best hyperparameter values are selected for each outer fold based on the performance on the validation set.

We perform the experiments using scikit-learn version 1.2.2 and PyTorch v2.0.0 in a machine equipped with a Nvidia Quadro RTX 6000 GPU (24 GB of GPU memory). We also provide the code to reproduce the Hybrid CNN-MLP and our experiments at https://github.com/cardiamc/ICANN2024.

7 Results and Discussion

We first analyze the results obtained by MLP, CNN, and LI-ESN using the absorbance, without considering additional features. Table 3 describes the performance of the ML models in predicting COD values in wastewater. The reported values represent the average of each evaluation metric across the 5 folds of the cross validation, along with their standard deviation. The values highlighted in bold indicate the best results among the different models. The R^2 values are higher than 0.87 for all models, indicating that the models can capture a significant portion of the variance in COD using the absorbance. The MLP, trained solely on the absorbance, exhibits the best performance across all metrics except for RMSE, where the LI-ESN achieves the lowest value of 2029.34.

Table 4 presents the validation and test results obtained by the the Hybrid CNN-MLP ensemble model. We compared our results with MLP and other ensemble models using all available inputs: the absorbance, the extracted features and the scalars pH, conductivity and dilution factor. Each row represents a different model configuration and columns denote different performance metrics. Each reported value is the average of the corresponding evaluation metric across the 5 folds, along with its standard deviation. In the case of ensemble models

Table 3. Results of the MLP, CNN, and LI-ESN applied to the UV-Vis absorbance spectrum.

	val logRMSE	test logRMSE	test RMSE	test R^2	test MAPE
MLP	0.47 (±0.03)	**0.47 (±0.10)**	2086.18 (±485.56)	**0.88 (±0.04)**	**4.88% (±1.00%)**
CNN	0.46 (±0.05)	**0.47 (±0.10)**	2350.55 (±671.53)	0.88 (±0.04)	5.03% (±1.05%)
RNN	0.52 (±0.04)	0.52 (±0.10)	2237.69(±379.10)	0.87 (±0.06)	5.51% (±1.16%)
LI-ESN	0.48 (±0.03)	0.51 (±0.12)	**2029.34 (±484.21)**	0.87 (±0.06)	5.44% (±1.20%)

based on the averaging method, we do not include results on the validation set as there is no training phase specific for the ensemble itself. Indeed, the ensemble combines the predictions of its constituent models. The focus is to present the performance metrics derived from the ensemble's predictions on an independent test set.

All ensemble models achieve excellent results, with a coefficient of determination exceeding 0.88 and a MAPE of less than 5%. The performance of the Hybrid CNN-MLP ensemble is slightly superior to that of other ensemble methods across all evaluation metrics considered. Specifically, we observe an R^2 of 0.89 and an RMSE of 1843. in the test set.

Table 4. Results of the Hybrid CNN-MLP. Its performance are compared against a MLP, the Hybrid MLP-MLP, and with averaging ensemble models.

	val logRMSE	test logRMSE	test RMSE	test R^2	test MAPE
Hybrid CNN-MLP	0.46 (±0.06)	**0.43 (±0.14)**	**1843.84 (±586.70)**	**0.89 (±0.07)**	**4.45% (±1.08%)**
MLP	0.50 (±0.05)	0.47 (±0.10)	1968.97 (±535.38)	0.88 (±0.05)	4.87% (±0.92%)
Hybrid MLP-MLP	0.40 (±0.05)	0.44 (±0.12)	1971.96 (±399.38)	**0.89 (±0.06)**	4.74% (±0.96%)
MLP MLP ensemble	-	0.44 (±0.13)	1984.15 (±511.71)	0.88 (±0.06)	4.75% (±1.18%)
CNN MLP ensemble	-	0.44 (±0.11)	2027.48 (±609.43)	0.88 (±0.05)	4.51% (±1.05%)
MLP RF ensemble	-	0.44 (±0.11)	1976.06 (±550.66)	**0.89 (±0.05)**	4.58% (±1.12%)
CNN RF ensemble	-	0.44 (±0.10)	2151.46 (±614.18)	**0.89 (±0.04)**	4.58% (±1.06%)
LI-ESN MLP ensemble	-	0.46 (±0.13)	1991.90 (±507.22)	0.88 (±0.06)	4.77% (±1.20%)
LI-ESN RF ensemble	-	0.46 (±0.12)	2115.75 (±542.74)	**0.89 (±0.06)**	4.88% (±1.12%)

Figure 3 compares the performance of the ensemble models with that of the non-ensemble models having the best RMSE. The results indicate that using all inputs leads to improved performance compared to the case in which only absorbance is input. This is already confirmed by comparing the RMSE obtained by the MLP in the two cases. Furthermore, an enhancement in performance is observed when employing hybrid models (CNN-MLP and MLP-MLP) compared to averaging ensembles.

To gain a deeper insight into feature relevance, we conducted a feature importance analysis using the Random Forest model. The features considered include

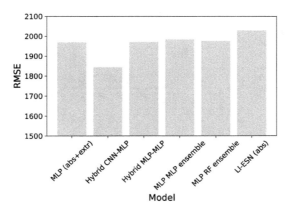

Fig. 3. Comparison of the models having the best performances in terms of RMSE.

the absorbance, extracted features, and the three scalar dilution factor, conductivity, and pH. Figures 4 and 5 illustrate the feature importance for COD prediction. Figure 4 depicts the importance of each wavelength in the absorbance, while Fig. 4 showcases the importance of the extracted features. The wavelengths between 225 nm and 275 nm of the absorbance emerge as the most important. Models confirms the relevance of the UV absorbance spectra in the estimation of COD. Indeed, organic compounds could absorb light in this area, and their presence could be related to higher COD values. Notably, among the additional features, pH, slope, AUC, and kurtosis are the most important. While the slope is well known in the literature for its relevance in the water quality estimation, the AUC is not as common as other features. This lays the basis for further investigation into the relevance of this feature from a chemical perspective.

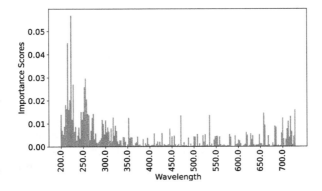

Fig. 4. Feature importance of the absorption spectrum

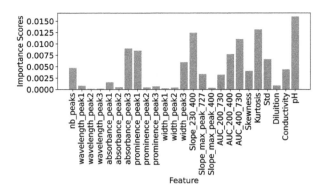

Fig. 5. Feature importance of the extracted features

In summary, the results underscore the slightly superior performance of hybrid ensemble models compared to ensemble models with averaging. Furthermore, the MLP incorporating additional features outperforms the MLP using only the absorbance as input. Particularly noteworthy is the hybrid CNN-MLP ensemble model, which stands out as the best-performing model across all the evaluation metrics considered.

8 Conclusion

In the context of wastewater quality assessment based on UV-Vis spectroscopy, we propose a novel soft sensor based on absorbance spectrum, features inspired to human-based interpretation, and additional data such as pH, dilution factor and conductivity. The soft sensor makes use of an hybrid ensemble models based on CNN and MLP.

We conducted experimental assessment using a double k-fold cross-validation approach to evaluate model performance over a real-world dataset collected to this purpose. The dataset collects data from various tanneries in Italy, overall comprising 244 samples (after outlier removal) with COD ranging from 21.2 mg/L to 92900 mg/L. An extensive preprocessing pipeline addresses the complexity and variability of sample composition and to extract new features.

Notably, the Hybrid CNN-MLP ensemble model achieves the best results with an RMSE of 1843.84, a R^2 of 0.89, and a MAPE of 4.45% on the test set. Furthermore, we investigated the most relevant features among the absorbance and the extracted features for COD prediction, by using the feature importance of the Random Forest model. Among these, we find a high importance of the UV absorbance spectra and other features such as the pH, AUC, kurtosis, and slope. In our evaluation and comparison among different solutions, we considered both models already tested in the literature and novel approaches to our specific case study. However, it is important to note that directly comparing our results with those reported in other works is challenging due to differences in dataset composition and data characteristics [7,25,26]. Our dataset comprises wastewater samples from tanneries, with COD values ranging from 21.2 mg/L to 92900 mg/L, contrasting with the literature where the maximum COD value is 500 mg/L [25,26].

Potential enhancements for this approach involve exploring the use of near-infrared spectroscopy. Additionally, extending these methodologies to drinking water analysis presents an interesting avenue for advancement.

References

1. Cao, H., Qu, W., Yang, X.: A rapid determination method for chemical oxygen demand in aquaculture wastewater using the ultraviolet absorbance spectrum and chemometrics. Anal. Methods **6**(11), 3799–3803 (2014)
2. Cardia, M., Chessa, S., Franceschi, M., Gambineri, F., Micheli, A.: Estimation of cod from UV-VIS spectrometer exploiting machine learning in leather industries wastewater. In: 8th World Congress on Civil, Structural, and Environmental Engineering. Avestia Publishing (2023). https://doi.org/10.11159/iceptp23.160
3. Cardia, M., Chessa, S., Franceschi, M., Gambineri, F., Micheli, A.: Machine learning for the estimation of cod from UV-VIS spectrometer in leather industries wastewater. Int. J. Environ. Pollu. Rem. **11**, 10–19 (2023)
4. Cardia, M., Chessa, S., Micheli, A., Luminare, A.G., Franceschi, M., Gambineri, F.: Multitarget wastewater quality assessment in a smart industry context. In: 20th International Conference on Intelligent Environments. IEEE (2024), to be published
5. Chen, H., Xue, H.Y., Liu, J., Li, Z., Hou, Y.: Research on COD detection method based on UV-VIS spectroscopy. In: International Symposium on Precision Mechanical Measurements, vol. 11343 (2019). https://doi.org/10.1117/12.2548522
6. Fortuna, L., Graziani, S., Rizzo, A., Xibilia, M.G., et al.: Soft Sensors for Monitoring and Control of Industrial Processes, vol. 22. Springer, London (2007). https://doi.org/10.1007/978-1-84628-480-9
7. Guan, L., Zhou, Y., Yang, S.: An improved prediction model for cod measurements using UV-VIS spectroscopy. RSC Adv. **14**, 193–205 (2024). https://doi.org/10.1039/D3RA05472A
8. Hartpence, B., Kwasinski, A.: CNN and MLP neural network ensembles for packet classification and adversary defense. Intell. Converged Netw. **2**(1), 66–82 (2021). https://doi.org/10.23919/ICN.2020.0023
9. ISO, C.S.: Water quality – Determination of the chemical oxygen demand index (ST-COD) – Small-scale sealed-tube method. Standard, International Organization for Standardization, Geneva, CH (2002)
10. Jaeger, H., Lukoševičius, M., Popovici, D., Siewert, U.: Optimization and applications of echo state networks with leaky- integrator neurons. Neural Netw. **20**(3), 335–352 (2007). https://doi.org/10.1016/j.neunet.2007.04.016
11. Kadlec, P., Gabrys, B., Strandt, S.: Data-driven soft sensors in the process industry. Comput. Chem. Eng. **33**(4), 795–814 (2009). https://doi.org/10.1016/j.compchemeng.2008.12.012
12. Korostynska, O., Mason, A., Al-Shamma'a, A.: Monitoring pollutants in wastewater: traditional lab based versus modern real-time approaches. Smart sensors for real-time water quality monitoring, pp. 1–24 (2013)

13. Larsen, D.A., Green, H., Collins, M.B., Kmush, B.L.: Wastewater monitoring, surveillance and epidemiology: a review of terminology for a common understanding. FEMS Microbes **2** (2021). https://doi.org/10.1093/femsmc/xtab011
14. Lyu, Y., Zhao, W., Kinouchi, T., Nagano, T., Tanaka, S.: Development of statistical regression and artificial neural network models for estimating nitrogen, phosphorus, cod, and suspended solid concentrations in eutrophic rivers using UV-vis spectroscopy. Environ. Monit. Assess.t **195** (2023). https://doi.org/10.1007/s10661-023-11738-0
15. Moon, T., Park, J., Son, J.E.: Prediction of the fruit development stage of sweet pepper (capsicum annum var. annuum) by an ensemble model of convolutional and multilayer perceptron. Biosystems Engineering **210**, 171–180 (2021). https://doi.org/10.1016/j.biosystemseng.2021.08.017
16. Mrkva, M.: Evaluation of correlations between absorbance at 254 nm and cod of river waters. Water Res. **17**(2), 231–235 (1983)
17. Pescod, M.: Wastewater characteristics and effluent quality parameters. Wastewater treatment and use in agriculture, Food and agriculture organization of the Unitied Nations (2013)
18. Qin, X., Gao, F., Chen, G.: Wastewater quality monitoring system using sensor fusion and machine learning techniques. Water Res. **46**(4), 1133–1144 (2012). https://doi.org/10.1016/j.watres.2011.12.005
19. Suetani, H., Kitajo, K.: Exploring individuality in human EEG using reservoir computing. In: Iliadis, L., Papaleonidas, A., Angelov, P., Jayne, C. (eds.) Artif. Neural Netw. Mach. Learn. ICANN 2023, pp. 551–555. Springer Nature Switzerland, Cham (2023)
20. Swinehart, D.F.: The beer-lambert law. J. Chem. Educ. **39**(7), 333 (1962). https://doi.org/10.1021/ed039p333
21. Tchobanoglus, G., Burton, F., Stensel, H.D.: Wastewater engineering: treatment and reuse. Am. Water Works Assoc. J. **95**(5), 201 (2003)
22. Van Den Broeke, J., Langergraber, G., Weingartner, A.: On-line and in-situ UV/VIS spectroscopy for multi-parameter measurements: a brief review. Spectrosc. Eur. **18**(4), 15–18 (2006)
23. Victor, R., Kotter, R., O'Brien, G., Mitropoulos, M., Panayi, G.: Who guidelines for the safe use of wastewater, excreta and greywater, vol. 1–4 (2008)
24. Vozzi, F., et al.: Echo state networks for the recognition of type 1 Brugada syndrome from conventional 12-lead ECG. Heliyon **10**(3) (2024)
25. Xu, Z., et al.: Data fusion strategy based on ultraviolet-visible spectra and near-infrared spectra for simultaneous and accurate determination of key parameters in surface water. Spectrochim. Acta Part A Mol. Biomol. Spectrosc. **302**, 123007 (2023). https://doi.org/10.1016/j.saa.2023.123007
26. Ye, B., et al.: Water chemical oxygen demand prediction model based on the CNN and ultraviolet-visible spectroscopy. Front. Environ. Sci. **10** (2022). https://doi.org/10.3389/fenvs.2022.1027693
27. Zhang, J., et al.: Multi-sensor fusion and feature selection in ultraviolet-visible spectrometry system for predicting chemical oxygen demand. In: 11th IEEE International Conference on Control & Automation (ICCA), pp. 904–907 (2014). https://doi.org/10.1109/ICCA.2014.6871041
28. Zhou, M., Li, Y., Lu, H.: Ensemble-based deep metric learning for few-shot learning. In: Farkaš, I., Masulli, P., Wermter, S. (eds.) ICANN 2020. LNCS, vol. 12396, pp. 406–418. Springer, Cham (2020). https://doi.org/10.1007/978-3-030-61609-0_32

Short-Term Forecasting of Wind Power Using CEEMDAN-ICOA-GRU Model

Yun Wu[1], Wei Zheng[1(✉)], Yongbin Zhao[2], Jieming Yang[1], Ning An[3], and Dan Feng[4]

[1] Department of Computer Science, Northeast Electric Power University, Jilin, China
19818987537@139.com
[2] State Grid Baicheng Power Supply Company, Jilin, China
[3] Northeast Branch of State Grid Corporation of China, Liaoning, China
[4] Chaoyang Service Center of Ecology and Environment, Liaoning, China

Abstract. Accurate wind power prediction plays a vital role in ensuring the safe operation of wind power connected to the grid. To improve the prediction accuracy of wind power, a short-term wind power prediction model (CEEMDAN-ICOA-GRU) based on a Complete Ensemble Empirical Mode Decomposition with Adaptive Noise (CEEMDAN) combined with improved Coati Optimization Algorithm(ICOA) to optimize Gated Recurrent Unit (GRU) is proposed in this paper. First, CEEMDAN was used to decompose the original wind power data, reduce its volatility, and reduce the lag of the forecast curve. Then, to solve the problem that the traditional Coati Optimization Algorithm(COA) is prone to locally optimal solutions, chaotic sequences are added to improve the initial population distribution to be more uniform. Finally, ICOA is used to optimize the hyperparameters of GRU to obtain the optimal prediction model, which is used to predict different sub-sequences, and the prediction results are superimposed to obtain the final prediction results. To verify the validity of the model, a large number of experiments were conducted using the data set of a wind power plant in Turkey in 2022. The results show that CEEMDAN-ICOA-GRU can effectively improve the accuracy of wind power prediction.

Keywords: Wind power forecasting · CEEMDAN · Coati optimization algorithm · Gated recurrent unit

1 Introduction

Due to the advantages of high environmental protection and low cost, wind energy has become the main force of new energy development [1]. At the same time, the intermittency and volatility of wind have brought great limitations to the utilization rate of wind power. Therefore, accurate and efficient short-term wind power forecasting is of great significance to the stable operation of the power system [2]. Wind power forecasting methods are mainly divided into physical methods, statistical methods, and combined forecasting methods [3].

The physical method require high accuracy of historical meteorological data and the model solution process is complex, which leads to poor performance of wind power prediction [4]; statistical methods, such as probabilistic autoregression, have simple model structure and fast calculation speed, and can obtain better prediction results through a large number of historical data, but their prediction accuracy decreases rapidly with the increase of time series [5]; the combined forecasting methods are the optimization and fusion improvement of the input, output and model itself of the first two methods, and have powerful nonlinear mapping function, so they are much more widespread now [6].

In recent years, many researchers have focused on the combined prediction model of decomposition-prediction-reconstruction. In literature [7], the original wind power sequence was decomposed into N sub-sequences with different center frequencies by Variational Mode Decomposition (VMD), and then each sub-sequence was divided into high-frequency and low-frequency components by frequency run discrimination. Finally, two different combined prediction methods are used to predict the high-frequency and low-frequency components respectively, to improve the power prediction accuracy; literature [8] proposes a time convolutional network model with improved Complete Ensemble Empirical Mode Decomposition with Adaptive Noise (CEEMDAN) and time-mode attention mechanism, which is also a decomposition-prediction-reconstruction wind power combination prediction model. First, CEEMDAN is used to decompose the original sequence of wind power, and then the combined model is used to predict each component, and good prediction results are obtained; literature [9] proposed a combined prediction model of Support Vector Machine(SVM) optimized by Whale Optimization Algorithm(WOA). The WOA was introduced to solve the difficult problem of learning parameter selection in SVM, and then the WOA-SVM prediction model was constructed for each subsequence. The results show that the wind power prediction accuracy is relatively high; in literature [10], a new probability density prediction model based on quantile regression and Temporal Convolutional Network (TCN) is established by combining quantile regression with TCN, and combining attention mechanism with Long Short-term Memory Networks(LSTM). It provides a new idea for the traditional short-term power forecast of wind power; literature [11] proposes a short-term wind power prediction method by decomposing the original wind power data by VMD technology and using the improved squirrel algorithm to optimize the quantile regression of Gated Recurrent Unit (GRU). This method reduces the randomness and uncertainty of wind power data and shortens the operation time.

The above combined forecasting method has improved the forecasting accuracy, but there are still some limitations: (1) When wind power data is forecast as time series data, there is often a lag in forecasting results. (2) It is still difficult to determine the hyperparameters of the traditional prediction model. (3) Optimization algorithms seeking hyperparameters are prone to problems such as local optimal solutions.

In conclusion, this paper proposes a prediction model based on CEEMDAN combined with an improved Coati Optimization Algorithm(ICOA) to optimize GRU (CEEMDAN-ICOA-GRU). Firstly, by CEEMDAN technology, wind power data is decomposed into sub-sequences of different modes, which can greatly reduce the influence caused by the randomness of wind power data and solve the lag caused by time series data prediction; then, the population initialization of the chaotic sequence is used to improve the COA, which helps the algorithm get rid of the local extreme value and expand the global optimization ability; finally, ICOA was used to optimize the two hyperparameters of GRU, the number of hidden layers and the number of batches, and the optimized GRU model was used to predict each sub-sequence, and the final wind power prediction result was obtained by superposition of the predicted results. The final prediction results of the CEEMDAN-ICOA-GRU model are compared with other excellent models to verify the validity of the model proposed in this paper.

2 Relevant Principles and Methods

2.1 CEEMDAN Decomposition Algorithm

Empirical Mode Decomposition (EMD) is an adaptive signal processing method for nonlinear and non-stationary signals. It does not need to set other basis functions to decompose the signal directly according to the characteristics of the data itself, but it has a mode aliasing phenomenon [12]; after signal decomposition, the Ensemble Empirical Mode Decomposition (EEMD) can obtain a high-frequency signal that reflects the fluctuation of the original sequence and a low-frequency residual signal that reflects its stability. Although the whole process can effectively overcome the mode aliasing defect of the EMD method, it can not effectively eliminate the white noise signal, so the reconstruction error is too large [13].

Compared with EEMD, CEEMDAN has a faster decomposition speed and greatly reduced computation amount. It performs average calculations after obtaining the first IMF. This processing method effectively avoids the transmission of white noise signals from high frequency to low frequency, thus affecting the final analysis and processing effect [14]. CEEMDAN adds a finite number of white noise signals in each subsequent decomposition stage, which increases the amount of computation to some extent but can effectively reduce the reconstruction error of EEMD [15].

2.2 Coati Optimization Algorithm

The COA is an optimization algorithm based on coatis' behavior proposed in 2022 [16], the algorithm is optimized by simulating the hunting behavior of coatis and has the characteristics of strong optimization ability and fast convergence speed. The specific process is as follows:

2.2.1 Initialization

In the optimization space, the formula (1) is used to randomly initialize the population:

$$x_{i,j} = lb_j + r \cdot (ub_j - lb_j) \tag{1}$$

where $x_{i,j}$ is an individual, lb_j is the lower boundary for optimization, ub_j is the upper boundary for optimization, and r is the random number between [0,1].

2.2.2 Hunting and Attacking Phases

The hunting and attacking phases of updating the coati population in the search space are modeled based on their strategy when attacking iguanas, with a group of coatis climbing a tree to scare one iguana and several others waiting under the tree until the iguana lands and the coati attacks and hunts it. This strategy causes the coati to move to a different location in the search space. Assuming that the location of the best member of the population is that of the iguana, and also assuming that half of the coatis climb the tree and the other half wait for the iguana to fall to the ground. Therefore, the position of the coati rising from the tree is mathematically simulated using Eq. (2).

$$X_i^{P1} : x_{i,j}^{P1} = x_{i,j} + r \cdot \left(\text{Iguana}_j - I \cdot x_{i,j} \right), \text{ for } i = 1, 2, \ldots, \left\lfloor \frac{N}{2} \right\rfloor \text{ and } j = 1, 2, \ldots, m. \tag{2}$$

After the iguana falls to the ground, it is placed in a random location in the search space. Based on this random location, coatis on the ground move through the search space, which is simulated using Eqs. (3) and (4).

$$\text{Iguana}^G : \text{Iguana}_j^G = lb_j + r \cdot (ub_j - lb_j), j = 1, 2, \ldots, m. \tag{3}$$

$$X_i^{P1} : x_{i,j}^{P1} = \begin{cases} x_{i,j} + r \cdot \left(\text{Iguana}_j^G - I \cdot x_{i,j} \right), & F_{\text{Iguana}}^G < F_i, \\ x_{i,j} + r \cdot \left(x_{i,j} - \text{Iguana}_j^G \right), & \text{else}, \end{cases}$$

$$for \ i = \left\lfloor \frac{N}{2} \right\rfloor + 1, \left\lfloor \frac{N}{2} \right\rfloor + 2, \ldots, N \text{ and } j = 1, 2, \ldots, m. \tag{4}$$

If the updated individual is better, then Eq. (5) is used to update the current individual; otherwise, keep it as it is:

$$X_i = \begin{cases} X_i^{P1}, & F_i^{P1} < F_i \\ X_i, & \text{else}. \end{cases} \tag{5}$$

Here X_i^{P1} is the updated position of the ith coati, $x_{i,j}^{P1}$ is its jth dimension, F_i^{P1} is its objective function value, F_i is the objective function value obtained based on the ith coati, r is a random real number between [0,1]. $Iguana$ is the position of the prey iguana, actually refers to the position of the best individual in the population, $Iguana_j$ is its jth dimension, j is an integer, randomly selected from the set {1,2}, $Iguana^G$ is the position on the ground, it is randomly generated, $Iguana_j^G$ is its jth dimension, F_{Iguana}^G is his objective function value.

2.2.3 Escape from Predators Phase

The escape from predators phase is used to update the coatis' location and is mathematically modeled based on the natural behavior of coatis when they encounter and flee a predator. When a predator attacks a coati, it flees from its position, putting it in a safe position close to its current location. To simulate this behavior, random positions are generated near the location of each coatis based on Eqs. (6) and (7).

$$lb_j^{local} = \frac{lb_j}{t}, ub_j^{local} = \frac{ub_j}{t}, \text{ where } t = 1, 2, \ldots, T \tag{6}$$

$$X_i^{P2} : x_{i,j}^{P2} = x_{x,j} + (1-2r) \cdot \left(lb_j^{local} + r \cdot \left(ub_j^{local} - lb_j^{local}\right)\right), \\ i = 1, 2, \ldots, N, j = 1, 2, \ldots, m. \tag{7}$$

where t is the number of iterations, lb_j^{local} and ub_j^{local} are the upper, and lower bounds of the jth dimensional variable updated with the number of iterations, if the updated individual is better, then Eq. (8) is used to update the current individual; otherwise, keep it as it is.

$$X_i = \begin{cases} X_i^{P2}, F_i^{P2} < F_i \\ X_i, \text{ else } . \end{cases} \tag{8}$$

2.3 Gated Recurrent Unit

GRU is an improvement of the LSTM [17]. It combines the forget gate and the input gate of the LSTM unit into an "update gate", while controlling the amount of information that the current state needs to retain from the previous state and the amount of information that needs to be received from the candidate state. The reset gate is used to control the degree of forgetting of the previous moment [18]. GRU is widely used because of its special gate structure, which makes GRU better than LSTM in prediction accuracy and prediction speed in some application scenarios [19].

The structure of GRU neurons is shown in Fig. 1. The values of the update and reset gates are determined by the previously hidden state h_{t-1} and the current input x_t.

The data transmission process in GRU neurons can be described as follows:

$$r_t = \sigma\left(W_r \cdot [h_{t-1}, x_t]\right) \tag{9}$$

$$z_t = \sigma\left(W_z \cdot [h_{t-1}, x_t]\right) \tag{10}$$

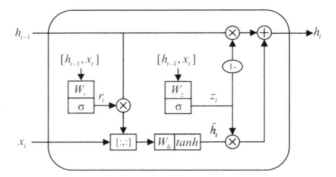

Fig. 1. The framework of GRU.

$$\widetilde{h}_t = tanh\left(W_h \cdot ([r_t \odot h_{t-1}, x_t])\right) \quad (11)$$

$$h_t = (1 - z_t) \odot h_{t-1} + \widetilde{h}_t \odot z_t \quad (12)$$

The two activations σ (sigmoid) and $tanh$ are used to simulate the gate and normalize the input, respectively. r_t and z_t are the outputs of the reset and update gates, respectively. $\widetilde{h}_t$ is the candidate hidden state, h_t is the hidden state, W_r, W_z and W_h are the weight matrices that will be updated by optimization.

3 CEEMDAN-ICOA-GRU Model

This paper presents a combined prediction model of wind power. First, the CEEMDAN method was used to decompose the original wind power data, and multiple IMF components were obtained. Secondly, the COA was improved by using a chaotic sequence to initialize the population. Finally, the ICOA was used to optimize the two hyperparameters of the GRU, the number of hidden layers and the number of batches, and the optimal parameters were obtained and input into the model to obtain the optimal model. The optimal model was used to predict each IMF component and the final wind power prediction result was obtained by superposition of the prediction results. This comprehensive model can avoid the predicted value lagging behind the actual value, effectively improve the prediction accuracy of the prediction algorithm, and provide a reliable tool for wind power prediction.

3.1 CEEMDAN Data Decomposition

Because of the strong fluctuation of wind power, the wind power sequence can be regarded as a superposition of subsequences of single frequency components. In this paper, the original wind power sequence is firstly extended by CEEMDAN decomposition to obtain the decomposed sub-sequence, and the original wind speed sequence is decomposed into a series of IMF ranked from high frequency to low frequency. The specific steps are as follows [20]:

(1)Add white noise to the original wind power sequence $C_{(t)}$, the formula is:

$$C_{j(t)} = C_{(t)} + \beta_k n_{j(t)} \tag{13}$$

In Eq. (13), t is the number of decomposition of the original power; $C_{j(t)}$ is the latest obtained power sequence; β_k is the kth signal-to-noise ratio; $n_{j(t)}$ is the white Gaussian noise added for the jth time.

(2)The residual signal $R_{1(t)}$ is calculated using the original power sequence, the formula is:

$$R_{1(t)} = C_{(t)} - imf_{1(t)} \tag{14}$$

In Eq. (14): $imf_{1(t)}$ is the first modal component obtained by decomposition.

(3)Add positive and negative pairs of white Gaussian noise in $R_{1(t)}$, the new signal $R_{1(t)} + \beta_1 E_1 \left[n_{j(t)} \right]$ is decomposed for N times, and the mean is calculated. The decomposed second modal component imf_2 and the second residual signal $R_{2(t)}$ are respectively obtained from Eqs. (15) and (16):

$$imf_2 = \frac{1}{N} \sum_{j=1}^{N} E_1 \left\{ R_{1(t)} + \beta_1 E_1 \left[n_{j(t)} \right] \right\} \tag{15}$$

$$R_{2(t)} = R_{1(t)} - imf_2 \tag{16}$$

(4)Then the calculation process of step 3 is repeated, and the k+1 modal component imf_{k+1} and the kth residual signal R_k are obtained by CEEMDAN.

(5)After repeating the above steps, several IMF components are obtained and the corresponding residual components satisfying the conditions are calculated, that is, no meaningful IMFs can be extracted anymore. The final signal is decomposed into:

$$C_{(t)} = R_{(t)} - \sum_{k=1}^{K} imf_{k(t)} \tag{17}$$

3.2 ICOA Optimization Algorithm

To alleviate the problems of complex parameter adjustment process and overfitting in GRU model. In this paper, the COA is used to optimize the model. At the same time, the traditional COA is easy to fall into the local optimal problem due to the decrease of population diversity in the late iteration period, and the chaotic sequence is proposed to improve COA. By introducing a chaotic sequence, the quality of the initial solution is improved, the location distribution of the initial population is more uniform, and the population diversity is increased.

The chaotic mapping expression is as follows:

$$z_{i+1} = (2z_i) \bmod 1 + \text{rand}(0,1) \times (1/NT) \qquad (18)$$

where NT is the number of particles in the chaotic sequence, rand(0,1) is the random number between [0,1], z_i and z_{i+1} are the original population and the updated population respectively, mod refers to the modular operation. According to the chaotic mapping characteristics, the steps to generate chaotic sequences in the domain are as follows:

(1) Randomly generate the initial value z_0 in (0,1), denoting $i = 0$;
(2) Using the above formula for iteration, the Z sequence is generated, and i increases by 1;
(3) If the number of iterations meets the maximum number, the program stops running, and the resulting Z sequence is saved and used as the initial population of the COA.

3.3 ICOA Optimizes GRU Prediction Model

To achieve better training efficiency, ICOA and GRU are integrated. The two hyperparameters, the number of hidden layers G and the number of batch processing A, are assumed to be the initial positions of iguanas. The improved COA is introduced to optimize G and A, where the number of hidden layer units G ∈ [20,100] and the number of batch processing A ∈ [10,100], ICOA optimizes the GRU process as follows:

(1) Set the initial population size of coati, search space dimension, and maximum number of iterations;
(2) Update the parameter positions(G,A) by simulating the hunting and aggressive behavior of coati;
(3) The GRU model is trained on the new location and its performance on the validation set is evaluated, then the new and old locations are compared and the better one is selected;
(4) Then simulate predator escape behavior to further adjust the position;
(5) Repeat this process until the termination condition is met, and finally output the optimal parameter value.

The optimal parameter values obtained are input into the GRU to obtain the optimal GRU, and then the optimal GRU is used as the main body of the model to respectively predict the wind power of the sub-sequences obtained by CEEMDAN decomposition. Finally, the obtained sub-sequence prediction results are superimposed to obtain the final wind power prediction results.

Figure 2 is the flow chart of the short-term wind power forecasting using CEEMDAN-ICOA-GRU model, IMF1~IMFn are the different components of decomposition, and the data are the original wind power.

4 Experimental Results and Analysis

4.1 Experimental Environment and Related Parameters

Details of the experimental environment in this paper are shown in Table 1.

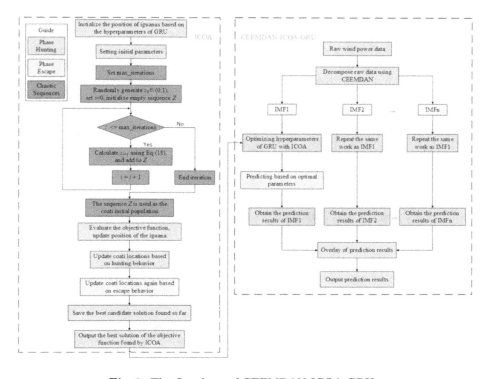

Fig. 2. The flowchart of CEEMDAN-ICOA-GRU.

Table 1. Experimental environment

Configuration item	Version
Python	3.8
Deep learning framework	Pytorch 1.8
Operating system	Windows 10
Memory size	16GB
CPU	Intel(R)Core(TM)i5-8700H
GPU	NVIDIA GeForce RT1050ti

4.2 Experimental Dataset

The dataset in this paper is from the wind power data of a wind power plant in Turkey in 2022. The time granularity is selected as the resource monitoring situation of the power plant every 10 min, with a total of 50531 (one year) data. The model is divided into a training set and a testing set at the ratio of 8:2 for training and testing. The basic parameters are set as follows: feature_size is set to 1; output_size is set to 1; the number of layers of the GRU, num_layers is set to 3; learning_rate is set to 0.01; the loss function is set to MSELoss().

4.3 CEEMDAN Results Analysis

Before the prediction began, the CEEMDAN decomposition algorithm was first used to decompose part of the original wind power data, as shown in Fig. 3. In Fig. 3, the original data is decomposed into 8 IMF components with different frequencies. IMF components are ranked from high frequency to low frequency and maintain a high degree of independence among components.

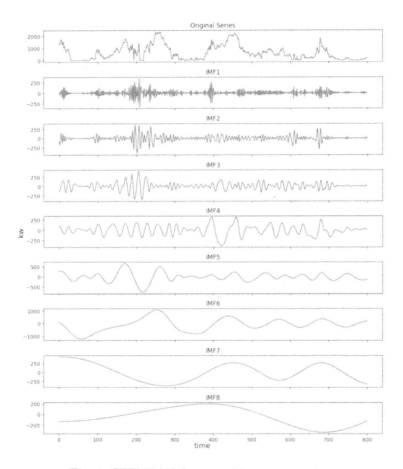

Fig. 3. CEEMDAN decomposition components.

Among them, the high-frequency signal component is more volatile, which reflects the greater fluctuation of the original power series, and will produce a large error in the prediction. The low-frequency signal component is relatively stable, close to a smooth curve, which reflects the stable change trend of the original sequence, and can fit the effect better in the prediction.

The prediction error of wind power is mainly determined by the high-frequency mode prediction error. Compared with the direct prediction using the original characteristics of wind power, CEEMDAN can reduce the influence of non-stationarity and complexity of wind power on the prediction accuracy, thus improve the prediction accuracy of the model.

4.4 Evaluation Index

In this paper, two indexes are mainly used to evaluate the performance of the prediction model when predicting wind power, namely mean square error (MSE) and mean absolute error (MAE) [2]. The calculation method is shown in Eqs. (19) and (20).

$$MSE = \frac{1}{n} \sum_{i=1}^{n} (\hat{y}_i - y_i)^2 \qquad (19)$$

$$MAE = \frac{1}{n} \sum_{i=1}^{n} |\hat{y}_i - y_i| \qquad (20)$$

where n is the prediction length, $\hat{y}_i$ is the actual value, and y_i is the predicted value. The lower the value of the two evaluation indexes, the better the model performance.

4.5 Comparative Experiments

The following comparison experiments will be carried out from two aspects, namely the analysis of the lag of the prediction results and the analysis of the prediction results of the combination model.

4.5.1 Analysis of the Lag of the Prediction Results

Due to the lack of readability of the first few sub-sequence images, to get a better view of the subsequence prediction, this paper selects the prediction curves of the last five subsequence models obtained by decomposition and lists them. The prediction situation is shown in Fig. 4.

It can be seen that the prediction error always exists in the places where the power change is more intense, the regularity of these places is weak, and the prediction of the model for these places is difficult to grasp. However, for paragraphs with moderate changes, the model shows a good fit.

To verify that the CEEMDAN method mentioned in this paper can improve the autocorrelation of GRU in predicting time series and reduce the lag of prediction curve, the prediction results of the four models of single prediction model GRU, single prediction model LSTM, CEEMDAN-GRU and CEEMDAN-LSTM were compared and analyzed with the actual value curve, and the prediction results are shown in Fig. 5 and Fig. 6.

Combining with Figs. 5 and 6, it is obvious that LSTM and GRU will have the problem of predictive value lag in prediction. After data decomposition,

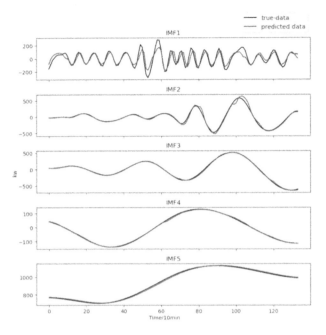

Fig. 4. The case of subsequence prediction.

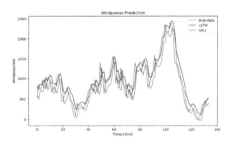

Fig. 5. Single model prediction results

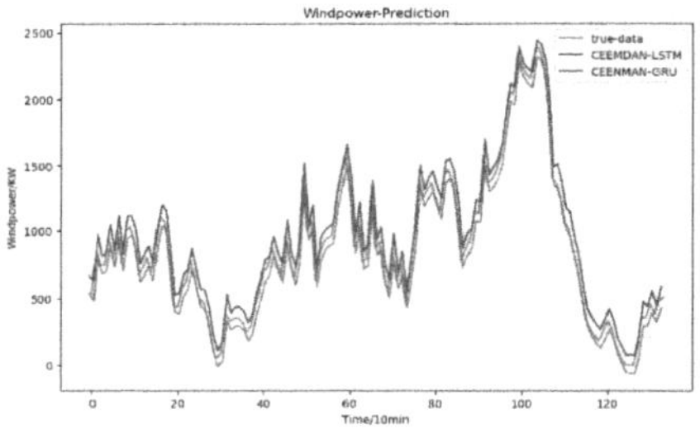

Fig. 6. Prediction results after decomposition

CEEMDAN-GRU and CEEMDAN-LSTM will solve part of the curve lag and appropriately improve the accuracy.

4.5.2 Analysis of the Prediction Results of the Combined Model
(1)Analysis of Experimental Results of COA
To verify that the COA can improve the prediction accuracy of the model, this paper compares and analyzes the prediction results of five models: GRU, LSTM, ANN, RNN, and COA-GRU, as shown in Fig. 7, and the error indicators are shown in Table 2.

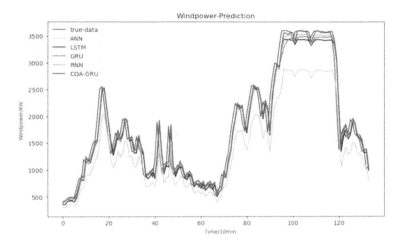

Fig. 7. The prediction results of COA-GRU and other models.

Table 2. Analysis of predictive performance of COA-GRU and other models

Model	MSE	MAE	Running time /s
ANN	1.035	1.305	195
RNN	0.688	0.739	189
LSTM	0.523	0.691	183
GRU	0.534	0.615	**165**
COA-GRU	**0.459**	**0.544**	286

Combining the above experimental data and figures, the following can be obtained: the basic GRU and LSTM are better than ANN and RNN in prediction accuracy, and the running time of the GRU is shorter than that of the LSTM. The MSE of the COA-GRU model is reduced by 14% compared with the basic GRU, which shows that COA has higher optimization accuracy for the number of hidden layers and batch processing of the GRU prediction model. The accuracy of the prediction model is improved.

(2) Analysis of Experimental Results of the CEEMDAN-ICOA-GRU

Since the basic COA uses a random generation method for population initialization, it will lead to an uneven distribution of the coati population. Chaotic map has the characteristics of regularity and ergodicity, so this paper uses chaotic sequences to initialize the initial position of the coatis to improve the quality of the initial solution. The chaotic sequence distribution is shown in Fig. 8, and it can be found that the initial population distribution is relatively uniform.

To prove that the proposed model has a great improvement in accuracy, the real data curve is compared with the prediction results of five models: COA-GRU, ICOA-GRU, CEEMDAN-WOA-GRU, CEEMDAN-COA-GRU, and

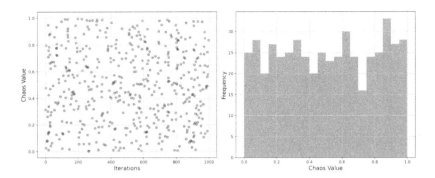

Fig. 8. Chaotic sequence distribution.

CEEMDAN-ICOA-GRU (the method in this paper). The results are shown in Fig. 9, and the error indicators are shown in Table 3.

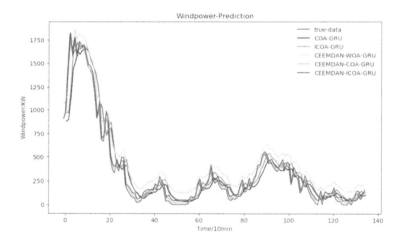

Fig. 9. The prediction results of different combined forecasting models.

As can be seen from the table, the results show that the MSE of ICOA-GRU and COA-GRU is reduced by 13%, indicating that the ICOA improves the prediction accuracy of the GRU more than the original COA in accuracy, that is, the introduction of chaotic sequence to optimize COA proposed in this paper has advantages; compared with COA-GRU, the MSE of CEEMDAN-COA-GRU is reduced by 34%, indicating that CEEMDAN can also improve the accuracy; compared with CEEMDAN-WOA-GRU, the MSE of CEEMDAN-COA-GRU is reduced by 29%, and the running time is reduced by 5%, which reflects the superiority of COA algorithm compared with other optimization algorithms in prediction accuracy.

Table 3. Analysis of predictive performance of combined models

Model	MSE	MAE	Running time/s
COA-GRU	0.443	1.526	283
ICOA-GRU	0.384	0.398	**279**
CEEMDAN-WOA-GRU	0.412	0.496	314
CEEMDAN-COA-GRU	0.293	0.342	298
CEEMDAN-ICOA-GRU	**0.236**	**0.227**	291

At the same time, the MSE of the CEEMDAN-ICOA-GRU joint prediction model and CEEMDAN-COA-GRU is reduced by 19%. In general, the experimental results show that the proposed combined prediction model has better prediction accuracy than the above prediction model, although the running time is slightly increased compared to the other models.

5 Conclusions

The following conclusions are obtained through comparative experiments:

(1) CEEMADAN decomposes the original wind power data to obtain the stable component, strengthens the ability of the model to learn and capture the time series, greatly solves the problem of large errors caused by the non-linear original data instability, and improves the data autocorrelation. Finally, it reduces the lag of the prediction curve and improves the accuracy of the prediction model.

(2) This paper proposes a short-term power prediction model for wind power based on CEEMDAN-IOA-GRU. The chaotic sequence is introduced to improve the COA to improve the quality of the initial population and overcome the shortcomings of the algorithm that is easy to fall into local extremum in the optimization process. At the same time, ICOA is used to optimize the relevant hyperparameters of the GRU model, which can effectively improve its accuracy. Compared with other prediction models, the test results show that the proposed method can make accurate and reliable judgments for wind power prediction, and has good application value.

In future work, we will further optimize the model structure, such as trying to use Bi-GRU to make it consider both past and future information, so as to better extract the underlying features of time series data and improve prediction accuracy. In addition, trying to add the self-attention mechanism to the model may further enhance model efficiency and computing time.

Acknowledgements. This work is supported by the Science and Technology Projects of Northeast Branch of State Grid Corporation of China(No:52992623001P).

References

1. Dai, Q.B., Huang, N.T.: Review of research on error correction of short-term wind power prediction. J. Northeast Electr. Power Univ. **43**(02), 1–7 (2023)
2. Zhang, H.F., Li, D.X., Li, X.J.: Impact of large-scale wind power integration on space-time distribution of power system dynamic frequency response. J. Northeast Electr. Power Univ. **42**(05), 74–82 (2022)
3. Yang, M., Yu, X.N.: An ultra-short term combined prediction considering wind farm power climbing. J. Northeast Electr. Power Univ. **42**(01), 63–70 (2022)
4. Liu, H.b., Gai, X.Y., Sun, L.: Ultra-short-term wind power prediction considering data reduction and data cleaning. J. Northeast Electr. Power Univ. **43**(04), 1–9 (2023)
5. Liu, Z., Liu, D.: Wind power ultra-short-term forecasting based on multiple regression-ARIMA-Kalman combination model. Wind Energy **06**, 74–78 (2018)
6. Wu, Y.H., Wang, Y.S., Xu, H.: Survey of wind power output power forecasting technology. Comput. Sci. Explor. **16**(12), 2653–2677 (2022)
7. Wang, R., Ran, F., Li, J.: Wind power prediction based on run discriminant method and VMD residual correction. J. Hunan Univ. (Nat. Sci. Ed.) **49**(08), 128–137 (2022)
8. Shi, J.R., Zhao, D.M., Wang, L.H.: Short-term wind power prediction based on RR-VMD-LSTM. Power Syst. Prot. Control **49**(21), 63–70 (2021)
9. Yue, X.Y., Peng, X.G., Lin, L.: Short-term wind power forecasting based on whales optimization algorithm and support vector machine. Proc. SCU-EPSA **32**(02), 146–150 (2020)
10. Pang, H., Gao, J.F., Du, Y.H.: A short-term load probability density prediction based on quantile regression of time convolution network. Power System Technol. **44**(04), 1343–1350 (1998)
11. Feng, S.C., Guo, J.C., Fu, H.: Wind power probability density prediction based on ISSA and GRU quantile regression. Adv. Technol. Electr. Eng. Energy **42**(10), 55–65 (2023)
12. Mohammad, A.: An efficient EMD-Based reversible data hiding technique using dual Stego images. Comput. Mater. Continua **75**(01), 1139–1156 (2023)
13. Ma, Q., Ye, R.X.: Short-term prediction of the intermediate point temperature of a supercritical unit based on the EEMD-LSTM method. Energies **17**(04), 949 (2024)
14. Huang, J.Z., Yin, Y.h., Tang, M.X: Short-term photovoltaic output prediction based on CEEMDAN and improved LSTM. Electr. Energy Mana. Technol. **10**, 36–43 (2023)
15. Hong, Y., Wang, D., Su, J.M.: Short-term power load forecasting in three stages based on CEEMDAN-TGA model. Sustainability **15**(14), 11123 (2023)
16. Dehghan, M., Montazeri, Z.: Coati optimization algorithm: a new bio-inspired metaheuristic algorithm for solving optimization problems. Knowl. Based Syst. **259**, 110011 (2023)
17. Liu, X.L., Lin, Z., Feng, Z.M.: Short-term offshore wind speed forecast by Seasonal ARIMA - A comparison against GRU and LSTM. Energy **227**, 120492 (2021)
18. Wang, Y.Q., Gui, R.Z.: A hybrid model for GRU ultra-short-term wind speed prediction based on Tsfresh and sparse PCA. Energies **15**(20), 7567 (2022)

19. Xiao, F., Ping, X., Li, Y.Y.: The short-term prediction of wind power based on the convolutional graph attention deep neural network. 1 State Grid Hubei Electric Power Research Institute ,Wuhan, 430077 ,China;2 College of Energy and Electrical Engineering, Hohai University ,Nanjing, 210098 ,China, **121**(02), 359–376 (2024)
20. Wang, D., Pan, C., Lu, L.: Source-storage staged planning strategy considering wind-photovoltaic timing related characteristics. J. Northeast Electr. Power Univ. **40**(04), 1–10 (2020)

City Planning

Predicting City Origin-Destination Flow with Generative Pre-training

Mingwei Zhang[1]([✉]), Lizhong Gao[1], Qiao Wang[1], and Weihao Gao[2]

[1] Southeast University, Nanjing, China
{zhmw,lzgao,qiaowang}@seu.edu.cn
[2] Tsinghua International Graduate School, Tsinghua University, Beijing, China

Abstract. Predicting Origin-Destination (OD) flow is a critical issue in the construction and management of smart cities, with challenges mainly stemming from the complex spatial-temporal dependencies in urban environments for three reasons: first, a considerable number of city grids are interrelated; second, the diversity of travel modes leads to highly imbalanced feature distributions in crowd flow data; lastly, crowd flow and geographical indicators have potential correlations. Inspired by natural language processing, we apply the concept of Generative Pre-training (GPT) to OD flow prediction. The proposed OD-GPT model frames grid sequence prediction as next token prediction in language models. Extensive testing on large-scale mobile signaling datasets from two real-world cities demonstrates the effectiveness of the model. In comparison to baseline models, it exhibits a notable enhancement in one-step prediction accuracy. Furthermore, we explore the model's multi-step prediction capability and research its adaptability to unconventional day through fine-tuning. The analysis indicates that our model can extract latent geographical features from OD data and generate reasonable embeddings with precise dynamic spatial-temporal dependencies.

Keywords: Origin-Destination flow prediction · generative pre-training · grid embedding · spatial-temporal data analysis

1 Introduction

In the context of smart cities, predicting crowd flow can effectively enhance municipal functions to serve the public. OD data is used to reveal trends in large-scale population movement [2]. OD flow prediction aims to forecast the flow from a specific origin to another specific destination in the next time period given historical OD data [8], which leads to efficient resource allocation, traffic management, and epidemic control, benefiting more people. Additionally, as there is a potential connection between crowd flow and static built environment information, exploring these connections can also contribute to related work on OD flow prediction. Several models have been proposed for predicting OD flow, such as the Long Short-Term Memory (LSTM) model focusing on time series analysis, or the Graph Neural Network (GNN) [9,13] model considering the inherent topological structure between grids. The main issues in this task are:

- **Issue 1: Global dynamic correlation.** Compared to predicting subway or taxi flows, which typically involve no more than a few hundred grids and relatively constant speeds, predicting general OD flows involving multiple modes of transportation is more challenging[2]. For the overall flow of a city, a vast urban area can lead to a significant increase in the number of grids, resulting in sparsity of OD pairs. Moreover, the correlations between grids dynamically change over time [3]. In most GNN models, the adjacency graph or node weight graph constructed from external data sources is usually static.
- **Issue 2: Complex flow distribution.** Firstly, the diversification of modes result in more imbalanced data distributions and frequent occurrences of extreme values, making it difficult for a simple model to capture global correlations [1,5]. Secondly, a finest model should be able to adapt to predicting unconventional days, especially during annual holidays, where the crowd flow exhibits different dynamics compared to regular days and lacks sufficient historical data.
- **Issue 3: Information fusion.** In fact, there is an inherent connection between crowd flow and the built environment, and exploring this can help in model construction [3]. One the one hand, we can use flow data to drive the acquisition of land use characteristics in a certain area; on the other hand, we can also inversely utilize geographic information to construct crowd flow prediction models. Traditional regression methods cannot solve the complex underlying relationship between regional indicators and OD flows [2], while GNNs waste a lot of resources for the computation of sparse OD matrices.

This paper proposes a novel model named OD-GPT, to address the aforementioned issues. Firstly, after utilizing the causal self-attention mechanism to learn the dynamic correlations in grid sequence data, the model is used for one-step prediction of historical OD data. Secondly, in the face of unconventional days with significantly different crowd flow trends, we evaluate the adaptability of the model using a strategy of freezing and fine-tuning. Finally, combining urban geographical indicators, we visualize and verify the effectiveness of the learned embedding layer of the model. We conduct evaluations on two large-scale real mobile signaling datasets from Nanjing and Quanzhou. Experimental results show that OD-GPT can effectively learn dynamic dependencies among massive grids and make accurate OD flow predictions.

2 Related Work

2.1 OD Flow Prediction

The prediction of OD flow is a crucial problem in modern urban management. A city is a complex and dynamic system, and predicting human mobility is a high-dimensional and multimodal problem [6,7]. To address this, there are traditional models and deep learning-based methods to capture the long-term dependencies in flow data. These studies mainly focus on predicting the flow distribution for each grid or the overall flow volume between origins and destinations at specific moments [4,5], while overlooking the rich historical information contained in each flow. However, these models often struggle to effectively capture the complex spatial-temporal relationships. Different from them, our model can consider individual OD flow sequences and their associated information, rather than just focusing on grid-level or aggregated flow predictions.

2.2 Location Representation Learning

The objective of location embedding is to encode locations in a way that the similarity of vectors in the embedding space approximates that in the original space [2,11]. The low-dimensional dense vectors can be used to reduce the computational cost of sparse matrix features [14,16]. When faced with diverse downstream tasks, pre-trained location embeddings can be leveraged and modified to alleviate the need for feature engineering in each task [15].

2.3 Generative Pre-training (GPT)

The excellent performance of the GPT model on large-scale unsupervised datasets has demonstrated its ability to effectively address the long-range dependency problem and its powerful feature representation capability to capture data characteristics [10,12].

Given an unsupervised sequence $X = \{x_1, ..., x_n\}$, the objective is to maximize the following likelihood using a standard language model:

$$L(X) = \sum_i \log P(x_i \mid x_{i-k}, ..., x_{i-1}; \Theta) \tag{1}$$

where k is the size of the context window, and the likelihood P is defined as the sum of logarithmic conditional probabilities. Each token x_i depends on the previous k tokens $(x_{i-k}, ..., x_{i-1})$, and is modeled using a neural network with parameters Θ. To train these parameters, stochastic gradient descent is employed. We use GPT as the basic framework, which applies a random multi-head attention mechanism to the input context tokens and then generates the output distribution of the target token through a feed-forward layer:

$$\begin{aligned} h_0 &= XW_e + W_t \\ h_j &= temporal\_attention\_block(h_{j-1}) \forall j \in [1, n] \\ P(x) &= softmax(h_n W_e^T) \end{aligned} \tag{2}$$

where $X = \{x_{-k}, ..., x_{-1}\}$ represents the vector of the context token, n represents the number of layers, W_e represents the token embedding matrix, and W_t represents the time embedding matrix. In the temporal attention block, the causal self-attention mechanism enhances the model's ability to learn the causal relationship between input and output, making it more suitable for combining the semantics of previous OD grids with timestamps to make predictions.

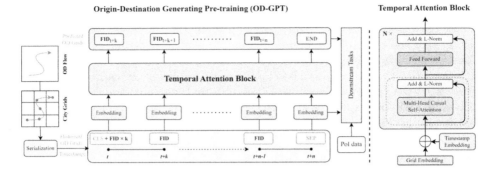

Fig. 1. An overview of the proposed Origin-Destination GPT (OD-GPT) model. Firstly, we map the mobile signaling data onto the city grids and serialize each OD flow at fixed time intervals. Secondly, we input the coordinate information of flows along with their corresponding timestamps, and use the model to capture the information contained in each sequence. Finally, we predict flows with the historical OD flows and combine embedding layers for further downstream work.

3 Methodology

Fig. 1 provides a detailed description of how to utilize the OD-GPT model for temporal modeling, aiming to obtain predictions for remaining timestamps and meaningful embeddings. We gridify the city area into $G = \{g_1, ..., g_N\}$, where g_i represents the Fid(Field id) of the i-th grid.

3.1 OD Flow Prediction

Definition 1 (OD Flow Prediction): Given historical OD sequence $X = \{x_1, ..., x_m\}$ and city grid information, we aim to predict $X' = \{x_{m+1}, ..., x_n\}$ [3].

Definition 2 (OD Matrix): In the OD matrix $A \in \mathbb{R}^{N \times N}$, N donates the numbers of grids. A_{ij} indicates the number of people from origin g_i to destination g_j [9].

As mentioned earlier, the geospatial information of grids is static. However, the OD sequences composed of grids are dynamic over time and contain inherent temporal dependencies. This means that the same grid often has different feature representations at different time points. To capture the dynamic changes, we extract semantic and spatial structural information from a large number of real OD flows and embed them separately for grids and timestamps. During the prediction phase, given a historical sequence as input, the model uses the learned parameters to generate the next most probable grid Fid. The main steps for crowd flow prediction using OD-GPT are outlined in Algorithm 1.

Algorithm 1. Prediction via OD-GPT
1: **Input:** Grid Fids, OD Flows
2: **Output:** Predictions, Timestamp, Grid Embeddings
3: $m, n \leftarrow$ size of OD flows, timestamps
4: **for** $i = 1$ to m **do**
5:     **for** $j = 1$ to n **do**
6:         Extract temporal features
7:         $ODSequences_{i,j} \leftarrow$ grid Fid of flows$_{i,j}$
8:     **end for**
9: **end for**
10: $T, V \leftarrow$ training, validation set of OD sequences S
11: **for** $batch_i$ in T **do**
12:     $Timestamp\ \&\ Grid\ Embedding \leftarrow$ compute embeddings($batch_i$)
13:     Train with $batch_i$ and $embeddings$
14: **end for**
15: **for** Y in V **do**
16:     $\hat{Y} \leftarrow Y[:k]$
17:     **while** length of $\hat{Y} < n$ **do**
18:         $Fid_{k+1} \leftarrow$ predict(Y)
19:         append Fid_{k+1} to $\hat{Y}$
20:     **end while**
21: **end for**
22: **return** Predictions, Timestamp, Grid Embeddings

Construct OD Sequences(Step 1–10). In this step, we set a fixed time interval, denoted as τ minutes. The total number of timestamps, denoted as n, within a day is 24×60 divided by τ. For each OD flow data, it is divided into segments based on τ (we set τ to 15). Then, each coordinate point is mapped to the corresponding grid Fid based on the time order, generating the OD sequence $X = \{x_1, ..., x_n\}$. For coordinates with a small number of missing grids (g^j) at the j-th timestamp, we use the method of calculating the center of coordinates before (g^i) and after (g^k) the timestamps to fill in the gaps.

$$g^j = g^i + \frac{g^k - g^i}{k - i} \qquad (3)$$

All generated X form the OD sequences and are divided into the training set T and the validation set V.

Obtain the Pre-Trained Model(Step 11–14). First, we need to split each X in the training set T using $[cls]$ and $[sep]$ tokens, and embed each pair of timestamp and Fid as tokens into vectors. In the autoregressive stage, for each batch of input sequences with a length of l, the input sequence consists of the first l-1 Fids of the current batch, and the target sequence consists of the last l-1 Fids. Using a feed-forward neural network, we generate the predicted Fid of the grid and calculate the cross-entropy loss function.

$$Loss = -\sum_{i=1}^{l-1}\sum_{j=1}^{N} p_{ij} log\left(\hat{p}_{ij}\right) \qquad (4)$$

where l-1 is the length of target sequence and N represents the size of grids. p_{ij} represents the probability of the i-th true grid Fid while $\hat{p}_{ij}$ represents the predicted. Model's parameters are updated through backpropagation with the loss, and the final grid embedding and timestamp embedding are obtained.

Predict OD Sequences(Step 15–22). For each Y in the validation set V, we select the first k Fids of Y as the historical OD sequence $\hat{Y}$. The model takes $\hat{Y}$ as input and predicts the $Fid_{k+1} = \text{OD-GPT}(Fid_1, ..., Fid_k)$. Then, Fid_{k+1} is appended at the end of $\hat{Y}$, and the updated sequence is input to the model again until the last timestamp of $\hat{Y}$ reaches the desired timestamp or the maximum prediction length (block size), which is the final prediction corresponding Y.

Calculate OD Matrix. Given OD sequences $S \in \mathbb{R}^{m \times n}$, one column o is chosen as the origin and another column d as the destination. We iterate through each row of S, incrementing A_{ij} accordingly. With the true sequence Y and the predicted sequence $\hat{Y}$, we can obtain OD matrix A and $\hat{A}$.

$$A_{ij} = \sum_{c=1}^{m}(Y_{co} = g_i \text{ and } Y_{cd} = g_j) \tag{5}$$

Since the value of some pairs is zero for certain time intervals, we only consider computing the error for non-zero OD pairs. We denote a binary matrix M of the same shape with A, where $M_{ij} = 1$ if A_{ij} is non-zero, otherwise 0.

3.2 Downstream Work

Fine Tuning On Few Samples. The pre-trained model based on weekday OD data performs well in predicting regular days. However, on unconventional days, especially during annual holidays, insufficient historical OD flow data for training makes it difficult for the model to learn the inter-grid correlations and spatial-temporal dependencies effectively. Considering the static nature of city grid features and the generic low-level features learned by shallow hidden layers, we choose to freeze them and fine-tune the deeper hidden layers containing more task-specific features using the limited OD data from unconventional days.

Embedding Analysis. Embedding is a low-dimensional vector representation of high-dimensional complex semantics. The embeddings generated by the model have a strong correlation with the attributes in the original space and can to some extent reflect the characteristics and connections of grids and timestamps in the real world. Cosine similarity is used to represent the similarity between grids at different timestamps. The higher the value, the higher the similarity. By reducing the dimensionality of the embedding layer with PCA or T-SNE and comparing it with various indicators of urban grids, we can discover certain underlying correlations between the two.

$$S \langle v_i, v_j \rangle = \frac{v_i \cdot v_j}{\| v_i \| \cdot \| v_j \|} \tag{6}$$

4 Evaluation

This section compares the proposed OD-GPT model with several baseline models for one-step OD flow prediction. In addition, we also explore the model's multi-step prediction capability and fine-tune it for unconventional days, and visualize the semantics learned by the model.

4.1 Experimental Setup

Dataset. We conduct extensive experiments on the mobile signaling data sets from the cities of Nanjing and Quanzhou. The data set consists of four weeks in 2019 and several holidays in 2019 and 2020. Unlike a single transportation mode scenario, this data includes almost all situations where mobile phone usage exists, such as pedestrians, trains, and ships. We divide the main areas of these two cities into grids of 500 m × 500 m, resulting in over 8,000 and 6,000 grids for Nanjing and Quanzhou respectively. After filtering out invalid grids, we have approximately 3.5 million usable OD flows for Nanjing and Quanzhou, with each flow containing 96 grids. The pre-trained model is trained on the workday portion of the dataset. The training and validation sets are randomly assigned in a 4:1 ratio.

This dataset covers four seasons, and is not limited to a single mode of transportation, including walking, taxis, subways, etc. With thousands of nodes in space, it presents a challenge for the model's performance.

Evaluation Metrics. Following previous manner, we use the root mean square error (RMSE) and mean average error ratio (MAE) to evaluate the performance of several models.

$$\text{RMSE}\left(A, \hat{A}\right) = \sqrt{\frac{\sum_{i=1}^{N}\sum_{j=1}^{N}\left(A_{ij} - \hat{A}_{ij}\right)^2 M_{ij}}{\sum_{i=1}^{N}\sum_{j=1}^{N} M_{ij}}} \quad (7)$$

$$\text{MAE}\left(A, \hat{A}\right) = \frac{\sum_{i=1}^{N}\sum_{j=1}^{N}\left|A_{ij} - \hat{A}_{ij}\right| M_{ij}}{\sum_{i=1}^{N}\sum_{j=1}^{N} M_{ij}} \quad (8)$$

where N denotes the number of OD pairs in the ground truth, and A represents the actual OD matrix while $\hat{A}$ represents the predicted. M is a binary matrix indicates whether the OD pair from g_i to g_j exists.

Baseline Models. We compare the proposed OD-GPT with several baseline models, ranging from traditional statistical methods to nonlinear regression models, and more recently, graph neural network methods. The details of the baseline models are as follows. **History Average(HA).** It calculates the OD pairs at t-th timestamp by averaging those with the same timestamp in previous days or weeks. **Linear Regression(LR)**. It is a regression model which exploits linear correlations between input and output. **Long Short-term Memory(LSTM)**. It utilizes historical data to input into the model, allowing it to make predictions for each individual OD grid sequence. **STGCN** [4]. It is a GCN based model for traffic forcasting which models the spatial and temporal dependencies. The difference is that we choose OD pairs as nodes and the distance between destination of one pair and origin of another as weights. **CMOD** [1]. It is based on modeling demand as a dynamic graph and utilizes a multi-level structure to adaptively exploit spatial dependencies between traffic nodes.

Experiment Settings. The proposed model is implemented with PyTorch and trained on one NVIDIA RTX GPU. In the temporal attention block, the number of hidden layers is set to 12, and each layer consists of 8 heads of multi-head casual self-attention, with 128-dimensional embedding layer. We set the learning rate to 0.0003 and use Adam optimizer for optimization. During the fine-tuning, we freeze the embedding layers and the first 10 temporal attention blocks in the pre-trained model, and train the model for fewer epochs with a learning rate of 0.00001. During the embedding analysis, PCA and T-SNE are used as dimensionality reduction tools.

4.2 Result And Discussion

OD Flow Prediction on Working Days. We first conduct experiments for one-step prediction. Table 1 presents the evaluation results. OD-GPT achieves the best metrics for the two indicators mentioned in Sect. 4.

Table 1. Performance comparison for one-step prediction

Model	Nanjing		Quanzhou	
	RMSE	MAE	RMSE	MAE
HA	2.815	1.730	3.569	1.704
LR	2.777	1.604	3.575	1.626
LSTM	2.545	1.615	3.339	1.521
STGCN	2.176	1.507	2.986	1.334
CMOD	1.854	0.925	2.615	1.012
OD-GPT	**1.106**	**0.553**	**1.555**	**0.586**

Generally, with strong expression ability, deep learning models are better than the traditional models. Due to the introduction of nonlinear expression, LSTM performs better than linear regression model(LR). STGCN is originally designed for traffic flow prediction, only capturing spatial dependencies on either the source or destination side. Although CMOD can capture the dependency

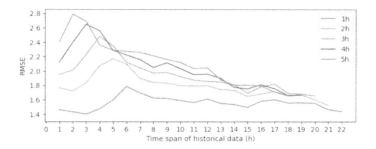

Fig. 2. Effect of historical data time span and generation length on Quanzhou multi-step prediction for a singal working day.

between two sides adaptively through a multi-level structure, it is better suited for dense OD matrices based on nodes such as subway stations, which typically have only a few dozen nodes. However, when dealing with a sparse matrix composed of thousands of nodes, the difficulty of dependency capture and the consumption of GPU memory increase sharply. Compared to them, OD-GPT can model the dynamic relationship to predict the possibility of the grids at the next timestamp by more outstanding capturing the whole spatial-temporal dependencies of historical OD flow and leverage rich past grid information.

Figure 2 explores the multi-step prediction capability of OD-GPT. The horizontal axis represents different time spans of historical OD data used as input to the model, while the vertical axis represents the RMSE of the prediction results with five different prediction steps. We observe that for all the multi-step prediction curves, the peak RMSE occurrs during the early peak time (7h). This is because during the early peak period, the crowd flow becomes more active, and the fluctuation of the OD matrix gradually increases, while the limited span of the input historical data can simply result in insufficient extraction of past information. Afterwards, as the length of the input increases, the gradual decrease in RMSE further supports this observation. Starting from the evening rush hour, there is a slight rebound in the indicators caused by the vibrant nightlife in city.

The findings suggest that in a multi-step prediction scenario, input data covering a longer time span captures more dependency information, providing an advantage for predictions. This flexible multi-step prediction capability reduces the model's sensitivity to the input data format, eliminating the necessity to train a separate model for each prediction step length. Furthermore, as the prediction steps increase, the prediction accuracy tends to decrease, although the rate of decline gradually diminishes.

Fine Tuning on Unconventional Days. The model, pre-trained on weekday data, exhibits reduced performance when predicting OD flow on unconventional days, especially during active holiday periods or COVID-19 lockdowns which lead to decreased mobility. This decline is attributed to significant changes in crowd activity. Given the limitations of historical data, training the model separately for these unconventional patterns can make it challenging for OD-GPT to capture the semantics and dependencies in the data. To address this, we fine-tune the deep hidden layers of the pre-trained model using appropriate historical data, considering the foundational nature of embedding and shallow hidden layers. For instance, during the Lantern Festival in 2020, which coincided with the early stage of the epidemic, we utilize historical data from the previous day, while for National Day, historical data from the same day of the previous year was employed.

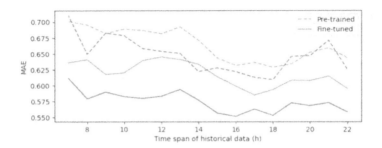

Fig. 3. Comparision between Pre-trained and Fine-tuned Models on the Lantern Festival (orange) and National Day(blue). (Color figure online)

Figure 3 illustrates the comparison of one-step predictions, demonstrating an enhancement in the model's predictive performance for unconventional OD flows following the fine-tuning process. Notably, the performance improvement observed during the early stages of the epidemic for the Lantern Festival is notably smaller in comparison to that of National Day. This discrepancy can be attributed to the epidemic's deceleration during the 2020 National Day, as evidenced by travel data reflecting a slightly lower figure of over 600 million trips, in contrast to the 780 million trips in 2019. The result shows that, through fine-tuning the deep layers of the network, we are able to more effectively adapt to the specific features of the task and data, consequently leading to improved predictive performance for unconventional OD flows.

Embedding Analysis. In Fig. 4(c), the cosine similarity between the selected grid and others in Quanzhou City is depicted. This particular grid encompasses parks, tourist areas, ancient temples, etc. Upon comparison with Fig. 4(a) and Fig. 4(b), it becomes evident that grids located near or containing these elements exhibit higher similarity on the heat map, while grids situated in oceanic areas or regions with minimal human presence display markedly lower similarity.

Figure 5(a) illustrates the outcome of the dimensionality reduction of timestamp embedding, portraying a day divided into 96 segments, forming a loop. Proximity between adjacent timestamps signifies higher correlation, underscoring the anticipated temporal pattern. In Fig. 5(b), we further divide two days

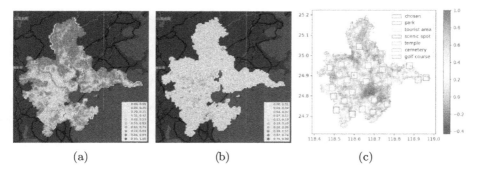

Fig. 4. Grid embedding analysis in Quanzhou City. (a)Proportion of ecological green land and waterways. (b)Proportion of land for commercial and service industries. (c)Grids including parks and scenic spots, etc. (Color figure online)

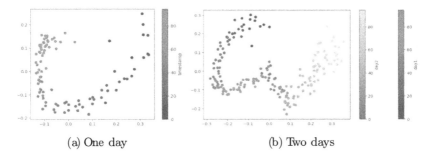

Fig. 5. Dimensionality reduction on timestamp embedding

into 196 segments, revealing dimensionally reduced timestamps that strikingly resemble the character 'ω', indicating periodicity of the timestamps.

Through the use of different timestamp embeddings (e.g., 8 a.m. and 6 p.m.) to shift grid embeddings and analyzing them in conjunction with T-SNE, we can glean insights into the temporal sensitivity of various grids, reflecting dynamic correlations. These grids with extremely small shifts(compared to the average ≈ 3) are extracted from comparision between Fig. 6(a) with Fig. 6(b) and displayed in Fig. 6(d), along with the information entropy shown in Fig. 6(c). Upon comparison, it can be seen that grids with higher information entropy tend to have smaller time shifts.

The analysis results demonstrate a significant correlation between grid embedding and land attribute categories. Simultaneously, timestamp embedding successfully captures the periodic characteristics of temporal changes. The dynamic integration of these two embeddings effectively reflects key indicators such as land-use structure information entropy, which reflects the diversity and balance of land use.

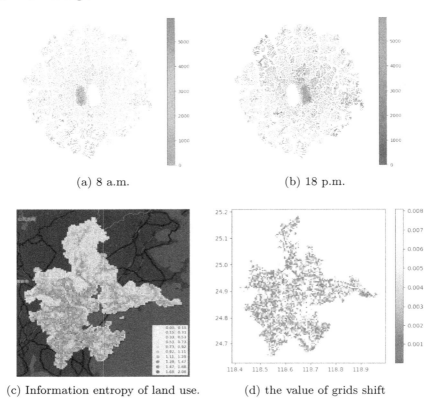

(a) 8 a.m. (b) 18 p.m.

(c) Information entropy of land use. (d) the value of grids shift

Fig. 6. Dynamic embedding analysis based on grid and timestamp.

5 Conclusion

This paper proposes a novel model called OD-GPT for city origin-destination flow prediction that can successfully capture spatial-temporal relationships and significantly improve one-step predictions. We extensively experiment with two large datasets from Nanjing and Quanzhou, further validating the model's multi-step prediction capability and adaptability to unconventional days. The results indicate that the combination of grid and timestamp embeddings can effectively express complex dynamic dependencies and indicators in cities.

Given that the embedding layers contribute significantly to the model, fine-tuning the embedding layers can effectively enhance the quality and convergence speed of the model. While pre-trained models are efficient in generating embeddings, gathering extensive crowd flow data poses a challenge. Consequently, we aim to investigate potential relationships between embeddings and various urban geographical indicators in the future. Reasonable embeddings derived from these indicators can be utilized in few-shot learning for other cities and tasks.

The proposed OD-GPT model is released through the link below:
https://github.com/scotlandowl/OD-GPT

References

1. Han, L., et al.: Continuous-time and multi-level graph representation learning for origin-destination demand prediction. In: Proceedings of the 28th ACM SIGKDD Conference on Knowledge Discovery and Data Mining (2022)
2. Cai, M., Pang, Y., Sekimoto, Y.: Spatial attention based grid representation learning for predicting origin-destination flow. In: Big Data (2022)
3. Shi, H., et al.: Predicting origin-destination flow via multi-perspective graph convolutional network. In: ICDE (2020)
4. Yu, B., Yin, H., Zhu, Z.: Spatio-temporal graph convolutional networks: a deep learning framework for traffic forecasting. In: IJCAI (2018)
5. Shao, Z., et al.: Decoupled dynamic spatial-temporal graph neural network for traffic forecasting. In: VLDB (2022)
6. Xu, J., Huang, J., Chen, Z., Li, Y., Tao, W., Xu, C.: ODNET: a novel personalized origin-destination ranking network for flight recommendation. In: ICDE (2022)
7. Zheng, G., Liu, C., Wei, H., Chen, C., Li, Z.: Rebuilding city-wide traffic origin destination from road speed data. In: ICDE (2021)
8. Guo, S., et al.: Self-supervised spatial-temporal bottleneck attentive network for efficient long-term traffic forecasting. In: ICDE (2023)
9. Geng, X., et al.: Spatiotemporal multi-graph convolution network for ride-hailing demand forecasting. In: AAAI (2019)
10. Radford, A., Narasimhan, K., Salimans, T., Sutskever, I.: Improving language understanding by generative pre-training (2018)
11. Hu, J., Yang, B., Guo, C., Jensen, C.S., Xiong, H.: Stochastic origin-destination matrix forecasting using dual-stage graph convolutional. recurrent neural networks. In: ICDE (2020)
12. Zou, A., et al.: Forecasting future world events with neural networks. In: NIPS (2022)
13. Monti, F., Bronstein, M., Bresson, X.: Geometric matrix completion with recurrent multi-graph neural networks. In: NIPS (2017)
14. Kuo, A.-T., Chen, H., Ku, W.-S.: BERT-Trip: effective and scalable trip representation using attentive contrast learning. In: ICDE (2023)
15. Ho, N.L., Hui Lim, K.: POIBERT: A Transformer-based model for the tour recommendation problem. In: Big Data (2022)
16. Bashir, S.R., Raza, S., Misic, V.B.: BERT4Loc: BERT for Location-POI Recommender System. In: Future Internet (2023)

Vehicle-Based Evolutionary Travel Time Estimation with Deep Meta Learning

Chenxing Wang[1], Fang Zhao[1(✉)], Haiyong Luo[2(✉)], Yuchen Fang[1], Haichao Zhang[1], and Haoyu Xiong[1]

[1] School of Computer Science, Beijing University of Posts and Telecommunications, Beijing, China
{wangchenxing,zfsse,haichaozhang,haoyuxiong}@bupt.edu.cn,
fyclmiss@gmail.com

[2] Institute of Computing Technology, Chinese Academy of Sciences, Beijing, China
yhluo@ict.ac.cn

Abstract. Vehicle-based travel time estimation is crucial for many travel scheduling and city planning applications in intelligent transportation systems. Since trajectories in different trips are affected by evolutionary Spatio-temporal dynamics (e.g., evolving travel patterns for different days of the week and varying road networks affected by traffic accidents or temporary restrictions, etc.), it is substantial to investigate these dynamics for accurate estimation. In this paper, we propose a novel deep learning model which fuses location features, distance features, and temporal features with meta learning-based neural networks, to implicitly learn path representations for evolving travel patterns in different days of week and road networks. Specifically, we utilize the meta learning-based optimization method to transfer the shared meta knowledge across trajectories in distinct Evolving-Tasks (i.e., a limited amount of trajectory data on different days of the week), which facilitates generalizing rapidly on evolving travel patterns for different days of the week. In addition, road network information is obviated in our model, which makes it a natural solution to tolerate evolving road networks while mitigating the computation burden contemporaneously. Comprehensive experiments on three real-world datasets demonstrate the superiority of our proposed model.

Keywords: Travel time estimation · Meta-learning · Spatio-temporal data mining

---

This work was supported in part by the National Natural Science Foundation of China under Grant 62261042, the Key Research Projects of the Joint Research Fund for Beijing Natural Science Foundation and the Fengtai Rail Transit Frontier Research Joint Fund under Grant L221003, Beijing Natural Science Foundation under Grant 4232035 and 4222034, the Strategic Priority Research Program of Chinese Academy of Sciences under Grant XDA28040500, the China Postdoctoral Science Foundation under Grant 2024M750200 and BUPT Excellent Ph.D. Students Foundation under Grant CX2022132.

© The Author(s), under exclusive license to Springer Nature Switzerland AG 2024
M. Wand et al. (Eds.): ICANN 2024, LNCS 15024, pp. 246–262, 2024.
https://doi.org/10.1007/978-3-031-72356-8_17

1 Introduction

(a) Spatial dimension.

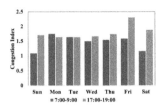

(b) Temporal dimension.

Fig. 1. The illustration for evolutionary Spatio-temporal dynamics. For spatial patterns, we present the partial area in Chengdu which has temporary restriction regions. For temporal dimension, we present the congestion indexes (Clear: $[1.00, 1.50)$, Slow: $[1.50, 1.80)$, Congestion: $[1.80, 2.00)$, Severe congestion: $[2.00, +\infty)$) of morning peak (i.e., 7:00-9:00) and night peak (i.e., 17:00-19:00) for each day of week in Chengdu from 2022-10-16 to 2022-10-22. We can observe that the different days of week and hours have distinct peak patterns. For instance, congestion indexes for morning peak are relatively low while that for night peak are relatively high on weekends.

Travel time estimation is crucial in smart city applications, especially for intelligent vehicle-based systems including navigation, route recommendation [22], and ride-hailing services [5]. However, it is still challenging to attain consistent performance for travel time estimation due to evolutionary Spatio-temporal dynamics, shown in Fig. 1, where several significant problems remain to be further addressed: (1) *Travel patterns evolve over time.* Travel patterns evolve during distinct periods (e.g., weekdays, weekends, traffic accidents, etc.). While most existing works treat trajectories in different periods as a whole, the evolutionary travel patterns are often neglected. (2) *Road networks evolve over time.* Road networks are evolving due to temporary restrictions for specific regions induced by accidents. Also, drivers may know new roads that haven't been addressed in current road networks, affecting the accuracy of travel time estimation. Most existing works treat road network information as static attributes integrated into models which have natural defects [5,6].

Recently, enormous efforts have been undertaken to address the problem of travel time estimation. Traditional statistics-based methods [1,13,24] are being replaced by end-to-end methods [6,11,20,23,25] because of the substantial error accumulation of successive road segments. Among the end-to-end methods, deep-learning-based methods can achieve good results by generalizing from many historical traffic data. However, these methods cannot provide accurate travel time estimation at consistent performance when traffic patterns and road networks evolve. To retain good performance, the pre-trained model has to be frequently finetuned with the latest traffic data and road networks, which brings tremendous computation burdens and requires careful labor-extensive processing [20].

Meta-learning has recently been suggested as an efficient approach to learning from the limited amount of data and attempts to optimize in a fast learning manner, using distinct tasks of training data [14]. This approach aims to produce an agent with an excellent average performance on the test set for various tasks. To this end, Fan et al. [5] construct tasks for meta-learning as distinct trajectories which support personalized adaptation to each trajectory. Moreover, Wang et al. [19] construct tasks as different regions to enhance the generalization ability on more sparse areas. Although the adaptation between different drivers or different regions is contemplated, the dynamics for evolutionary travel patterns are ignored. In this paper, we propose a novel end-to-end meta-learning-based Spatio-temporal prediction approach to provide accurate estimations based on elaborately designed Evolving-Tasks. It entails nontrivial challenges to train such a robust model, which can be mainly summarized as two aspects: **(1) Evolutionary Spatio-temporal Dynamics:** Most of the studies neglect the Spatio-temporal dynamics since: (i) spatial dimension: they utilize static road networks which may change sporadically due to accidents or exigencies; (ii) temporal dimension: they model different time horizons (e.g., different days of week or hours) by same parameter initialization which cannot obtain consistent performance under different horizons. How to elaborately tackle this issue is substantial for models to obtain gratified performance. **(2) Fast Learning for Generalization:** Since the evolutionary Spatio-temporal dynamics shift rapidly affected by peak or off-peak factors in distinct days of the week or exigencies, how to adapt to recent trajectory data in a fast learning manner becomes more substantial.

To address the above challenges, we exploit deep neural networks to implicitly learn the evolutionary Spatio-temporal dynamics in this paper. Moreover, a meta-learning-based optimization algorithm is adopted to allow DMTTE to generalize from the limited amount of data into distinct Evolving-Tasks in a fast learning manner. Specifically, the main contributions of this paper can be summarized as: **First**, we propose a novel vehicle-based evolutionary travel time estimation model with meta learning, termed DMTTE, which captures evolutionary Spatio-temporal dynamics (i.e., evolving travel patterns and road networks) to estimate the travel time given the query path. *For spatial dimension*, the road network information is obviated in DMTTE and competitive results are obtained comparing to those road network aware methods (see Sect. 5.3). *For temporal dimension*, we construct Evolving-Tasks and utilize meta learning to generalize from limited amount of data which show superior results comparing to baselines (see RQ7). **Second**, we extract Evolving-Tasks from three real-world datasets, each of which contains trajectory data from an individual day of the week (e.g., Monday etc.). To this end, DMTTE can generalize well for daily evolving travel patterns and mitigates computation burdens when adapting DMTTE to recent trajectory data. **Third**, we elaborately design an efficient Spatial-Temporal Modelling network (i.e., STMNet) to implicitly learn evolving road network dynamics in a data-driven way which obviates infusing static road network information. **At last**, We conduct extensive experiments on three real-world, large-scale datasets to demonstrate the effectiveness of our method.

2 Related Work

2.1 Travel Time Estimation

Recent studies on travel time estimation can be mainly categorized into **origin-destination-based methods** and **path-based methods**.

Origin-destination-based methods can be applied when only origin and destination information is provided in the query data. TEMP [21] attempts to find neighbors of the query path in historical trajectories to estimate travel time, which cannot obtain good results due to data sparsity issues. MURAT [12] leverages the underlying road network and the Spatio-temporal prior knowledge based on a multi-task learning scheme for travel time estimation. STNN [9] first predicts the travel distance between an origin and a destination GPS coordinate and then combines the prediction with the time of day to predict the travel time. These methods predict travel time without online route information, which reduces the time consumption compared to path-based methods. However, the performance of these models is limited due to the ignorance of complex Spatio-temporal information in the whole path [8].

Path-based methods exploit the whole query path to estimate travel time considering complex Spatio-temporal factors. Among these methods, Deep-Travel [25] automatically extracts different features with the auxiliary supervision model. WDR [23] jointly trains wide linear models, deep neural networks, and recurrent neural networks together to take full advantage of all models and provide estimation results. ConSTGAT [6] adopts a graph attention mechanism and fully exploits the joint relations of spatial and temporal information. DeepIST [7] converts 1D query paths to 2D images of paths and applies multiple convolutional layers to extract implicit features for travel time estimation. STTE [8] leverages multiple features, including semantic representations in non-Euclidean space and Euclidean space, of a given path to estimate travel time. Though some studies [20] [11] utilize historical trajectory data without road network information to estimate travel time, they fail to capture evolutionary travel patterns. Specifically, they cannot adapt to recent trajectory data in a fast learning scheme, which not only limits the performance currently but also limits the performance in the future.

3 Preliminary

In this section, we present definitions and the objective of our proposed DMTTE. Note that we use $V.r_i$ to denote r_i where $r_i \in V$ for simplification in the following.

Definition 1 (Trip of the vehicle V): Given a trip of the vehicle with L data points, denoted by $V = \{r_i\}, 1 \le i \le L$, each of which is arranged in order and represents the i-th location in trip V. Let $r_i = (p_1, p_2, d, w, h, t)$, of which $r_i.p_1$ and $r_i.p_2$ denote the latitude and longitude of the GPS coordinates, $r_i.d$ denotes the great-circle distance [17] from current location r_i to last location r_{i-1}, $r_i.w$ denotes the long term attribute (i.e. the day of week), $r_i.h$ denotes the short term attribute (i.e. the hour of the day) and $r_i.t$ denotes the timestamp of the current location r_i.

Definition 2 (DMTTE trajectory G): Let $T = L-1$. Given a DMTTE trajectory G with T data points, denoted by $G = \{g_j : j = 1, 2, \ldots, T\}$, each of which is arranged in order and represents the j-th location in DMTTE trajectory $G \in \mathbb{R}^{T \times 6}$. Let $g_j = (\Delta p_1, \Delta p_2, d, w, h, t)$, of which $g_j.\Delta p_1$ and $g_j.\Delta p_2$ denote the Δ of latitudes and longitudes between $r_{j+1}.p_1, r_j.p_1$ and $r_{j+1}.p_2, r_j.p_2$, respectively and $g_j.d = r_j.d, g_j.w = r_j.w, g_j.h = r_j.h, g_j.t = r_j.h$. We regard g_1 as the origin of this trajectory and g_T as the destination. Hence let $y = |g_T.t - g_1.t|$ be the travel time for G.

Definition 3 (Evolving-Task $\mathbb{G}$): Given a set of DMTTE trajectories $\mathbb{G}_i = \{G \in \mathbb{G}_i : g \in G \wedge g.w = i\}, i = 1, 2, \ldots, 7$ as the i-th Evolving-Task, which indicates that the week starts on Monday when $i = 1$ and ends on Sunday when $i = 7$. Note that we use $N_i = |\mathbb{G}_i|$ to denote the number of trajectories in the i-th Evolving-Task.

Definition 4 (TTE Dataset $\mathcal{D}$): Given a TTE Dataset $\mathcal{D}$, denoted by $\mathcal{D} = \{\boldsymbol{X}_i \in \mathbb{X}, \boldsymbol{Y}_i \in \mathbb{Y}\}$, of which $\boldsymbol{X}_i(:,:,:) = \{G \in \mathbb{G}_i\}$ and $\boldsymbol{Y}_i(:,:,:) = \{|G.g_T.t - G.g_1.t| : G \in \mathbb{G}_i\}$ denote a set of DMTTE trajectories in the i-th Evolving-Task and the corresponding ground-truth (i.e. travel time), respectively. Note that in this paper, $\boldsymbol{X}_i(:,:,:) \in \mathbb{R}^{N_i \times T \times 6} \wedge \boldsymbol{Y}_i(:,:,:) \in \mathbb{R}^{N_i} \wedge \mathbb{X} \in \mathbb{R}^{7 \times N_i \times T \times 6} \wedge \mathbb{Y} \in \mathbb{R}^{7 \times N_i}$ and we denote DMTTE trajectories and the corresponding ground-truth in i-th Evolving-Task as $\boldsymbol{X}, \boldsymbol{Y}$ by ignoring the subscript i for convenience.

Definition 5 (TTE Problem): Given a query trip V, the objective of this paper is to firstly transform V to G and then predict the travel time $\hat{y}$ corresponding to this query.

4 DMTTE: Deep Meta Learning-Based Travel Time Estimation

As illustrated in Fig. 2, we utilize a meta learning based optimization algorithm to iteratively adapt parameter initializations with Evolving-Tasks during meta training in a fast learning scheme (see Sect. 4.6). More specifically, as illustrated in Fig. 3, our DMNet consists of two main components: *Spatio-temporal modeling* and *fused context modeling*. Note that we elaborate the structure of DMNet based on the i-th Evolving-Task in the following for simplification.

The *Spatio-temporal modeling* component utilizes four types of time series data in a matrix of DMTTE trajectories as input: (i) GPS sequence, which is time-ordered sequence matrix $\boldsymbol{X}(:,:,:2)$; (ii) distance sequence, which is

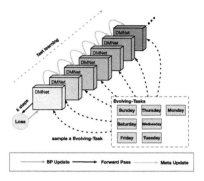

Fig. 2. The overall framework of DMTTE. The DMNet is meta learning-based neural networks, marked with different depths of color, indicating the different evolving states of parameter initialization.

time-ordered sequence matrix $\boldsymbol{X}(:,:,3)$; (iii) week sequence, which is time-ordered sequence matrix $\boldsymbol{X}(:,:,4)$; (iv) hour sequence, which is time-ordered sequence matrix $\boldsymbol{X}(:,:,5)$. These different types of time series data are then fed into four networks, i.e., GPSNet, DisNet, WeeNet and HouNet, respectively. Each network firstly embeds the corresponding sequence matrix from numeric sequence matrix to categorical sequence matrix except for GPSNet, which calculates the difference between adjacent original GPS coordinates and then learns implicit spatio-temporal feature representations for each type of time series data, respectively, to model the Spatio-temporal correlations of the historical trajectory data in a Spatial-Temporal Modelling network component (i.e., STMNet) with the same structure, including a recurrent network module to capture complex Spatio-temporal dependencies. Let D denote the last dimension of learned Spatio-temporal features. The *fused context modeling* components utilizes attention-based fusion to fuse four learned spatio-temporal features $\boldsymbol{Z}^P, \boldsymbol{Z}^S, \boldsymbol{Z}^W, \boldsymbol{Z}^H \in \mathbb{R}^{N \times D}$ and obtains the fused context feature $\boldsymbol{Z}^F \in \mathbb{R}^{N \times D}$. Finally, we utilize a Fully Connected Modelling network component (i.e. FCMNet) to further learn implicit features representations from $\boldsymbol{Z}^F$.

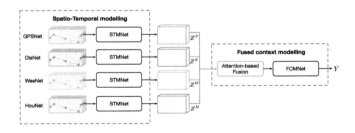

Fig. 3. The structure of DMNet.

4.1 GPSNet and DisNet

As mentioned in Sect. 3, we utilize the difference between two adjacent GPS coordinates to construct our DMTTE trajectory data G in the GPSNet. This approach of embedding to obtain embedded GPS sequence matrix $\boldsymbol{X}(:,:,:2)$ has the following pros. On the one hand, raw GPS data have to be resampled to gratify a relatively fixed pattern for better performance [20] which adds to computation complexity. On the other hand, intuitively, the driving distance between two locations with the similar speed patterns may contain the implicit time feature representations for accurately estimating the travel time for most vehicles. Since the driving distance may contain the implicit time features in its numerical form, the raw distance sequence matrix $\boldsymbol{X}(:,:,:3)$ are utilized in this stage. To align the last dimension of learned Spatio-temporal features D, we exploit two separate fully connected layers to project $\boldsymbol{X}(:,:,:2)$ and $\boldsymbol{X}(:,:,3)$ to ε^W and ε^S, formulated as:

$$\varepsilon^P = \max(0, \boldsymbol{W_1}\boldsymbol{X}(:,:,:2) + \boldsymbol{b_1}), \tag{1}$$

$$\varepsilon^S = \max(0, \boldsymbol{W_2}\boldsymbol{X}(:,:,:2) + \boldsymbol{b_2}), \tag{2}$$

where $\boldsymbol{W_1} \in \mathbb{R}^{2 \times D}, \boldsymbol{W_2} \in \mathbb{R}^{1 \times D}, \boldsymbol{b_1} \in \mathbb{R}^D, \boldsymbol{b_2} \in \mathbb{R}^D$ are trainable parameters.

4.2 WeeNet and HouNet

The week sequence matrix $\boldsymbol{X}(:,:,4)$ and hour sequence matrix $\boldsymbol{X}(:,:,5)$ are embedded into categorical sequence matrix $\varepsilon^W \in \mathbb{R}^{N \times D}, \varepsilon^H \in \mathbb{R}^{N \times D}$ in WeeNet and HouNet, respectively. To accomplish this, one possible solution is to utilize one-hot encoding to transform a day of a week and hour into D-dimensional vector. However, one-hot representation is too sparse, and the distance between any two one-hot codes are identical, so the distance between different days of the week or hours cannot be distinguished. Hence, to tackle this issue, we firstly use a one-hot encoding to transform $\boldsymbol{X}(:,:,4), \boldsymbol{X}(:,:,5)$ into $\boldsymbol{O}^W \in \mathbb{R}^{N \times T \times E^W}, \boldsymbol{O}^H \in \mathbb{R}^{N \times T \times E^H}$. Then we use a fully connected layer to embed them, formulated as:

$$\varepsilon^W = \boldsymbol{W_3}\boldsymbol{O}^W, \tag{3}$$

$$\varepsilon^H = \boldsymbol{W_4}\boldsymbol{O}^H, \tag{4}$$

where $\boldsymbol{W_3} \in \mathbb{R}^{E^W \times D}, \boldsymbol{W_4} \in \mathbb{R}^{E^H \times D}$ are trainable parameters, and $E_W = 7, E_H = 24$ are the number of categorical values in week and hour, respectively. To enhance the scalability of our DMTTE, we embed different factors, including week, hour, and other external features (e.g. driver ID) in different networks, respectively, which allows users to add or remove several embeddings without significant modifications to neural networks.

4.3 STMNet

In STMNet, we utilize a sequence of recurrent neural network units that can be customized by users (e.g., GRU, LSTM, etc.). Let $\boldsymbol{U}^{(t)}, \boldsymbol{Z}^{(t)}$ denote the output matrix and hidden state matrix at time step t, respectively. Note that we use $\boldsymbol{U}^{(0)}, \boldsymbol{Z}^{(0)}$ to denote the initial output and initial hidden state, and they are all zero matrices in our experimental settings. Specifically, we denote the output and state matrix at timestep T as $\boldsymbol{U}, \boldsymbol{Z}$ by ignoring the superscript for convenience. Then we formulate the calculation of RNN as:

$$(\boldsymbol{U}^{(t)}, \boldsymbol{Z}^{(t)}) = F_{RNN}(\boldsymbol{X}(:,t,:), \boldsymbol{U}^{(t-1)}, \boldsymbol{Z}^{(t-1)}), \tag{5}$$

where $F_{RNN}(\cdot)$ is the mapping function of a type of recurrent network (e.g. GRU [4], LSTM [16], and BiLSTM [3]). In this paper, we utilize the hidden state matrix $\boldsymbol{Z}^P, \boldsymbol{Z}^S, \boldsymbol{Z}^W, \boldsymbol{Z}^H$ at time step T as the four types of learned Spatiotemporal features, respectively.

4.4 Attention-Based Fusion

In the attention-based fusion component, inspired by a type of self-attention mechanism [2], we design an attention-based fusion method to learn the significance of different features in $\boldsymbol{Z}^P, \boldsymbol{Z}^S, \boldsymbol{Z}^W, \boldsymbol{Z}^H$ and fuse them to obtain a context feature $\boldsymbol{Z}^F$. We concatenate $\boldsymbol{Z}^P, \boldsymbol{Z}^S, \boldsymbol{Z}^W, \boldsymbol{Z}^H$ and then calculate the significant scores automatically for different features using a score function, formulated as:

$$f_S = \max(0, \boldsymbol{W}_5(\boldsymbol{Z}^P||\boldsymbol{Z}^S||\boldsymbol{Z}^W||\boldsymbol{Z}^H) + \boldsymbol{b}_3), \tag{6}$$

where $\cdot||\cdot$ denotes the concatenation function, $\boldsymbol{W}_5 \in \mathbb{R}^{4\times4}, \boldsymbol{b}_3 \in \mathbb{R}^4$ are trainable parameters. We then normalize the learned scores f_S^i for i-th feature to obtain the attention values, formulated as:

$$f_{SN}^i = \frac{\exp(f_S^i)}{\Sigma^i \exp(f_S^i)}, \tag{7}$$

And we calculate the fused context feature $\boldsymbol{Z}^F$, formulated as:

$$\boldsymbol{Z}^F = \Sigma_{i=1}^4 f_{SN}^i \odot (\boldsymbol{Z}^P||\boldsymbol{Z}^S||\boldsymbol{Z}^W||\boldsymbol{Z}^H)^i, \tag{8}$$

where $\odot$ represents the Hadamard product.

4.5 FCMNet

In FCMNet, we utilize multiple fully connected layers (i.e., FCs) to learn implicit feature representations. Since the residual mechanism can expedite the convergence speed [18], we design residual blocks with four FCs for feature learning which enhances the performance without consuming much time. The formula of each FC k ($k=1,2,\ldots,4$) is defined as follows:

$$\boldsymbol{z}^{'(k)} = \boldsymbol{W}^{(k)} \boldsymbol{z}^{'(k-1)} + \boldsymbol{b}^{(k)}, \tag{9}$$

where $\boldsymbol{W}^{(k)}, \boldsymbol{b}^{(k)}$ are trainable parameters in the layer. Let $\boldsymbol{Z}^{'(0)} = \boldsymbol{Z}^F$, the final estimation output is defined as:

$$\hat{\boldsymbol{Y}} = \boldsymbol{W}_6(\boldsymbol{Z}^F + \boldsymbol{Z}^{'(4)}) + \boldsymbol{b}_4, \tag{10}$$

where $\boldsymbol{W_6} \in \mathbb{R}^{D \times 1}, \boldsymbol{b}_4 \in \mathbb{R}^1$ are trainable parameters.

4.6 Optimization Algorithm

We design an efficient optimization algorithm to optimize DMTTE. Algorithm 1 shows the pseudocode of the optimization algorithm for our DMTTE. The input of the proposed algorithm is TTE Dataset $\mathcal{D}$, model $\mathcal{M}$ which contains the trainable parameters θ, loss function for i-th task, and the maximum iteration times J. We need to set a proper value for J first to declare the stopping criteria for meta learning. In practice, DMTTE can reach gratified MAE (i.e., Mean Average Error) in 1000 iterations and thus let $J = 1000$. For each outer iteration, we randomly select an Evolving-Task $\mathbb{G}_i$ at first and then define current model parameters as θ_1. Then we repeat k times to update model parameters on the selected Evolving-Task in the inner iteration. After the k time optimization, we define the optimized model parameters as θ_2 and derive the model's final parameters θ_f, formulated as:

$$\theta_f = \beta(1 - \frac{j}{J})(\theta_2 - \theta_1) \tag{11}$$

where β is the hyperparameter for learning rate scheduling, j is the current iteration, J is the maximum iteration times, θ_1, θ_2 are the parameters before and after k-time inner optimization steps, respectively.

Algorithm 1: Optimization algorithm for DMTTE.

Input : TTE Dataset $\mathcal{D}$, model $\mathcal{M}$, loss function for i-th task: $\mathcal{L}_i$, the maximum iteration times J.
Output: DMTTE Model.

1: $j = 1$;
2: **repeat**
3:     Randomly sample an Evolving-Task $\mathbb{G}_i$ from $\mathcal{D}$;
4:     Define model parameters as θ_1 ;
5:     **for** j in $[1, k]$ **do**
6:         Randomly select a batch of DMTTE trajectories $\boldsymbol{X}'$ and ground-truth $\boldsymbol{Y}'$ from $\mathbb{G}_i$;
7:         $\hat{\boldsymbol{Y}} \leftarrow \mathcal{M}(\boldsymbol{X}')$;
8:         Optimize model parameters based on Adam optimizer [10] and $\mathcal{L}_i$;
9:     **end**
10:    Define model parameters as θ_2 ;
11:    Derive model parameters θ_f based on θ_1, θ_2, j, J;
12:    Set model parameters to θ_f ;
13: **until** *stopping criteria is met*;
14: Derive optimized DMTTE;

5 Experiments

In this section, we mainly investigate the effectiveness of DMTTE to answer six research questions. RQ1: does our proposed DMTTE outperform the baselines? RQ2: how do hyper-parameters affect DMTTE? RQ3: how do different components of DMTTE affect its performance? RQ4: how does DMTTE perform under different cases? RQ5: is our proposed DMTTE more efficient than baselines? RQ6: how does meta-learning affect DMTTE? RQ7: how does the baselines perform on limited amount of data?

5.1 Datasets and Experimental Settings

Datasets. We utilize three real-world datasets to evaluate the performance of DMTTE, shown as the following:
(i) Chengdu$_{DC}$[1]: Chengdu$_{DC}$ dataset is collected from ride-hailing services in Chengdu, China which contains over 1.4 billion GPS records.
(ii) Porto[2]: Porto dataset is collected from taxi dispatch central from Jul 1st, 2013 to Jun 30th, 2014 involving 442 taxis running in Porto, Portugal.
(iii) Chengdu$_{DD}$[3]: Chengdu$_{DD}$ dataset is collected from ride-hailing services in Chengdu, China with a more regular sampling rate compared to Chengdu$_{DC}$ dataset and contains 2,918,946 trips in total.
Implementation Details. The implementation details include batch size (32), inner optimizer Adam [10] with its learning rate α (0.01), step size β (0.3), inner number of shots k (10), maximum iteration times J (1000), the dimension number of embedding D (64), and the number of units in FCMNet (1024, 512, 256, 64), training epochs (100), and train-val-test split rate (0.7, 0.2, 0.1).

5.2 Baselines

AVG [11]. **AV**era**G**e speed firstly calculates the average speed using GPS trajectories on the training set and then estimates travel time by calculating the average speed of those trajectories with similar origins and destinations.
LR [11]. **L**inear **R**egression models the relation between geographical locations for origin-destination and travel time to estimate travel time.
GBM [11]. Light**GBM** exploits temporal features from historical trajectories, to estimate travel time.
TEMP [21]. **TEMP**orally weighted neighbors utilize the average travel time based on neighbors from historical trajectories on training data.
WDR [23]. **W**ide-**D**eep-**R**ecurrent network extracts handcrafted features from historical trajectory data and aggregates road segment information to jointly estimate travel time.

---
[1] https://www.dcjingsai.com/v2/cmptDetail.html?id=175.
[2] https://www.kaggle.com/crailtap/taxi-trajectory.
[3] https://outreach.didichuxing.com/research/opendata/en/.

DeepTTE [20]. DeepTTE learns feature representations from historical trajectories and external information using 1-D convolutional based neural networks.
STNN [9]. Spatio-Temporal Neural Network firstly predicts the travel distance between an origin and a destination, and then combines this prediction with the temporal factors to predict travel time.
MURAT [12]. MUlti-task Representation learning model for Arrival Time estimation predicts travel time and travel distance in multi-tasking scheme for performance enhancement.
Nei-TTE [15]. Deep learning method based on Neighbors for Travel Time Estimation takes the entire trajectory as series of segments based on road network information and predicts the travel time.
DMTTE$_{WA}$. DMTTE$_{WA}$ selects LSTM as the recurrent cell in STMNet without fused context feature.
DMTTE$_{LSTM}$. DMTTE$_{LSTM}$ selects LSTM as the recurrent cell in STMNet.
DMTTE$_{BiLSTM}$. DMTTE$_{BiLSTM}$ selects BiLSTM as the recurrent cell in STMNet.
DMTTE$_{GRU}$. DMTTE$_{GRU}$ selects GRU as the recurrent cell in STMNet.

5.3 Results

Table 1. Performance comparison results for different baselines. The best of all baselines is marked with **bold font** and the best of each part is marked with underline.

Baselines	Chengdu$_{DC}$			Porto			Chengdu$_{DD}$		
	MAE	MAPE (%)	RMSE	MAE	MAPE (%)	RMSE	MAE	MAPE (%)	RMSE
AVG	442.20	39.71	8443.60	182.64	26.66	1128.21	135.40	21.97	451.31
LR	516.23	49.09	1204.99	194.40	33.90	279.20	189.98	36.11	244.41
GBM	454.50	41.67	1121.32	148.53	24.59	209.07	194.42	39.85	227.68
TEMP	334.60	39.70	761.05	174.44	28.73	260.81	142.30	25.38	220.41
WDR[a]	433.99	29.74	1024.92	164.04	22.84	244.41	136.66	22.14	194.52
DeepTTE	413.09	24.22	926.04	84.29	14.79	90.29	121.64	19.55	128.29
STNN[a]	427.33	30.08	1011.88	226.30	35.44	331.75	173.67	30.90	231.08
MURAT[a]	439.19	28.54	1102.14	165.91	27.10	177.83	79.93	11.25	138.15
Nei-TTE[a]	414.16	30.04	1038.71	106.30	15.23	183.03	130.73	20.05	192.26
DMTTE$_{WA}$	265.98	26.00	751.75	4.44	0.62	6.53	32.19	5.86	68.85
DMTTE$_{LSTM}$	246.52	24.49	746.56	4.30	0.57	5.97	29.76	5.27	60.95
DMTTE$_{BiLSTM}$	292.01	31.45	803.21	6.82	0.81	10.85	32.66	5.83	62.04
DMTTE$_{GRU}$	**231.96**	**22.44**	**729.85**	**2.07**	**0.26**	**3.83**	**24.37**	**4.13**	**51.73**

[a] These baselines have road network information.

Performance Comparison (RQ1). Table 1 compares the performance of different methods on three real-world datasets. We can observe that: (i) most deep learning-based methods outperform other traditional time series modelling methods in MAPE, which indicates the superior of their ability to learn dynamic

Spatio-temporal features for travel time estimation; (ii) variants of DMTTE outperforms other deep learning-based methods on three datasets, which indicates the ability of DMTTE to capture the evolutionary Spatio-temporal dynamics; (iii) among all the variants of DMTTE, DMTTE$_{GRU}$ outperforms other baselines (*even those baselines with static road network information*) on all the datasets which indicates that DMTTE$_{GRU}$ has the most robust generalization ability among all the variants of DMTTE.

Parameter Sensitivity Analysis (RQ2). Figure 4 depicts the results of parameter sensitivity analysis of our proposed DMTTE. We search the dimension number of embedding D and step size β from a search space of $[32, 64, 128]$ and $[0.1, 0.3, 0.5]$. For the dimension number of embedding D, the best performance is achieved with 64. Increasing the model size is capable of endowing our predictive model with better representation ability, while increasing more dimension number of embedding D may involve noise in refining learned representations. For the step size β, the best performance is achieved with 0.3. Increasing the step size can expedite the convergence of our model, while increasing more step size may hinder our model from converging to the local optimum.

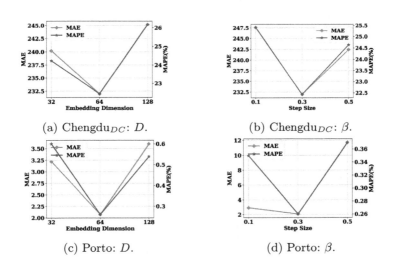

Fig. 4. Hyper-parameter study on Chengdu$_{DC}$ and Porto datasets.

Ablation Study (RQ3). To investigate the effectiveness of the attention mechanism, we compare the performance between DMTTE$_{WA}$ and DMTTE$_{LSTM}$. For Chengdu$_{DC}$ dataset, the MAE, MAPE, and RMSE are reduced by approximately 7.3%, 5.8%, and 0.69%, respectively. For Porto dataset, the MAE, MAPE, and RMSE are reduced by approximately 3.2%, 8.1%, and 8.6%, respectively. For Chengdu$_{DD}$ dataset, the MAE, MAPE, and RMSE are reduced by approximately 7.5%, 10.1%, and 11.5%, respectively. Hence the effectiveness of the attention mechanism has been verified.

Case Studies (RQ4). Figure 5 depicts four different cases of travel time estimation on Chengdu$_{DC}$ dataset. We show different fractions and MAPEs under different circumstances in this figure and elaborate on the following:

Case 1: Travel Time. As illustrated in Fig. 5a, DMTTE$_{GRU}$ significantly surpasses other baselines on most fractions of travel time on Chengdu$_{DC}$ dataset. Even for those trajectories that occupied less than 10% of all the trajectories, of which travel time is beyond 14 min, DMTTE$_{GRU}$ still outperforms other baselines.

Case 2: Travel Distance. As illustrated in Fig. 5b, DMTTE$_{GRU}$ surpasses other baselines on all fractions of travel distance. Since there are a limited amount of trajectories in which travel distances are less than 2 km or more than 7 km, it is challenging for all the baselines to estimate travel time accurately.

Case 3: Day of the Week. As illustrated in Fig. 5c, DMTTE$_{GRU}$ surpasses other baselines on each day of the week. Due to significant differences in traffic patterns between workdays and weekends and data-sparse issues on weekends, the performance of estimation on weekends for all the baselines is inferior. Especially, performance on Tuesday for all the baselines is inferior due to its fraction of all trajectories, which is less than 10 %.

Case 4: Hour. As illustrated in Fig. 5d, DMTTE$_{GRU}$ surpasses other baselines on all fractions of hours. Since there are a limited amount of trajectories of which hours are between 0:00 to 9:00, it is challenging for all the baselines to estimate travel time accurately.

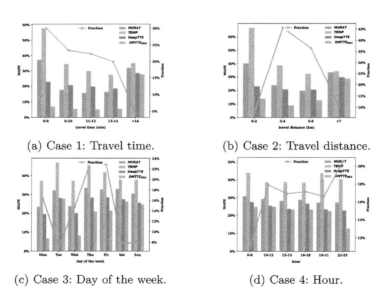

(a) Case 1: Travel time.   (b) Case 2: Travel distance.

(c) Case 3: Day of the week.   (d) Case 4: Hour.

Fig. 5. Case studies on Chengdu$_{DC}$.

Computation Cost (RQ5). To show the performance of the computation of our model, we compare the variants of DMTTE with other baselines. The training time for WDR, DeepTTE, STNN, MURAT, Nei-TTE, DMTTE$_{WA}$, DMTTE$_{LSTM}$, DMTTE$_{BiLSTM}$ and DMTTE$_{GRU}$ are 1.3, 160.7, 2.4, 5.0, 1.7, 3.1, 3.2, 5.2, and 3.1 hrs, respectively. The training time of variants of DMTTE are analogous, ranging from 3.1 to 5.2 hrs. Among all the baselines, DeepTTE performs well on the MAPE metric compared to DMTTE$_{LSTM}$ at high computation cost. WDR, STNN, and Nei-TTE are faster than the variants of DMTTE but show poor performance on all metrics. In a nutshell, DMTTE seems to be able to make a better trade-off in terms of training time and performance.

Meta-learning Study (RQ6). To investigate the impact of the introduction of meta-learning, we conduct experiments by training DMTTE$_{LSTM}$ using vanilla deep learning techniques for 100 epochs (denoted by DMTTE$_{WOM}$) on the Chengdu$_{DC}$ dataset. The MAE, MAPE and RMSE for DMTTE$_{WOM}$ are 329.02, 48.23% and 828.89, respectively, which takes training time 97.7 hrs comparing to 3.2 hrs for DMTTE$_{LSTM}$. We can conclude that: (1) the MAE, MAPE, and RMSE metrics are significantly diminished by introducing meta-learning techniques, which proves that the meta learning-based optimization method is essential in DMTTE; (2) the training time significantly diminishes by introducing meta-learning techniques which verifies that meta learning-based optimization algorithm enhances DMTTE's ability of fast learning.

Table 2. Scalability of test result on Chengdu$_{DC}$.

Baselines	MAPE (%) on Chengdu$_{DC}$				
	20%	40%	60%	80%	100%
AVG	56.62	43.00	41.29	39.89	39.71
LR	49.54	49.44	49.22	49.10	49.09
GBM	50.82	44.99	42.10	41.76	41.67
TEMP	43.19	41.96	40.80	40.60	39.70
WDR	45.85	33.19	32.38	32.99	29.74
DeepTTE	28.89	26.38	25.90	25.70	24.22
STNN	41.33	36.62	34.61	30.31	30.08
MURAT	44.93	35.86	30.65	29.24	28.54
Nei-TTE	52.96	40.26	33.74	32.62	30.04
DMTTE$_{GRU}$	**24.23**	**23.69**	**23.48**	**23.25**	**22.44**

Scalability Comparision (RQ7). To investigate the scalability, we conduct experiments by varying the training dataset size of Chengdu$_{DC}$. More specifically, we sample 20%, 40%, 60%, 80% and 100% from the Chengdu$_{DC}$ dataset to train these baselines and then present the MAPE metrics in Table 2. We can observe that: (1) All methods perform better when more training data are utilized which cover more diverse patterns. (2) Our DMTTE$_{GRU}$ has more stable and effective performance than other methods. On one hand, the MAPE metric only increases 8% when the sampling size of dataset varies from 20% to 100%. On the other hand, the MAPE metric (24.23%) of DMTTE$_{GRU}$ trained on 20% of the Chengdu$_{DC}$ dataset is almost the same compare to that (24.22%) of DeepTTE trained on 100% of the Chengdu$_{DC}$ dataset. This two-fold observation demonstrates the stability and effectiveness of our proposed DMTTE$_{GRU}$ which supports our first contribution (i.e., generalize well from limited amount of data).

6 Conclusion

In this paper, we propose a novel meta-learning-based travel time estimation framework, namely DMTTE. Specifically, DMTTE is optimized based on the meta-learning algorithm, which captures evolutionary Spatio-temporal dynamics to provide accurate travel time estimation over time. Evolving-Tasks are extracted from three real-world datasets, supporting DMTTE to generalize better for daily evolving travel patterns and mitigates computation burdens when adapting DMTTE to recent trajectory data. An efficient Spatial-Temporal Modelling network (i.e., STMNet) is elaborately designed for learning evolving road networks in a data-driven way. Extensive experiments on three real-world, large-scale datasets have demonstrated the effectiveness of our DMTTE.

References

1. Asif, M.T., et al.: Spatiotemporal patterns in large-scale traffic speed prediction. IEEE Trans. Intell. Transp. Syst. **15**(2), 794–804 (2013)
2. Bahdanau, D., Cho, K., Bengio, Y.: Neural machine translation by jointly learning to align and translate. arXiv preprint arXiv:1409.0473 (2014)
3. Chiu, J.P., Nichols, E.: Named entity recognition with bidirectional lstm-cnns. Trans. Assoc. Comput. Linguist. **4**, 357–370 (2016)
4. Chung, J., Gulcehre, C., Cho, K., Bengio, Y.: Empirical evaluation of gated recurrent neural networks on sequence modeling. arXiv preprint arXiv:1412.3555 (2014)
5. Fan, Y., Xu, J., Zhou, R., Li, J., Zheng, K., Chen, L., Liu, C.: Metaer-tte: an adaptive meta-learning model for en route travel time estimation. In: Proceedings of the Thirty-First International Joint Conference on Artificial Intelligence (IJCAI-ECAI 2022) (2022)
6. Fang, X., Huang, J., Wang, F., Zeng, L., Liang, H., Wang, H.: Constgat: contextual spatial-temporal graph attention network for travel time estimation at baidu maps. In: Proceedings of the 26th ACM SIGKDD International Conference on Knowledge Discovery & Data Mining, pp. 2697–2705 (2020)
7. Fu, T.y., Lee, W.C.: Deepist: deep image-based spatio-temporal network for travel time estimation. In: Proceedings of the 28th ACM International Conference on Information and Knowledge Management, pp. 69–78 (2019)

8. Han, L., Du, B., Lin, J., Sun, L., Li, X., Peng, Y.: Multi-semantic path representation learning for travel time estimation. IEEE Trans. Intell. Transp. Syst. (2021)
9. Jindal, I., Chen, X., Nokleby, M., Ye, J., et al.: A unified neural network approach for estimating travel time and distance for a taxi trip. arXiv preprint arXiv:1710.04350 (2017)
10. Kingma, D.P., Ba, J.: Adam: a method for stochastic optimization. arXiv preprint arXiv:1412.6980 (2014)
11. Lan, W., Yanyan, X., Zhao, B.: Travel time estimation without road networks: an urban morphological layout representation approach. In: Proceedings of the Twenty-Eighth International Joint Conference on Artificial Intelligence (IJCAI-2019) (2019)
12. Li, Y., Fu, K., Wang, Z., Shahabi, C., Ye, J., Liu, Y.: Multi-task representation learning for travel time estimation. In: Proceedings of the 24th ACM SIGKDD International Conference on Knowledge Discovery & Data Mining, pp. 1695–1704 (2018)
13. Lv, Y., Duan, Y., Kang, W., Li, Z., Wang, F.Y.: Traffic flow prediction with big data: a deep learning approach. IEEE Trans. Intell. Transp. Syst. **16**(2), 865–873 (2014)
14. Nichol, A., Achiam, J., Schulman, J.: On first-order meta-learning algorithms. arXiv preprint arXiv:1803.02999 (2018)
15. Qiu, J., Du, L., Zhang, D., Su, S., Tian, Z.: Nei-tte: intelligent traffic time estimation based on fine-grained time derivation of road segments for smart city. IEEE Trans. Industr. Inf. **16**(4), 2659–2666 (2019)
16. Sak, H., Senior, A., Beaufays, F.: Long short-term memory based recurrent neural network architectures for large vocabulary speech recognition. arXiv preprint arXiv:1402.1128 (2014)
17. Sinnott, R.W.: Virtues of the haversine. Sky and telescope **68**(2), 158 (1984)
18. Wang, C., Luo, H., Zhao, F., Qin, Y.: Combining residual and lstm recurrent networks for transportation mode detection using multimodal sensors integrated in smartphones. IEEE Trans. Intell. Transp. Syst., 1–13 (2020)
19. Wang, C., Zhao, F., Zhang, H., Luo, H., Qin, Y., Fang, Y.: Fine-grained trajectory-based travel time estimation for multi-city scenarios based on deep meta-learning. IEEE Trans. Intell. Transp. Syst. **23**(9), 15716–15728 (2022)
20. Wang, D., Zhang, J., Cao, W., Li, J., Zheng, Y.: When will you arrive? estimating travel time based on deep neural networks. In: AAAI, vol. 18, pp. 1–8 (2018)
21. Wang, H., Tang, X., Kuo, Y.H., Kifer, D., Li, Z.: A simple baseline for travel time estimation using large-scale trip data. ACM Trans. Intell. Syst. Technol. (TIST) **10**(2), 1–22 (2019)
22. Wang, Z., Peng, Z., Wang, S., Song, Q.: Personalized long-distance fuel-efficient route recommendation through historical trajectories mining. In: Proceedings of the Fifteenth ACM International Conference on Web Search and Data Mining, pp. 1072–1080 (2022)
23. Wang, Z., Fu, K., Ye, J.: Learning to estimate the travel time. In: Proceedings of the 24th ACM SIGKDD International Conference on Knowledge Discovery & Data Mining, pp. 858–866 (2018)

24. Yang, B., Guo, C., Jensen, C.S.: Travel cost inference from sparse, spatio temporally correlated time series using Markov models. Proc. VLDB Endowment **6**(9), 769–780 (2013)
25. Zhang, H., Wu, H., Sun, W., Zheng, B.: Deeptravel: a neural network based travel time estimation model with auxiliary supervision. arXiv preprint arXiv:1802.02147 (2018)

Machine Learning in Engineering and Industry

APF-DQN: Adaptive Objective Pathfinding via Improved Deep Reinforcement Learning Among Building Fire Hazard

Ke Zhang[1]($\boxtimes$), Dandan Zhu[1]($\boxtimes$), Qiuhan Xu[1], Hao Zhou[1], and Xuemei Peng[2]

[1] China University of Petroleum, Beijing, China
2021211702@student.cup.edu.cn, zhu.dd@cup.edu.cn
[2] The Hong Kong University of Science and Technology (Guangzhou), Guangzhou, China
xpeng558@connect.hkust-gz.edu.cn

Abstract. Evacuation path planning is a critical task to enable the safety of individuals in a fire hazard. Current evacuation planning approaches mainly calculate a fixed optimal path given a deterministic task. Nevertheless, fire evacuation guidance confronts some vital challenges including multi-exits existing in building and unstable evacuation path caused by dynamic fire spread. To resolve these issues, this paper proposes an evacuation agent which possesses a novel Artificial Potential Field Deep Q-Learning (APF-DQN) algorithm to calculate an evacuation route which evacuation agent enables choosing an appropriate exit and plan a dynamic evacuation path. A concept called artificial potential field is introduced into deep Q-learning architecture to lead agent adaptively choose targeted exit and avoid damage from fire spread. Meanwhile, deep Q-learning framework ensures evacuation agent plan a dynamic path. Then, APF-DQN is estimated in a proposed simulation experiments and compared with several conventional path-finding methods. Our APF-DQN reduces time-step cost by 18.7% and increases distance to closest fire by 20.1% compared with classical A star and APF methods. Our code can be downloaded from URL: https://github.com/ColaZhang22/APFDQN-Indoor-fire-hazard-path-planning.

Keywords: Deep reinforcement learning · Intelligent agent · Adaptive Objective · Pathfinding · Fire evacuation

1 Introduction

Pathfinding is a vital component of evacuation system inside building when confronting with fire hazard. Nevertheless, with architecture of building increasingly

---

Supported by the Science Foundation of China University of Petroleum, Beijing (No. 2462020YXZZ024).

becoming more intricate and larger, planning a rational evacuation path enables to efficiently reduce threat caused by fire hazard (e.g., fire by-product and bomb threat). Beside, due to multi-exits and dynamic fire hazard in a building, a fixed evacuation path is not enough to tackle multi-exits in building. Thus, a rapid rising focus in fire hazard evacuation is to build a dynamic, reliable and safety route.

Conventional pathfinding method [18] is time-consuming caused by large-scale architecture of indoor building and thus unacceptable in emergency scenario. Although numerous prior research [1] has been proposed to solve this issue based on some modifications, like improved ant colony algorithm [18] and Hierarchical A-star [3]. But there still exists two challenges in current methods as shown in Fig. 1. The first issue is negative impact from variation of dynamic fire hazard spread. With development of fire hazard, some path in the environment cannot pass and meanwhile some room become an obstacle in evacuation way. The second problem is there exists multi-exits in a large building. Therefore, choosing an appropriate exits according to fire hazard is another vital issue.

Fig. 1. One exit and multi-exits evacuation guidance pathfinding. In the left side (a), current pathfinding methods enable to tackle dynamic fire and arrive at targeted exit. But in some certain scenario, there exists multi-exits so that evacuation agent needs to consider both influence of damage from fire and multi-exits.

Few researchers have addressed the lack of variation of circumstance of building within development of fire hazard and meanwhile consider adaptive choice due to multi-exits. In some certain scenarios, an exit will be invalid because of locked or obstacle of fire and more than one place outbreaks of fire simultaneously. Therefore, there remains a need for an extensible pathfinding method which can dynamically plan an optimal evacuation path in fire hazard and adaptively choose an appropriate exit among multi-exits.

To solve mentioned issues, this study aims to perform a method called Artificial Potential Field Deep Q-Learning (APF-DQN) to train an evacuation agent to navigate individual to exits in indoor building which has lower egress time and safer distance. We examine our method in two building scenarios and compare with two typical pathfinding algorithms. Our agent is able to plan a dynamic evacuation path considering fire spread and choose a rational exit as target to avoid damage from fire.

Our main contributions can be described as follows:

1. Considering the influence factors of fire spread and multi-exits in building, Artificial potential fields are introduced into conventional DRL method and thus we perform a novel algorithm architecture called **Artificial Potential Field Deep Q-Learning** (APF-DQN) to solve variation of environment in fire hazard and adaptively choose a targeted exit.
2. Our research applies Deep Reinforcement Learning (DRL) into indoor building fire evacuation path planning circumstance to tackle high recalculation expense and fixed routes in comparison with conventional path planning methods.
3. Our evacuation agent model can be applied in different fire evacuation building circumstances included multi-exits and fires. Evacuation path planned by APF-DQN enables to adaptively choose an appropriate exit among multi-exits and take lowest time-consuming and avoiding damage of fire meanwhile.

The rest of this paper is organized as follow: Sect. 2 describes related work concerning evacuation path plan in fire hazard and motivation. Section 3 presents architecture of APF-DQN and detail components of evacuation agent. Section 4 introduces the simulated environment which tests our proposed methodology. The result and conclusion of experiment is described in Sect. 5. Finally, Sect. 6, we consider future pathfinding in multi-agent pathfinding scenario.

2 Related Work

Current pathfinding studies, such as heuristic A star [9] and improved ant colony algorithm [18], modify conventional path planning methods to dynamically calculate an evacuation path. Nevertheless, computation time in these methods explosively increases when the size of building grows. In an urgent evacuation scenario, computing paths that take a lot of time reduces evacuation efficiency.

In recent years, deep reinforcement learning has obtained enormous success and shown potential in various fields, ranging from electric games [16], robotics control [2] and go [12]. Therefore, some researchers [8,10,15] has adapted reinforcement learning (RL) to dynamic calculate a route in real-time. Reinforcement learning-based approaches can guarantee the dynamic feasibility and be extensible with different building scenarios (e.g. different building scenarios and various exits). [8] indicates that heuristic search algorithms, like A*, do not take the obstacles into account, and proposes instance-dependent heuristic proxies to learn a correction factor and path probability so that to increase the efficiency of the search.

To solve pathfinding problems with many conflicting objectives, [15] introduces a new model-free many objective reinforcement learning algorithm, called Voting Q-learning, that is capable of finding a set of optimal policies in an initially unknown, stochastic environment with several conflicting objectives. [4] presents zone-based path finding (or ZBPF) where agents move among zones and agents' movements require uncertain travel time. Meanwhile, [4] present a multi-agent credit assignment scheme(DC) that helps pathfinding learning approach converge faster.

Nevertheless, current reinforcement learning based pathfinding approaches exit two drawbacks to tackle fire hazard evacuation scenario. The first is due to development of fire hazard, fire in the building is not like a static obstacle but a dynamic diffused obstacle. Some exits in the building are destroyed by fire and thus become invalid for evacuation. Spread of fire violates time-invariant of MDP and is not only decided by current state of fire and building information but also time-scale. To solve this problem, our paper introduces the artificial potential field (APF) into DQN to handle instability of dynamic fire environment. Development of fire is considered as a repulsive field and add into state of agent. By calculating field produced by fire, development of fire hazard is transferred as part of state so that lead evacuation agent to re-satisfy MDP.

The second drawback is that each agent has individual one targeted objective in conventional path planning methods. However, in fire evacuation scenario, there exits multi-exits in a large building so that each individual needs to choose an appropriate exit as target and adjust targeted exit with development of fire. Therefore, our APF-DQN calculate an attractive field to estimate attractive forces from various exits. In the evacuation process, agent adjusts targeted goal among multi-exits based on variation of attractive forces. Next section specifically explains framework and process of APF-DQN method.

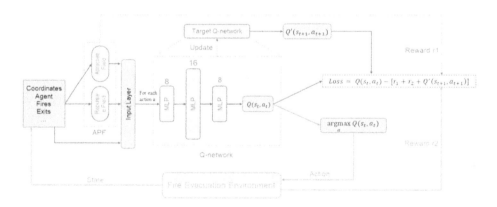

Fig. 2. Network structure of artificial potential field-deep Q learning (APF-DQN). For each fire and exit in the building, evacuation agent calculates repulsive field and attractive field and adds into states. Meanwhile, variation of field is set as reward of action. Therefore, our method enable evacuation agent to adaptively choose exit and avoid damage of fire.

Few researchers have addressed the lack of variation of circumstance of building within development of fire hazard and meanwhile consider adaptive choice due to multi-exits. In some certain scenarios, an exit will be invalid because of locked or obstacle of fire and more than one place outbreaks of fire simultaneously. Therefore, there remains a need for an extensible pathfinding method which can dynamically plan an optimal evacuation path in fire hazard and adaptively choose an appropriate exit among multi-exits.

To solve mentioned issues, this study aims to perform a method called Artificial Potential Field Deep Q-Learning (APF-DQN) to train an evacuation agent to navigate individual to exits in indoor building which has lower egress time and safer distance. We examine our method in two building scenarios and compare with two typical pathfinding algorithms. Our agent is able to plan a dynamic evacuation path considering fire spread and choose a rational exit as target to avoid damage from fire.

3 Methodology

Artificial potential field deep Q-learning (APF-DQN) is divided into two parts included reinforcement learning (RL) and artificial potential field. In comparison with classical pathfinding approaches, RL based path planning enables the provision of a real-time dynamic evacuation route so that avoid computing overhead. Artificial potential field leads agent to adaptively choose an appropriate targeted exit as goal and adjust targeted exit according to variation of fire hazard in the building.

3.1 Reinforcement Learning Framework for Fire Evacuation Guidance Agent

In order to reduce the time consumption of calculating the optimal path, DQN, a decision-making model, is adapted as path planning framework in fire evacuation scenario. In this research, fire evacuation guidance system is regarded as an evacuation agent in DQN and interacted with fire evacuation environment. Evacuation agent plans next evacuation coordinate and adjusts targeted exit according to self-state and fire hazard.

The entire DQN process is illustrated in Fig. 2. The whole learning process of evacuation agent is defined as a Markov decision process (MDP), which can be described as a five-tuple $(s_t, a_t, P_t, r_t, \gamma_t)$. s_t is state of evacuation agent at time t and a_t is the action adapted by agent at time t. P_t represents state transition function for agent from s_t to s_{t+1}. r_t represents reward function that agent receives from environment at time t and γ is discount parameter. The goal of evacuation agent aims to maximize the accumulated discounted reward:

$$maximize \underset{s}{\overset{S}{\mathbb{E}}}(\gamma^1 r_1 + \gamma^2 r_2 + \gamma^3 r_3 + ... + \gamma^t r_t)$$

where t is the expect time horizon in an evacuation episode and S represents state distribution.

In fire evacuation scenario, evacuation agent senses surrounding attributions such as current coordinates and positions of fire points from environment as state s_t of agent. Then fire evacuation guidance agent chooses next escape action based on sensed information. At last the fire evacuation environment feeds back reward to agent and transfers state s_t into next state s_{t+1}. The goal of evacuation agent aims to accumulate and maximize these rewards. Classical RL method Q-learning [17] introduces a function called Q value function to estimate the value of evacuation agent's current paired action-state as follow:

$$Q(s_t, a_t) = \mathbb{E}\{\sum_t^T (\gamma^t r_t | s = s_t, a = a_t)\} \quad (1)$$

where $Q(s_t, a_t)$ denotes the value of evacuation agent in state s_t took action a_t. Higher Q value means current escaped location is safer and evacuation agent prefers choosing this escaped location as optimal behavior. Besides, in training process, to keep balance between exploration and exploitation, evacuation agent adapts ε-greedy strategy to choose next action:

$$\pi(s_t, a_t) = \begin{cases} \text{random } a & \text{with probability } \varepsilon \\ \arg\max_a Q(s_t, a_t) & \text{with probability} 1 - \varepsilon \end{cases} \quad (2)$$

Q value function is a critical component in aforementioned method. Conventional method [17] always adapts a table to storage Q value regarding current state. However, with explosion of state space, there is not enough space to storage each state Q value. Thus, [6] proposed a novel method to, Deep Q learning, using deep neural network to approximate Q value function and transfer Eq. (1) into novel formulation:

$$Q(s_t, a_t) = \gamma * r_t + Q(s_{t+1}, a_{t+1}, \omega) \quad (3)$$

Which ω denotes parameters of deep neural network. r_t and γ denote reward evacuation agent received from fire hazard indoor building environment and discount factor. $Q(s_t, a_t)$ in Eq. (3) is regarded as target to calculate gradient of parameters w in Q value function. The optimization object is:

$$\underset{\omega}{minimize} \quad Q(s_t, a_t) - Q(s_t, a_t, \omega) \quad (4)$$

To get an optimal solution in Eq. (4), parameters ω in deep neural network is updated by rule:

$$w_{t+1} = w_t + \nabla_w (Q(s_t, a_t) - (\gamma * r_t + Q(s_{t+1}, a_{t+1}, w_t))) \quad (5)$$

In fire evacuation environment, development of fire disturbs assumption of MDP process, thus leads agent to make mistake. In some scenarios, agent judges and chooses the next action with closest distance to secure exit but this choice may be wrong caused by by-product and spread of fire. To eliminate impact from development of fire to DQN in fire evacuation, our research then introduces artificial potential field (APF) into DQN and performs APF-DQN method in next subsection.

3.2 Artificial Potential Field for Adaptive Targeted Exit

Classical RL method cannot tackle development of fire hazard in large building. Therefore, artificial potential field (APF) is utilized to transfer development of fire into a field and add this field into state of evacuation agent to eliminate instability of environment. Meanwhile, based on different attractive field from exits, evacuation agent enables to adaptively choose an exit as targeted goal.

The APF algorithm generates two forces field included both attractive force field and repulsive force field. In a fire evacuation scenario, fire and obstacle produces a repulsive field and guide evacuation agent to stay away from fire hazard. In comparison with fire, multi-exits in building separately produce an attractive field and lead agent towards the targeted exit. Therefore, APF enables to accurately reflect variation information of fire and dissipates instability of building environment caused by development of fire and different scenarios.

Attractive field in APF represents intensity of attractive force by multi-exits and is formulated as:

$$U_{attr,i}(x) = \begin{cases} \frac{1}{2}\lambda(x - x_{goal,i})^2 & (x - x_{goal,i})^2 \leq x_{thres} \\ 0 & (x - x_{goal,i})^2 > x_{thres} \end{cases} \quad (6)$$

x and x_{goal} represent position of evacuation agent and exits in our building environment. $(x - x_{goal})^2$ is distance between agent and exit. λ represents impact factor to illustrate importance of attractive field, our paper sets λ as 0.8. i denotes the number of exits and x_{thres} indicates effective range of exits.

Equation (6) means if x in range of x_{thres}, agent will be driven to targeted exit by attractive force. Thus, attractive field becomes larger With distance closer and evacuation agent would be guided into closest exit among all exits in the building. Nevertheless, closest exit is not a ideal target when fire hazard happened near the closest exit. Therefore, repulsive field in APF is also used to guide path planning of evacuation agent. Compared with attractive field, repulsive field represents intensity of repulsive force by obstacle or fire point. Repulsive field is designed as:

$$U_{req,i}(x) = \begin{cases} \frac{1}{2}\mu(\frac{1}{d(x-x_{fire,i})} - \frac{1}{x_{thres_f}})^2 & (x - x_{fire,i})^2 \leq x_{thres_f} \\ 0 & (x - x_{fire,i})^2 > x_{thres_f} \end{cases} \quad (7)$$

where $U_{req}(x)$ represents repulsive force from fire and obstacle in evacuation scenario. μ is hyper-parameter to control importance of repulsive and our paper set μ as 1.2. $d(x - x_{fire,i})$ represents distance between $fire_i$ and agent. When X_{thres_f} is within range of fire effective range, Eq (7) illustrates that agent will get larger repulsive force if agent becomes closer to fire point. Therefore, evacuation agent inclines to keep a safe distance with fire hazard and is guided toward an appropriate exit by attractive field at the same time. These both fields compose into artificial potential fields and drive evacuation agent to targeted exit.

$$U(x) = U_{attr}(x) + U_{rep}(x) \quad (8)$$

Furthermore, in conventional reinforcement learning, DQN exists sparse reward and limited state problem. Sparse reward prevents evacuation agent from learning efficiently from experience samples. Therefore, variation of APF is regarded as reward function between two states to enhance quality of training. Our method uses:

$$r = \begin{cases} \alpha(U_{attr,t+1} - U_{attr,t}) + \beta(U_{attr,t+1} - U_{attr,t}) & \text{else} \\ 10 & \text{arrive goal} \end{cases} \quad (9)$$

to represent reward function of APF-DQN. α and β is positive impact factor and negative impact factor to control importance of attractive field and repulsive field. When evacuation agent arrived exit, it acquires a fixed reward 10. And APF-DQN introduces attractive force and repulsive force which been represented as gradient of field into state of agent (Fig. 3).

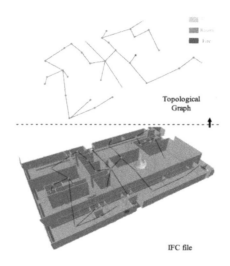

Fig. 3. Simulated Evacuation Building Environment based on IFC file. Extract topological information about indoor space from IFC fired and then construct S-graph based on topological. Add fire spread model into graph to construct a fire evacuation environment.

Adjacent tuples in one episode always have relation in time dimension and this relation always reduce efficiency of training. To erase relationship between two tuples in one episode, Deep Q-learning builds a replay buffer to store tuples. When train our model, agent samples batch size number of tuples in replay buffer.

4 Indoor Building Fire Hazard Evacuation Experiment

4.1 Fire Evacuation Environment

Considering domain of DRL, a high-quality environment can facilitate relevant research into RL application. For building a high quality environment, simulated fire evacuation environment used in our paper bases on the International Foundation Class (IFC) format [14] to construct our fire evacuation environment. This process can be divided into two parts include indoor building model and fire hazard model.

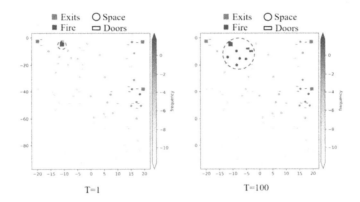

Fig. 4. Fire spread model in building environment. Red square represent place where original fire hazard happened. With time T increasing, range of fire space becomes larger so that part of evacuation space becomes invalid and cannot pass. (Color figure online)

Firstly, to simulate real indoor building environment, as shown in Fig. 4, a building IFC file is chosen to build a simulated environment. IFC fire is a standard file format in building engineer and compose all information about building. S-graph is simple representation to evaluate relationship and distance between two rooms and has been used in [11]. Therefore, our paper utilizes information from IFC file to build a S-graph to represent relative relationship and distance between various space in the building.

Furthermore, to evaluate negative impact from fire hazard, a fire spread model is introduced into our simulated building environment. Our environment uses model proposed in [7] which considers material, wind velocity and direction as our fire spread model. Development of fire from current node to neighborhood of nodes satisfies:

$$F = \theta * S * W^\rho * p \tag{10}$$

where S is a building structure parameter, various materials of building has a significant effect in fire spread. W is a parameter determined by the wind velocity and direction, and p is the ability of current state to cause the spread of fire. The variable θ is a coefficient to tune the degree of slowdown in spreading caused by the conditions except wind velocity, and ρ is a coefficient to adjust the range and direction of spreading. Based on this model, our environment enables to provide a dynamic fire evacuation simulated environment to test our method.

4.2 Experiment Settings

The whole APF-DQN method is implemented based on Python and Pytorch package. And the experimental hardware platform is an Intel(R) Core(TM) i7-10750H CPU, NVIDIA RTX 2080Super GPU, and 32 GB memory. Our Q network, as shown in Fig. 2, comprises three fully linear layers and one dropout layer. Meanwhile, fire evacuation simulation mentioned in last section is used as interactive environment to train our agent. Our experiment considers variation of Exits and fire points to verify our expansion capability. For instance, some exits are unable to pass because of broken or locked and in some extreme situation there exists several fire points at same time. At time t, agent senses APF field and self-attribution as current state. For each next possible action, agent has ϵ probability to choose action randomly and $1-\epsilon$ probability to choose next action which has maximum Q-value. Relevant parameter of our model was set as in Table 1.

Table 1. Relevant parameter in our model

Name	Value	Name	Value
Episode	10000	α	0.8
Episode length	100	Hidden layers	4
Optimizer	RMSprop	Activation Function	ReLU
Buffer size	100000	β	1.2
Batch size	32	γ	0.9
Lr	0.0025	λ	0.8
Update freq	100	μ	1.2

The goal of agent is to minimize the overall evacuation time steps and avoid damage from fire at same time. For instance, the position of one agent is closed to A exit but far away from B exit, meanwhile closed to fire point. When agent arrived at A exit, fire maybe has spread so that A exits has occupied by fire so that cannot pass. By calculating field, agent would choose the exit has largest attractive field as targeted objective and plan a route far away from fire hazard. Thus, agent based on our APF-DQN will learn how to avoid damage and are willing to choose B exist in fire evacuation process. After training, our agent enables to find an efficient path which possesses shortest evacuation time and safest distance to fire hazard.

5 Performance

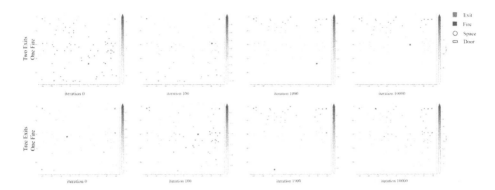

Fig. 5. Q value function for every state in various condition at $t = 0$. x, y coordinates represent position of space and color depth of circle represent value of state in building when fire happened. Value of state is expectation of Q value. At the training beginning, circle color represents ruleless and evacuation agent is hard to choose right way. With training process, state which is closed distance to exits and far away to fire points has a high value. Thus, agent is willing to choose those high value state far away from dangerous fire and closed to exits.

Two environments, one consists of two exits and one fire and the other considers three exits, is utilized to text our proposed method. Considering various condition and calculate every Q value at time $t = 0$ for every state, the result of experiment is shown in Fig. 5. Deeper color represents a higher Q value and means evacuation agent inclines to take action which has a higher Q value. At beginning of training, Q value of each state is irregular and thus agent cannot choose an action correctly. With training process, states near the exits has higher Q value and near fire hazard possesses lower value in the environment. Therefore, evacuation agent enables to adaptively choose an appropriate exit among multi-exits and keep far away from far hazard at the same time.

As for evacuation path, our experiment aims to simulate real fire evacuation situation, so agent is initially set in position randomly and fire hazard also happens in random place. As shown in Fig. 6(f)(b), our model enables to find optimal evacuation path to escape from fire in various circumstance and choose a route which far away from fire to avoid damage from fire hazard. Meanwhile, evacuation planned by APF-DQN shows when faced multi-exits at same time, agent is willing to choose exit away from fire point. In (c)(d) of Fig. 6, planned evacuation route shows although some exits close to current state of agent, agent prefers to plan a relatively far but safer path to ensure escape of fire. Therefore, APF-DQN enables to dynamically choose an appropriate exit as target to escape among multi-exits in the building.

Then, we evaluate our proposed APF-DQN from two criteria, egress time and shortest distance from agent to fire hazard. Egress time estimates time-step agent spend to arrive at an appropriate exit from initial coordinate. In fire hazard evacuation, lower egress time means agent efficiently arrive at an exit so that avoid damage from fire hazard. Shortest distance represents distance from agent to the closet fire.

To estimate performance of our model, our method take ablation experiment with two classical path planning methods, A-star and APF. A-star [13] is a classical shortest path planning which has been applied in electric games. A-star chooses next action based on current position and heuristic function like Euclidean distance. APF [5] also is common shortest path planning algorithm, which considers influence of obstacles in environment. In our experiment, for each agent, difference of initial positions significantly effect evacuation time-step and closest distance to fire. Thus, our experiments begin 100 games and calculate their average distance and time-steps to eliminate error per episode. As shown in Table 2, our method(APF-DQN) has a shortest egress time and largest distance from fire hazard.

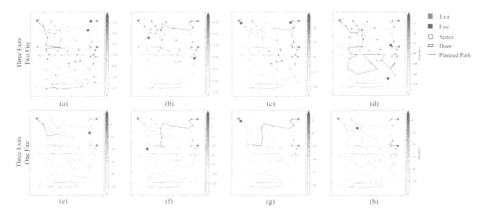

Fig. 6. Fire evacuation path planned by APF-DQN. Red line is the evacuation path planned by our model. (a)(b) show that agent would choose exit away from fire rather than exists fire points near about exits. (c)(d) illustrates agent choose safer path to exit despite there exists closer path to exit. (Color figure online)

As shown in Fig. 7, several classical approaches are stable method and thus has a horizontal line. As for our method, in the beginning of training, APF-DQN has a longer time-step compared with classical path finding method. However, with the training progressing, fire evacuation routes planned by our evacuation agent are progressively shorter and the distance to nearest fire point gradually increase. Average of closed distance base on APF-DQN method is 16.6 m, compared to 10.2 m of A-star and 4.8 m of APF, our APF-DQN (red line) can provides a safer distance to fire point. Meanwhile, average evacuation time-steps

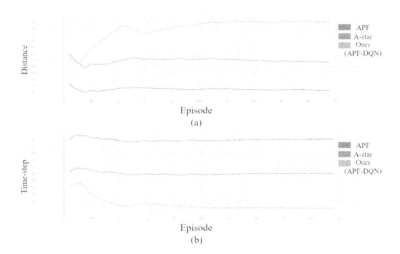

Fig. 7. Distance from fire and Evacuation time-steps of three methods. x axis means episode. y axis in (a) is agent closest distance from fire. y axis in (b) is agent time-steps from initial position to exit. Compared to APF and A star, our method keeps safer distance and plans a faster path.

Table 2. Experimental results

Method	Optimal Found Ratio	Egress time	Shortest distance
A*	100	17.23 ± 5.23	10.2 ± 2.32
APF+A*	100	21.53 ± 3.45	8.8 ± 2.03
APF	100	23.42 ± 4.24	4.8 ± 3.34
DQN	85.3	13.21 ± 2.32	0.41 ± 1.31
APF-DQN(**Ours**)	89.4	**12.21 ± 4.83**	**16.6± 2.12**

of our method are 12.2 but A-star are 17.2 and APF are 23.4. Our method always has lower time-steps expense so that evacuation agent enables to plan a shortest evacuation route for individual. At the same time, distance to fire of evacuation agent gradually increases so that can avoid damage from fire. Therefore, it demonstrates our APF-DQN method can provide highly efficient fire evacuation path guidance. The experimental result illustrates that evacuation agent can plan a rational path for evacuation while avoiding damage from fire compared to classical methods.

6 Conclusion and Future Work

Our paper focuses on evacuation path planning in fire hazard building environment. There exits multi-exit and dynamic fire hazard in the building so that evacuation agent needs to choose an appropriate targeted exits to escape and avoid damage from fire hazard. Conventional methods require recalculation for each circumstance and just give a static path for a fixed exit. However, place which fire hazard happened is random and multi-exits cause various evacuation route. To tackle these issues, our paper proposes an APF-DQN to plan evacuation path to an exit. APF-DQN fuses artificial potential field methods into deep q-learning to calculate shortest evacuation path and keep safe distance from fire point simultaneously. Beside, our method enables to adaptively choose an appropriate exit among multi-exits in environment and reshape reward function based on variation of attractive field and repulsive field to lead agent sense development of fire and exits.

To test availability of our method, we construct a fire indoor environment based on our building model and set a fire spread model in this indoor environment. Result of our experiment shows that our method increase 38.55% and 71% closest distance to make individual safer compared to A-star and classical APF methods. Meanwhile, evacuation time-steps APF-DQN consumed decreases 29.06% and 47.86% compared with above function. Figure 6 illustrates that agent based on APF-DQN enable to find a optimal evacuation path and keep away from fire or choose shortest path with closer to fire point but have a safe distance. Besides, evacuation route planned by APF-DQN enables to adaptively choose an appropriate exit to guide individual to avoid damage from fire.

Nevertheless, some limitation still exits in our paper. This work just currently considers one agent scenario and provides a simple training environment. Congestion caused by crowd and complexity of multi-layer building have a negative impact for evacuation efficiency. Therefore, future research will focus on crowded fire evacuation in multi-layer complex building scenario. In this scenario, restriction of space will cause congestion and decay rate of evacuation. To overcome these problems, we will focus on cooperation of multi-agent in fire evacuation circumstance to solve potential congestion in fire hazard evacuation.

References

1. Cuesta, A., Abreu, O., Balboa, A., Alvear, D.: Real-time evacuation route selection methodology for complex buildings. Fire Saf. J. **91**(April), 947–954 (2017). https://doi.org/10.1016/j.firesaf.2017.04.011
2. Gu, S., Holly, E., Lillicrap, T., Levine, S.: Deep Reinforcement Learning for Robotic Manipulation with. ICRA, July 2018
3. Holte, R.C., Perez, M.B., Zimmer, R.M., MacDonald, A.J.: Hierarchical A *: searching abstraction hierarchies efficiently. In: Proceedings of the thirteenth national conference on Artificial intelligence - Volume 1, pp. 530–535. AAAI Press, Portland, Oregon (1996)
4. Ling, J., Gupta, T., Kumar, A.: Reinforcement learning for zone based multiagent pathfinding under uncertainty. In: Proceedings of the International Conference on Automated Planning and Scheduling **30**(1), 551–559, June 2020. https://doi.org/10.1609/icaps.v30i1.6751, https://ojs.aaai.org/index.php/ICAPS/article/view/6751

5. Ma, B., Wei, C., Huang, Q., Hu, J.: Apf-rrt*: an efficient sampling-based path planning method with the guidance of artificial potential field. In: 2023 9th International Conference on Mechatronics and Robotics Engineering (ICMRE), pp. 207–213 (2023). https://doi.org/10.1109/ICMRE56789.2023.10106516
6. Mnih, V., et al.: Playing Atari with Deep Reinforcement Learning, pp. 1–9 (2013)
7. Ohgai, A., Gohnai, Y., Watanabe, K.: Cellular automata modeling of fire spread in built-up areas-A tool to aid community-based planning for disaster mitigation. Comput. Environ. Urban Syst. **31**(4), 441–460 (2007). https://doi.org/10.1016/j.compenvurbsys.2006.10.001
8. Panov, A.I., Yakovlev, K.S., Suvorov, R.: Grid path planning with deep reinforcement learning: Preliminary results. Procedia Comput. Sci. **123**, 347–353 (2018). https://doi.org/10.1016/j.procs.2018.01.054, https://www.sciencedirect.com/science/article/pii/S1877050918300553, 8th Annual International Conference on Biologically Inspired Cognitive Architectures, BICA 2017 (Eighth Annual Meeting of the BICA Society), held August 1-6, 2017 in Moscow, Russia
9. Rios, L.H.O., Chaimowicz, L.: A survey and classification of A* based best-first heuristic search algorithms. In: da Rocha Costa, A.C., Vicari, R.M., Tonidandel, F. (eds.) SBIA 2010. LNCS (LNAI), vol. 6404, pp. 253–262. Springer, Heidelberg (2010). https://doi.org/10.1007/978-3-642-16138-4_26
10. Sartoretti, G., Kerr, J., Shi, Y., Wagner, G., Kumar, T.K.S., Koenig, S., Choset, H.: Primal: pathfinding via reinforcement and imitation multi-agent learning. IEEE Robot. Automation Lett. **4**(3), 2378–2385 (2019). https://doi.org/10.1109/LRA.2019.2903261
11. Shaheer, M., Bavle, H., Sanchez-Lopez, J.L., Voos, H.: Robot Localization using Situational Graphs and Building Architectural Plans (2022)
12. Silver, D., et al.: A general reinforcement learning algorithm that masters chess, shogi, and Go through self-play. Science **362**(6419), 1140–1144 (2018). https://doi.org/10.1126/science.aar6404
13. Tang, G., Tang, C., Claramunt, C., Hu, X., Zhou, P.: Geometric A-star algorithm: an improved A-star algorithm for AGV path planning in a port environment. IEEE Access **9**, 59196–59210 (2021). https://doi.org/10.1109/ACCESS.2021.3070054
14. Thein, V.: Industry foundation classes (ifc)-BIM interoperability through a vendor-independent file format (2011)
15. Tozer, B., Mazzuchi, T., Sarkani, S.: Many-objective stochastic path finding using reinforcement learning. Expert Syst. Appl. **72**, 371–382 (2017). https://doi.org/10.1016/j.eswa.2016.10.045. https://www.sciencedirect.com/science/article/pii/S0957417416305863
16. Vinyals, O., et al.: Grandmaster level in StarCraft II using multi-agent reinforcement learning. Nature **575**(7782), 350–354 (2019). https://doi.org/10.1038/s41586-019-1724-z
17. Watkins, C.J., Dayan, P.: Technical note: Q-learning. Mach. Learn. **8**(3), 279–292 (1992). https://doi.org/10.1023/A:1022676722315
18. Xu, L., Huang, K., Liu, J., Li, D., Chen, Y.F.: Intelligent planning of fire evacuation routes using an improved ant colony optimization algorithm. J. Build. Eng. **61**(August), 105208 (2022). https://doi.org/10.1016/j.jobe.2022.105208

DDPM-MoCo: Enhancing the Generation and Detection of Industrial Surface Defects Through Generative and Contrastive Learning

Xiaozong Yang[1], Huailiang Tan[1(✉)], and Xinyan Wang[2]

[1] Hunan University, Changsha 410082, China
{yangxiaozong,tanhuailiang}@hnu.edu.cn
[2] Qilu University of Technology (Shandong Academy of Sciences), Jinan 250353, China

Abstract. The task of industrial detection based on deep learning often involves solving two problems: (1) obtaining sufficient and effective data samples, (2) and using efficient and convenient model training methods. DDPM-MoCo, a novel processing model, is proposed in this paper to address these issues. Firstly, Denoising Diffusion Probabilistic Model (DDPM) is utilized to generate high-quality defect data samples, overcoming the problem of insufficient sample data for model learning. Secondly, we introduces the unsupervised learning momentum contrast model (MoCo) to train the model with unlabeled sample data, addressing efficiency and consistency challenges in large-scale negative sample encoding during diffusion model training. The experimental results demonstrates a complete visual detection solution for metal surface defects from unlabeled sample data generation to model training and detection, providing practical guidance and application value for industrial visual detection in the metal processing industry.

Keywords: Defect generation · Diffusion model · Momentum contrast learning

1 Introduction

The defect detection [1-4] of industrial products requires a large number of feature samples to train the network so that the model can distinguish individuals of the same type as the training samples. This presents two main challenges: acquiring a sufficient amount of sample data and annotating the distinguishing features in the data. In reality, obtaining a large dataset of product defects is difficult, and manually labeling each sample is tedious and monotonous. This study focuses on defect detection in industrial products such as precision Aluminum alloy plate surfaces which are particularly challenging to obtain due to production enterprises' efforts to minimize product defects. In general, metal surface defects include corrosion, cracks, dents, scratches, ink marks, and brittleness. These defects not only affect the appearance of the product but also indicate potential quality issues [3]. In this research, we mainly focus on three surface defects that impact high-end Aluminum alloy products: dents, scratches, and corrosion. Dent is a single or multiple non-smooth depression defects produced by mechanical collision of Aluminum alloy plate, which is very destructive to the surface of the material. The scratch is a common defect in the processing, storage, and transportation of Aluminum alloy plates, caused by friction or scratching on the surface, and scratches damage the oxide film and the Aluminum alloy cladding layer, reducing the corrosion resistance of the material. The specific characteristics of corrosion are a little bit of white or black spots on the surface of the Aluminum alloy plate, which is caused by the production, packaging, transportation, storage process contact with acid or water. If the corrosion defects are not found and treated in time, they will not only make the surface of the Aluminum alloy plate lose its luster, but also reduce the corrosion resistance and comprehensive performance of the material [2]. as shown in Fig. 1.

The proposed method DDPM-MoCo, potentially greatly mitigates the above issues. To be more specific, firstly, we utilize the probabilistic diffusion model (DDPM) [19] to generate defect samples. Then, we employ the MoCo model with momentum contrast learning [20] for efficient training without annotations. Specifically, only a single sample is required for each type of defect in the diffusion model. The newly generated samples from the model are then mapped to the real data space. Once a large dataset is obtained, this paper utilizes the MoCo model for training without labeling data. The model compares each individual sample with all other samples and continuously iterates and optimizes to capture inherent high-level structural characteristics of each sample data. Subsequently, it classifies samples of similar types by training a linear classifier. Our contribution includes:

- We developed the DDPM-MoCo model, combining Denoising Diffusion Probabilistic Models and Momentum Contrast for enhanced defect detection [14,19].
- We addressed sample data scarcity by generating high-quality defect data and utilizing unlabeled samples for training [13].
- Our experimental results demonstrate effectiveness in metal surface defect generation, offering a practical solution for industrial visual inspection tasks [3].

2 Related Work

Tomer Amit, et al. proposed an image segmentation algorithm based on probabilistic diffusion models [5], which can automatically select segmentation points and has high robustness and reliability. Jascha Sohl-Dickstein proposed an unsupervised learning method based on probabilistic diffusion model [6], which can be used for data clustering, feature learning and dimensionality reduction tasks.

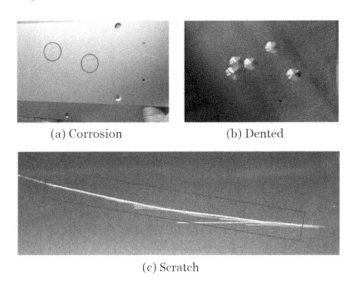

Fig. 1. Example of three most common defect types of our metal surfaces. (a) They are Original Aluminum alloy plate with corrosion sampling images . (b) They are Self-made dented aluminum alloy plate sampling images . (c) Shows Homemade scratched aluminum alloy plate sampling images .

Mengwei Ren, et al. proposed a probabilistic model of multiscale diffusion process [7], and conducted experiments on image classification and text classification using this model. Tianfei Zhou, et al. proposed a nonparametric probabilistic diffusion model [8], which can better adapt to the actual data, and used this model to conduct text classification and image segmentation experiments. R Yamini, et al. proposed a convolutional neural network (CNN) framework based on probabilistic diffusion model [9], which can be used for image denoising tasks and achieved good results. X. Zhang, et al. proposed a data clustering method based on probabilistic diffusion model [10], which uses diffusion embedding to map data into low-dimensional space and clusters data through spectral clustering algorithm. Y. Luo, et al. proposed a probabilistic encoder based on Gaussian diffusion processes [11], which can be used for unsupervised learning and feature extraction. Changde Du, et al. proposed a Multimodal deep generative adversarial models for scalable doubly semi-supervised learning in [12], which can improve the classification accuracy in the case of very little labeled data. Andrey Voynov, et al. proposed an Object Segmentation Without Labels with Large-Scale Generative Models [13], which demonstrated that large-scale unsupervised models can also perform a more challenging object segmentation task, requiring neither pixel-level nor image-level labeling. Kaiming He, et al. initially proposed an unsupervised learning method called momentum contrast in their work "Momentum contrast for unsupervised visual representation learning" [14]. This approach leverages online updating and momentum technology to maximize similarity and learn visual features. In the follow-up paper "MoCo v2: Improved Baselines with Momentum Contrastive Learning" [15], MoCo was further enhanced by incorporating more sophisticated data augmentation techniques to enhance the quality of feature representation. Tongzhou Wang and Phillip Isola investigated the contrastive loss function in MoCo, exploring it through alignment and uniformity on the hypersphere [16]. They proposed a novel contrastive loss function called NT-Xent, which optimizes alignment and uniformity in high-dimensional space to enhance learning quality. Zhirong Wu, et al. proposed another self-supervised learning method called instance discrimination [17]. The method learns discriminative features by comparing cropped regions within the same image. Ting Chen, et al. studied the learning ability of large self-supervised models in Big self-supervised models are strong semi-supervised learners [18] and found that these models can achieve good performance with a small amount of labeled data. MoCo was evaluated as one of the self-supervised learning methods and achieved good results. The above works has unique characteristics, but fails to effectively integrate the generation of large-scale sample data, preservation of high-quality features from the original data, and improvement in model training efficiency. These issues are addressed by the proposed DDPM-MoCo technology model. Experimental results suggest that DDPM-MoCo can efficiently generate large amounts of defects, and obtain richer diversity and better quality than other methods.

3 Methodologies

Our work is based on [15] [19]. The model we are using is U-Net architecture from guided diffusion, and the attention mechanism is also applied in DDPM. In this section, we illustrate step by step how we apply a diffusion process to metallic surface defect generation, and we demonstrate the application of a MoCo process for model training performed with unlabeled sample data.

3.1 Evolving Image Generation: From Deep Models to DDPM

Currently, the prevailing deep generative models can be broadly categorized into four distinct groups, as illustrated in well-known Fig. 2. These models primarily serve the purpose of generating or synthesizing various forms of content such as images, audio, video, text, and more. Diffusion models [5,7,8,19] have emerged as the new state-of-the-art family of deep generative models. They have broken the long-time dominance of generative adversarial networks (GANs) in the challenging task of image synthesis and have also shown potential in a variety of domains, ranging from computer vision, natural language processing, temporal data modeling ,multi-modal modeling to interdisciplinary applications.

Denoising Diffusion Probabilistic Models (DDPM) [19,26] is a very popular class of deep generative model that have been successfully applied to a diverse range of problems including image and video generation, protein and material synthesis, and neural surrogates of partial differential equations. DDPM can be

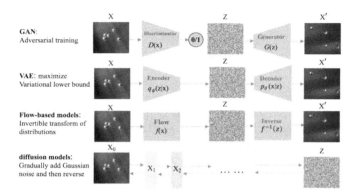

Fig. 2. Overview of classic deep generative models defect generation process.

used to generate images from random noise, and its performance is comparable to or better than GAN, VAE and other models. Many red-hot diffusion models in recent years are based on this. DDPM Model Highlights are the forward process, the backward pass, how to train and how to use.

(a) Forward Process (Diffusion Process). The original defect image x_0 is transformed into x_T by gradually adding Gaussian noise, so as to achieve the purpose of destroying the image. The forward process can be formulated as follows:

$$x_t = \sqrt{a_t} x_{t-1} + \sqrt{1-a_t} \epsilon_{t-1} \tag{1}$$

where $\{a_t\}_{t=1}^T$ is a pre-defined hyper parameter, called Noise schedule, which often includes columns with very small values. $\epsilon_{t-1} \sim N(0,1)$ is Gaussian noise. Equation (2) can be iteratively deduced by Equation (1). $\bar{a}_t$ is also a hyper parameter set with Noise schedule, $\epsilon \sim N(0,1)$ is Gaussian noise too. So the forward process can be depicted by Equation (1) or (2), and the (1) is used to destroy an input image gradually, but the (2) can make it in one step.

$$x_t = \sqrt{\bar{a}_t} x_0 + \sqrt{1-\bar{a}_t} \epsilon \tag{2}$$

(b) The Backward Pass (Denoising Process). The reverse process is to gradually restore the damaged x_T to x_0 by estimating the noise and iterating many times. The backward process can be formulated as follows:

$$x_{t-1} = \frac{1}{\sqrt{a_t}} x_t - \frac{\sqrt{1-a_t}}{\sqrt{a_t}} \epsilon_\theta(x_t, t) + \sigma_t \tag{3}$$

Since the real noise ϵ in Equation (3) is not allowed to be used in the restoration process, the key to DDPM is to train a model $\epsilon_\theta(x_t, t)$ that estimates the noise from x_t and t. θ is the training parameter of the model. σ_t is also Gaussian noise and $\sigma_t \sim N(0,1)$ which is used to indicate the difference between the estimate and the actual. In DDPM, U-Net serves as a framework for estimating noise.

(c)How to Train (To Obtain the Noise Estimation Model). From the above, we know the key to DDPM is to train a model $\epsilon_\theta(x_t, t)$ and it should be made to predict $\hat{\epsilon}$ close to the ϵ that is actually used for destruction. So L2 distance is a good way to describe the similarity and the Loss is formulated as:

$$Loss = \left\| \epsilon - \epsilon_\theta(x_t, t) \right\|^2 = \left\| \epsilon - \epsilon_\theta(\sqrt{\bar{a}_t} x_{t-1} + \sqrt{1 - \bar{a}_t}\epsilon, t) \right\|^2 \quad (4)$$

The Fig. 3 illustrates the training process of DDPM with enhanced MoCo.

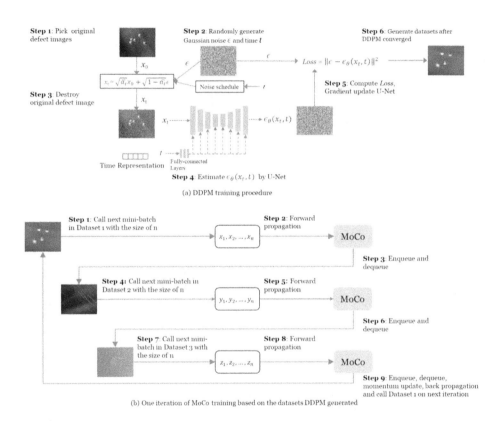

Fig. 3. DDPM training process with enhanced MoCo representation

3.2 Enhancing Data Representation and Extraction with MoCo

(a) Contrastive Learning and Momentum Contrast (MoCo). MoCo [14] refines contrastive learning by creating dynamic dictionaries for pretext tasks in unsupervised learning. This approach focuses on enhancing data representation through pretext tasks and optimizing loss functions, particularly employing contrastive loss. MoCo innovates by adjusting the loss function for instance discrimination tasks that use Noise-Contrastive Estimation (NCE) [14], facilitating learning from matched and unmatched image views. NCE loss, given by Eq. 5 and regulated by a temperature parameter τ, measures the similarity of a query q to a positive key k_+ over K negative keys, classifying q among k_+ out of $k+1$ options.

$$\mathcal{L}_q = -log \frac{exp(q * k_+/\tau)}{\Sigma_{i=0}^{k} exp(q * k_i/\tau)} \quad (5)$$

The MoCo algorithm, is an unsupervised technique that employs contrastive loss for training a visual representation encoder. This encoder matches a query q with a dynamically evolving dictionary of encoded keys $\{k_0, k_1, k_2, \ldots\}$, functioning via a queue mechanism to manage entry and exit of mini-batches, thus decoupling their size scalability. Key encoding leverages a momentum-based updating mechanism, ensuring synchronization with the query encoder's progression. Such an expansive dictionary facilitates robust visual representation learning by sampling a vast, continuous visual space. The efficacy of this approach hinges on the dictionary's size and the congruence between keys and query, necessitating comparable encoding mechanisms.

(b) Enhanced Batch Contrastive Representation Loss. Our DDPM model, designed for identifying three types of surface defects, advances the field by innovating upon the NCE contrastive loss function through "Dataset Discrimination." Training our DDPM across three defect classes yields three specialized models. From these, we generate corresponding datasets, laying the groundwork for a novel approach to contrastive learning. Utilizing a "for loop" across three PyTorch DataLoader instances, one for each class, allows us to refine the instance discrimination loss into what we term "Dataset Discrimination", as formalized in Eq. 6.

$$\mathcal{L}_q = -\frac{1}{n}\sum_{j=1}^{n} \log \frac{\exp(q \cdot k_j/\tau)}{\sum_{i=0}^{K} \exp(q \cdot k_i/\tau)} \quad (6)$$

Mirroring the structure of Eq. 5, our dataset-level discrimination (Eq. 6), with n representing the mini-batch size, enhances dataset representation through a "for loop" over DataLoader instances. This process ensures each query engages in an inner product with every positive key within its batch, significantly enriching the dataset representation. With Dataset Discrimination Loss, we underscore its ability to select negative samples of same labels as the anchor. With stronger negative samples, Dataset Discrimination Loss learns better representation. In a parallel study on representation learning, this batch-by-label fashion demonstrates its robustness against data corruption and reduction in the training dataset. Consequently, it proves particularly suitable for our scenario where the number of defective images is limited.

MoCo distinguishes itself through two primary advantages: it expands the size of negative sample pairs, more accurately mimicking real-world distributions for enhanced training efficacy; and it ensures consistent representation of identical data samples, significantly boosting the stability and performance of the training process.

(c) Queue and Momentum Enhancement in MoCo. In MoCo, the dictionary is conceptualized as a queue of data samples; the incorporation of queues enables the dictionary size to be decoupled from the mini-batch size, thereby enhancing flexibility and allowing it to be set as a hyperparameter; upon entry of each new mini-batch into the queue, the oldest mini-batch in the queue is dequeued, ensuring that the data within the dictionary always represents a sampled subset of all available samples (with each element in the queue representing an individual mini-batch).

The inclusion of queues increases the dictionary size, but it hinders updating the key encoder during backpropagation (as the gradient needs to propagate across all samples in the queue). To address this issue, a straightforward solution is to disregard the gradient and directly copy query encoder f_q to key encoder f_k. However, this approach exhibits poor performance due to its potential impact on reducing consistency in query features caused by rapid changes in encoders. So MoCo proposes Momentum update to solve the above problems, shown in Equation (7) where θ_k and θ_q are parameters of the encoder f_k and f_q respectively, m is Momentum coefficient and the larger the value of m is, the better it is, that is, learning θ_q slowly is more effective, and only θ_q updates with back propagation. The momentum updating strategy in Equation (6) ensures smoother learning of θ_k compared to θ_q Therefore, despite the extraction of key features by encoders from different mini-batches, the differences between these features are minimal. The kernel of MoCo used for our model is shown in Algorithm 1.

$$\theta_k \leftarrow m\theta_k + (1-m)\theta_q, \quad m \in [0,1) \tag{7}$$

Through MoCo, we can learn an encoder that effectively extracts the essential features of the dataset in an unsupervised manner. By fine-tuning the encoder while retaining its main weight parameters and adjusting the classification number of its last fully connected layer, we can achieve effective classification of the dataset. Specifically, our fine-tuned model is capable of accurately classifying data based on their respective characteristics. This allows dictionary size to be unlimited while maintaining a high degree of dictionary consistency.

4 Experiments

In this section, we aim to investigate whether our proposed framework can generate higher-quality defect images compared to other state-of-the-art methods and if the detection module trained on our generated dataset can boosts its performance.

4.1 Experiment Setup

(a)Laboratory Workbench. As shown in Fig. 4, the camera we used was GS3-U3-50S5M-C, which was used to take pictures of the surface of the Aluminum alloy plate from top to bottom on the experimental platform. We used an Aluminum alloy plate with a size of 280mm*200mm, and divided it into six rectangular blocks with a size of 80mm*80mm. Then, we manually made the features of two types of defects, namely depression and scratch, on each grid of the surface with an electric drill, shown as Fig. 1 (b) and (c).And we also find some defective products with corrosion spots, which is appeared in the process of production, packaging, transportation and storage, shown as Fig. 1 (a).

Algorithm 1. Simplified MoCo Kernel for Our Model

1: Initialize a dictionary queue of K keys and encoder networks f_q and f_k with identical parameters.
2: For each subset x in the queue with mini-batch size N, apply distinct random augmentations to generate positive sample pairs x_q and x_k.
3: Perform forward inference to obtain encodings $f_q(x_q)$ and $f_k(x_k)$ for the positive pairs.
4: Compute the alignment of $f_q(x_q)$ with $f_k(x_k)$ to determine positive pairs logits.
5: Calculate the alignment with the queue (negative pairs logits) using equation 6, considering the queue has K samples.
6: Combine positive and negative logits into a $(N, K + 1)$ matrix, marking 1 for the positive sample and 0 for the K negative samples.
7: Prepare labels for computing cross-entropy loss, noting a strategy to handle the element with an n by n matrix filled with zeros.
8: Update f_q using backpropagation based on the computed loss (equation 6).
9: Update f_k using a momentum method with $m = 0.999$.
10: Adjust the queue size by enqueuing and dequeuing each mini-batch of N samples.
11: Note: Apply the same augmentation method to other keys in the queue as used for x_k; this can be preprocessed offline.

Fig. 4. Laboratory workbench setup for data collection .

(b)**Dataset.** The purpose of this experiment is to generate a large number of samples from a small set in order to meet the requirements of industrial detection. Initially, we only had 120 original sampling images (Fig. 1). To ensure stable model training and prevent overfitting, targeted data augmentation techniques were applied to these samples. We performed random rotation, random horizontal flip, and random contrast augmentation on each of the 120 samples for a total of 10 repetitions, resulting in a dataset containing 1200 samples. The purpose of these is to train the model on a specific dataset size, minimizing overfitting. The input data will undergo common preprocessing operations.

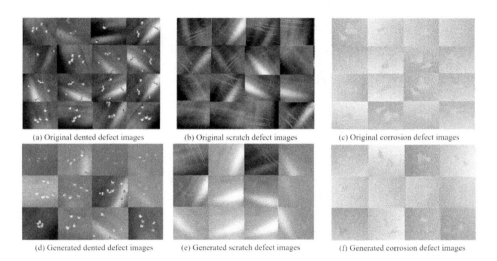

Fig. 5. Sampling images generated by the simulation of the (a) Original dented defect, (b) Original scratch defect, and (c) Original corrosion defect.

4.2 Implementation of DDPM-MoCo for Defect Generation

The experiment uses Phil Wang's Pytorch-based diffusion probability model code [24] and adjusts the parameters accordingly. The images from our dataset are cropped to small patches (640^2) and then resized to a smaller resolution (512^2) as input to the diffusion model. The processed data x_t and *time* will be input into the U-Net main network of the diffusion model. The module Attn utilizes a linear attention mechanism [25] to interact and reorganize input data, extracting key information while maintaining the original data dimension. This aims to generate more representative output The output of the model under the current batch input, ϵ_θ in Equation (4), can be obtained after multiple feature extraction and time information fusion.

(a)Improved Training Loss Schedule. The problem of infinite loss in the training process is addressed by proposing a dynamic model learning rate adjustment plan based on the training step, and we propose a learning rate adjustment scheme based on the cosine transformation, shown as Equation (8), where $current\_steps$ is the number of epochs currently trained, and $total\_steps$ is the total number of epochs (epochs) of training. Obviously, due to the sine function, the learning rate of model training will gradually decrease from 1 to 0 with the progress of training, which ensures the stability of training to the greatest extent.

$$lr = \frac{1}{2}\left(1 + \cos\left(\pi \frac{current\_steps}{total\_steps}\right)\right) \tag{8}$$

(b)Generated Images Quality Evaluation. Table 1 shows the performance improvements brought by the generated defect samples with DDPM when the training dataset is very small, which indicates those fake defects can be valid inputs for training unsupervised deep learning models.

Table 1. Performance comparison using Yolov5-mobileone on real data and generated data (scaled)

Case	Real data	Generated data	Real + generated data
Number of samples	120	1500	1620
Recall	0.35	0.68	0.662
Precision	0.25	0.55	0.891

The real and generated dented defects in Fig. 5 are visually indistinguishable. To compare the quality of different models, we use quantitative metrics for evaluation:

- **Inception Score (IS)** [27], measures how well a model capture the full ImageNet class distribution while still producing individual samples. The IS metric, although commonly used, has limitations as it only considers image content quality and diversity, neglecting other factors like image details and clarity, and models which memorize a small subset of the full dataset will still have high **IS**.
- **Fréchet Inception Distance (FID)** [27], was proposed to better capture diversity than IS, which is more consistent with human judgment than Inception Score. FID provides a symmetric measure of the distance between two image distributions in the Inception-V3 latent space [29]. Recently, sFID [28]

was proposed to replace spatial features with standard pooled features. We use FID as our default metric for overall quality comparisons as it captures both diversity and fidelity and has been adopted by generative modeling work [14,19].

4.3 Implementation of Momentum Comparison Model (MoCo)

The model utilizes three key hyperparameters: dictionary size K, momentum m, and temperature τ for the contrastive loss. Encoders f_q and f_k, based on an 18-layer ResNet [23], process matching samples and dictionary entries, respectively. Our dataset comprises 1500 samples, detailed in Table 2. Training parameters were set to $K = 16384$, $m = 0.999$, $\tau = 0.07$, batch size of 32, and an initial learning rate of 0.03, adjusted dynamically via cosine transformation (Eq. 8). Conducted over 200 iterations on an NVIDIA GeForce RTX 3090 GPU, the refined ResNet model adeptly discriminates between defects, as shown in Fig. 5. The model's precise identification of defects show the effective representation learning by f_q, as depicted in Fig. 6. This demonstrates the model's capability to accurately classify defect types with minimal input.

As Table 3 illustrates, we have conducted a comprehensive evaluation of image quality generated by our proposed method. Yolo-v5 serves as the principal detection framework in our study, where the model is trained on various synthetic datasets produced by different generative approaches. However, for a consistent assessment, the evaluation is performed on the same real dataset. We meticulously collect and compare the F1 scores for each category. Our model not only exhibits superior image generation quality but also enhances the performance of the detection model when trained on our synthetically generated dataset. This indicates that our method's controlled generation process significantly benefits the training phase, optimizing the model's ability to generalize from synthetic to real-world data. The results underscore the effectiveness of our approach, with the DDPM-MoCo method we developed outperforming others in terms of F1 score, Inception Score (IS), and Fréchet Inception Distance (FID) across various categories.

Table 2. The number of defect images in each category(original/constructed).

Defects	Dataset
	Aluminum alloy plate
scratch	50 / 500
dented	50 / 500
corrosion	20 / 500

Table 3. The Anomaly Detection and Image Quality Evaluation, F1 score/IS/FID, the best and the second-best numbers are shown in red and blue, respectively.

Method	Corrosion	Dent	Scratch	(Average)
DDPM ([19])	71.37/11.45/15.74	65.73/11.49/12.38	79.47/18.01/14.29	72.19/13.65/14.14
Image Synthesis ([30])	65.97/14.17/17.13	66.15/14.78/12.73	78.94/15.45/19.13	70.35/14.80/16.33
AdaBLDM ([31])	75.33/11.39/16.20	74.67/18.65/19.13	78.92/15.32/17.93	76.31/15.12/17.75
DDPM-MoCo (ours)	85.64/22.34/18.15	86.55/15.46/20.16	79.47/19.97/21.91	83.89/19.26/20.07

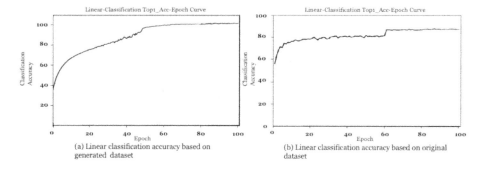

Fig. 6. Instance discrimination and Classification results.

5 Conclusion

The paper proposed a novel and effective solution DDPM-MoCo, addressing the issues of few unlabeled Aluminum alloy plate data samples and inefficient model training in surface defect detection of industrial workpieces using deep learning. Our method can effectively solve the problems of lacking accurate pixel-level annotations, poor diversity, and interference from non-defective information in industrial complex defect synthesis. Experimental results demonstrate that the probability diffusion model enables extensive simulations on the original defect samples obtained from images, while the momentum contrast learning model extracts feature representations for classification purposes. The dictionary encoder of the MoCo model in our experiment uses an 18-layer residual neural network as the backbone classifier, enabling almost real-time sample classification and providing valuable engineering guidance. These findings have practical implications for engineering applications. In future works, we will continue to explore the application value of our methods in other scenarios.

References

1. Dong, H., Song, K., He, Y., Xu, J., Yan, Y., Meng, Q.: PGA-Net: pyramid feature fusion and global context attention network for automated surface defect detection. IEEE Trans. Ind. Inform. **16**(12), 7448–7458 (2020)
2. Tong, X., Huang, Y., Xiao, L., Chen, X., Shen, R.: Surface defect detection method based on improved faster-RCNN. In: Proceedings of 2021 4th International Conference on Information Communication and Signal Processing (ICICSP 2021), pp. 357–362. IEEE, Shanghai (2021)
3. Tao, X., Zhang, D., Ma, W., Liu, X., Xu, D.: Automatic metallic surface defect detection and recognition with convolutional neural networks. Appl. Sci. **8**(9), 1575–1589 (2018)
4. Luo, Q., Fang, X., Liu, L., Yang, C., Sun, Y.: Automated visual defect detection for flat steel surface: a survey. IEEE Trans. Instrum. Measur. **69**(3), 626–644 (2020)
5. Amit, T., Shaharbany, T., Nachmani, E., Wolf, L.: SegDiff: Image Segmentation with Diffusion Probabilistic Models. arXiv preprint arXiv:2112.00390 (2022)
6. Sohl-Dickstein, J., Weiss, E., Maheswaranathan, N., Ganguli, S.: Deep unsupervised learning using nonequilibrium thermodynamics. In: Proceedings of the 32nd International Conference on Machine Learning, ICML 2015,37, pp. 2256–2265. JMLR: W&CP, Lille (2015)
7. Ren, M., Delbracio, M., Talebi, H., Gerig, G., Milanfar, p.: Multiscale structure guided diffusion for image deblurring. In: Proceedings of 2023 IEEE/CVF International Conference on Computer Vision (ICCV), pp. 3143–3154. IEEE, Paris (2023)
8. Zhou, T., Wang, W., Konukoglu, E., Van Gool, L.: Rethinking semantic segmentation: a prototype view. In: Proceedings of 2022 IEEE/CVF Conference on Computer Vision and Pattern Recognition (CVPR), pp. 2572–2583. IEEE, New Orleans (2022)
9. R. Yamini.,R. Ratha Jeyalakshmi.,R. Sowmini.: PDE based Diffusion Filters for Image Denoising. In: Proceedings of 2022 OPJU International Technology Conference on Emerging Technologies for Sustainable Development (OTCON). https://doi.org/10.1109/OTCON56053.2023.10114016. IEEE, Raigarh, Chhattisgarh(2022)
10. Xu, Y., Wu, P.: Multiscale clustering based diffusion representation learning method. In: Proceedings of 2021 IEEE/ACM 8th International Conference on Big Data Computing, Applications and Technologies (BDCAT '21), pp. 46–51. Association for Computing Machinery, Leicester (2021)
11. Luo, Y., Gong, Y., Liu, Y.: Probabilistic encoder based on gaussian diffusion processes. IEEE Trans. Neural Netw. Learn. Syst. **30**(6), 1846–1858 (2019)
12. Du, C., Du, C., He, H.: Multimodal deep generative adversarial models for scalable doubly semi-supervised learning. J. Inf. Fusion, **68**, 118–130 (2021)
13. Voynov, A., Morozov, S., Babenko, A.: Object segmentation without labels with large-scale generative models. In: Proceedings of the 38th International Conference on Machine Learning, ICML 2021, pp. 10596–10606. PMLR 2021, (Online virtual conference) (2021)
14. He, K., Fan, H., Wu, Y., Xie, S., Girshick, R.: Momentum contrast for unsupervised visual representation learning. In: Proceedings of 2020 IEEE/CVF Conference on Computer Vision and Pattern Recognition (CVPR2020), pp. 9729–9738. IEEE, (Online virtual conference) (2020)
15. Chen, X., Fan, H., Girshick, R.B., He, K.: MoCo v2: Improved baselines with momentum contrastive learning. arXiv preprint arXiv:2003.04297v1 (2020)

16. Wang, T., Isola, P.: Understanding contrastive representation learning through alignment and uniformity on the hypersphere. In: Proceedings of the 37th International Conference on Machine Learning, ICML 2020, PMLR 119, pp. 9929–9939 (2020)
17. Wu, Z., Xiong, Y., Yu, S.X., Lin, D.: Unsupervised feature learning via non-parametric instance discrimination. In: Proceedings of 2018 IEEE/CVF Conference on Computer Vision and Pattern Recognition (CVPR), pp. 3733–3742. IEEE, Salt Lake City (2018)
18. Chen, T., Kornblith, S., Swersky, K., Norouzi, M., Hinton, G.: Big self-supervised models are strong semi-supervised learners. In: Proceedings of 34th Conference on Neural Information Processing Systems, pp. 3143–3154. NeurIPS 2020, (Online conference) (2020)
19. Ho, J., Jain, A., Abbeel, P.: Denoising Diffusion Probabilistic Models. arXiv preprint arXiv:2006.11239 (2020)
20. Hadsell, R., Chopra, S., LeCun, Y.: Dimensionality reduction by learning an invariant mapping. In: Proceedings of the 2006 IEEE Computer Society Conference on Computer Vision and Pattern Recognition, CVPR 2006, 2, pp. 1735–1742. IEEE, New York (2006)
21. van den Oord, A., Li, Y., Vinyals, O.: Representation learning with contrastive predictive coding. arXiv preprint arXiv:1807.03748 (2018)
22. Song, K., Yan, Y.: NEU surface defect database. http://faculty.neu.edu.cn/songkechen/ Accessed 4 Dec 2023
23. He, K., Zhang, X., Ren, S., Sun, J.: Deep residual learning for image recognition. In: Proceedings of the 2016 IEEE Computer Society Conference on Computer Vision and Pattern Recognition, pp. 770–778. IEEE, Las Vegas (2016)
24. Wang, P.: denoising-diffusion-pytorch [Software]. https://github.com/lucidrains/denoising-diffusion-pytorch. Accessed 4 Nov 2023
25. Vaswani, A., et al.: Attention is all you need. In: Proceedings of the 31st Conference on Neural Information Processing Systems, pp. 6000–6010. NeurIPS 2017, Long Beach (2017)
26. Turner, R.E.: Denoising Diffusion Probabilistic Models in Six Simple Steps. arXiv preprint arXiv:2402.04384 (2024)
27. Barratt, S., Sharma, R.: A Note on the Inception Score. CoRR abs/1801.01973 (2018). https://arxiv.org/abs/1801.01973
28. Nash, C., Menick, J., Dieleman, S., Battaglia, P.W.: Generating images with sparse representations. In: Proceedings of the 38th International Conference on Machine Learning, ICML 2021, pp. 7958–7968. PMLR, (Online virtual conference) (2021)
29. Szegedy, C., Vanhoucke, V., Ioffe, S., Shlens, J., Wojna, Z.: Rethinking the inception architecture for computer vision. In: Proceedings of 2016 IEEE Conference on Computer Vision and Pattern Recognition, CVPR 2016, pp. 2818–2826. IEEE, Las Vegas (2016)
30. Fulir, U., Bosnar, L., Hagen, H., Gospodnetic, P.: Synthetic data for defect segmentation on complex metal surfaces. In: Proceedings of the IEEE/CVF Conference on Computer Vision and Pattern Recognition, CVPR 2023, pp. 4423–4433. IEEE, Vancouver (2023)
31. Li, H.: A novel approach to industrial defect generation through blended latent dif- fusion model with online adaptation. arXiv preprint arXiv:2402.19330(2024)

Detecting Railway Track Irregularities Using Conformal Prediction

Andreas Plesner[1]([✉]) and Allan P. Engsig-Karup[2] and Hans True[2]

[1] ETH Zurich, Zurich, Switzerland
aplesner@ethz.ch
[2] Technical University of Denmark (DTU), Lyngby, Denmark
{apek,htru}@dtu.dkand

Abstract. This study addresses the challenge of assessing railway track irregularities using convolutional neural networks (CNNs) and conformal prediction techniques. Using high-fidelity sensor data from high-speed trains, the study proposes a CNN model that outperforms state-of-the-art results, achieving a mean unsigned error of 0.31 mm on the test set. Incorporating conformal prediction with the CV-minmax method, the model delivers prediction intervals with 97.18% coverage, averaging 2.33 mm in width, ensuring reliable uncertainty estimation. The model also exhibits impressive computational efficiency, processing data at a rate suitable for real-time applications, with the capacity to evaluate over 2,000 km of track data per hour. These advances demonstrate the potential of the model for practical implementation in continuous monitoring systems, providing a contribution to the field of predictive maintenance within the railway industry.

Keywords: Railway track integrity · convolutional neural networks · conformal prediction · predictive maintenance · sensor data analysis · machine learning

1 Introduction

In the evolving landscape of transportation, railways play a pivotal role, offering a blend of efficiency, reliability, and environmental sustainability. As rail networks burgeon, paralleled by an upsurge in speed and passenger expectations, the imperatives of track safety and maintenance have ascended to the forefront of railway operations. The integrity of railway tracks, susceptible to irregularities due to wear and external forces, directly influences the safety, comfort, and operational efficiency of rail services. Traditional track inspection methodologies, although precise, grapple with limitations such as high operational costs, limited coverage, and high latency between inspections. The advent of machine learning and sensor technology indicates a transformative approach that allows continuous, real-time monitoring of track conditions by collecting data from in-service railway vehicles to generate predictive machine learning models.

This study delves into applying convolutional neural networks (CNNs) and conformal prediction methods to preemptively identify track irregularities from dynamic responses of in-service railway vehicles. The research is based on the use of high-fidelity sensor data from a high-speed train, embodying a shift from conventional reactive maintenance strategies to a predictive maintenance paradigm. The fusion of CNNs with conformal prediction offers a robust way to quantify the uncertainty of predictions, improving the reliability of the predictive framework. This integration not only showcases the potential of deep learning to decipher complex patterns from high-dimensional data but also underscores the importance of conformal prediction in providing robust, uncertainty-aware inferences.

Focusing on the prowess of the convolutional neural network, the study highlights the architecture, training, and optimization decisions that underpin the successful application of CNNs to the task at hand. CNNs are used because of their ability to handle spatial hierarchies in data, making them especially suitable for analyzing the nuanced dynamics captured by onboard sensors. The investigation extends to the realm of conformal prediction, highlighting its utility in giving prediction intervals that encapsulate the expected deviations with a quantifiable confidence level. The results derived from this application of CNNs and conformal prediction not only demonstrate a marked advancement in the accuracy and reliability of track irregularity prediction but also show the way for operationalizing these insights in real-world railway maintenance operations.

The prediction errors of the models, i.e., the difference between the predictions and the actual values, will have to be low enough so that it can be determined if the operating limits are exceeded. The EN:13848–5 standard [8] is used to establish a benchmark. This document contains operating limits for the track measurements at various speeds. The strictest limits are deviations of 1 mm for 100-meter running means and standard deviations. Based on this, a 0.1 mm benchmark will be chosen for the mean unsigned error, ME. Furthermore, to ensure that the predictions can be safely used to assess operating limits, further benchmarks are established to say that the maximum of the unsigned errors is below 0.5 mm. Since the limit values in [8] are only specified to the nearest millimeter, a maximum unsigned error of less than 0.5 mm would imply that all running means are within 0.5 mm of the actual value.

Kawasaki and Youcef-Toumi [11] give error ranges that would accept maximum unsigned errors of less than 4 mm and an ME of around 1 mm, while Hao et al. [9] set a benchmark of 0.25 mm and 0.45 mm of the mean unsigned error in the wavebands [3 m, 42 m] and [42 m, 120 m], respectively. These are less strict than our benchmarks mentioned above. We therefore set additional benchmarks of an ME of 0.35 mm (the mean of 0.25 mm and 0.45 mm) and a maximum unsigned error of 4 mm and call these the "satisfying" levels.

2 Related Work

Vehicle dynamics Traditionally, the integrity assessment of railway infrastructure has relied heavily on periodic inspections using specialized measurement vehicles, a process that, while accurate, suffers from limitations such as high costs, limited coverage, and the potential for subjective error. Ravitharan [18] highlight the operational benefits of proactive maintenance strategies, advocating for continuously monitoring track conditions using in-service railway vehicles, a concept explored over the past two decades [11,26]. Lee et al. [12] underscores the direct correlation between vehicle dynamics and track conditions, laying the foundation for the use of vehicle dynamics as a means of assessing track quality.

Classic Methods Prior works have largely focused on classical mathematical analysis tools, such as Kalman filters, system identification techniques, digital and analog processing, and other signal processing methods [3,5,11,12,15,16,22,25]. These works have prioritized interpretable models over complex data-driven solutions. These traditional methods often encounter mathematical difficulties, such as the issue of double integration of accelerations to obtain positions, which complicates their application in real-world scenarios [26].

Deep Learning Recent advances in machine learning, particularly in the application of convolutional neural networks (CNNs), present promising alternatives to traditional methods. Data-driven machine learning models have begun to shift the paradigm in various domains, demonstrating superior performance in fields such as image analysis [1,17]. In the context of monitoring the condition of the railway track, initiatives have explored the use of cameras on board and binary classification techniques to differentiate between good and bad track conditions [7,14,21,27]. However, the adoption of machine learning in this domain is not without its challenges. Despite their promise, these approaches face their own set of limitations, including computational demands and the lack of severity assessment in track irregularities [14].

Research by Hao et al. [9] presents a notable advancement, which showcases the potential of deep learning approaches to predict vertical track irregularities with a high degree of precision. However, this method does not address lateral irregularities and is based on simulated data, which may not fully capture the complexity of real-world track conditions. Similarly, the use of autoencoders to compress irregularity data presents innovative solutions but is again limited to simulated environments and specific types of irregularities [13].

Exploring data-driven methods for road quality monitoring has also yielded encouraging results, suggesting that similar approaches could be beneficial for the maintenance of railway tracks [23].

3 Methodology and Data

This section outlines the methodology employed to predict railway track irregularities using Convolutional Neural Networks (CNNs) complemented by conformal prediction techniques to estimate the uncertainties of these predictions. The section also includes a bit of background for these methods, how the CNN is designed specifically for the task at hand, and training the model. In addition, we will look at the data used for this project.

Table 1. Features in a sample of geometry dataset – With labels used in this project.

Label	Unit	Description	Notes
Position	km	Position along the track	
Lateral left D1	mm	Lateral irregularities of the left and right rail in the D1 wavelength domain	D1 is the first frequency band with wavelengths in [3 m, 25 m]
Lateral right D1	mm		
Vertical left D1	mm	Vertical irregularities of the left and right rails in the D1 wavelength domain	
Vertical right D1	mm		
Lateral left D2	mm	Lateral irregularities of the left and right rail in the D2 wavelength domain	D2 is the second frequency band with wavelengths in [25 m, 70 m]
Lateral right D2	mm		
Vertical left D2	mm	Vertical irregularities of the left and right rails in the D2 wavelength domain	
Vertical right D2	mm		
Lateral left D3	mm	Lateral irregularities of the left and right rail in the D3 wavelength domain	D3 is the third frequency band with wavelengths in [70 m, 200 m]
Lateral right D3	mm		
Vertical left D3	mm	Vertical irregularities of the left and right rails in the D3 wavelength domain	
Vertical right D3	mm		

3.1 Data Collection and Preprocessing

The data used in this study comprise high-fidelity sensor readings from a high-speed train, capturing various dynamic responses under operating conditions. The preprocessing steps involved the removal of outliers and normalization and segmentation to ensure compatibility with the CNN architecture. This preprocessing facilitated the transformation of raw sensor data into a structured format conducive to machine learning models.

Data are collected using multiple accelerometers located at various points on the railway vehicle; see Fig. 1 for locations. The input data consist of time series with measurements from each accelerometer. The output data consist of the irregularities of the track in the lateral and vertical directions for the left and right rails. These have been split into three frequency domains D1, D2, and D3 with wavelengths of [3 m, 25 m], [25 m, 70 m], and [70 m, 200 m], respectively, thus giving 12 output series. An outlier analysis found that five of the sensors had regions where they were faulty; in these regions, the faulty data were zeroed. Data were collected with a sampling frequency of $\approx$ 1000 Hz, and this was interpolated to have a constant sample spacing of 0.167 m. The train did not drive at a constant speed, so the spacing was irregular in the positional domain before interpolation. The features in the track geometry (output) data can be seen in Table 1 with a large table of all dynamics (input) features in Appendix Table 4.

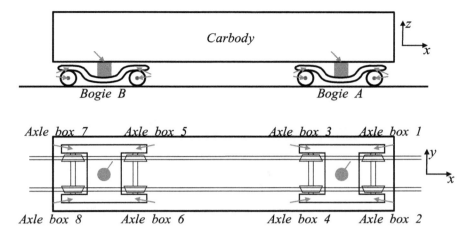

Fig. 1. Data measurement locations of the vehicle dynamics. The red arrows indicate the placement of the accelerometers on the axle boxes, bogies, and car body. (Color figure online)

We then split the data into training and testing regions. The data have 287,827 observations in total, and the 1st to 23,827th sample, the 94,001st to 117,827th sample, and the 188,001st to 211,827th sample are used as the test data. The training data consist of the 23,828th to 94,000th sample, the 117,828th to 188,000th sample, and the 211,827th to 281,827th sample. These regions have been shown in Fig. 2. The training data are further divided into training and validation segments by splitting it into 9 regions and using 1 for validation and the remaining 8 for training. Six of the nine regions are used for validation; a separate model is trained for each of the six validation regions, and the results are the mean across the six models.

3.2 Convolutional Neural Network (CNN) Architecture

The CNN architecture was designed to process time-series data, capturing spatial and temporal dependencies inherent in the train's dynamic responses. The model comprises multiple convolutional layers, each followed by pooling layers to reduce dimensionality and enhance feature extraction. Dropout layers were incorporated to mitigate overfitting, ensuring the model's generalizability across different track conditions. The model consists of batch normalization of the input and then 3 hidden CNN layers using batch normalization, the ELU activation function, and dropout of 60 %, with a final CNN layer to obtain the output [6,10,20]. The first convolutional layer uses very large kernels to ensure that features with 300 m wavelengths can be captured. This is relevant as the irregularities can exhibit wavelengths up to 200 m. A diagram of the final network has been shown in Fig. 4.

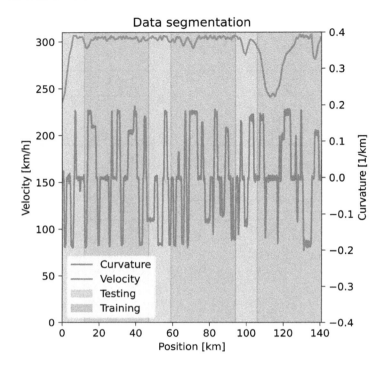

Fig. 2. Training and testing regions of the data. During the training of the models, the training data is divided into training and validation segments by splitting it into 9 regions and using 1 for validation and the remaining 8 for training.

Hyperparameters, including the learning rate, number of convolutional layers, kernel size, and dropout rate, were tuned using a combination of grid search and cross-validation to find the optimal model configuration by comparing the validation losses Fig. 3.

3.3 Conformal Prediction Framework

To quantify the uncertainty of CNN predictions, we applied conformal prediction methods. These methods use residuals from the training dataset to construct prediction intervals for new observations. Different variants of conformal prediction, Naïve, Holdout, and Cross-Validation, were evaluated to determine the most effective approach for this application. The Cross-Validation variant can further be split into three versions, CV, CV+, and CV-minmax [2,4,19,24]. The best intervals were produced by CV+ and CV-minmax, so the results will include only these. These methods have assumption-free theoretical guarantees that the α level interval contains $> 1 - \alpha$ of the samples [4]. For our results, we will use $\alpha = 0.05$ intervals. The focus on CV+ and CV-minmax is due to the better theoretical guarantees of these methods [4].

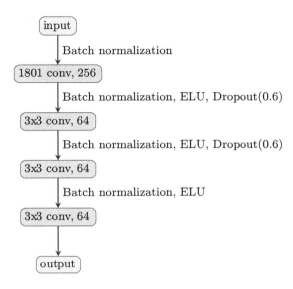

Fig. 3. Depiction of the CNN after tuning the hyperparameters.

3.4 Evaluation Metrics

The performance of the CNN model and the effectiveness of the conformal prediction intervals were evaluated using a few metrics. For CNN, the metrics were the mean and maximum of unsigned errors and the compute time. For conformal prediction, the focus was on the accuracy of the prediction intervals measured through the coverage probability (how often the true values were inside the interval) and the width of the interval assessed through the mean and maximum width.

4 Results

This section will present the results of this project for the best CNN model constructed, the use of conformal predictions, and, lastly, the compute time required to evaluate the model and produce prediction intervals.

4.1 CNN Predictions

The convolutional neural network (CNN) model showcased proficiency in predicting track irregularities from dynamic responses of in-service railway vehicles. The model, after rigorous tuning, achieved a satisfactory mean unsigned error (ME) by beating the "satisfying" benchmark for the ME. This significant achievement is depicted in Fig. 4, illustrating the training and validation mean errors across epochs, where the model's performance is notably highlighted by its capacity to maintain errors below the "satisfying" benchmark level.

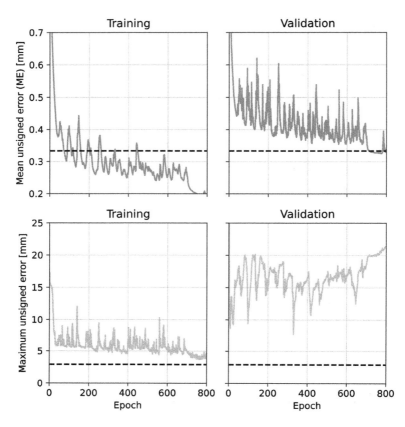

Fig. 4. The training and validation mean and maximum unsigned error during training for the best performing CNN model. The black dashed lines are the "satisfying" benchmark levels. We can see that the model gets a satisfactory mean unsigned error (ME), but the maximum is still off.

Architecture and hyperparameter optimization played a pivotal role in enhancing the model's accuracy. The final CNN model utilized a sophisticated arrangement of convolutional layers coupled with dropout regularization and batch normalization techniques. These elements collectively contributed to a robust model capable of discerning the intricate patterns associated with track irregularities from the vast and complex data derived from railway dynamics.

An extensive error analysis was conducted to dive into the predictive capabilities of the model and areas of improvement. This analysis was crucial in understanding the nuances of the model's performance, including the instances where it deviated from expected outcomes. Despite achieving high accuracy, the model faced challenges with maximum errors, especially in the validation data, prompting a detailed examination of error characteristics to identify potential model enhancements. The result of this analysis showed that the model makes the largest errors in regions with faulty sensor data. This is highlighted in Table 2, which shows key statistics for the model evaluated on the test data. The test data did not contain faulty sensor data. From the table, we see that the model also gets a satisfactory mean, but not a satisfactory maximum, on the test data. However, the aggregated maximum unsigned error is much smaller in the test data compared to the validation data errors seen in Fig. 4. Additionally, the model beats the state-of-the-art results from [9] for short wavelengths and the aggregated mean unsigned errors (ME). However, the model falls short of the 0.1 and 0.5 mm benchmarks for, respectively, the mean and maximum of unsigned errors.

Thus, to further improve the model, the focus should be on the data used to train the model. This might then eliminate the issues with missing sensor data.

Table 2. Mean and maximum of the unsigned test errors for each of the 12 output features. Values highlighted in red and in bold are those that exceed satisfactory levels. Recall that D1, D2, and D3 correspond to wavelengths of [3 m, 25 m], [25 m, 70 m], and [70 m, 200 m], respectively. From this, the mean unsigned errors for the three wavelength regions are 0.205 mm, 0.248 mm, and 0.486 mm, respectively.

	Mean [mm]	Maximum [mm]
Lateral left D1	0.14	2.54
Lateral right D1	0.13	2.62
Vertical left D1	0.27	2.23
Vertical right D1	0.28	2.56
Lateral left D2	0.20	3.55
Lateral right D2	0.18	3.49
Vertical left D2	0.30	2.28
Vertical right D2	0.31	2.63
Lateral left D3	0.35	**7.33**
Lateral right D3	0.33	**7.61**
Vertical left D3	**0.63**	**6.58**
Vertical right D3	**0.64**	**6.19**
Aggregates	0.31	**7.61**

4.2 Conformal Prediction

Integration of conformal prediction methods notably enhanced the predictive capabilities of the CNN model. The CV+ and CV-minmax methods were used to calculate the prediction intervals for the test data, which would improve the confidence in the model outputs for new predictions. The predictions are made as a mean aggregate of the six model instances trained for each validation segment. As depicted in Fig. 5, the CV+ and CV-minmax methods achieved high true value coverage rates, illustrating their effectiveness in encompassing data variability. The CV+ intervals are slightly narrower than those for CV-minmax but are overall very similar.

Table 3 presents the aggregate statistics for the $\alpha = 0.05$ intervals. It reveals that although CV-minmax offers higher coverage at 97.18 % compared to CV+'s 95.76%, it produces wider intervals on average (2.33 mm for CV-minmax versus 1.78 mm for CV+). Notably, as will be shown later, the CV-minmax method's prediction intervals are computed faster than those of CV+, an advantage for real-time applications.

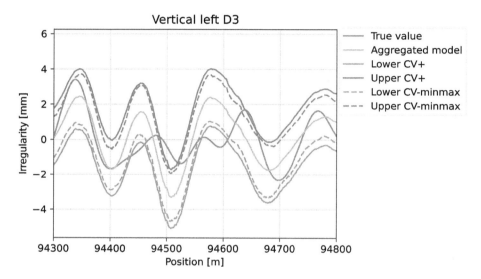

Fig. 5. Comparison of prediction intervals from CV+ and CV-minmax methods against the true values of track irregularities for the vertical D3 (long wavelengths) irregularities of the left rail. We see that the intervals often capture the true value, but there are instances, where they fail.

4.3 Compute Time

A crucial aspect of the CNN model's design was its ability to process and make predictions at a rate that exceeds the operational speeds of high-speed railway

Table 3. Aggregate statistics for conformal prediction intervals using the CV+ and CV-minmax methods. CV+ produces narrower intervals, but they have a slightly lower coverage.

Measure	CV+	CV-minmax
True value coverage (%)	95.76	97.18
Average interval width (mm)	1.78	2.33
Maximum interval width (mm)	4.54	5.25

vehicles, which can reach speeds over 300 km per hour. The CNN model could process 35.30 km of test data in 60.72 s using a GTX 970. Thus, the model can process track data at a rate of 2093 km per hour. Meanwhile, for conformal predictions, the CV-minmax and CV+ methods require 0.12 s and 12 min, respectively, to process the test data. This means they can process track data at rates of over 1,000,000 km and 176.5 km per hour, respectively.

Thus, the CV+ method cannot process sufficiently fast. However, the CNN model and CV-minmax demonstrated exceptional efficiency, capable of evaluating substantial lengths of track data within a constrained timeframe, thereby ensuring its applicability in real-time monitoring systems. We propose a CNN-based model with uncertainty quantified via conformal predictions as a solution for continuous real-time monitoring of railway track conditions.

5 Conclusion

This research ventured into the domain of using data-driven machine learning methods, with a focus on convolutional neural networks (CNNs) and conformal prediction, to predict railway track irregularities from the observed dynamics of in-service railway vehicles. The core achievement was the development of a predictive model that not only delivered satisfactory accuracy in detecting track irregularities but also incorporated conformal prediction to estimate the uncertainty of these predictions reliably. Satisfactory results are set as a mean unsigned error (ME) of 0.35 mm based on state-of-the-art results from related work. Our model has a mean unsigned error of 0.31 mm on the test set, thus improving the state-of-the-art results of [9]. Interestingly, the conformal prediction methodology achieved a high coverage of 97.18 % of the true values, with prediction intervals of an average width of 2.33 mm, thus ensuring a robust and reliable predictive framework.

However, it was noted that, while the prediction coverage was impressively high, the width of the intervals, though relatively small, indicates room for optimization to refine the precision further. These intervals were derived using the CV-minmax method, highlighting the potential for real-time application of this approach, given its ability to evaluate more than 1M km of track data per hour. Additionally, the CNN model could process track data at a rate of more than 2,000 km per hour. This efficiency underscores the feasibility of deploying this methodology in real-world settings, where it can serve as a cornerstone for continuous real-time monitoring of railway track conditions using in-service high-speed vehicles.

The journey to improve the accuracy and reliability of track irregularity detection through machine learning is far from over. Future endeavors can pivot around several key areas to push the boundaries of current achievements. Primarily, addressing the identified data issues will be crucial. This includes refining sensor data quality by removing or correcting data from faulty sensors and handling outliers more effectively. The model could, for instance, be made more robust so that it can allow for faulty sensors.

Further exploration of vehicle modeling offers a promising avenue for advancement. Transitioning the codebase to Julia has opened up new possibilities for using scientific computing methods. For example, delving into the domain of scientific machine learning, specifically through the lens of Neural Ordinary Differential Equations (NODEs), presents an exciting frontier. This approach could fundamentally change the way we model vehicle dynamics by integrating data-driven insights directly into the differential equations governing these dynamics.

We tried using transfer learning to simulate vehicle dynamics and pre-training the model on these simulated data. However, this did not produce the expected benefits in this study, suggesting a potential misalignment in data formatting or a lack of representation in the ODE system. Future research could aim to refine these aspects, potentially leading to breakthroughs in model performance and generalizability. Similarly, future work could try using physics-informed neural networks (PINNs) to enhance the model by incorporating physical laws directly into the learning process.

In sum, the groundwork laid by this project not only contributes to the current body of knowledge but also charted a course for future research to explore uncharted territories in railway track maintenance and safety through the lens of advanced machine learning techniques.

A Appendix

Table 4. Features in a sample of dynamics dataset – With provided labels and labels used in this project.

ID	Label	Unit	Description
0	Position	km	Position along the track
1	Velocity	km/h	Velocity of the railway vehicle
2	AccB1Y	m/s^2	Lateral acceleration of axle box 1
3	AccB1Z	m/s^2	Vertical acceleration of axle box 1
4	AccCR1Y	m/s^2	Lateral acceleration of bogie A at axle box 1
5	AccCR1Z	m/s^2	Vertical acceleration of bogie A at axle box 1
6	AccB2Y	m/s^2	Lateral acceleration of axle box 2
7	AccB2Z	m/s^2	Vertical acceleration of axle box 2
8	AccCR2Y	m/s^2	Lateral acceleration of bogie A at axle box 2
9	AccCR2Z	m/s^2	Vertical acceleration of bogie A at axle box 2
10	AccB3Y	m/s^2	Lateral acceleration of axle box 3
11	AccB3Z	m/s^2	Vertical acceleration of axle box 3
12	AccCR3Y	m/s^2	Lateral acceleration of bogie A at axle box 3
13	AccCR3Z	m/s^2	Vertical acceleration of bogie A at axle box 3

(*continued*)

Table 4. (*continued*)

ID	Label	Unit	Description
14	AccB4Y	m/s^2	Lateral acceleration of axle box 4
15	AccB4Z	m/s^2	Vertical acceleration of axle box 4
16	AccCR4Y	m/s^2	Lateral acceleration of bogie A at axle box 4
17	AccCR4Z	m/s^2	Vertical acceleration of bogie A at axle box 4
18	AccB5Y	m/s^2	Lateral acceleration of axle box 5
19	AccB5Z	m/s^2	Vertical acceleration of axle box 5
20	AccCR5Y	m/s^2	Lateral acceleration of bogie B at axle box 5
21	AccCR5Z	m/s^2	Vertical acceleration of bogie B at axle box 5
22	AccB6Y	m/s^2	Lateral acceleration of axle box 6
23	AccB6Z	m/s^2	Vertical acceleration of axle box 6
24	AccCR6Y	m/s^2	Lateral acceleration of bogie B at axle box 6
25	AccCR6Z	m/s^2	Vertical acceleration of bogie B at axle box 6
26	AccB7Y	m/s^2	Lateral acceleration of axle box 7
27	AccB7Z	m/s^2	Vertical acceleration of axle box 7
28	AccCR7Y	m/s^2	Lateral acceleration of bogie B at axle box 7
29	AccCR7Z	m/s^2	Vertical acceleration of bogie B at axle box 7
30	AccB8Y	m/s^2	Lateral acceleration of axle box 8
31	AccB8Z	m/s^2	Vertical acceleration of axle box 8
32	AccCR8Y	m/s^2	Lateral acceleration of bogie B at axle box 8
33	AccCR8Z	m/s^2	Vertical acceleration of bogie B at axle box 8
34	AccCSAY	m/s^2	Lateral acceleration of car body at bogie A
35	AccCSAZ	m/s^2	Vertical acceleration of car body at bogie A
36	AccCSBY	m/s^2	Lateral acceleration of car body at bogie B
37	AccCSBZ	m/s^2	Vertical acceleration of car body at bogie B
38	Curvatura	$1/m$	Curvature of the circle the track is forming

References

1. Aslam, Y., Santhi, N.: A review of deep learning approaches for image analysis. In: 2019 International Conference on Smart Systems and Inventive Technology (ICSSIT), pp. 709–714, 10.1109/ICSSIT46314.2019.8987922 (2019)
2. Balasubramanian, V., Ho, S.S., Vovk, V.: Conformal Prediction for Reliable Machine Learning: Theory. Elsevier Inc., Adaptations and Applications (2014). https://doi.org/10.1016/C2012-0-00234-7
3. Balouchi, F., Bevan, A., Formston, R.: Development of railway track condition monitoring from multi-train in-service vehicles. Veh. Syst. Dyn. **59**(9), 1397–1417 (2021)
4. Barber, R.F., Candés, E.J., Ramdas, A., Tibshirani, R.J.: Predictive inference with the jackknife+. Ann. Stat. **49**(1), 486–507 (2021). https://doi.org/10.1214/20-AOS1965

5. Chudzikiewicz, A., Bogacz, R., Kostrzewski, M., Konowrocki, R.: Condition monitoring of railway track systems by using acceleration signals on wheelset axle-boxes. Trans. (Vilnius, Lithuania) **33**(2), 555–566 (2017)
6. Clevert, D.A., Unterthiner, T., Hochreiter, S.: Fast and accurate deep network learning by exponential linear units (elus). arxiv 2015. arXiv preprint arXiv:1511.07289 (2020)
7. De Rosa, A., et al.: Monitoring of lateral and cross level track geometry irregularities through onboard vehicle dynamics measurements using machine learning classification algorithms. Proc. Inst. Mech. Eng. Part F J. Rail Rapid Trans. **235**(1), 107–120 (2021)
8. En, C.: 13848–5, Railway Applications-track-track Geometry Quality-part 5: Geometric Quality Levels-plain Line, Switches and Crossings. European Committee for Standardization, Brussels (2017)
9. Hao, X., Yang, J., Yang, F., Sun, X., Hou, Y., Wang, J.: Track geometry estimation from vehicle-body acceleration for high-speed railway using deep learning technique. Veh. Syst. Dyn. **61**(1), 239–259 (2023)
10. Ioffe, S., Szegedy, C.: Batch normalization: accelerating deep network training by reducing internal covariate shift. In: International conference on machine learning, PMLR, pp. 448–456 (2015)
11. Kawasaki, J., Youcef-Toumi, K.: Estimation of rail irregularities. In: Proceedings of the American Control Conference **5**, 3650–3660 (2002). https://doi.org/10.1109/acc.2002.1024495
12. Lee, J.S., Choi, S., Kim, S.S., Kim, Y.G., Kim, S.W., Park, C.: Waveband analysis of track irregularities in high-speed railway from on-board acceleration measurement. J. Solid Mech. Mater. Eng. **6**(6), 750–759 (2012)
13. Li, C., He, Q., Wang, P.: Estimation of railway track longitudinal irregularity using vehicle response with information compression and Bayesian deep learning. Comput. Aided Civ. Infrastruct. Eng. **37**(10), 1260–1276 (2021)
14. Mittal, S., Rao, D.: Vision based railway track monitoring using deep learning. arXiv preprint arXiv:1711.06423 (2017)
15. Muñoz, S., Ros, J., Urda, P., Escalona, J.L.: Estimation of lateral track irregularity through Kalman filtering techniques. IEEE Access **9** 60010–60025 (2021)
16. Naganuma, Y., Kobayashi, M., Okumura, T.: Inertial measurement processing techniques for track condition monitoring on shinkansen commercial trains. J. Mech. Syst. Trans. Logistics **3**(1), 315–325 (2010)
17. O'Mahony, N., et al.: Deep learning vs. traditional computer vision. In: Advances in Computer Vision: Proceedings of the 2019 Computer Vision Conference (CVC), Vol.1 1, Springer, pp. 128–144 (2020)
18. Ravitharan, R.: Safer rail operations: reactive to proactive maintenance using state-of-the-art automated in-service vehicle-track condition monitoring. In: 2018 International Conference on Intelligent Rail Transportation, ICIRT 2018, pp. 8641587, 10.1109/ICIRT.2018.8641587 (2019)
19. Shafer, G., Vovk, V.: A tutorial on conformal prediction. J. Mach. Learn. Res. **9**, 371–421 (2008)
20. Srivastava, N., Hinton, G., Krizhevsky, A., Sutskever, I., Salakhutdinov, R.: Dropout: a simple way to prevent neural networks from overfitting. J. Mach. Learn. Res. **15**(1), 1929–1958 (2014)
21. Tsunashima, H.: Condition monitoring of railway tracks from car-body vibration using a machine learning technique. Appl. Sci. **9**(13), 2734 (2019)
22. Tsunashima, H., Hirose, R.: Condition monitoring of railway track from car-body vibration using time-frequency analysis. Veh. Syst. Dyn. **60**(4), 1170–1187 (2022)

23. Varona, B., Monteserin, A., Teyseyre, A.: A deep learning approach to automatic road surface monitoring and pothole detection. Pers. Ubiquit. Comput. **24**(4), 519–534 (2020)
24. Vovk, V., Gammerman, A., Shafer, G.: Algorithmic learning in a random world. Springer, US, (2005). https://doi.org/10.1007/b106715
25. Wei, X., Liu, F., Jia, L.: Urban rail track condition monitoring based on in-service vehicle acceleration measurements. Measurement **80**, 217–228 (2016)
26. Weston, P., Roberts, C., Yeo, G., Stewart, E.: Perspectives on railway track geometry condition monitoring from in-service railway vehicles. Veh. Syst. Dyn. **53**(7), 1063–1091 (2015)
27. Yang, C., Sun, Y., Ladubec, C., Liu, Y.: Developing machine learning-based models for railway inspection. Appl. Sci. **11**(1), 13 (2020)

Identifying the Trends of Technological Convergence Between Domains Using a Heterogeneous Graph Perspective: A Case Study of the Graphene Industry

Shan Jiang, Yuan Meng[(✉)], and Danni Zhou

Shanghai University of International Business and Economics, Shanghai, China
nancymeng@suibe.edu.cn

Abstract. Technological convergence can lead to the emergence of new products and disruptive technologies within industries, offering both opportunities and challenges. Therefore, it is crucial for enterprises to timely recognize the trends of technology convergence to make informed business decisions. In this study, we propose a deep learning method within a heterogeneous graph framework to identify technological convergence trends between domains. The approach involves collecting patent texts and metadata, constructing a heterogeneous graph network based on IPC patent, and learning representations of IPC nodes based on the IPCvec model. Using the graphene industry as a case study, we validate the effectiveness of our method by comparing it with other graph neural network models. Moreover, we estimate the trends of technological convergence across different domains in the industry at a broader data level and confirm the presence of integration over time. This method can assist enterprises in identifying technological convergence trends, uncovering new business opportunities, fostering cross-domain collaboration and innovation, guiding strategic decision-making and resource allocation, and promoting technological and industrial development. Ultimately, it enables enterprises to drive innovation, achieve growth, and enhance their competitiveness.

Keywords: Heterogeneous graph · Technology convergence · Graphene industry

1 Introduction

The formation of emerging industries is a long-term development process, where convergence plays a crucial role as a driving factor. Convergence involves a sequential process across four levels: science, technology, market, and industry. Firstly, scientific convergence narrows the gap between different domains, facilitating the exchange and integration of knowledge. Secondly, technological convergence combines previously disparate technological knowledge to address specific technical challenges. Subsequently, technology overlaps in the market convergence, resulting in new product-market combinations. Finally, industry convergence occurs, giving rise to new industries.

However, for existing enterprises, the formation of emerging industries presents both an opportunities and a challenge. The emergence of these industries brings with it new market shares and the entry of new competitors, significantly impacting the market position and profitability of existing enterprises. Therefore, it becomes crucial for companies to identify early signals of convergence and make timely strategic deployments before the formation of emerging industries.

Technological convergence is described as a phenomenon that blurs the boundaries between different technologies and involves the mutual inclination of two or more technological domains (Curran and Leker, 2011) [1]. The intersection of two or more technological domains can generate new technologies and even form a new industry sector (such as bioinformatics). Therefore, it is a scientific approach to identify signals of technological convergence at the technical level. Analyzing the development trends of technological convergence and clarifying the role and significance of key technological nodes in the process are crucial for gaining insights into future opportunities and planning research and development activities. They also play a significant role in promoting technological innovation and guiding the formation and development of emerging industries. Based on these factors, this paper proposes a study on technological convergence between domains from a heterogeneous graph perspective.

2 Literature Review

The primary sources of data for research on the convergence of industrial technologies are patent data. Research methods in this field can be categorized into citation analysis, co-classification methods, and semantic analysis.

Citation analysis utilizes the citation information between patents to analyze technology convergence and knowledge flow between technology fields. Kim et al. [2] constructed a dependency structure matrix (DSM) using patent citation relationships to identify knowledge flow and integration trends. Smojver et al. [3] applied link prediction algorithms to co-citation networks to forecast future knowledge flow.

The co-classification method involves analyzing the co-occurrence relationships between patent classification codes to uncover fusion relationships between technologies. Wang et al. [4] determined the degree of technology convergence by co-classifying patents and calculating their novelty scores. Wu Xiaoyan et al. [5] conducted empirical research on synthetic biology patents using the ISI OST-INPI classification system.

The semantic method focuses on analyzing the content of technologies. Preschitschek et al. [6] used the DSS-Jaccard measure to quantify textual similarity between documents. Kim et al. [7] used the Doc2Vec method to generate vector representations for technology fields and predicted technology convergence using bibliometric indicators. Zhu et al. [8] proposed a graph convolutional network (GCN)-based method to generate technology keywords and patent representations for assessing the semantic proximity between patents and technology fields.

Integrating citation, co-classification, and semantic methods has become a research hotspot in identifying technology convergence. Zhang Jinzhu et al. [9] represented patent classification sequences and texts semantically, using machine learning methods to obtain comprehensive representations for predicting technology convergence.

In conclusion, research on technology convergence has made significant progress by integrating various methods. However, it is important to consider the limitations of each method, such as delays in citation data and the granularity of patent classification codes. Integrating multiple methods can enhance the accuracy and comprehensive understanding of technology convergence.

3 Proposed Methodology

3.1 General Logic

The overall research approach is as follows: Firstly, a patent dataset is acquired, and a patent heterogeneous graph network centered around IPC is constructed, taking into account the citation relationships, ownership relationships, application relationships, and inclusion relationships between patents. IPCs are also mapped to 35 technology domains. Secondly, an IPC representation learning model is constructed by designing different types of meta-paths to integrate the semantic information from multiple nodes. Node-level and meta-path-level attention mechanisms are utilized to learn the importance of nodes and meta-paths, generating comprehensive IPC representations. Then, the effectiveness of the model is validated by using the learned IPC representations for technology domain classification and comparing them with other graph embedding algorithms. Finally, the phenomenon of technology convergence among domains is explored by analyzing the similarity between patents and IPCs, aggregating information at the technology domain level.

3.2 Heterogeneous Graph Network

The article introduces a patent heterogeneous graph network utilizing the IPC system. The network consists of patents, inventors, assignees, and IPC codes (first four digits) as nodes. Connections in the network represent patent-patent citations, patent-inventor ownership, patent-assignee applications, and patent-IPC code inclusion. Please refer to Fig. 1 for a visual representation of the IPC heterogeneous graph network.

3.3 IPCvec Model

The IPCvec model is trained using the IPC patent heterogeneous graph network to generate patents and IPC vectors. The model consists of three main

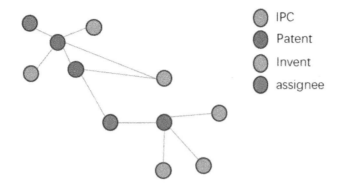

Fig. 1. The IPC heterogeneous graph network.

components: (1) Initial node embeddings, (2) Node-level aggregation, and (3) Meta-path aggregation. The overall architecture of the model is illustrated in Fig. 2.

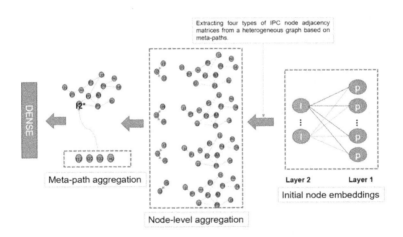

Fig. 2. IPCvec Model.

IPC Node Initial Embeddings. Starting from the first layer, we utilize Doc2Vec to train abstract and title data of patent to initialize each patent vector. In the second layer, the IPC node initial embeddings are obtained by linearly transforming and aggregating neighboring patents, with the inclusion of learning coefficients.

$$h_i^{(2)} = \sigma \left(W^{\text{encoder}} \sum_{\text{Patent } n \in N(i)} \frac{1}{|N(i)|} h_n^{(1)} + b^{\text{encoder}} \right) \quad (1)$$

Node-Level Aggregation. A self-attention mechanism is utilized to determine the significance of neighboring nodes in the meta-path. The mechanism learns the influence of different neighboring nodes on the target node. Using these learned importance weights, the information from the neighboring nodes is then merged into the target node. This aggregation process facilitates the acquisition of a comprehensive representation for the target node within the IPC patent heterogeneous network.

Given a meta-path Φ, the importance of node j to node i in the node pair (i,j) can be calculated using the following formula:

$$e_{ij}^{\Phi} = a_{\text{node}}\left(\mathbf{h}_i, \mathbf{h}_j; \Phi\right) \tag{2}$$

In equation (2), anode represents a multilayer neural network. j is a neighboring node that represents the set of neighboring nodes of node i on meta-path Φ. After obtaining the importance score e_{ij}^{Φ}, we use the Softmax function to calculate the weight coefficients for each neighboring node according to equation (3).

$$\alpha = \left(\alpha_{ij}^{\Phi}\right) = \text{softmax}_j\left(e_{ij}^{\Phi}\right) = \frac{\exp\left(\sigma\left(\mathbf{a}_{\Phi}^T \cdot [\mathbf{h}_i \| \mathbf{h}_j]\right)\right)}{\sum_{k \in \mathcal{N}_i^{\Phi}} \exp\left(\sigma\left(\mathbf{a}_{\Phi}^T \cdot [\mathbf{h}_i \| \mathbf{h}_k]\right)\right)} \tag{3}$$

In equation (3), the node-level attention vector on the meta-path Φ is represented, which can be obtained automatically through model training. σ denotes an activation function. The $\|$ symbol represents the concatenation operation. Finally, based on the node-level attention on the meta-path σ, the representation of node i can be summarized using its neighboring node representations and their corresponding weight coefficients. Then, we utilize K attention heads to obtain the representation.

$$\mathbf{z}_i^{\Phi} = \prod_{k=1}^{K} \sigma\left(\sum_{j \in \mathcal{N}_i^{\Phi}} \alpha_{ij}^{\Phi} \cdot \mathbf{h}_j'\right) \tag{4}$$

Within the meta-path set $\{\Phi_1, \Phi_2, \ldots, \Phi_n\}$ based on the node-level attention, we can obtain n node representations for each node in the meta-paths. Each node representation is denoted as $\{Z_{\Phi_1}, Z_{\Phi_2}, \ldots, Z_{\Phi_n}\}$.

Meta-path Aggregation. In this step, we utilize self-attention mechanism to automatically learn the importance of different meta-paths and explore the semantic information in the IPC patent heterogeneous network. The weight coefficients of meta-paths are calculated using equation (5)

$$\left(\beta_{\Phi_1}, \beta_{\Phi_1}, \ldots, \beta_{\Phi_n}\right) = a_{tt_{sem}}\left(Z_{\Phi_1}, Z_{\Phi_1}, \ldots, Z_{\Phi_n}\right) \tag{5}$$

In equation (5), $a_{tt_{sem}}$ represents a multi-layer neural network. To obtain the importance of each meta-path, we first perform a non-linear transformation on the node representations. Then, we measure the importance of meta-path Φ_i, denoted as w_{Φ_i}, by calculating the similarity (dot product) between the node representations and the meta-path attention vector q^T. The importance of meta-paths is calculated using equation (6).

$$w_{\Phi_i} = \frac{1}{|\mathcal{V}|} \sum_{i \in \mathcal{V}} q^T \cdot \tanh\left(W \cdot z_i^\Phi + b\right) \tag{6}$$

In equation (6), W is a weight matrix, b is a bias term, and q^T is the semantic attention vector, which is automatically obtained through model training. Therefore, the weight coefficient w_{Φ_i} for meta-path Φ_i can be computed using the Softmax function.

$$\beta = (\beta_{\Phi_i}) = \frac{\exp(w_{\Phi_i})}{\sum_{i=1}^{n} \exp(w_{\Phi_i})} \tag{7}$$

Based on the obtained weight coefficients of the meta-paths, we integrate the semantic information contained in different meta-paths to obtain the final node representation Z.

3.4 Technological Convergence

Threshold Setting: To determine an appropriate threshold for distinguishing patents close to the IPC classification code from those that are far from it, we employ a criterion based on the reflection of patents within their own IPC domain. The criterion states that the majority of patents should surpass the threshold compared to the IPC codes related to their respective domains. We consider this criterion met if it applies to at least 75% of the patents. After testing various thresholds ranging from 0.01 to 0.05, we found that 0.02 is the highest value that satisfies this criterion.

Single Patent Level: The level of proximity between an individual patent m in domain i and domain j. At the level of a single patent, it is necessary to select a threshold that represents the degree of proximity to a semantic anchor. This threshold distinguishes patents that are close to the semantic anchor from those that are far from it.

$$IJ_m : I-J \text{ closeness centrality } m = \max_{j(\text{ ipc code }) \text{ set}} s_{mj} \quad m \in i(pn) \text{ set} \tag{8}$$

Technical Field Level: The strength of technological convergence from domain i to domain j, indicating the degree of proximity from domain i to domain j.

This measure of strength serves as the weights for the connections between nodes in the technological convergence network.

$$\text{intensity}_{ij} = \sum_{m \in i(pn) \text{ set}} \sum_{j(\text{ipc code}) \text{ set}} \frac{C_{mj}}{n} \text{ where } i \neq j C_{mj} \begin{cases} 1 & \text{if } IJ_m \geq \theta \\ 0 & \text{otherwise} \end{cases} \tag{9}$$

θ represents the minimum threshold for establishing a connection between an individual patent in domain i and the technological field j. n denotes the total number of patents in domain i.

$$Z = \sum_{i=1}^{n} \beta_{\Phi_i} \cdot Z_{\Phi_i} \tag{10}$$

4 Experiments

4.1 Data Collection

In this study, the Derwent Patent Database was chosen as the data source. Through consultation with experts and literature review, a thematic search method was employed for patent retrieval.

The search strategy was as follows: (TS=(Graphene)) OR TS=(Graphenes) AND TM = (2004-01-01,2022-12-31)

Considering that graphene was first extracted and produced in 2004, the starting point for patent retrieval was set as 2004. The 99,916 graphene patents were mapped to one or more of the 35 technical subclasses of ISI-OST-INPI based on their respective IPC classification codes. Each technical subclass represents a corresponding technological field. We exclude technical subclasses with a low number of patent data, resulting in a final dataset of 81,737 patents, as shown in Table 1.

Table 1. ISI-OST-INPI Mapping Table

Field	Area	IPC code	PN count
Mechanical Engineering	Textile and paper machines	['D06M', 'D02J', 'B31F', 'D02H', 'D21J', 'C14B', 'D01G', 'D21F', 'B41J', 'D05C', ...]	6958
	Other special machines	['A23N', 'F42D', 'F41J', 'A01B', 'B28D', 'A01K', 'B29C', 'F42B', 'B28C', 'B29D', ...]	7330
	Thermal processes and apparatus	['F28D', 'F24C', 'F25B', 'F24D', 'F28B', 'F28C', 'F24T', 'F27B', 'F24V', 'F27D', ...]	2850
Chemistry	Macromolecular chemistry, polymers	['C08G', 'C08H', 'C08C', 'C08B', 'C08F', 'C08K', 'C08L', ...]	25071
	Chemical engineering	['B01L', 'B05B', 'D06B', 'F25J', 'D06C', 'H05H', 'B03D', 'B03B', 'B01J', 'B07C', ...]	13677
	Basic materials chemistry	['C05D', 'C10B', 'C09F', 'C10N', 'C05G', 'C09D', 'C11D', 'C09J', 'A01P', 'C10M', ...]	14581
	Environmental technology	['C02F', 'F23G', 'F23J', 'B09B', 'G01T', 'B65F', 'C02B', 'F01N', 'B09C', 'A62C','A62D']	4491
	Materials, metallurgy	['C01B', 'C01D', 'C01G', 'B22C', 'C21B', 'C01F', 'C21D', 'C03C', 'C01C', 'C22F', ...]	26535
	Surface technology, coating	['C25C', 'B05D', 'C23D', 'C25D', 'B32B', 'C23F', 'C25F', 'C23C', 'C25B', 'B05C','C23G', 'C30B']	10578
	Micro-structure and nano-technology	['B81C', 'B81B', 'B82Y', 'B82B']	7397
Instruments	Measurement	['G04G', 'G01M', 'G01R', 'G04B', 'G01S', 'G01J', 'G04D', 'G01D', 'G01H', 'G01V', ...]	7553
	Medical technology	['A61C', 'A61J', 'A61F', 'A61H', 'A61N', 'A61L', 'A61B', 'A61G', 'A61D', 'H05G', 'A61M']	5069
	Audio-visual technology	['G09F', 'H04S', 'H04R', 'G09G', 'H05K', 'G11B']	3007
Electrical Engineering	Electrical machinery, apparatus, energy	['H01B', 'F21W', 'H02H', 'H02J', 'H02S', 'F21V', 'H02N', 'H02B', 'H01G', 'H01K', ...]	23811
	Semiconductors	['H01L']	7648

4.2 Heterogeneous Graph of Patents in the Graphene Field

This paper establishes a heterogeneous graph network by extracting relationships between four nodes: classification code, patent, rights holder, and assignee. The heterogeneous graph information is shown in Table 2 as follows:

Table 2. Graphene Heterogeneous Graph Information

Data	Node Type	Edge Type
graphne	'ipc': 258, 'pn': 81737 'ae': 27386, 'au': 24761	('pn', 'cite', 'pn'): 136159, ('pn', 'contain', 'ipc'): 146069, ('pn', 'own', 'ae'): 98637, ('pn', 'write', 'au'): 300014

Establishment of Meta-paths: In a heterogeneous network, meta-paths are paths defined on the network pattern that connect different nodes. A meta-path is represented as $M_p : V_1 \xrightarrow{K_1} V_2 \xrightarrow{R_2} \cdots \xrightarrow{R_l} V_{l+1}$, indicating the semantic relationship between nodes V_i and V_{i+1} through the relationship R_i. Table 3 provides the specific semantic details represented by each meta-path (Table 4).

Table 3. Meta-Path Designs based on Heterogeneous Graph

Index	Meta-Path	Semantic Information of Meta-Path
M1	IPC → Patent → Inventor ← Patent ← IPC	Same inventor
M2	IPC → Patent → Assignee ← Patent ← IPC	Same assignee
M3	IPC → Patent ← IPC	Co-occurrence of IPC in the same patent
M4	IPC → Patent → Patent ← IPC	Mutual citation of patents

Table 4. Some examples of meta-paths

Meta-Path	Example
M1	D06M → CN113929090-A → YANG S ← CN114314570-A ← A61N
M2	C01B → CN113929090-A → QUJING HUAJIN YULIN TECHNOLOGY CO LTD ← CN114314570-A ← C25B
M3	C02F → WO2022261706-A1 ← C01B
M4	C01B → CN113929090-A - CN108190868-A ← C01B

4.3 Model Evaluation

Visualization of IPC Representation. In order to visually demonstrate the effectiveness of the model training, we employ the TSNE algorithm to map the learned IPC representations onto a two-dimensional space. This approach allows us to present a more intuitive visualization of the IPC representations.

The graph clearly demonstrates noticeable distances between embeddings from different technological domains, providing further evidence of the discriminative IPC representations learned by our model. Additionally, embeddings within the same domain exhibit a smaller coverage range. This finding indicates a certain level of correlation among embeddings within the same domain (Fig. 3).

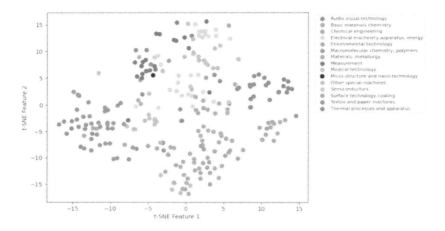

Fig. 3. The IPC representations of IPCvec Model.

Classification Evaluation of IPC Representation. When comparing different approaches, two factors are considered for model selection: (1) Graph Neural Network (GNN) models, and (2) initial embeddings for patent text. The GNN algorithms evaluated are GCN, GAT, and GraphSAGE, while the initial embeddings for patent text included Doc2Vec and PatentBERT (Table 5).

Table 5. IPC Classification Performance Comparison for Different Models

Model	Mirco-F1	Top-1	Top-2	Top-3
GCN	0.6226	0.6226	0.7547	0.8679
GAT	0.5849	0.5849	0.7170	0.8491
Graphsage	0.6981	0.6981	0.8491	0.9245
Doc2vec-IPCvec	0.6981	0.6981	0.8679	0.9434
Patentbert-IPCvec	0.5283	0.5283	0.6415	0.7547

a) The Doc2vec-IPCvec model demonstrates significant improvements in multiple evaluation metrics compared to other models, indicating the effectiveness of IPC vector representations learned by our model for IPC classification.
b) Both the GraphSAGE and Doc2vec-IPCvec models perform similarly in Micro-F1 and Top-1 metrics. However, the Doc2vec-IPCvec model slightly outperforms the GraphSAGE model in Top-2 and Top-3 metrics. This suggests that they have similar overall prediction accuracy, but the Doc2vec-IPCvec model excels in considering a broader range of candidate categories. Its ability to capture crucial information between nodes using heterogeneous graphs and attention mechanisms contributes to this advantage over Graph-SAGE, which may have limitations in predicting more diverse categories.

c) The use of the PatentBert pretraining model for computing IPC initial embeddings performs poorly in the IPC classification task, indicating its failure to capture relevant information for IPC classification. Therefore, choosing the Doc2vec algorithm for IPC initial embeddings appears to be more suitable.

4.4 Technology Convergence

Estimating One-Way and Two-Way Technology Convergence. A technological fusion network graph is utilized to evaluate the overall level of technological fusion between different domains. This network graph serves as a foundation for measuring the degree of fusion and predicting trends. The figure below illustrates the network, with nodes representing 15 technology domains. The connection weights between nodes represent the intensity of fusion between the corresponding domains ($intensity_{ij}$). Fusion levels are categorized as follows: strong fusion (0.75–1), moderate fusion (0.5–0.75), weak fusion (0.25–0.5), and no fusion (0–0.25) (Fig. 4).

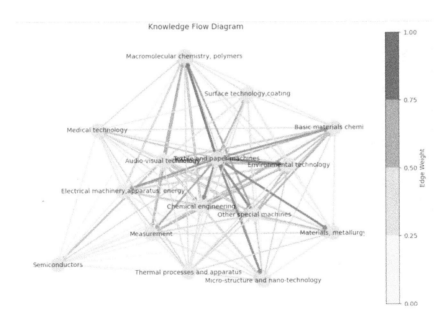

Fig. 4. The IPC heterogeneous graph network.

The technological convergence network can be regarded as a complex network, enabling the application of various metrics from complex network science. The table below illustrates the utilization of weighted degree centrality to identify important nodes representing significant technology domains.

Table 6. Identification of important technology domains

Field	Area Node	Weighted Degree Centrality
Mechanical engineering	Textile and paper machines	13.57403
Chemistry	Chemical engineering	13.14107
Mechanical engineering	Other special machines	12.66718
Electrical engineering	Audio-visual technology	11.42802
Chemistry	Environmental technology	11.41023
Electrical engineering	Electrical machinery, apparatus, energy	11.23777

Table 6 shows that the graphene industry involves important technology domains such as Textile and paper machines, Chemical engineering, Other special machines, Audio-visual technology, Environmental technology, and Electrical machinery, apparatus, energy. Next, we will delve into identifying the technological fusion phenomena within these six specific technology domains (Fig. 5).

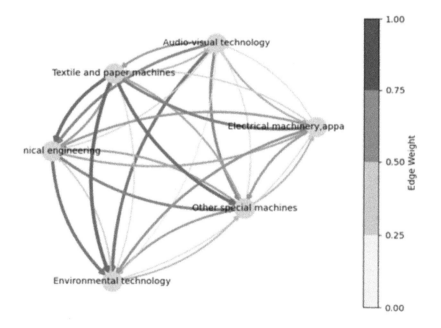

Fig. 5. The first six technical areas of convergence.

a) Audio-visual technology demonstrates strong convergence with multiple fields, including Other special machines, Chemical engineering, and Textile and paper machines.

b) Environmental technology exhibits weak convergence with other fields. It shows a strong one-way fusion with Audio-visual technology but weak fusion with Chemical engineering, Electrical machinery, apparatus, energy, and Textile and paper machines.
c) Other special machines exhibit strong fusion with multiple fields, including Environmental technology, Chemical engineering, and Textile and paper machines.
d) Chemical engineering displays evident convergence with various fields, showing strong two-way fusion with Audio-visual technology, Other special machines, Electrical machinery, apparatus, energy, and Textile and paper machines.
e) Electrical machinery, apparatus and energy have weak fusion with other fields, including Audio-visual technology, Other special machines, and Chemical engineering.
f) Textile and paper machines have significant two-way fusion with Audio-visual technology, Other special machines, and Chemical engineering, but weaker fusion with Electrical machinery, apparatus, energy.

Verification of One-Way and Two-Way Technology Convergence. This convergence trend is a long-term process that takes time to gradually form and develop. Therefore, it is necessary to examine the technology convergence trends in different fields over specific time periods.

The Tables (7, 8, 9, 10, 11 and 12) show relative patent counts between a specific technical field and another field in four timeframes. Continuous patent counts (exceeding a threshold in two or more timeframes) or a significant number of patents indicate unidirectional or bidirectional technological convergence.

a) Bidirectional technological convergence between CE and TP is observed, with a stronger convergence from TP to CE. Knowledge flow from CE to TP is significant, while the convergence from TP to CE is weaker.
b) OSM and TP exhibit high bidirectional technological convergence.
c) Unidirectional technological convergence is seen from TP to AT, but limited convergence is observed from AT to TP.
d) Bidirectional technological convergence between ET and TP, with a stronger knowledge flow from ET to TP.
e) Bidirectional technological convergence between EM and TP, with a stronger knowledge flow from EM to TP.
f) Bidirectional technological convergence between OSM and CE, with decreasing knowledge flow from CE to OSM and increasing flow from OSM to CE.
g) Unidirectional technological convergence from CE to AT, with weak convergence observed.
h) Bidirectional technological convergence between ET and CE, although the degree of convergence decreases over time.
i) Unidirectional technological convergence from CE to EM, with limited continuity in patent counts.

Table 7. IPC Field: TP

		Timeframe			
		T1	T2	T3	T4
Patent set	CE	0.17	0.39	0.58	0.44
	OSM	0.22	0.80	0.53	0.65
	AT	0.06	0.26	0.19	0.38
	ET	0.22	0.53	0.47	0.44
	EM	0.11	0.70	0.26	0.41

Table 8. IPC Field: CE

		Timeframe			
		T1	T2	T3	T4
Patent set	CE	0.14	0.78	0.76	0.49
	OSM	0.14	0.83	0.58	0.35
	AT	0	0.38	0.24	0.14
	ET	0.29	0.64	0.55	0.18
	EM	0	0.71	0.37	0.16

Table 9. IPC Field: OSM

		Timeframe			
		T1	T2	T3	T4
Patent set	TP	0.23	0.60	0.76	0.58
	CE	0.23	0.31	0.64	0.44
	AT	0.08	0.29	0.31	0.32
	ET	0.38	0.50	0.57	0.41
	EM	0.38	0.84	0.42	0.35

Table 10. IPC Field: AT

		Timeframe			
		T1	T2	T3	T4
Patent set	TP	0	0.51	0.71	0.60
	CE	0	0.26	0.50	0.42
	OSM	0	0.88	0.54	0.44
	ET	0	0.50	0.60	0.29
	EM	0.67	0.87	0.39	0.22

Table 11. IPC Field: ET

		Timeframe			
		T1	T2	T3	T4
Patent set	TP	0	0.77	0.80	0.46
	CE	1.00	0.54	0.66	0.22
	OSM	0	0.81	0.64	0.27
	AT	0	0.33	0.27	0.10
	EM	0	0.65	0.43	0.11

Table 12. IPC Field: EM

		Timeframe			
		T1	T2	T3	T4
Patent set	TP	0.28	0.71	0.79	0.66
	CE	0.24	0.50	0.61	0.34
	OSM	0.31	0.85	0.64	0.34
	AT	0.24	0.26	0.27	0.17
	ET	0.41	0.59	0.63	0.19

CE: Chemical engineering; OSM: Other special machines; AT: Audio-visual technology;
ET: Environmental technology; EM: Electrical machinery, apparatus, energy; TP: Textile and paper machines

j) Bidirectional technological convergence between AT and OSM, with decreasing knowledge flow from AT to OSM and consistently weak flow from OSM to AT.
k) Bidirectional technological convergence between ET and OSM, with decreasing convergence from OSM to ET and weak but stable convergence from ET to OSM.
l) Bidirectional technological convergence between EM and OSM, with decreasing convergence over time.
m) Unidirectional technological convergence from ET to AT, with weakening convergence, but no convergence observed from AT to ET.

n) Unidirectional technological convergence from EM to AT, with weakening convergence, but no convergence observed from AT to EM.
o) Unidirectional technological convergence from ET to EM, with weakening convergence, but no convergence observed from EM to ET.

Based on the provided information, the following conclusions can be drawn:

The Audio-Visual Technology (AT) field has limited patent applications in other domains, indicating fewer external market threats and expansion opportunities. Weak convergence towards the Other Special Machines (OSM) field suggests potential for further development in this direction.

Textile and Paper Machines (TP) field is active and exhibits strong convergence with other domains, indicating significant prospects and a broader range of action for companies in this field.

Other Special Machines (OSM) field shows a diminishing convergence trend with other domains, but there is sustained demand and application opportunities from other domains. This suggests potential for technology and knowledge application in OSM field, despite lagging behind in fusion compared to other domains.

Chemical Engineering (CE) field maintains a stable convergence trend with other domains, despite a decrease in convergence with Environmental Technology (ET). Collaboration opportunities with other domains can still be sought, indicating good development prospects for CE field.

Environmental Technology (ET) field shows a decrease in technology application from Electrical Machinery, Apparatus, and Energy (EM) field and CE field. This could be attributed to emerging technologies and energy transitions, resulting in reduced connections between ET and EM fields.

Electrical Machinery, Apparatus, and Energy (EM) field does not exhibit convergence towards ET and CE fields. Convergence towards AT field weakens, but it maintains stable convergence towards OSM and TP fields. Companies in EM field should focus on understanding the needs of OSM and TP fields, engage in technological innovation, seek collaborations, enhance product diversity, and apply digital technologies to maintain competitiveness and seize new opportunities.

5 Conclusions

A new research method for technology convergence is proposed using patent data and the classification of patents into technological fields based on the ISI-OST-INPI table provided by WIPO. Each technological field is associated with specific IPC codes. The proximity between different fields is calculated by measuring the closeness between patents and IPC codes, enabling the assessment of the degree of convergence between fields. In this study, we construct a patent heterogenous graph centered around IPC codes and utilize the IPCvec model to learn IPC node representations by considering both patent text and structural information, thereby reconstructing patent representations. We apply this method to the graphene field and compare it with other graph embedding algorithms. The results demonstrate that the IPCvec model outperforms other models in terms of representation effectiveness. Finally, we calculate the degree of technological convergence among subfields in the graphene industry and validate the convergence trends over time. By combining the IPC-centered patent heterogenous graph with the HAN algorithm to construct the IPCvec model, this study provides a new approach for patent analysis. It also offers a method for analysts in companies to fully integrate semantic and structural information from patents to identify unidirectional and bidirectional technological convergence phenomena for early warning signals. The limitation of this research lies in the analysis being based solely on existing technologies in the graphene field, without the ability to predict other potential technologies that may emerge. To enhance the generality of the technology forecasting framework, future research can incorporate multidisciplinary studies, cross-sector collaboration, market intelligence and trend analysis, expert consultation and think tank research, as well as advanced technological forecasting tools and algorithms. This would address the limitation of solely analyzing existing technologies and improve the generality and accuracy of the technology forecasting framework. Such an integrated approach can help businesses and decision-makers gain a more comprehensive understanding of and respond to the ever-changing technological landscape.

References

1. Curran, C.-S., Leker, J.: Patent indicators for monitoring convergence-examples from NFF and ICT. Technol. Forecast. Soc. Chang. **78**(2), 256–273 (2011)
2. Kim, J., Lee, S.: Forecasting and identifying multi-technology convergence based on patent data: the case of it and BT industries in 2020. Scientometrics **111**(1), 47–65 (2017)
3. Smojver, V., Štorga, M., Zovak, G.: Exploring knowledge flow within a technology domain by conducting a dynamic analysis of a patent co-citation network. J. Knowl. Manag. **25**(2), 433–453 (2021)
4. Wang, Z., Porter, A.L., Wang, X., Carley, S.: An approach to identify emergent topics of technological convergence: a case study for 3D printing. Technol. Forecast. Soc. Chang. **146**, 723–732 (2019)
5. Xiaoyan, W.: Research on technology convergence analysis framework based on patent co-classification: taking synthetic biology as an example. Inf. Stud. Theory Appl. **44**(10), 179–184 (2021)
6. Preschitschek, N., Niemann, H., Leker, J., Moehrle, M.G.: Anticipating industry convergence: semantic analyses vs IPC co-classification analyses of patents. Foresight **15**(6), 446–464 (2013)
7. Kim, J., Yoon, J., Park, E., Choi, S.: Patent document clustering with deep embeddings. Scientometrics **123**(2), 563–577 (2020)
8. Zhu, C., Motohashi, K.: Identifying the technology convergence using patent text information: a graph convolutional networks (GCN)-based approach. Technol. Forecast. Soc. Chang. **176**, 121477 (2022)
9. Zhang, J., Li, Y.: Technology convergence prediction by the semantic representation of patent classification sequence and text. J. China Soc. Sci. Tech. Inf. **41**(6), 609–624 (2022)

Machine Learning Accelerated Prediction of 3D Granular Flows in Hoppers

Duy Le[1,2], Linh Nguyen[2], Truong Phung[2], David Howard[1], Gayan Kahandawa[2], Manzur Murshed[3], and Gary W. Delaney[1(✉)]

[1] CSIRO Data61, Sydney, Australia
{duy.le,david.howard,gary.delaney}@data61.csiro.au
[2] Federation University Australia, Victoria, Australia
{l.nguyen,t.phung,g.appuhamillage}@federation.edu.au
[3] Deakin University, Victoria, Australia
m.murshed@deakin.edu.au

Abstract. Granular materials are crucial components in a broad variety of industrial and natural processes. However, despite their widespread importance, predicting their complex flow behaviour under different conditions remains extremely challenging. The Discrete Element Method (DEM) is the primary computational technique used to simulate granular flows, and while it can produce highly accurate predictions, it is also inherently computationally expensive. We look to overcome this computational bottleneck through the use of a neural network surrogate model for 3D granular flow simulation, and consider the industrially important use case of flow in grain hoppers. We investigate our model performance across a range of different time scales, quantifying its accuracy and generalizability to different hopper geometries. The use of deep learning techniques for the prediction of granular flow dynamics offers an excellent opportunity for providing step change increases in computational efficiency for industrial decision-making and potential application in real-time decision making in diverse manufacturing settings.

Keywords: Discrete Element Method · Deep Neural Networks · Granular Flow · Industrial Machinery

1 Introduction

Granular materials are critical components in a broad range of systems that we observe throughout both the natural world and in industrial processes such as pharmaceutical manufacturing, food processing, additive manufacturing, minerals processing and chemical production [5,21]. Despite this importance, the detailed complex behavior of granular material flow remains difficult and frequently computationally expensive to predict [22].

The ability to simulate and accurately predict the complex granular flow dynamics within industrial processes offers the opportunity to optimise both the design of the equipment being used and also the process parameters - reducing production costs, minimising down time and enhancing product quality [26]. Prediction of granular flow in natural systems can also be of great benefit, and enable deeper understanding of phenomena such as avalanches and landslides, and prediction of the severity of these events under different conditions [12].

Industrial hoppers are one of the most commonly used pieces of apparatus for storage and controlled discharge of granular materials [3]. Understanding and accurately predicting the flow behaviour within hoppers is crucial in optimisation of their design to ensure effective storage, consistent and controllable discharge and predictable flow behaviour of different types of granular materials within the hopper. This is highly important for example in pharmaceutical manufacturing where product consistency from constituent powders inputted from multiple dispensers is critical for maintaining product consistency [10], or in metal additive manufacturing where a consistent and uniform spreading of powder dispensed from hoppers is required to maintain the quality of the final 3D printed part [19].

The current state-of-the-art simulation methods commonly employed are time-consuming and computationally expensive due to the need to model the detailed particle-particle and particle-boundary interactions of the granular material [7]. There is thus a strong need for faster prediction methods that sufficiently preserve the granular flow characteristics to enable accurate predictions of relevant quantities of interest.

The discrete element method (DEM) [4] is the most widely used technique for accurate modelling of granular flows in industrial devices. This technique requires modeling the individual particle-particle and particle-boundary interactions of each individual grain within the system and often requires use of very small timesteps to resolve the very short timescales over which collisions occur (typically using a timestep of around 10^{-5} s) [1]. Deep learning methods have been explored as a means of training surrogate models that are capable of operating using much larger timesteps. Recent results [14,18] have suggested their potential to be applied as fast and accurate surrogate models, opening new possibilities to replace traditional physics-based simulations.

Studies have demonstrated the effectiveness of the use of deep learning for predicting granular flows [16,28], achieving results with an accuracy close to DEM simulations. These results demonstrate how multi-layered neural networks are capable of learning aspects of the fine intricacies and complex inter dependencies within granular flow systems, offering computational efficiency advantages over more traditional simulation approaches.

In this work, we focus on developing a surrogate model based on a deep neural network approach to rapidly and accurately predict granular flow for an industrially relevant use case of flow of material in hoppers. We aim for the surrogate model to be able to provide a computationally efficient alternative to DEM simulations, while maintaining sufficient accuracy in predicting the flow behaviour within the hopper and allowing for accurate estimation of industrially relevant quantities such as the mass discharge rate of a granular material from within the hopper.

To guide our exploration, we pose two key research questions:

1. **Sensitivity to temporal resolutions**: How well can the model predict particle motions using significantly higher timesteps than those employed in DEM simulations, while still ensuring sufficient accuracy and numerical stability?
2. **Generalization to unseen contexts**: To what extent does the trained model generalize to scenarios and new hopper geometries that are not present in the training data?

The remainder of this paper is organized as follows: Sect. 2 provides a brief overview of related work in granular flow simulation and surrogate modeling. In Sect. 3, we describe the technical details of our approach, the structure of our neural network model and the training process. Section 4 presents the results of our experiments and discusses their implications. In the conclusion, we summarize our findings and outline the potential avenues for future research.

2 Related Work

Granular flow simulations have been a topic of extensive research due to their importance in both understanding the underlying physics of granular materials and their application in optimizing industrial processes. Traditional approaches often employ the discrete element methods (DEM) [2,5,6] to model granular interactions. This method primarily requires solving Newton's equations of motion for every particle within the system. This is achieved by exhaustively calculating all the individual interactions and forces on all particles within the system and integrating their motions through time. Such methods are extremely powerful, but they suffer from being highly computational expensive, limiting their applicability in real-time industrial scenarios [30].

Surrogate modeling, leveraging machine learning techniques, has recently emerged as a promising alternative to traditional simulation methods. In the context of industrial applications, previous work has successfully employed surrogate models to accelerate simulations and make predictions with reduced computational costs [9]. Deep neural networks (DNNs), in particular, have shown great potential in capturing complex relationships within data, making them suitable candidates for surrogate modeling in diverse industrial domains [24].

For granular flow simulations, Andreas et al. [17] proposed a graph neural network (GNN) model, building upon Sanchez-Gonzalez et al.'s work [23], for 3D granular flow simulations with complex geometric boundaries. This model handles triangularization of 3D objects by introducing virtual nodes near boundaries and estimating proximity to adjacent triangles. However, precise hyperparameter tuning is necessary for stable predictions; improper settings lead to instability. Yang et al. [28] introduced a Graph-based Physics Engine (GPE) that simulates diverse physical systems using a unified model architecture with various graph structures. Despite its versatility, this approach does not significantly improve computational speed or memory usage compared to traditional physics engines.

In terms of learning properties of granular material, Mengmeng et al. [27] used an artificial neural network (ANN) to predict the anisotropy of contact force chains (CFCs) in heterogeneous granular materials. Raj Kumar [10] predicted the mass discharge rate of multi-component particle systems from conical hoppers using a neural network and feedforward backpropagation. Yaoyu et al. [13] introduced a supervised learning model trained on DEM simulation data to predict particle flow inside a drum. They employed a Support Vector Machine for Regression (SVR) to predict the angle of repose and collision energy, showing good performance for the angle of repose due to its simple relationship with operational conditions but less accuracy for collision energy due to complex parameter dependencies.

Existing studies in surrogate modeling have primarily focused on applications such as structural analysis [25], fluid dynamics [15], and material design [29]. Surrogate modelling of granular flow simulations for industrial applications is a relatively less well studied topic. It is also an ideal candidate for such an approach due to the computational challenges posed by irregular geometries and dynamic particle interactions that are prevalent in such systems and may necessitate tailored surrogate models that can offer improved efficiency without compromising accuracy.

This gap motivates our research to develop and evaluate a deep neural network-based surrogate model, aiming to address the challenges posed by granular flow simulations for industrial devices. In this study we will consider such a system in a hopper with a worn internal geometry consisting of a complex rough surface with numerous non-convexities.

3 Methodology

3.1 Boundary Representation

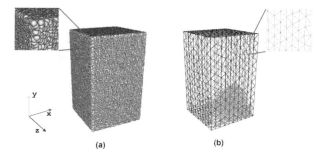

Fig. 1. Two types of boundary representation used in DEM simulations: (a) Overlapping spherical caps; (b) Non-overlapping triangular mesh

We employ the discrete element method (DEM) to generate our ground truth granular flow data. This is one of the most widely used simulation methods for granular materials. A detailed description of the DEM model used and examples of its application to industrially relevant problems can be found in Ref. [2]. We have considered two distinct boundary representations (as shown in Fig. 1), the first being a traditional mesh boundary and the second being formed by sets of overlapping spherical shells distributed randomly on the surface of the hopper. We will focus here on this second case as it offers two distinct advantages. This first is that it allows us to readily model the rough internal surface of a realistically worn industrial hopper, which will contain surface defects and non-convexities across its internal structure. The fully symmetric nature of the interaction potential with the moving particles also simplifies the required network structure and accelerates the learning process when training the artificial neural network (ANN).

3.2 Neural Network Structure

In a granular flow model such as the Discrete Element Method, the simulation proceeds by iteratively calculating all the particle interactions within the system to determine the forces and acceleration of each particle. This information is then used to update the velocities and positions of the particles. Building upon this concept and the deep learning model introduced in [11], in our model we initially compute intermediate positions and velocities, x^* and v^*. This serves as a preliminary prediction neglecting particle interactions:

$$\mathbf{v}_i^* = \mathbf{v}_i^n + \mathbf{g}\Delta t \qquad (1)$$

$$\mathbf{x}_i^* = \mathbf{x}_i^n + \frac{\mathbf{v}_i^* + \mathbf{v}_i^n}{2}\Delta t \qquad (2)$$

These values are then passed as inputs to a neural network to produce a correction term, $\Delta \mathbf{x}_i$, which is used to predict the updated particle positions:

$$\mathbf{x}_i^{n+1} = \mathbf{x}_i^* + \Delta \mathbf{x}_i \qquad (3)$$

Using the updated positions, we can then calculate the updated velocities as:

$$\mathbf{v}_i^{n+1} = \frac{\mathbf{x}_i^{n+1} - \mathbf{x}_i^n}{\Delta t} \qquad (4)$$

where the subscript i is the particle index, superscript n represents the nth step in the simulation, and Δt is the integration time step.

To determine the position correction, $\Delta \mathbf{x}_i$, we employ a 3-D convolutional neural network based on an extension to the structure originally proposed by Ummenhofer et al. [24] (as shown in Fig. 2). For a set of N particles with feature values f_i at position $\mathbf{x}_i$, the convolution at position $\mathbf{x}$ is calculated as:

$$(f * g)(\mathbf{x}) = \sum_{i \in \mathcal{N}(\mathbf{x}, R)} a(\mathbf{x}_i, \mathbf{x}) f_i g(\Lambda(\mathbf{x}_i - \mathbf{x})) \qquad (5)$$

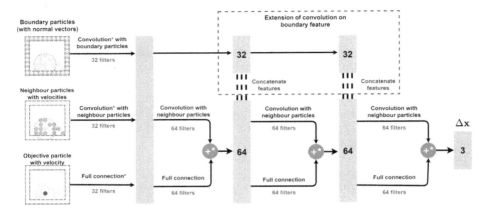

Fig. 2. The architecture of our convolutional neural network. Operations annotated with * are followed by the ReLU activation. All convolutional and fully-connected operations use an additive bias

where f_i is the input features of the neighbour particle i. The key input features to the network are the the positions in 3D space and the intermediate velocities of all of the particles within the system. $\mathcal{N}(\mathbf{x}, R)$ is defined as the set of particles within a radius R around $\mathbf{x}$, where the convolution is applied. The importance of each neighboring particle is weighted using a window function, which allows a smooth response of the convolution operation with different numbers of neighbor particles:

$$a(\mathbf{x}_i, \mathbf{x}) = \begin{cases} \left(1 - \frac{\|\mathbf{x}_i - \mathbf{x}\|_2^2}{R^2}\right)^3 & \text{for } \|\mathbf{x}_i - \mathbf{x}\|_2 < R \\ 0 & \text{else} \end{cases} \quad (6)$$

To represent three-dimensional features, a spherical filter with radius R is employed. This filtering function, denoted as g, is continuous and utilizes linear interpolation on a structured grid. A mapping function $\Lambda(r)$ [8] is used to map a unit sphere to a unit cube, simplifying filter value storage and lookup processes. This configuration allows for the spherical filter to process spatial information while retaining filter function values on a structured grid.

To capture the additional complexity of our rough surface description, we modify the previously proposed architecture by Ummenhofer [24], and add an additional component to the hidden layers within the network that is composed of a convolution on the boundary particles. This addition is highlighted in the red dotted box in Fig. 2. We found from testing that this addition increases the accuracy of the predictions of collisions with the boundary surface.

3.3 Training Procedure

The neural network is trained in a supervised manner, with data from the DEM simulation serving as the ground truth. The model reads ground-truth data at frame n and predicts particle movements for subsequent frames. Given the large number of particles and varying error magnitudes in individual particle positions between training epochs, a multi-scale loss function is employed [16] to aid the optimization process as follows

$$L^{n+1} = \alpha \frac{1}{N} \sum_{i=1}^{N} \|\widehat{x}_i^{n+1} - x_i^{n+1}\|_2$$
$$+ (1-\alpha) \left\| \frac{1}{N} \sum_{i=1}^{N} \widehat{x}_i^{n+1} - \frac{1}{N} \sum_{i=1}^{N} x_i^{n+1} \right\|_2 \quad (7)$$

where N denotes the total number of particles, x_i^{n+1} represents the predicted positions for particle i at frame $n+1$, and $\widehat{x}_i^{n+1}$ is the ground-truth positions. This loss function comprises of two components: the micro-scale loss, which quantifies individual errors between predicted and ground-truth positions, and the macro-scale loss, measuring the error between the centers of ground truth and predicted particles. Both terms are weighted by a factor $\alpha = 0.5$.

Predictions for frame $n+1$ are used iteratively to forecast subsequent frames up to a predetermined limit, denoted as F_r. To optimize memory usage, rather than updating model parameters once per epoch, we partition the training dataset into small batches (e.g., $B = 16$ in our case) and update parameters after processing each batch. The total batch loss is a weighted sum of losses across the predicted frames:

$$L = \sum_{i=1}^{F_r} w^i L^{n+i} \quad (8)$$

where $F_r = 5$ is the number of predicted frames used to calculate the loss, w^i is the weight of the loss at frame $n+i$.

3.4 Dataset Preparation

This work utilizes the CSIRO DEM solver [2] to create a dataset for the training and testing phases. The dataset is deliberately constructed to maximize particle interactions for both training and testing with the hopper. Particles are initially arranged in a cluster with some randomness in their locations. The initial velocities are set between 1 m/s and 2 m/s. The DEM simulation uses a time step of $\Delta t = 1.77 \times 10^{-5}$ s corresponding to a material stiffness of $k_n = 20000$ N/m.

3.5 Simulation Setup

The training environment is a hopper with dimensions shown as in Fig. 3. The hopper surface consists of overlapping spherical hemisphere sections with a radius of $r_{boundary} = 0.003$ m to represent a highly worn and rough surface. Moving particles within the hopper have a radius of $r_{particle} = 0.003$ m. Key

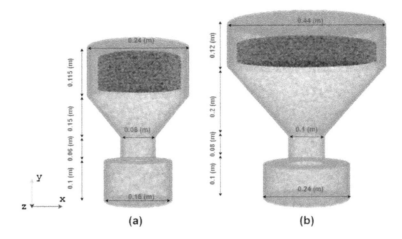

Fig. 3. Geometry of the hopper in simulation: (a) training data; (b) testing data

parameters for the DEM simulation are given in Table 1. The training simulations include cases with multiple different fill levels, with particle numbers in the range of $N = 3000$ to 4000. We measure the flow rate of material through the orifice of the hopper over time, and compare our surrogate model predictions to the DEM simulation ground truth results.

Table 1. DEM simulation parameters

Parameters	Value
Particle diameter (m)	0.006
Mass density (kg/m^3)	2400
Normal spring stiffness (N/m)	20000
Coefficient of restitution	0.5
Friction coefficient	0.5

In the subsequent section, we present and discuss the results of these experiments, offering a better understanding of the capabilities and limitations of our trained model.

4 Results and Discussion

4.1 Hopper with the Same Geometry as in Training Data

In the first test case, we assess our model's predictive performance by applying it to predict granular flow in a hopper with the same geometry used in

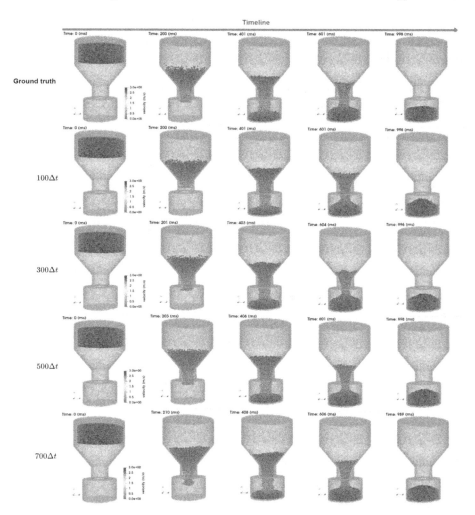

Fig. 4. Prediction of granular flow in a hopper. The simulation starts with 5000 stationary particles in the top section of the hopper, which fall under gravity and flow through the hopper orifice into the lower chamber. Results are shown for the ground truth simulation and our model predictions for 100, 300, 500, and 700Δt. All models give reasonably accurate predictions of the granular flow, with very similar material states and flow profiles at the time snapshots shown.

the training dataset but with an increase in the number of particles, from the maximum of 4000 particles considered in the training set to 5000 particles in the testing set. Figure 4 demonstrates the model's accuracy in predicting granular flow using a much larger timesteps than that required for DEM simulations to remain numerically stable. The model predictions were performed with time steps varying from $100\Delta t$ to $700\Delta t$ (results shown from the second row to the fifth row in Fig. 4. In this configuration, 5000 stationary particles are placed in the hopper's top section. The particles fall due to gravity, colliding with the walls of the hopper and flowing through the central orifice within the hopper. The model accurately predicts the granular flow behaviour and both the particle-particle and particle-boundary collision behaviour, with no penetration of particles beyond the boundary during the simulation or other obvious errors in the flow behaviour.

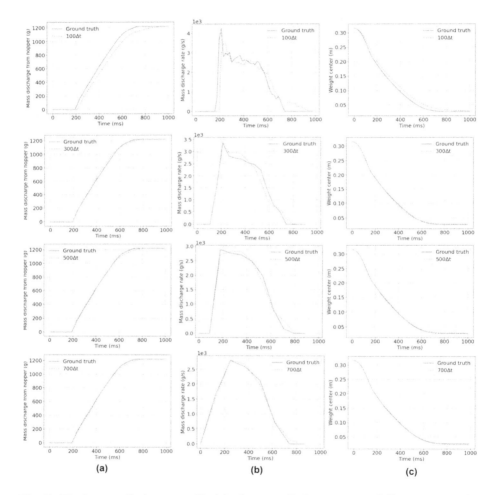

Fig. 5. The hopper discharge profile (a), the mass discharge rate at different time scales (b), the evolution of weight centers (c).

To assess the model's accuracy, quantitative results are collected and presented in Fig. 5. Hoppers are designed to control particle movement under gravity (represented on the y-axis), with the mass discharge rate being a crucial parameter that is optimised for a particular industrial application.

Figure 5 shows total mass discharged over time, along with the mass discharge rate and evolution of the particle weight center with time.

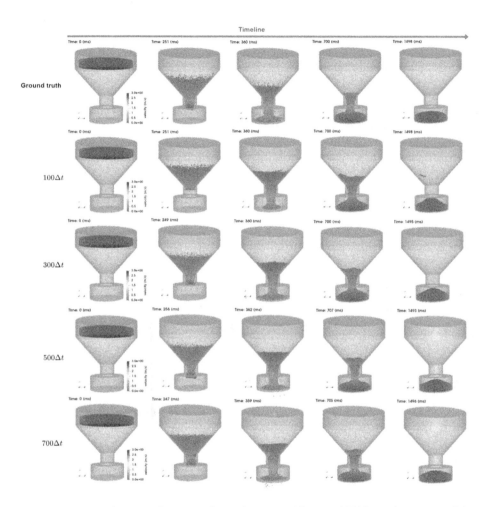

Fig. 6. Granular flow simulation within a hopper with over 10000 stationary particles at the top. Discharge begins around 250 ms (when particles fall below a reference height of 0.1 m), continuing until simulation completion at approximately 1500 ms.

We see in Fig. 5(b), that the material begins to flow through the hopper orifice at $t = 200$ ms. The mass discharge rate has non-zero values in the period from 200 ms to about 750 ms with a continuous flow observed inside the hopper. The flow stabilises to a near-constant discharge rate as shown by the plateau region in the graph between 250 – 550 ms. All material has been discharged through the orifice by $t = 800$ ms, and the flow rate falls to zero.

At this point, the total mass that has been discharged through the hopper is slightly more than 1200 g.

The results from this test case shows our model's consistent ability to accurately predict the flow within the hopper right up to very large timesteps of $700\Delta t$, suggesting that our surrogate model has the ability to perform simulations at much larger timesteps while achieving similar predictive power to the equivalent DEM simulation.

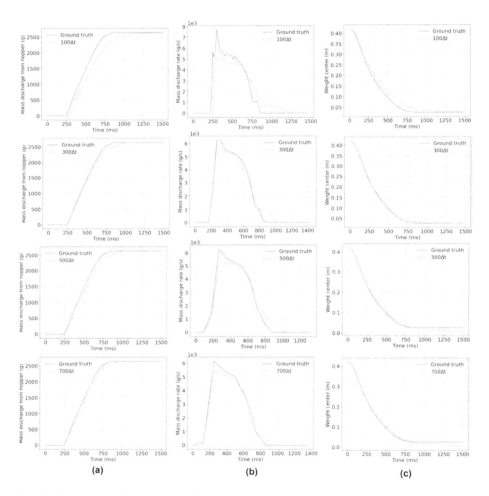

Fig. 7. The hopper discharge profile (a), the mass discharge rate at different time scales (b), the evolution of weight centers (c) of granular flow in a hopper with different geometry.

4.2 Hopper with Different Geometry

To assess our model's generalizability, we deploy it to predict granular flow in a hopper with a different geometry to that used in the training data, and featuring an increased particle count exceeding 10000 to additional increased complexity to the flow within the hopper.

Figure 6 shows the granular flow in the new hopper geometry at different time points in the simulation. The results indicate accurate predictions of overall granular motion by our surrogate model, with the particles stabilized at the end of the simulation, even with large timesteps. For our predictions with a timestep of $100\Delta t$, where we observe the strongest damping effects, the slower granular flow leads to a small number of particles remaining sitting on the rough shallower surface of the upper section of the hopper at the end of the simulation. (A phenomena frequently observed in equivalent realworld granular systems which involve flow over shallow rough sloping surfaces). For timesteps of $500\Delta t$ and $700\Delta t$, we also observe a few particles go through the boundary around 360 ms, an artifact which has been seen in similar models applied previously [16].

Figure 7 shows comparisons of the mass discharge for the DEM simulation and surrogate models using different timesteps.

Despite the surrogate model for this case exhibiting larger prediction errors in the fine details of the flow, the model again generally gives reasonably accurately estimates for the macroscopic behavior of the granular flow within the hopper and provides good predictions of the mass discharge rate within the hopper up to timesteps of $700\Delta t$.

4.3 Run-Time Comparison

We have evaluated the runtime for both the DEM model and the ANN surrogate model. For the hopper case shown in Fig. 6, the surrogate model running with a $700\Delta t$ has a 70x faster runtime compared to the DEM model. This demonstrates the excellent potential of these surrogate models to be used to give predictions with an acceptable degree of accuracy with significantly reduced computation runtimes.

5 Conclusions

This study presents a deep neural network approach for creating accurate surrogate models to speed up granular flow simulations in hoppers. Our findings and contributions include:

- **Efficient model performance**: Our surrogate model accelerates granular flow simulations while maintaining a sufficient degree of accuracy in predicting granular flow dynamics
- **Non-convex boundary representation**: Our use of overlapping spherical shells in our boundary description allows for the representation of highly non-convex realistic rough surfaces and can thus capture the complex granular

flow dynamics with such irregular surfaces commonly found in real industrial hoppers.
- **Practical industrial applicability**: Our model is demonstrated to be able to predict key industrial quantities of interest, including the mass discharge rate from the hopper.

Future improvements to our model may expand the applicability to other industrial scenarios and address challenges such as granular materials composed of multiple species with diverse particle properties. The speed-up achieved with machine learning surrogates over traditional modelling techniques also opens up the possibility of applications in new areas including real-time prediction and optimisation of complex industrial processes [19,20].

References

1. Podlozhnyuk, A., Pirker, S., Kloss, C.: Efficient implementation of superquadric particles in discrete element method within an open-source framework. Comput. Part. Mech. **4**(1), 101–118 (2016). https://doi.org/10.1007/s40571-016-0131-6
2. Cleary, P.W.: Industrial particle flow modelling using discrete element method. Eng. Comput. **26**(6), 698–743 (2009)
3. Cleary, P.W., Sawley, M.L.: DEM modelling of industrial granular flows: 3D case studies and the effect of particle shape on hopper discharge. Appl. Math. Model. **26**(2), 89–111 (2002)
4. Cundall, P.A., Strack, O.D.L.: A discrete numerical model for granular assemblies. Geotechnique **1**, 47–65 (1979)
5. Delaney, G.W., Morrison, R.D., Sinnott, M.D., Cummins, S., Cleary, P.W.: DEM modelling of non-spherical particle breakage and flow in an industrial scale cone crusher. Miner. Eng. **74**, 112–122 (2015). https://doi.org/10.1016/j.mineng.2015.01.013
6. Delaney, G.W., Cleary, P.W., Hilden, M., Morrison, R.D.: Testing the validity of the spherical DEM model in simulating real granular screening processes. Chem. Eng. Sci. **68**(1), 215–226 (2012)
7. Ge, W., et al.: Discrete simulation of granular and particle-fluid flows: from fundamental study to engineering application. Rev. Chem. Eng. **33**(6), 551–623 (2017)
8. Griepentrog, J.A., Höppner, W., Kaiser, H.C., Rehberg, J.: A bi-Lipschitz continuous, volume preserving map from the unit ball onto a cube. Note di Mat. **28**(1), 177–193 (2008)
9. Harmon, D., Zorin, D.: Subspace integration with local deformations. ACM Trans. Graph. **32**(4), 1–10 (2013)
10. Kumar, R., Patel, C.M., Jana, A.K., Gopireddy, S.R.: Prediction of hopper discharge rate using combined discrete element method and artificial neural network. Adv. Powder Technol. **29**(11), 2822–2834 (2018)
11. Ladický, L.U., Jeong, S., Solenthaler, B., Pollefeys, M., Gross, M.: Data-driven fluid simulations using regression forests. ACM Trans. Graph. **34**(6), 1–9 (2015)
12. Lemiale, V., et al.: Combining statistical design with deterministic modelling to assess the effect of site-specific factors on the extent of landslides. Rock Mech. Rock Eng. **55**(1), 1–15 (2021). https://doi.org/10.1007/s00603-021-02674-x

13. Li, Y., Bao, J., Yu, A., Yang, R.: A combined data-driven and discrete modelling approach to predict particle flow in rotating drums. Chem. Eng. Sci. **231**, 116251 (2021)
14. Ling, J., Kurzawski, A., Templeton, J.: Reynolds averaged turbulence modelling using deep neural networks with embedded invariance. J. Fluid Mech. **807**, 155–166 (2016)
15. Loy, Y.Y., Rangaiah, G.P., Lakshminarayanan, S.: Surrogate modelling for enhancing consequence analysis based on computational fluid dynamics. J. Loss Prev. Process Ind. **48**, 173–185 (2017)
16. Lu, L., Gao, X., Dietiker, J.F., Shahnam, M., Rogers, W.A.: Machine learning accelerated discrete element modeling of granular flows. Chem. Eng. Sci. **245**, 116832 (2021)
17. Mayr, A., Lehner, S., Mayrhofer, A., Kloss, C., Hochreiter, S., Brandstetter, J.: Learning 3D granular flow simulations. arXiv preprint arXiv:2105.01636 (2021)
18. Morton, J., Jameson, A., Kochenderfer, M.J., Witherden, F.: Deep dynamical modeling and control of unsteady fluid flows. Adv. Neural Inf. Process. Syst. **31** (2018)
19. Phua, A., Cook, P.S., Davies, C.H., Delaney, G.W.: Smart recoating: a digital twin framework for optimisation and control of powder spreading in metal additive manufacturing. J. Manuf. Process. **99**, 382–391 (2023)
20. Phua, A., Delaney, G.W., Cook, P.S., Davies, C.H.: Intelligent digital twins can accelerate scientific discovery and control complex multi-physics processes. In: ICML 2022 2nd AI for Science Workshop (2022)
21. Phua, A., Doblin, C., Owen, P., Davies, C.H.J., Delaney, G.W.: The effect of recoater geometry and speed on granular convection and size segregation in powder bed fusion. Powder Technol. (2021). https://doi.org/10.1016/j.powtec.2021.08.058
22. Rao, K.K., Nott, P.R., Sundaresan, S.: An Introduction to Granular Flow, vol. 490. Cambridge University Press, Cambridge (2008)
23. Sanchez-Gonzalez, A., Godwin, J., Pfaff, T., Ying, R., Leskovec, J., Battaglia, P.: Learning to simulate complex physics with graph networks. In: International Conference on Machine Learning, pp. 8459–8468. PMLR (2020)
24. Ummenhofer, B., Prantl, L., Thuerey, N., Koltun, V.: Lagrangian fluid simulation with continuous convolutions. In: International Conference on Learning Representations (2019)
25. Wang, Y.Z., Zheng, X.Y., Lu, C., Zhu, S.P.: Structural dynamic probabilistic evaluation using a surrogate model and genetic algorithm. In: Proceedings of the Institution of Civil Engineers-Maritime Engineering, vol. 173, pp. 13–27. Thomas Telford Ltd. (2020)
26. Windows-Yule, C., Tunuguntla, D.R., Parker, D.: Numerical modelling of granular flows: a reality check. Comput. Part. Mech. **3**, 311–332 (2016)
27. Wu, M., Wang, J.: Prediction of 3D contact force chains using artificial neural networks. Eng. Geol. **296**, 106444 (2022)
28. Yang, C., Gao, W., Wu, D., Wang, C.: Learning to simulate unseen physical systems with graph neural networks. In: NeurIPS 2021 AI for Science Workshop (2021)
29. Ye, F., Wang, H., Li, G.: Variable stiffness composite material design by using support vector regression assisted efficient global optimization method. Struct. Multidiscip. Optim. **56**, 203–219 (2017)
30. Yeom, S.B., Ha, E.S., Kim, M.S., Jeong, S.H., Hwang, S.J., Choi, D.H.: Application of the discrete element method for manufacturing process simulation in the pharmaceutical industry. Pharmaceutics **11**(8), 414 (2019)

RD-Crack: A Study of Concrete Crack Detection Guided by a Residual Neural Network Improved Based on Diffusion Modeling

Yubo Huang[2(✉)], Xin Lai[2], Zixi Wang[3], Muyang Ye[4], Yinmian Li[2], Yi Li[5], Fang Zhang[2], and Chenyang Luo[1(✉)]

[1] School of Computer, Electronics and Information, Guangxi University, Nanning 530004, China
sherlock511@126.com
[2] School of Civil Engineering, Southwest Jiaotong University, Chengdu 611756, Sichuan, China
ybforever@my.swjtu.edu.cn
[3] School of Computing and Artificial Intelligence, Southwest Jiaotong University, Chengdu 611756, Sichuan, China
[4] SWJTU-Leeds Joint School, Southwest Jiaotong University, Chengdu 611756, Sichuan, China
[5] Computing and Communications, Lancaster University, Lancaster, UK

Abstract. Automated crack detection in concrete structures is an important aspect of structural health monitoring (SHM) to ensure safety and durability. Traditional methods mainly rely on manual inspection, which suffers from subjectivity and inefficiency challenges. To address these issues, machine learning, especially deep learning techniques, has been gradually adopted to improve accuracy and reduce reliance on large amounts of labeled data. This paper introduces RD-Crack, an innovative concrete crack detection method. Our RD-Crack framework combines the encoder with ResNeXt and extrusion excitation modules for feature extraction and uses a diffusion model for parameter optimization to achieve accurate crack detection in complex engineering environments. Experimental results show that our RD-Crack outperforms other state-of-the-art methods in comprehensive performance.

Keywords: Crack Detection · Structural Health Monitoring · Diffusion Model · Unsupervised Learning · Neural Network Optimization

1 Introduction

Crack detection is the process of detecting cracks in a structure using various processing techniques. Structural health monitoring (SHM) is a technique

---

Y. Hauang, X. Lai, Z. Wang, M. Ye—These authors contributed equally to this work.

© The Author(s), under exclusive license to Springer Nature Switzerland AG 2024
M. Wand et al. (Eds.): ICANN 2024, LNCS 15024, pp. 340–354, 2024.
https://doi.org/10.1007/978-3-031-72356-8_23

for assessing the health of structures by acquiring, processing, and interpreting data from structural engineering systems in real-time or periodically [8,10,34]. Cracks have a significant impact on the mechanical behavior of structures and their identification and detection can reveal structural stress mechanisms, which is an important part of SHM to ensure structural safety and durability [21]. Conventional methods for concrete crack detection and evaluation typically rely on manual visual inspection, usually performed by experienced inspectors, and present challenges such as subjectivity, labor intensity, and inability to detect subtle or early cracks [7,26].

To overcome these challenges, automated crack detection has been proposed to replace subjective and inefficient manual visual inspections. Barazzetti et al. [3] proposed an image-based crack analysis method (IMCA) for processing digital image sequences to perform crack boundary extraction and crack deformation measurements. Nishiyama et al. [25] proposed a new digital photogrammetry-based crack monitoring method for measuring crack displacements in aged structures with high precision and accuracy. Jia et al. [18] investigated the application of vibro-thermography in the crack detection of concrete components and developed an acoustic excitation device to enhance crack detection using high-power ultrasound. Yehia et al. [39] used embedded Fabry-Pérot fiber optic sensors to monitor the concrete structure at different stages of construction behavior to investigate their performance in strain measurement, dynamic/cyclic loading, and fatigue testing. However, vision-based crack detection is limited by the abundance of tagged data sets, and tagging a large number of untagged initial crack data sets is one of the challenges faced by civil engineers. How to bridge the traditional civil engineering structural health inspection with the emerging computer technology is a handicap.

The proposal of machine learning algorithms presents new ideas to address the challenges of structural health monitoring (SHM). Zhang et al. [42] proposed a real-time crack detection method based on a 1D convolutional neural network (1D-CNN) and a long-short-term memory network (LSTM). Geetha et al. [9] proposed a method that combines image binarisation and a Fourier-based 1D DL model framework for real-time detection and classification of crack and non-crack features on concrete surfaces. Baduge et al. [2] proposed a framework for asphalt pavement health monitoring based on 1D-DFTCNN and CCR, and a geospatial mapping methodology for visualization of crack locations for health monitoring and maintenance purposes. Supervised neural networks require extensive human intervention for feature identification and extraction, and have poor generalization performance.

Therefore, semi-supervised and unsupervised algorithms are preferred by researchers in today's concrete crack detection. Jian et al. [19]. proposed a new semi-supervised algorithm to improve the performance of crack segmentation by means of a cross-teacher pseudo-supervised framework and a cross-enhancement strategy. Zheng et al. [43] proposed a multi-stage application-oriented semi-supervised algorithm. proposed a multi-stage application-oriented semi-supervised active learning framework (CAL-ISM) for bridge crack identification and measurement. Xiang et al. [36] proposed SemiCrack, a semi-supervised learning framework based on contrast learning and cross-pseudo-supervision, which can effectively reduce the dependence on labels in crack segmentation.

Diffusion modeling [31], as a new unsupervised method, is proposed to give new consideration to concrete crack detection. Although concrete crack detection is a supervised task, the optimization of various parameters of the supervised neural network in it cannot be derived in a supervised manner, which leads to poor detection results. In addition, the supervised approach requires higher image quality than the unsupervised approach, otherwise, the accuracy and reliability of detection will be affected. Therefore, this paper proposes an improved concrete crack detection method based on the diffusion model, namely RD-Crack. Specifically, RD-Crack uses the idea of diffusion model guidance to generate high-performance neural network parameters to optimize the performance of our neural network. Meanwhile, the embedded diffusion model can improve the image to make its crack detection more accurate and stable. Our contributions are mainly as follows:

- We introduce the RD-Crack framework, an innovative methodology for detecting cracks in concrete structures. By integrating sophisticated image processing algorithms with a diffusion model, the framework enhances the neural network's parameter optimization, leading to more accurate and stable crack detection.
- We delineate the implementation of an encoder-decoder architecture, which incorporates ResNeXt and squeeze-excitation modules. This integration aids in superior feature extraction and analysis, enabling nuanced segmentation of crack features within concrete structures.
- A pivotal contribution of Our RD-Crack is applying a diffusion model to refine the neural network parameters. This approach facilitates the generation of optimized parameters, substantially improving the model's efficacy in identifying and analyzing concrete cracks. The findings indicate the method's potential to transform current practices in crack detection and analysis, promoting safer and more efficient monitoring and maintenance of concrete infrastructures.

2 Related Work

2.1 Concrete Crack Detection

Concrete crack detection is an assessment technique used to identify and quantify the characteristics of cracks in concrete structures. Due to the specificities of the real-world environment, this detection relies on the combined application of visual inspection, digital image processing, acoustic emission analysis, and other non-destructive assessment methods. Methods grounded in traditional image processing employ computer vision techniques to extract crack-related information from digital images. Tiny cracks, often elusive to visual inspection and conventional non-destructive testing (NDT), can be rapidly and accurately identified using image processing methods, facilitated by visual sensors and computers [4,13].

The image processing workflow typically encompasses image acquisition, feature extraction, and crack segmentation. Acquisition can be performed using various modalities, such as photographic [24], infrared [30], ultrasonic [6], and laser imaging [28]. Adhikari et al. [1] proposed a model that integrates digital image processing to formulate numerical representations of concrete defects. Iyer et al. [16] developed segmented binary crack maps to detect crack patterns in pipeline images using alternating filters. However, these methods have limited functionality and often necessitate human intervention in complex scenarios, hindering full automation and efficiency, especially in large-scale or real-time applications.

The advent of machine learning has introduced novel approaches to addressing the challenge. Kaseko et al. [20] devised an automated method that merges artificial neural network models with traditional image processing to classify cracks in road video imagery. Moreover, support vector machines have been widely employed in conventional machine learning. O'Byrne et al. [27] proposed a semi-automated, augmented texture segmentation method employing defined feature vectors to detect and classify surface damage on infrastructure components.

Deep learning, surpassing traditional machine learning, excels in extracting complex and subtle features from images, garnering significant attention in research. Concrete crack detection using deep learning is categorized into image classification, object recognition, and semantic segmentation. Yokoyama et al. [40] applied deep learning in concrete crack detection, developing classifiers from image datasets. Song et al. [32] designed a Fully Attentional Network (FLANet) to address the limitations of non-local self-attention methods in semantic segmentation. These studies underscore the profound impact of deep learning in revolutionizing concrete crack detection.

2.2 Diffusion Modeling

Diffusion modeling is a methodology that commences with the training of an energy model using techniques such as score matching, followed by sampling from the energy model via the Langevin equation [31]. This method facilitates the generation and sampling of various continuous entities, such as speech and images. Ho et al. [12] expanded this framework by iteratively degrading data samples to random noise and then reconstructing them, enabling the production of high-resolution images, thereby garnering significant research interest.

Over time, the versatility and generalizability of diffusion models have established them as potent tools for addressing intricate problems, demonstrating exceptional performance across diverse application areas. Song et al. [33] introduced a comprehensive framework that integrates score-based generative modeling and Denoising Diffusion Probabilistic Models (DDPMs). This framework facilitates the transformation of complex data distributions into prior distributions through stochastic differential equations (SDEs) and then reverts these prior distributions to the original data distributions using backward SDEs. Radford [29] introduced the CLIP model, based on ConVIRT, capable of predicting new image classification tasks without specific dataset training.

Diffusion model-enhanced methods have found widespread application in various disciplines. Wang et al. [35] developed a high-fidelity technique for generating speckle patterns in Digital Image Correlation (DIC), enhancing the accuracy of crack-containing DIC image segmentation. Li et al. [22] introduced the DIM-UNet model, which enhances medical image segmentation performance. Additionally, Luo et al. [23] proposed a deep generative model that combines a diffusion probabilistic model with a covariant neural network, enabling the joint modeling of Complementarity Determining Regions (CDRs) based on their sequences and structures.

Despite the advancements in computer vision and multimodal learning, diffusion models' application in concrete crack detection remains underexplored. The subtlety, complexity, and irregularity of cracks can compromise the precision and accuracy of conventional target detection models. Therefore, the principles of diffusion modeling are particularly well-suited for enhancing the effectiveness of concrete crack detection methodologies.

3 Methodology

3.1 The Framework of RD-Crack

Figure 1 shows our framework for crack detection, RD-Crack. RD-Crack still mainly consists of an encoder and a decoder. The encoder mainly uses ResNeXt [37] as the main feature extraction network and uses the squeeze and excitation module [14] to learn the correlation between channels for better crack segmentation. We flatten the parameters of this model into one-dimensional vectors, and then introduce a new autoencoder that imports the one-dimensional vectors into the standard diffusion model for training, extracts potential representational information from random noise, and then reconstructs the parameters from the new representational information through the decoder, which is used to update the parameters of the model of RD-Crack. At this point, the model parameters of RD-Crack are all generated by the diffusion model and are more suitable for concrete crack detection.

Main Body Construction of RD-Crack Encoder. We consider the most classical residual neural network [11] to extract the crack features of concrete.

$$f(u) = \mathcal{F}(u, \{T_i\}) + u; \qquad (1)$$

Where u and $f(u)$ are input and output variables. The function $\mathcal{F}(\cdot)$ represents the residual mapping to be learned.

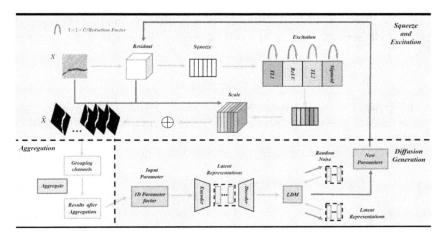

Fig. 1. The framework of RD-Crack. In the process of extrusion and excitation, the feature maps segmented by the residual neural network are given different weights and fed into the aggregation module for group aggregation. After the initial solution is obtained, parameter flattening is introduced into the autoencoder of the diffusion model. The parameters are adjusted through Gaussian noise interference, and the new parameters are output into the residual neural network. Iteratively solve until reaching the threshold of LDM. Please note that in our RD-Crack, the initial parameters are the parameters of the classic paper, which can save a large part of the convergence time of the diffusion model.

In the convolution process, global information is lost, and certain feature maps identified by crack detection may prove more valuable for subsequent layers. To address this, we preserve the global information from crack detection through squeeze and excitation operations [14], which dynamically adjust the dependencies among the feature maps. Denote the output of the preceding convolutional layer in the crack detection framework as Global information is $R = [r_1, r_2, \ldots, r_p]$ then extracted by applying a global average pooling operation to each channel r_q resulting in the channel statistics $T = [t_1, t_2, \ldots, t_p]$, where the ith element of T is calculated as follows:

$$T_i = \frac{1}{H * W} \sum_{j=1}^{H} \sum_{k=1}^{W} r_i(j, k); \qquad (2)$$

Equation 2 is the extrusion process. Having obtained the global information from the squeeze step, the next step is to embed the global information in the output. To recalibrate the feature map interdependencies, the excitation step will use a gating mechanism with the following output equation:

$$g(u) = TL_2(TL_1 f(u)) = sigmoid(w_2(\text{Re}LU(w_1 f(u)))); \qquad (3)$$

where TL_1 and TL_2 are fully connected layers with weights w_1 and w_2. In this case, $g(u)$ is a probability vector with the same shape as the input to the network. The final output of the Squeeze and Excitation module is obtained by scaling each element of $f(u)$ to the corresponding element of $g(u)$, denoted as follows:

$$\tilde{X}_l = f_l(u) * g_l(u); \qquad (4)$$

At this point, the feature maps are dynamically recalibrated. We then partitioned the feature maps n output from the Squeeze and Excitation module into groups using m. Each group was downsampled to a single feature map using the sum of aggregation operations [37], represented as follows:

$$\begin{aligned} D_1 &= [d_1, d_2, ..., d_k] \\ D_2 &= [d_{m+1}, d_{m+2}, ..., d_{2m}] \\ &\vdots \\ D_i &= [d_{((i-1)m)+1}, d_{((i-1)m)+2}, ..., d_{im}] \\ &\vdots \\ D_{n/m} &= [d_{(((n/m)-1)m)+1}, d_{(((n/m)-1)m)+2}, ..., d_m] \end{aligned} \qquad (5)$$

$$G(u) = \sum_{j=((i-1)m)+1}^{im} d_j; \qquad (6)$$

Automatic Construction of New Encoder Parameters. Using the above approach, we successfully constructed a comprehensive encoder. However, the parameters of this model are completely customized, and continuous debugging is required to determine the parameter value range with better results. At the same time, since there may be potential representations between feature maps, the set parameters may not be able to find their specific relationships. Simply relying on the stacking and aggregation of a large number of feature maps will significantly reduce performance. Therefore, after establishing the encoder with the initial parameter set, this study imported these parameters into the diffusion model, used random noise to optimize the parameters, and then fed it back to the original encoder as the output to find the potential in the crack feature map. characterization.

The encoder generates a parameter set that is then flattened into a one-dimensional vector $V = [v_1, v_2, ..., v_n]$. This study reconstructs V through training an autoencoder. Subsequently, a Variational AutoEncoder (VAE) is trained without conditioning. The VAE is trained to ensure that the decoded weights achieve similar performance to the input weights when evaluated on the test set. Specifically, we minimize the following objective function:

$$\mathcal{L} = -E_{q_\phi(z|V)}[\log p_\theta(V|z)] + KL[q_\phi(z|V) \parallel p(z)] \qquad (7)$$

where V is a one-dimensional vector, z is its latent representation, p_θ and q_ϕ are the reconstruction and approximate posterior terms, respectively, and $p(z)$ is the prior distribution. We use the standard Gaussian distribution for the prior.

The first term here is the log-likelihood, which represents the reconstruction loss. Since crack detection is still essentially a classification problem, we use the classic cross-entropy loss to measure it. The second term is the KL divergence between the encoder output and the prior probability, used for regularization.

The trained VAE is then utilized to train a latent diffusion model for dataset-conditioned weight sampling. At this stage, we have access to a pretrained VAE for encoding neural network weights and a pretrained Set Transformer module for encoding entire datasets. The next step involves defining a model to generate latent representations of weights conditioned on dataset representations, achieved through Diffusion Denoising Probabilistic Models (DDPM).

Given a weight embedding z obtained from the encoder of the pretrained VAE, the forward diffusion process involves successive Gaussian noise perturbations of z over T time steps. At time step t, the distribution is given by:

$$p(z_t|z_{t-1}) = \mathcal{N}(z_t; \mu_t = \sqrt{1-\beta_t}z_{t-1}, \beta_t I) \tag{8}$$

where $\beta_t \in (0,1)$ is the noise variance and $p(z_{1:T}|z_0) = \prod_{i=1}^{T} p(z_t|z_{t-1})$.

Decoder. When the diffusion model is applied to the autoencoder, the goal of the decoder is to reconstruct the parameter variables from the latent space through the inverse diffusion process. Specifically, the decoder formulation relies on the inverse process of the diffusion model, which progressively generates data from samples of some noise distribution (here a Gaussian noise distribution is used). It can be expressed as:

$$p_\theta(z_{t-1}|z_t) = \mathcal{N}(z_{t-1}; \mu_\theta(z_t, t), \Sigma_\theta(z_t, t)) \tag{9}$$

where μ_θ and Σ_θ are neural networks.

4 Experiments

4.1 Experimental Setup

Datasets and Settings. Caffe [17] was utilized to implement our RD-crack. The original encoder was initialized using the Kaiming [11] method on ResNet50, with an initial learning rate set to 0.1 and decreased by a factor of 10 every 150 epochs. The training duration was 600 epochs. Other hyperparameters such as weight decay, momentum, and optimizer were set to 0.0005, 0.85, and stochastic gradient descent (SGD), respectively. The batch size was fixed at 64. In the parameter reconstruction of the original encoder, both the self-encoder and the latent diffusion model include an encoder and decoder based on a 4-layer 1D CNN. We collected 600 training data points by default for all architectures. During the final training phase, we continued training the last two normalization layers while keeping other parameters fixed. We saved 600 checkpoints of the original model in the last epoch. In the inference process of the diffusion model, we generated 100 new parameters by inputting random noise to the potential diffusion model and the decoder. These synthesized parameters were then combined with the fixed parameters mentioned above to create our generated models. From these models, we selected the one that performed best on the training set, evaluated its accuracy on the validation set, and reported the results. This approach ensures a fair comparison with the model trained using the SGD optimization method. The test data set includes the SDDNET [5] (about 56,000 256*256 pixel images of concrete walls, bridge decks, and pavement with and without cracks.), CrackTree260 [44] (260 pavement images, 512*512 pixels, in order to expand the data set Scale, we crop the image center symmetrically into 4 sub-images, and rotate each sub-image once at 10-degree intervals, expanding the total to 37440 images), Crack500 [38] (500 road images of 2000*1500 pixels, in order to expand the data set Scale, we crop the image center symmetrically into 16 sub-images, and rotate each sub-image once at an interval of 60°C, expanding the total to 48,000 images), as shown in Fig. 2. We adjust all images to the standard 256*256 pixels for experiments. All experiments were conducted on an NVIDIA GeForce RTX 4090 and 12th Gen Intel(R)Core(TM) i7-12700KF.

Fig. 2. Concrete crack detection test data set. The negative pictures is a picture of concrete without cracks, and the positive picture is pictures of concrete with cracks.

Baselines. To verify the effectiveness of our RD-Crack, we compare it with several state-of-the-art concrete crack detection methods. These include a traditional digital image processing method [3], a machine learning method [9], and three deep learning-based methods, Deepcrack [45], Mostafa et.al [15], and CrackDiff [41]. For a fair comparison, we retrained other models using the same training dataset.

Evaluation Indicators. 4 indicators can be used to measure the performance of concrete crack detection. By comparing the detected cracks with manually annotated real data, the *precision*, and *recall* are calculated, and then the $F1score$ is obtained. In addition, $mIoU$ (average intersection greater than union) is used to measure the accuracy of pixel positioning, which can be used to comprehensively evaluate the performance of the model for image segmentation, and is also widely used in the evaluation of concrete crack detection.

4.2 Comparative Experimental Results with SOTA Method

Quantitative Results. Table 1 presents the numerical results of our RD-Crack and IMCA, Geetha et al., Deepcrack, Mostafa et al., and CrackDiff. As can be seen from Table 1, for concrete crack detection, our RD-Crack performs well in two comprehensive performance indicators, F1 score and mIoU, in the three data sets, both of which are better than other optimal methods. In addition, it can be seen from Table 1 that different precision pictures and crack complexity of different data sets will affect the performance of existing concrete detection methods by more than 5%, and the parameter fine-tuning of the supervised neural network by the diffusion neural network makes Its fluctuation does not exceed 3% in the selected data set. This shows the importance of parameter adjustment under actual engineering environmental conditions. This also provides a new idea for future researchers to improve the detection model in actual engineering. Figure 3 is a violin plot of various indicators detected by our RD-Crack in the test data set. As can be seen from Fig. 3, the highest value density of our RD-Crack in recall, precision, F1 score and exceeds 85%, and the highest value density of mIoU exceeds 80%. This shows that our proposed RD-Crack is reliable in performance.

Table 1. Comparison of the indicators of various results of different crack detection methods on different data sets. In the table, F1 represents F1 score, and TP represents true precision.

Methods	SDDNET			CrackTree260			Crack500		
	F1 ↑	mIoU↑	TP ↑	F1 ↑	mIoU↑	TP ↑	F1 ↑	mIoU↑	TP ↑
IMCA [3]	65.24	67.26	61.26	68.79	71.10	62.32	70.13	72.47	65.84
Geetha et al. [9]	81.72	84.36	74.37	79.68	82.95	76.31	83.36	87.14	77.18
Deepcrack [45]	78.62	81.71	76.26	75.34	83.32	71.29	79.23	84.13	73.94
Mostafa et al. [15]	83.69	86.27	79.86	84.24	85.72	78.64	88.17	86.91	85.32
CrackDiff [41]	82.32	84.69	79.34	85.76	88.23	81.63	87.93	89.71	84.26
RD-Crack	88.97	90.64	86.72	89.59	92.03	88.17	92.24	92.72	88.95

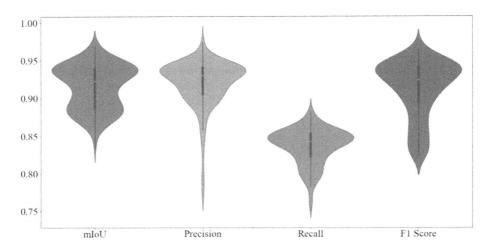

Fig. 3. Segmentation performance evaluation for crack detection using RD-Crack and CrackDiff.

Qualitative Results. Figure 4 shows the crack recognition images of our RD-Crack and CrackDiff, which perform best in the three datasets. It can be clearly seen that both CrackDiff and our RD-Crack perform well in more obvious data cracks. Our RD-Crack can more completely identify the severity of cracks because the diffusion model identifies the characteristics of cracks in more detail. Instead of ordinary supervised neural networks that can only identify cracks. Furthermore, we were pleasantly surprised to find that the diffusion model attempts to make its own judgments about complex actual engineering situations. During the recognition of the first original image, we were surprised to find that almost all neural networks represented by CrackDiff identified some tiny traces as the extension of cracks. However, our RD-Crack does not think this is a crack but believes that the location of some edges may be an extension of a crack that was not photographed. Therefore, our RD-Crack has good prospects for crack detection in actual engineering environments.

4.3 Ablation Experiment

To verify the effectiveness of the diffusion model and the squeeze, excitation, and aggregation modules in RD-Crack, this study gradually removed the diffusion model (DM), squeeze excitation, and aggregation modules (SEA) from our RD-Crack until only the most classic residual neural network (ResNet) and training on Crack500 dataset. The obtained precision, recall, F1 score, accuracy, and mIoU results are shown in Table 2:

As can be seen from the Table 2, compared with directly using supervised residual neural network for crack detection, the extrusion, excitation, and aggregation modules have improved in all indicators, which shows that the actual

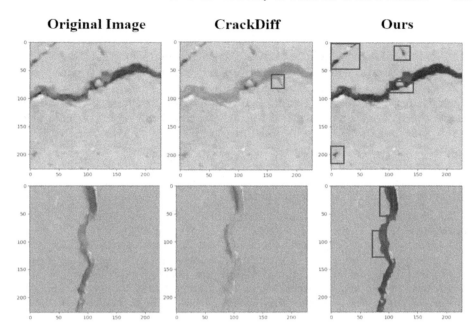

Fig. 4. Segmentation evaluation for crack detection using RD-Crack and CrackDiff.

engineering environment has a great impact on the accuracy of crack detection is extensive, it is necessary to consider global factors and the weight issues between different feature maps. Furthermore, the diffusion model improves precision, F1, and mIoU with a slight decrease in recall. This shows that diffusion models require more time and resources to find missing information than ordinary supervised neural networks. However, since our model parameters are not directly generated by the diffusion model, but are obtained by adjusting the classical model parameters using the diffusion model, better classical model parameters at this time indirectly speed up the solution of the diffusion model. This means that a good initial value can make the decrease in the recall rate of the diffusion model negligible.

Table 2. Ablation study results on RD-Crack

	Recall	Precision	F1 score	mIoU
RD-Crack	85.35	88.95	92.24	92.72
Remove DM	88.06	81.93	86.91	84.07
Remove SEA	84.29	85.34	88.73	84.62
ResNet	87.59	74.67	80.20	80.41

5 Conclusions

RD-Crack represents a significant advancement in concrete crack detection technology, incorporating diffusion models to refine neural network parameters, thereby improving detection accuracy. Experimental results demonstrate RD-Crack's superiority over traditional and contemporary methods across various datasets, evidenced by its higher F1 score and mIoU metrics. The integration of diffusion models offers a promising direction for enhancing crack detection in engineering practices, affirming the potential of combining traditional engineering approaches with advanced computational models. The method's success in handling different data complexities and conditions suggests its applicability in real-world structural health monitoring.

References

1. Adhikari, R., Moselhi, O., Bagchi, A.: Image-based retrieval of concrete crack properties for bridge inspection. Autom. Constr. **39**, 180–194 (2014)
2. Baduge, S.K., et al.: Assessment of crack severity of asphalt pavements using deep learning algorithms and geospatial system. Constr. Build. Mater. **401**, 132684 (2023)
3. Barazzetti, L., Scaioni, M.: Crack measurement: development, testing and applications of an automatic image-based algorithm. ISPRS J. Photogramm. Remote. Sens. **64**(3), 285–296 (2009)
4. Broberg, P.: Surface crack detection in welds using thermography. Ndt E Int. **57**, 69–73 (2013)
5. Choi, W., Cha, Y.J.: SDDNet: real-time crack segmentation. IEEE Trans. Industr. Electron. **67**(9), 8016–8025 (2019)
6. Dhital, D., Lee, J.R.: A fully non-contact ultrasonic propagation imaging system for closed surface crack evaluation. Exp. Mech. **52**, 1111–1122 (2012)
7. Dong, C.Z., Catbas, F.N.: A review of computer vision-based structural health monitoring at local and global levels. Struct. Health Monit. **20**(2), 692–743 (2021)
8. Farrar, C.R., Worden, K.: An introduction to structural health monitoring. Philos. Trans. R. Soc. Math. Phy. Eng. Sci. **365**(1851), 303–315 (2007)
9. Geetha, G.K., Sim, S.H.: Fast identification of concrete cracks using 1D deep learning and explainable artificial intelligence-based analysis. Autom. Constr. **143**, 104572 (2022)
10. Gharehbaghi, V.R., et al.: A critical review on structural health monitoring: definitions, methods, and perspectives. Arch. Comput. Methods Eng. **29**(4), 2209–2235 (2022)
11. He, K., Zhang, X., Ren, S., Sun, J.: Deep residual learning for image recognition. In: Proceedings of the IEEE Conference on Computer Vision and Pattern Recognition, pp. 770–778 (2016)
12. Ho, J., Jain, A., Abbeel, P.: Denoising diffusion probabilistic models. Adv. Neural. Inf. Process. Syst. **33**, 6840–6851 (2020)
13. Ho, S., White, R., Lucas, J.: A vision system for automated crack detection in welds. Meas. Sci. Technol. **1**(3), 287 (1990)
14. Hu, J., Shen, L., Sun, G.: Squeeze-and-excitation networks. In: Proceedings of the IEEE Conference on Computer Vision and Pattern Recognition, pp. 7132–7141 (2018)

15. Iraniparast, M., Ranjbar, S., Rahai, M., Nejad, F.M.: Surface concrete cracks detection and segmentation using transfer learning and multi-resolution image processing. In: Structures. vol. 54, pp. 386–398. Elsevier (2023)
16. Iyer, S., Sinha, S.K.: A robust approach for automatic detection and segmentation of cracks in underground pipeline images. Image Vis. Comput. **23**(10), 921–933 (2005)
17. Jia, Y., et al.: Caffe: convolutional architecture for fast feature embedding. In: Proceedings of the 22nd ACM International Conference on Multimedia, pp. 675–678 (2014)
18. Jia, Y., Tang, L., Xu, B., Zhang, S.: Crack detection in concrete parts using vibrothermography. J. Nondestr. Eval. **38**, 1–11 (2019)
19. Jian, Z., Liu, J.: Cross teacher pseudo supervision: enhancing semi-supervised crack segmentation with consistency learning. Adv. Eng. Inform. **59**, 102279 (2024)
20. Kaseko, M.S., Ritchie, S.G.: A neural network-based methodology for pavement crack detection and classification. Transp. Res. Part C: Emerg. Technol. **1**(4), 275–291 (1993)
21. Koch, C., Georgieva, K., Kasireddy, V., Akinci, B., Fieguth, P.: A review on computer vision based defect detection and condition assessment of concrete and asphalt civil infrastructure. Adv. Eng. Inform. **29**(2), 196–210 (2015)
22. Li, G., Zheng, Y., Cui, J., Gai, W., Qi, M.: DIM-UNet: boosting medical image segmentation via diffusion models and information bottleneck theory mixed with MLP. Biomed. Signal Process. Control **91**, 106026 (2024)
23. Luo, S., Su, Y., Peng, X., Wang, S., Peng, J., Ma, J.: Antigen-specific antibody design and optimization with diffusion-based generative models for protein structures. Adv. Neural. Inf. Process. Syst. **35**, 9754–9767 (2022)
24. Moon, H., Jung, H.K., Lee, C., Park, G.: Camera image processing for automated crack detection of pressed panel products (conference presentation). In: Active and Passive Smart Structures and Integrated Systems 2017, vol. 10164, pp. 53–53. SPIE (2017)
25. Nishiyama, S., Minakata, N., Kikuchi, T., Yano, T.: Improved digital photogrammetry technique for crack monitoring. Adv. Eng. Inform. **29**(4), 851–858 (2015)
26. O'Brien, D., Osborne, J.A., Perez-Duenas, E., Cunningham, R., Li, Z.: Automated crack classification for the CERN underground tunnel infrastructure using deep learning. Tunn. Undergr. Space Technol. **131**, 104668 (2023)
27. O'Byrne, M., Schoefs, F., Ghosh, B., Pakrashi, V.: Texture analysis based damage detection of ageing infrastructural elements. Comput.-Aid. Civil Infrastruct. Eng. **28**(3), 162–177 (2013)
28. Rabah, M., Elhattab, A., Fayad, A.: Automatic concrete cracks detection and mapping of terrestrial laser scan data. NRIAG J. Astron. Geophys. **2**(2), 250–255 (2013)
29. Radford, A., et al.: Learning transferable visual models from natural language supervision. In: International Conference on Machine Learning, pp. 8748–8763. PMLR (2021)
30. Rodríguez-Martín, M., Lagüela, S., González-Aguilera, D., Martínez, J.: Thermographic test for the geometric characterization of cracks in welding using IR image rectification. Autom. Constr. **61**, 58–65 (2016)
31. Sohl-Dickstein, J., Weiss, E.A., Maheswaranathan, N., Ganguli, S.: Deep unsupervised learning using nonequilibrium thermodynamics. In: Proceedings of the 32nd International Conference on International Conference on Machine Learning-vol. 37, pp. 2256–2265 (2015)

32. Song, Q., Li, J., Li, C., Guo, H., Huang, R.: Fully attentional network for semantic segmentation. In: Proceedings of the AAAI Conference on Artificial Intelligence, vol. 36, pp. 2280–2288 (2022)
33. Song, Y., Sohl-Dickstein, J., Kingma, D.P., Kumar, A., Ermon, S., Poole, B.: Score-based generative modeling through stochastic differential equations. arXiv preprint arXiv:2011.13456 (2020)
34. Sun, L., Shang, Z., Xia, Y., Bhowmick, S., Nagarajaiah, S.: Review of bridge structural health monitoring aided by big data and artificial intelligence: from condition assessment to damage detection. J. Struct. Eng. **146**(5), 04020073 (2020)
35. Wang, X., Yue, Q., Liu, X.: Conditional diffusion model-based generation of speckle patterns for digital image correlation. Opt. Lasers Eng. **175**, 107997 (2024)
36. Xiang, C., Gan, V.J., Guo, J., Deng, L.: Semi-supervised learning framework for crack segmentation based on contrastive learning and cross pseudo supervision. Measurement **217**, 113091 (2023)
37. Xie, S., Girshick, R., Dollár, P., Tu, Z., He, K.: Aggregated residual transformations for deep neural networks. In: Proceedings of the IEEE Conference on Computer Vision and Pattern Recognition, pp. 1492–1500 (2017)
38. Yang, F., Zhang, L., Yu, S., Prokhorov, D., Mei, X., Ling, H.: Feature pyramid and hierarchical boosting network for pavement crack detection. IEEE Trans. Intell. Transp. Syst. **21**(4), 1525–1535 (2019)
39. Yehia, S., Landolsi, T., Hassan, M., Hallal, M.: Monitoring of strain induced by heat of hydration, cyclic and dynamic loads in concrete structures using fiber-optics sensors. Measurement **52**, 33–46 (2014)
40. Yokoyama, S., Matsumoto, T.: Development of an automatic detector of cracks in concrete using machine learning. Proc. Eng. **171**, 1250–1255 (2017)
41. Zhang, H., Chen, N., Li, M., Mao, S.: The crack diffusion model: an innovative diffusion-based method for pavement crack detection. Remote Sensing **16**(6), 986 (2024)
42. Zhang, Q., Barri, K., Babanajad, S.K., Alavi, A.H.: Real-time detection of cracks on concrete bridge decks using deep learning in the frequency domain. Engineering **7**(12), 1786–1796 (2021)
43. Zheng, Y., Gao, Y., Lu, S., Mosalam, K.M.: Multistage semisupervised active learning framework for crack identification, segmentation, and measurement of bridges. Comput.-Aid. Civil Infrastruct. Eng. **37**(9), 1089–1108 (2022)
44. Zou, Q., Cao, Y., Li, Q., Mao, Q., Wang, S.: CrackTree: automatic crack detection from pavement images. Pattern Recogn. Lett. **33**(3), 227–238 (2012)
45. Zou, Q., Zhang, Z., Li, Q., Qi, X., Wang, Q., Wang, S.: DeepCrack: learning hierarchical convolutional features for crack detection. IEEE Trans. Image Process. **28**(3), 1498–1512 (2018)

Applications in Finance

Anomaly Detection in Blockchain Using Multi-source Embedding and Attention Mechanism

Ao Xiong[1], Chenbin Qiao[1(✉)], Baozhen Qi[2], and Chengling Jiang[3]

[1] State Key Laboratory of Networking and Switching Technology, Beijing University of Posts and Telecommunications, Beijing 100876, China
qiaochenbin@bupt.edu.cn
[2] Jiangsu Electric Power Company Suzhou Power Supply Company, Suzhou 215008, China
[3] Jiangsu Provincial Electric Power Corporation, Nanjing 210024, China

Abstract. Due to the lack of effective regulatory mechanisms, many risks and offences have emerged in the blockchain trading market. Therefore, in order to achieve anomaly detection for blockchain networks, this paper abstracts blockchain transaction data as a graph structure and proposes GraphAEAtt, a deep learning model based on multi-source embedding and attention mechanism. GraphAEAtt uses two encoders to generate structure embeddings and feature embeddings respectively, and utilizes attention mechanisms to generate composite embeddings. By using multiple embeddings and attention mechanisms, the GraphAEAtt model can integrate the structural information and feature information of the graph, while also learning the relationships between nodes to reduce the impact of abnormal nodes on the learning process. Experimental results on several datasets show that the deep learning model proposed in this paper can better explore the implicit information in blockchain transaction graphs compared to other methods, thereby more accurately identifying abnormal transactions on the blockchain.

Keywords: Blockchain · Anomaly detection · Machine learning

1 Introduction

Cryptocurrency is one of the most successful applications based on blockchain technology, and the identity information of each node in cryptocurrency does not need to be disclosed or verified, and the information transfer can be carried out anonymously, which also facilitates money laundering, fraudulent transactions, and other illegal acts. The significant changes brought about by blockchain technology have posed many challenges to financial security [1].

Due to the unique nature of blockchain technology, traditional regulatory systems are not applicable to the regulation of this field, so it is necessary to adopt

---

Supported by the National Key R&D Program of China(2022YFB2703400).

more advanced technologies to optimize regulatory methods. Firstly, since all transaction data is recorded on the blockchain in a publicly transparent manner, regulatory authorities can use big data technology to analyze this data and establish multi-dimensional data models to achieve data-driven intelligent supervision. Secondly, deep learning technology has powerful advantages in data processing and analysis. Deep learning mimics the neural architecture of the human brain, excelling at identifying complex patterns and representations in massive datasets, enabling prediction and information classification. Therefore, it is worth considering integrating deep learning-related technologies into regulatory methods [2].

In order to regulate blockchain, the primary goal is to detect anomalies in blockchain transactions and accurately identify abnormal transactions on the blockchain [3]. Anomaly detection technology is a technology used to identify behaviors that do not conform to normal patterns, aiming to discover abnormal connections and nodes in the network. Anomaly detection can assess potential risks to facilitate regulatory authorities in taking timely measures.

Many common relationships, such as social networks, financial transaction networks, and academic citation networks, can be represented using graph structures. The blockchain transaction network can also be seen as a type of graph structure, where nodes represent addresses and edges represent transaction behaviors. We can utilize graph structures to better understand the information contained in blockchain transaction data [4]. After constructing the transaction graph, deep learning models can be used to extract high-dimensional features from the graph structure, mine deeper information in the graph, and then identify abnormal connection structures, abnormal behavior nodes, and nodes containing abnormal information in the graph. The problem of detecting anomalies in blockchain transactions can be abstracted as a graph anomaly detection problem, and more efficient anomaly detection can be achieved by using methods related to graph representation learning.

However, based on the above ideas, detecting anomalies in blockchain transactions will also face multiple challenges. Existing anomaly detection methods have some limitations, with one main issue being the neglect of deep connections between the graph structure and node features in the graph, while the connections between information from different sources are crucial for anomaly detection. Additionally, current methods typically only use a single embedding representation to process multiple pieces of information, but this approach has limitations in extracting data features and cannot effectively capture the complex interaction information between graph structures and node features. Furthermore, due to the presence of anomalous nodes, traditional feature extraction methods are easily disturbed, leading to a significant decrease in the effectiveness of anomaly detection.

To address the above challenges, this paper proposes a deep learning model GraphAEAtt based on multi-source embeddings and attention mechanism. GraphAEAtt utilizes two encoders to generate structural embeddings and feature embeddings respectively, while using attention mechanism to generate composite embeddings. By utilizing multiple embeddings and attention mechanisms, the GraphAEAtt can integrate the structural and feature information of the graph, and learn the relationships between nodes to mitigate the impact of abnormal nodes on node embedding learning. In this model, the structural encoder is responsible for learning the structural embeddings of the graph, the graph attention layer combines the graph structure and node features to obtain composite embeddings, thereby better revealing the inherent relationships between nodes. The feature encoder maps feature vectors to feature embeddings. After generating multiple embeddings, the structural decoder is used to reconstruct the adjacency matrix, while the feature decoder reconstructs the feature matrix. Finally, abnormal situations in the graph can be judged by reconstruction errors.

Experimental results on several datasets show that the deep learning framework proposed in this paper can better uncover hidden information in graph data compared to other methods, and thus more accurately identify abnormal transactions on the blockchain.

2 Related Work

As one of the most successful applications of blockchain technology, cryptocurrencies represented by Bitcoin [5] demonstrate tremendous potential for development. However, with the rapid expansion of the market, a series of security-related issues have emerged, making it urgently necessary to regulate blockchain-related applications [6]. Currently, many countries and regions have begun to formulate relevant regulations and policies to include blockchain technology within the regulatory scope. For example, the Financial Crimes Enforcement Network (FinCEN) under the U.S. Department of the Treasury has established strict regulations to curb illegal activities in the cryptocurrency field, requiring institutions holding Money Services Business (MSB) licenses to fully understand their customers and assess the financial crime risks their customers may pose.

In various regulatory measures, it is crucial to implement anomaly detection for blockchain transactions. Anomaly can be termed as a specific pattern in the collected or transmitted data which does not show a well-suited regular behaviour [7]. Identification of abnormal activities within Bitcoin transactions has been intensively addressed in the past [8].

Early anomaly detection methods largely rely on statistical models constructed by experts, which often require a large amount of manpower and have limited capabilities in detecting unknown anomalies. For example, Demetis et al. [9] studied the transaction records of a bank in the UK, extracted a set of evaluation indicators for making decisions on anti-money laundering activities, and this method is suitable for the anti-money laundering system in the banking industry.

The anomaly detection methods can be roughly divided into three categories based on different training sets: fully supervised anomaly detection, semi-supervised anomaly detection, and unsupervised anomaly detection. For example, researchers like Jullum et al. [10] trained an XGBoost supervised prediction model using information such as sender/receiver background and transaction behavior to identify potential fraudulent transaction behavior, and applied it to bank regulation. On the other hand, researchers like Paula et al. [11] combined autoencoder algorithms to train an unsupervised model for detecting money laundering activities.

Graph representation learning is a field of artificial intelligence dedicated to developing algorithms that can learn graph data. In traditional machine learning tasks, data is typically represented in structured formats such as matrices or tensors. However, many data in the real world exhibit complex relationships that cannot be represented using traditional methods. The graph provides a way to simulate the relationships between entities in various fields. In these graphs, entities are represented as nodes, and the relationships between entities are represented as edges. Transactions on the blockchain can also be viewed as a graph structure with users as nodes and transactions between users as edges, providing a new direction for anomaly detection in transactions. For example, Weber et al. [12] introduced the idea of graph structure in their article "Anti-Money Laundering in Bitcoin: Experimenting with Graph Convolutional Networks for Financial Forensics", which transforms Bitcoin transactions into graphs and uses graph convolutional networks to learn the characteristics of transactions and determine their legitimacy.

Through graph representation learning, we can better understand and analyze complex network relationships. The method based on graph representation learning provides new ideas and technical means for security and regulation in the financial field. Martin et al. [8] use graph representation learning to study the interactions between blockchain nodes and enhance the analysis of graph structures by applying advanced modern machine learning techniques to detect anomalous transactions in various digital currency markets using multiple machine learning models.

Through the above introduction to related work, it can be seen that researchers worldwide have already achieved many research results. However, there are still many problems that need to be improved in the anomaly detection methods for blockchain transactions. Firstly, existing methods often ignore the deep connections between the structure of the graph and the features of the nodes in the graph. At the same time, existing methods usually only use one embedding to represent multiple pieces of information, which limits the extraction of features and fails to obtain complex interaction information between structure and features effectively. In addition, due to the presence of anomalous nodes, traditional feature extraction methods are easily interfered with, thus affecting the effectiveness of anomaly detection. Therefore, in order to better address these issues, we need to seek more innovative methods.

3 Methodology

3.1 Definition

A graph can be represented using $G = (V, E, X)$, where $V = \{V_1, V_2...V_N\}$ is a set of nodes $N = |V|$, $E \in V \times V$ is a set of $M = |E|$ edges between the nodes, and $X \in R^{N \times N}$ is the features of the N nodes. The adjacency matrix of the graph is represented using $A \in R^{N \times N}$, where $A_{i,j} = 1$ if a pair of nodes have an edge between them. Given a graph, the purpose of anomaly detection is to discover rare nodes that have significant differences in structure or features compared to the majority of nodes.

3.2 The Architecture of the Model

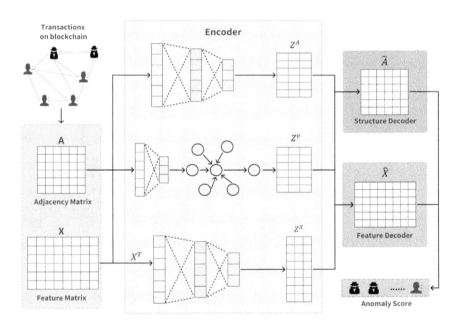

Fig. 1. The Architecture of GraphAEAtt.

The architecture of the GraphAEAtt model mentioned in this article is shown in Fig. 1. The model consists of two parts: an encoder and a decoder. The encoder includes a structure encoder, a graph attention layer, and a feature encoder, while the decoder includes a structure decoder and a feature decoder. The input to the encoder is the adjacency matrix of the graph and the feature matrix of the nodes in the graph, and the output of the decoder is the reconstructed matrix of the adjacency matrix and the feature matrix.

First, the structure encoder learns the embedded representation of the global structure based on the adjacency matrix of the graph, obtaining structure embeddings; the graph attention layer integrates the adjacency matrix and feature matrix, merging structural information and feature information to obtain composite embeddings, and the feature encoder learns feature embeddings based on the feature matrix. Subsequently, based on these encoders, multiple embeddings are generated, and two decoders respectively select different embeddings for deep fusion. The structure decoder reconstructs the adjacency matrix, and the feature decoder reconstructs the feature matrix. Finally, the training effectiveness is evaluated by calculating the reconstruction errors of the adjacency matrix and the feature matrix.

Existing methods use a single embedding to reconstruct various information, but this makes it difficult to effectively capture the deep interaction information that may exist between the structure and node features in the graph. At the same time, abnormal nodes in the graph can interfere with the process of extracting features from normal nodes. The deep learning model GraphAEAtt proposed in this paper is designed to address the above issues. This model utilizes multi-source embedding and attention mechanisms to generate multiple embeddings, learning graph features from both structural and feature perspectives, integrating information between the structure and features in the graph, minimizing the interference of abnormal nodes on feature extraction, and achieving an effective representation of graph information.

3.3 Encoder

Structure Encoder. The structure encoder is designed to learn the structure of the graph without considering the features of nodes. This article adopts two feature transformation layers to extract structural features. Taking the adjacency matrix $A \in R^{N \times N}$ as input, after the first layer of non-linear feature transformation, the hidden layer output $\tilde{Z}^A$ is obtained, as shown in Eq. (1).

$$\tilde{Z}^A = \sigma(AW^{A_1} + b^{A_1}) \qquad (1)$$

where σ is the activation function, W^{A_1} and b^{A_1} are the weight and bias learned by encoder. Then input $\tilde{Z}^A$ into the next feature transformation layer, as shown in formula Eq. (2).

$$Z^A = \tilde{Z}^A W^{A_2} + b^{A_2} \qquad (2)$$

where W^{A_1} and b^{A_1} are the weight and bias learned by encoder, Z^A is the structure embedding obtained after two layers of feature transformation.

Graph Attention Layer. The graph attention layer is designed to integrate the structural information of the graph and the feature information of the nodes to obtain composite embeddings. In anomaly detection tasks, anomalous nodes can affect the feature extraction of normal nodes, so the graph attention layer is introduced. The embeddings generated by the graph attention layer summarize the features of the nodes themselves, the features of neighboring nodes, and the network connection relationships, excluding the interference of anomalous information. The graph attention layer enables higher quality embeddings to be obtained, thereby optimizing the effectiveness of data analysis. The input of the attention layer includes the feature matrix and the adjacency matrix. The first step is to use the non-linear feature transformation layer to convert the feature matrix X into a low-dimensional embedding matrix, as shown in Eq. (3).

$$X' = \{x'_1, x'_2, ..., x'_N\} = \sigma(XW^{V_1} + b^{V_1}) \qquad (3)$$

where σ is the activation function, W^{V_1} and b^{V_1} are the weight and bias learned by encoder. X' represents the low-dimensional embedding obtained after feature transformation. N represents the number of nodes.

After obtaining the embedding X', the next step is to use the attention mechanism to obtain the weight coefficients.

First, use the following equation to calculate the relevance between two nodes, as shown Eq. (4).

$$e_{i,j} = attn(Wx'_i, Wx'_j) = LeakyReLU(a^T \cdot [Wx'_i||Wx'_j]) \qquad (4)$$

where $e_{i,j}$ is the importance weight of node i to node j, $attn()$ represents a neural network with parameters W and a, all nodes share the parameters of this network. $||$ denotes the concatenate operation. First, the node features are dimensionally augmented based on linear mapping. Then, the transformed features of node i and node j are concatenated and fed into a feed-forward neural network, which is activated using the LeakyReLU function.

In order to obtain the correlation coefficient, the importance weights of the node and its neighboring nodes are normalized. The standardization process uses the softmax function, as shown in Eq. (5).

$$\alpha_{i,j} = Softmax(e_{i,j}) = \frac{exp(e_{i,j})}{\sum_{k \in N_i} exp(e_{i,k})} \qquad (5)$$

where N_i denotes the first order neighbours of node i. After obtaining the weight coefficients of each neighbour of the target node i, the attention computation can be performed. A single computation is performed by weighting and summing the low-dimensional embeddings of the output from the previous step using the weight coefficients, which gives a new vector representation of node i. The multiple attention mechanism, on the other hand, performs K independent attention computations for each target node, such that K vector representations of the target vertex are obtained, and then an averaging operation is performed on these K vectors to obtain the output vector, as shown in Eq. (6).

$$\hat{x}_i = \frac{1}{K} \sum_{k=1}^{K} \sum_{j \in N_i} \alpha_{i,j}^k W^k x'_j \qquad (6)$$

where $\alpha_{i,j}^k$ denotes the normalisation factor of the kth attention mechanism and W^k corresponds to the weight matrix of the kth linear transformation. After obtaining the new vector representation of node i, the composite embedding is obtained, as shown in Eq. (7).

$$Z^V = \{\hat{x}_1, \hat{x}_2, ...\hat{x}_N\} \quad (7)$$

Feature Encoder. The responsibility of the feature encoder is to obtain feature embeddings for each node. In the feature encoder, two nonlinear feature transformation layers are also used to transform the feature data into low-dimensional feature embeddings, as shown in Eqs. (8) and (9).

$$\tilde{Z}^X = \sigma(X^T W^{X_1} + b^{X_1}) \quad (8)$$

$$Z^X = \tilde{Z}^X W^{X_2} + b^{X_2} \quad (9)$$

where W^{X_1} and b^{X_1} represent the parameters of the first transformation layer, while W^{X_2} and b^{X_2} represent the parameters of the second feature transformation layer. The feature matrix X includes the feature information of all nodes. $\tilde{Z}^X$ is the output of the hidden layer, and Z^X is the final obtained feature embedding.

3.4 Decoder

Structure Decoder. The responsibility of the structure decoder is to reconstruct a new adjacency matrix. In order to integrate multiple types of information, it utilizes the inner product operation between the composite embeddings from the graph attention layer outputs and the structure embeddings from the structure encoder outputs. Z^A contains global structural features, while Z^V combines graph structural information and node feature information. By using the two embeddings mentioned above, the structure identical to the original graph is reconstructed. This is shown in Eq. (10).

$$\hat{A} = Sigmoid(Z^V (Z^A)^T) \quad (10)$$

In the above equation, Z^V represents composite embedding, while Z^A represents structural embedding. Finally, the Sigmod function is used for normalization to make the associations between nodes closer to 0 and 1, so that the reconstructed matrix can be closer to the initial matrix.

Feature Decoder. The task of the feature decoder is to reconstruct a new feature matrix. In order to integrate various information, its input includes composite embedding Z^V and feature embedding Z^X, This is shown in Eq. (11).

$$\hat{X} = Z^V (Z^X)^T \quad (11)$$

In the above equation, Z^V represents composite embedding, while Z^X represents feature embedding.

3.5 Loss Function

The loss function of the model, as shown in Eq. (12), aims to minimize the reconstruction errors of the adjacency matrix and feature matrix to the greatest extent during training: where α, θ, η are hyperparameters that adjust the weights of reconstructing errors of the adjacency matrix and feature matrix, and $\odot$ represents the Hadamard product.

$$Loss = \alpha||(A - \hat{A}) \odot \theta||_F^2 + (1 - \alpha)||(X - \hat{X}) \odot \eta||_F^2 \tag{12}$$

The structure and features of abnormal nodes usually differ significantly from other nodes, therefore, the anomaly score of a node can be evaluated through reconstruction error. Nodes are sorted from high to low based on their anomaly scores, with nodes having higher scores possibly indicating anomalies. For nodes with anomaly scores exceeding a certain threshold, they are labeled as anomalies and compared with their true labels. Finally, the algorithm's anomaly detection performance is evaluated using multiple metrics.

4 Experiments

4.1 Datasets and Evaluation Metric

In this paper, we conduct an experimental study using two blockchain transaction datasets. The first dataset is the Bitcoin public transaction dataset. The data collection method refers to [26]. The dataset was published by Omer (https://ieee-dataport.org/open-access/bit-coin-transactions-data-2011--2013), extracted from the bitcoin core client's blk.dat file, and contains transaction information from 2011 to 2013, the dataset field information is shown in Table 1.

Existing Bitcoin transaction data usually lacks complete labelling information, and most of the time the data can only be labelled manually. Bitcoin Forum, an authoritative paper on Bitcoin, published a history of illegal transactions between 2011 and 2013, which covers detailed information on a variety of cases such as BTC theft, BTC hacking, and BTC loss, including the amount of money involved, the date, and the transaction ID. In this paper, illegal transactions are collected in this forum with the help of python crawler, and illegal transactions and their sub-transactions are labelled as anomalous (1) and the rest are labelled as normal (0). After counting, the dataset contains 108 abnormal classes and 30248025 normal classes.

Using transaction flow information, this paper uses the python-igraph package to construct a directed acyclic network graph with transactions as nodes, and based on this, we extract seven representative features related to abnormal behavioural patterns of nodes:

(1)Characteristics of transaction graph: degree of entry, degree of exit.

(2)Characteristics of transaction funds: gross receipts, gross dispatches, average receipts, average dispatches, net balance.

Table 1. Feature List.

Name of the feature	Content
tx_hash_from	The hash value of the transaction
tx_hash_to	The hash value of the transaction
detetime	Trading time
amount_bitcoins	Amount of transaction

The second dataset is a blockchain transaction dataset: the Elliptic dataset. The Elliptic dataset maps the raw bitcoin transaction information into a graph structure, where nodes represent bitcoin transactions and edges represent the flow of bitcoins between transactions. The Elliptic dataset consists of 203,769 nodes and 234,355 edges, and the nodes carry timestamps as well.

In the Elliptic dataset, Bitcoin transaction subjects are included as legitimate entities (e.g., exchanges, wallet providers, legitimate services, etc.) and illegitimate entities (e.g., scams, malware, ransomware, etc.). Transactions initiated by legitimate entities are labelled as legal transactions, and transactions initiated by illegal entities are labelled as illicit transactions. Using a heuristic inference process, the dataset is labelled with a total of 4,545 illegal transactions and 42,019 legal transactions, while the labelling of the remaining transactions is unknown. Each transaction contains 166 features, of which 94 features are local information of Bitcoin transactions (nodes), labelled as local features, such as time step, transaction fee, number of inputs/outputs, amount of outputs, and some aggregate data (e.g., the average number of BTC received/transferred, etc.), and the remaining 72 features are maximum, minimum, standard deviation, and correlation coefficient, etc., labelled as aggregated features. The experiment uses 166-dimensional features in the Elliptic dataset to differentiate between normal and abnormal transactions.

In this paper, the same evaluation metrics as other deep learning anomaly detection algorithms are used. Three main evaluation metrics are used to measure the performance of each anomaly detection algorithm, which are accuracy, precision rate and recall rate. The statistical source is the classification confusion matrix of anomaly detection. The precision rate indicates the proportion of all nodes predicted to be anomalous that are actually anomalous among all predicted results. The recall rate indicates the proportion of successfully detected anomalous nodes occupying all actual anomalous nodes in the sample.

4.2 Models for Comparison

In this paper, several blockchain transaction anomaly detection models are selected for experimentation.

Aiming at the problem that the performance of intelligent detection models is limited by the expressiveness of raw data (features), Zhu et al. [27] designed a residual network structure ResNet-32 for mining the implicit associations among blockchain transaction features. By automatically learning high-level abstract features containing rich semantic information, the ResNet-32 model can adaptively bridge the gap between high-level abstract features and original features, automatically remove redundant information, and mine the cross-feature information of the two to obtain the most distinguishable features.

Aiming at the existing machine learning-based anomaly transaction detection methods that are difficult to accurately generalize multiple anomaly types and have insufficient generalization ability, Liao et al. [26] construct a network structure and extract features related to anomalous behavioural patterns on bitcoin transaction data, apply the parallel integration algorithm based on the combination of local dynamic choices (LSCP) to construct a detection model, and incorporate seven classical anomaly detection algorithms into the algorithm to take advantage of the sensitivity of the base learner to different types of anomalies is used to improve the reliability and stability of the detection model.

Weber et al. [12] mapped Bitcoin transactions into a large, complex graph structure and extracted relevant features such as the number of transactions, transaction amounts, etc., and then used a graph convolutional network algorithm to distinguish between illegal and legal transactions, so in this paper, skip-gcn is also used as a comparison model.

4.3 Experimental Results

In the experiment, we train GraphAEAtt with 100 iterations for two datasets respectively. Through extensive experimental comparisons, the best performing hyperparameters and activation functions are selected in this paper. In the non-feature transformation layer, the dimension of the hidden layer is chosen to be 256, while the dimension of the output is 128, the Relu function is used as the activation function, and the Sigmoid function is chosen as the activation function for the decoder. The reconstruction error parameters vary for different datasets. The parameters α, θ, η are set as (0.9,5,50), (0.9,2,40) for two datasets respectively. Adam algorithm is utilized for optimization with learning rate as 0.001.

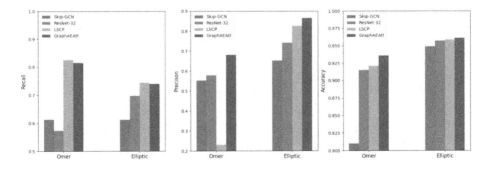

Fig. 2. Experimental results on two datasets.

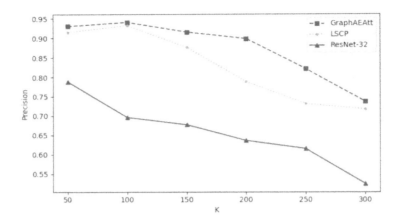

Fig. 3. The precision result of Omer dataset.

In the experiments, the model is first trained using the training set, and after the model is trained, the reconstruction error of each node in the reconstructed graph and the original graph is calculated separately in the way defined in Equation (12). And according to the way defined in Equation (12), the top K nodes in descending order of reconstruction error are marked as the predicted anomalous nodes.

As can be seen from Fig. 2, comparing the values of the GraphAEAtt model and the other algorithms in the three evaluation metrics, GraphAEAtt is not far from the highest level even when it is not at the highest level. Although it is not always better than the other methods, the overall performance is better than the other methods, and it has better reliability and stability for the detection of abnormal bitcoin transactions.

Through the analysis of precision and recall, two evaluation metrics, a line chart was used to demonstrate the detection effects of three methods on abnormal nodes. The comparison results of the experiments on the two datasets for the two metrics of precision and recall are shown in Figs. 3, 4, 5 and 6.

It can be observed that when K is set to 100, the precision rate of each method is higher. When K is set to greater than 100, the performance of the models decreases significantly. Compared with the LSCP model, which is the best performer among the compared models, the precision rate of the proposed method in this paper is improved by 13.58% and 19.05% in the two datasets, respectively.

From Figs. 5 and 6, it can be seen that the traditional methods have low recall and limited ability to detect anomalies, while the algorithm proposed in this chapter improves the recall by 23.63% and 27.08% in the two datasets, and the effectiveness of anomaly detection is significantly improved.

Based on the comprehensive analysis of all data, conclusions can be drawn. The GraphAEAtt framework, through the use of multiple embeddings and attention mechanisms, integrates the structural information and feature information

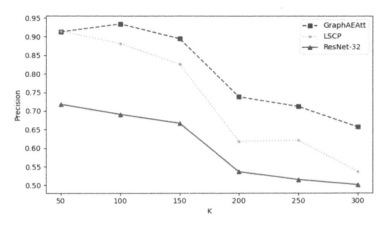

Fig. 4. The precision result of Elliptic dataset.

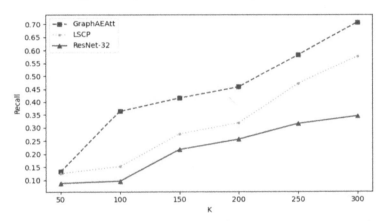

Fig. 5. The recall rate result of Omer dataset.

of the graph, while also learning the relationships between nodes to reduce the impact of abnormal nodes on node embedding learning. This enables a more accurate depiction of the essence of transaction data, effectively enhancing the quality of learned embeddings and facilitating the analysis of exceptional cases. Therefore, the deep learning model proposed in this study, based on multi-source embeddings and attention mechanisms, demonstrates outstanding performance in anomaly detection, successfully improving the effectiveness and efficiency of anomaly detection.

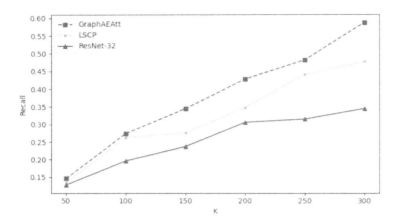

Fig. 6. The recall rate result of Elliptic dataset.

5 Conclusion

With the development of the Internet and blockchain technology, the cryptocurrency market has achieved rapid development, and the scale of transactions on the blockchain continues to expand. The scale of Bitcoin users is dynamic and the identity of participants is anonymous, which will lead to more hidden, complex and intelligent financial crimes. Therefore, in order to achieve intelligent blockchain transaction supervision, it is necessary to accurately identify abnormal transactions on the blockchain.

Based on this, this paper proposes a deep learning model GraphAEAtt based on multi-source embedding and attention mechanism. The autoencoder in this model helps to fully explore hidden patterns and features in the data, extract deeper network features, while the attention mechanism can minimize the interference of abnormal nodes on feature extraction. Compared with traditional models, GraphAEAtt can generate higher quality data embedding representations, while preserving the features of normal nodes and reducing the impact of abnormal nodes on the generated embeddings. In experiments on multiple datasets, the results show that compared to other methods, the proposed method in this study can more fundamentally describe transaction data, more accurately identify abnormal transactions on the blockchain, and improve the performance of abnormal transaction detection. This article has strong practical significance for research on anomaly detection in blockchain, and it can provide many possibilities for future research directions.

The method proposed in this paper can provide new ideas and methods for intelligent supervision of blockchain, and provide strong support for building a secure and transparent blockchain transaction environment. In the next step of our work, we need to further optimize the algorithm to improve its performance. Then, we can further enhance the scalability of the model, expand the application of the model to a wider range of fields, such as node classification, link prediction, and clustering.

References

1. Patel, V., Pan, L., Rajasegarar, S.: Graph deep learning based anomaly detection in Ethereum blockchain network. In: Kutyłowski, M., Zhang, J., Chen, C. (eds.) Network and System Security: 14th International Conference, NSS 2020, Melbourne, VIC, Australia, November 25–27, 2020, Proceedings, pp. 132–148. Springer International Publishing, Cham (2020). https://doi.org/10.1007/978-3-030-65745-1_8
2. Hasan, M., et al.: Detecting anomalies in blockchain transactions using machine learning classifiers and explainability analysis. ArXiv abs/2401.03530 (2024)
3. Zhang, R., Zhang, G., Liu, L., Wang, C., Wan, S.: Anomaly detection in bitcoin information networks with multi-constrained meta path. J. Syst. Archit. **110**, 101829 (2020). ISSN 1383-7621. https://doi.org/10.1016/j.sysarc.2020.101829
4. Pham, T., Lee, S.: Anomaly Detection in the Bitcoin System - A Network Perspective (2016)
5. Nakamoto, S.: Bitcoin: a peer-to-peer electronic cash system. Decentralized Bus. Rev., 21260 (2008)
6. Pocher, N., Zichichi, M., Merizzi, F., et al.: Detecting anomalous cryptocurrency transactions: an AML/CFT application of machine learning-based forensics. Electron Markets **33**, 37 (2023). https://doi.org/10.1007/s12525-023-00654-3
7. Ul Hassan, M., Rehmani, M.H., Chen, J.: Anomaly detection in blockchain networks: a comprehensive survey. IEEE Commun. Surv. Tutorials **25**(1), 289–318, Firstquarter (2023). https://doi.org/10.1109/COMST.2022.3205643
8. Martin, K., et al.: Anomaly detection in blockchain using network representation and machine learning. Secur. Priv., 5 (2021)
9. Demetis, D.S.: Fighting money laundering with technology: a case study of Bank X in the UK. Decis. Support Syst. **105**, 96–107 (2018)
10. Jullum, M., Lland, A., Huseby, R.B., et al.: Detecting money laundering transactions with machine learning. J. Money Laundering Control **23**, 173–186 (2020)
11. Paula, E.L., Laderia, M., Carvalho, R.N., et al.: Deep learning anomaly detection as support fraud investigation in Brazilian exports and anti-money laundering. In: Proceedings of 2016 15th IEEE International Conference on Machine Learning and Applications, pp. 954–960. Anaheim: IEEE (2016)
12. Weber, M., Domeniconi, G., Chen, J., et al.: Anti-Money Laundering in Bitcoin: experimenting with graph convolutional networks for financial forensics. arXiv, (2019). arXiv: 1908.02591
13. Vermander, P., Mancisidor, A., Cabanes, I., et al.: Intelligent systems for sitting posture monitoring and anomaly detection: an overview. J. Neuroengineering Rehabil. **21**(1), 28–28 (2024)
14. Ding, K., Li, J., Bhanushali, R., et al.: Deep anomaly detection on attributed networks. In: Proceedings of the 2019 SIAM International Conference on Data Mining, Society for Industrial and Applied Mathematics, pp. 594–602 (2019)
15. Wang, X., Cui, P., Wang, J., Pei, et al. Community preserving network embedding. In: Proceedings of 31st AAAI Conference on Artificial Intelligence, pp. 203–209. AAAI, San Francisco (2017)
16. Kipf, T.N., Welling, M.: Semi-supervised classification with graph convolutional networks. In: Proceedings of the International Conference on Learning Representations, Toulon: arXiv (2017)
17. Hamilton, W., Ying, Z., Leskovec, I.: Inductive representation learning on large graphs. In: Proceedings of the 31st Annual Conference on Neural Information Processing Systems, pp. 1024–1034. Long Beach (2017)

18. Li, J., Dani, H., Hu, X., et al.: Radar: residual analysis for anomaly detection in attributed networks. In: Proceedings of the 26th International Joint Conference on Artificial Intelligence, pp. 2152–2158. Melbourne: International Joint Conferences on Artificial Intelligence (2017)
19. Peng, Z., Luo, M., Li, J., et al.: ANOMALOUS: a joint modelling approach for anomaly detection on attributed networks. In: Proceedings of the 27th International Joint Conference on Artificial Intelligence, pp. 3513–3519. Stockholm: International Joint Conferences on Artificial Intelligence (2018)
20. Ding, K., Li, J., Bhanushali, R., et al.: Deep anomaly detection on attributed networks. In: Proceedings of the 19th SIAM International Conference on Data Mining, pp. 594–602. Calgary: Society for Industrial and Applied Mathematics Publications (2019)
21. Ma, X., et al.: A comprehensive survey on graph anomaly detection with deep learning. IEEE Trans. Knowl. Data Eng. **35**(12), 12012–12038 (2023). https://doi.org/10.1109/TKDE.2021.3118815
22. Perozzi, B., Akoglu, L.: Discovering communities and anomalies in attributed graphs: interactive visual exploration and summarization. ACM Trans. Knowl. Discov. Data (TKDD) **12**(2), 1–40 (2018)
23. Manessi, F., Rozza, A., Manzo, M., et al.: Dynamic graph convolutional networks. Pattern Recogn. **97**, 107000 (2020)
24. Zhou, R., Zhang, Q., Zhang, P., et al.: Anomaly detection in dynamic attributed networks. Neural Comput. Appl. **33**, 2125–2136 (2021)
25. Yang, R., Yu, F.R., Si, P., et al.: Integrated blockchain and edge computing systems: a survey, some research issues and challenges. IEEE Commun. Surv. Tutorials **21**(2), 1508–1532 (2019)
26. Qian, L.I.A.O., Yijun, G.U.: Bitcoin network abnormal transaction detection based on LSCP algorithm. Comput. Eng. Appl. **58**(15), 117–123 (2022)
27. Huijuan, Z.H.U., Jinfu, C.H.E.N., Zhiyuan, L.I., Shangnan, Y.I.N.: Block-chain abnormal transaction detection method based on adaptive multi-feature fusion. J. Commun. **42**(5), 41–50 (2021)

Beyond Gut Feel: Using Time Series Transformers to Find Investment Gems

Lele Cao[1(✉)], Gustaf Halvardsson[1,2], Andrew McCornack[1], Vilhelm von Ehrenheim[1,3], and Pawel Herman[2]

[1] Motherbrain, EQT Group, Regeringsgatan 25, 11153 Stockholm, Sweden
caolele@gmail.com,
{gustaf.halvardsson,andrew.mccornack,vilhelm.vonehrenheim}@eqtpartners.com
[2] KTH Royal Institute of Technology, Lindstedtsvagen 5, 11428 Stockholm, Sweden
paherman@kth.se
[3] QA.tech, Gävlegatan 16, 11330 Stockholm, Sweden

Abstract. This paper addresses the growing application of data-driven approaches within the Private Equity (PE) industry, particularly in sourcing investment targets (i.e., companies) for Venture Capital (VC) and Growth Capital (GC). We present a comprehensive review of the relevant approaches and propose a novel approach leveraging a Transformer-based Multivariate Time Series Classifier (TMTSC) for predicting the success likelihood of any candidate company. The objective of our research is to optimize sourcing performance for VC and GC investments by formally defining the sourcing problem as a multivariate time series classification task. We consecutively introduce the key components of our implementation which collectively contribute to the successful application of TMTSC in VC/GC sourcing: input features, model architecture, optimization target, and investor-centric data processing. Our extensive experiments on two real-world investment tasks, benchmarked towards three popular baselines, demonstrate the effectiveness of our approach in improving decision making within the VC and GC industry.

Keywords: Company success prediction · Venture capital · Growth equity · Private equity · Investment · Multivariate time series

1 Introduction

Private Equity (PE) is a rapidly growing segment of the investment industry that manages funds on behalf of institutional and accredited investors. PE firms acquire and manage companies with the goal of achieving high, risk-adjusted

---

Gustaf Halvardsson (currently works for BCG GAMMA, part of BCG X) contributed (equally as Lele Cao) to this work as part of his Master's Thesis project carried out in EQT Motherbrain supervised by Lele Cao (industrial supervisor) and Pawel Herman (academic supervisor).

© The Author(s), under exclusive license to Springer Nature Switzerland AG 2024
M. Wand et al. (Eds.): ICANN 2024, LNCS 15024, pp. 373–388, 2024.
https://doi.org/10.1007/978-3-031-72356-8_25

returns through subsequent sales [8]. These acquisitions can involve majority shares of private or public companies, or investments in buyouts as part of a consortium. Common PE investment strategies, as identified by [6], include Venture Capital (VC), Growth Capital (GC), and Leveraged Buyouts (LBO). These strategies offer varying degrees of risk and return potential, depending on the investment objectives and time horizon of the PE fund. The ability to accurately assess the likelihood of company success is crucial for PE firms in identifying attractive investment targets. Traditional evaluation of company performance often relies on manual analysis of financial statements or proprietary information, which may not be sufficient for capturing the dynamic nature of companies, especially those in early-stage or high-growth industries. This evaluation approach is often time consuming and as a result not every potential company can be properly evaluated. Therefore, there is a growing interest in leveraging data-driven methods to (1) debias decisions, so that the individual investment decision made for a particular deal is expected to drive lower risk and higher ROI (return on investment); and (2) enable automation, so that more companies can be evaluated without the need for additional resources [9].

For LBO in the PE industry, data-driven approaches may be less relevant due to the combination of two reasons: (1) LBO professionals often track and maintain in-depth knowledge of late-stage companies[1] in a few focus sectors, resulting in unique knowledge and understanding that can hardly be entirely replaced by public (or even proprietary) data; (2) the number of LBO investments is usually less than VC and GC leading to a lower sourcing frequency. VC investments often involve early-stage companies with prone-to-change business models and limited revenue, making data-driven approaches valuable for evaluating their growth potential. Additionally, VC investors typically manage larger portfolios with higher investment frequency, necessitating the use of data-driven models for efficient decision-making in identifying and evaluating investment opportunities. In practice, historical financial data (e.g., revenue) of startup[2] or scaleup[3] companies are commonly perceived as a good approximation of their true valuations [10]. The financial information of GC targets (scaleups) is much more accessible than that of VC targets (startups). Therefore, GC practitioners often directly use financial metrics to calculate the company's valuation for sourcing, which is why the adoption of big data in GC sourcing may not be as intensive as in VCs. However, data-driven approaches may still provide additional insights in assessing the growth potential and financial performance of the GC targets.

---

[1] Generally, a company is considered late-stage when it has proven that its concept and business model work, and it is out-earning its competitors.

[2] A startup is a dynamic, flexible, high risk, and recently established company that typically represents a reproducible and scalable business model. It provides innovative products or services, and has limited funds and resources [5,9,35,37].

[3] A startup moves into scaleup territory after proving the scalability and viability of its business model and experiencing an accelerated cycle of revenue growth. This transition is usually accompanied by the fundraising of outside capital [11].

Our contributions significantly advance data-driven strategies for sourcing investment opportunities in the VC and GC sectors by predicting the potential success of companies. These advancements include:

- We formally define the sourcing problem for VC/GC investments as a multivariate time series classification task and propose to employ a Transformer-based Multivariate Time Series Classifier (TMTSC) to address it.
- We introduce key components of our implementation, including input features, model architecture and optimization target, which all contribute to the successful application of TMTSC in VC/GC sourcing.
- We carry out extensive experiments, comparing TMTSC with widely adopted baselines on two real-world tasks, and demonstrate the effectiveness of our approach using a diverse set of evaluation metrics and strategies.

2 Related Work

Over the past two decades, data-driven approaches have been dominating research on deal sourcing for VC, i.e. identifying startups that eventually turn into unicorns[4]. In recent years, however, research has begun to intensify on GC deal sourcing, transforming the way scaleup companies are identified and assessed. Based on our extensive literature survey, data-driven methods for VC/GC deal sourcing can be broadly categorized into Statistical and Analytical (S&A) methods, conventional Machine Learning (ML) methods, and Deep Learning (DL) methods. S&A work [22,27,32,35] typically starts with defining some hypotheses followed by testing them using statistical tools. However, developing effective hypotheses for S&A approaches is a challenging task that requires simplicity, conciseness, precision, testability, and most importantly, a grounding in existing literature or established theory, as emphasized in [41]. It is worth mentioning that while DL methods technically fall under the broader umbrella of ML, we discuss DL work separately in recognition of its increasing popularity and relevance to our research.

2.1 Conventional Machine Learning Methods

Over the last few years, there has been a growing interest in leveraging ML algorithms for *hypothesis mining* from data, as an alternative to manually defining hypotheses upfront. Hypothesis mining involves conducting explainability analysis on trained ML models to summarize, rather than explicitly define, hypotheses [24]. For instance, by training an ML model on a labeled dataset containing features of various companies, and quantifying how changes in these features impact the prediction target (i.e. success probability), one can distill hypotheses that describe the relationships between the relevant features and the prediction

---

[4] Unicorn and near-unicorn startups are private, venture-backed firms with a valuation of at least $500 million at some point [14].

target. Compared to S&A, hypothesis mining is a much more structured procedure that trains an ML model using the entire dataset at hand. In general, ML based approaches, as demonstrated in previous works such as [3,7,27,29,43], typically require practitioners to define the input data $\mathbf{x}$ and annotation y (labeling "good" or "bad" investment according to some criteria) before training a model $f(\cdot)$ that maps $\mathbf{x}$ to y, i.e., $y = f(\mathbf{x})$. With the rapid growth of dataset size and diversity (origin and modality), conventional ML models[5] sometimes struggle to fit the large and unstructured[6] data due to lack of *capacity* and *expressivity*[7].

2.2 Deep Learning Methods

Most recently, DL algorithms have attracted an increasing number of researchers hunting for good VC/GC investment targets. DL is implemented (entirely or partly) with ANNs (artificial neural networks) that utilize at least two hidden layers of neurons. The *capacity* of DL can be controlled by the number of neurons (width) and layers (depth) [23]. Deep ANNs are exponentially *expressive* with respect to their depth [34]. While structured data is commonly used in many DL methods, such as [2,4,18], unstructured data is increasingly recognized as an important complement to structured data in recent studies [12,20,26,31,38], or even as a standalone input to the model [39,45]. Unstructured data often contains large-scale and intact-yet-noisy signals, which may result in superior performance when a proper DL approach is applied [19].

The main types of unstructured data seen include text [12], graph [1], image [13], video [39], audio [36] and time series [12]. Among these, fine-grained multivariate time series, which encompass various aspects of a company over time, hold particular significance for deal sourcing in the VC/GC domain. Some examples of these aspects include financial performance, team dynamics, funding rounds, market conditions, and other key indicators. Especially for GC, financial time series become highly relevant for evaluating scaleup companies whose periodical financial data points are usually available to the potential investor [10]. Due to the proprietary, costly, and scarce nature of multivariate time series company data, there is a limited number of DL based approaches in the literature that utilize time series as model input. To the best of our effort, we identified only three such studies [12,26,38], highlighting the challenges associated with utilizing multivariate time series data to source investment targets for Venture and Growth Capital. Inspired by [44], we frame the problem as a multivariate time series classification task and propose a solution that leverages a Transformer model. Our approach also incorporates carefully designed input features, optimization target, and investor-centric data processing [9].

---

[5] The frequently applied conventional ML models include many such as decision tree [3], random forest [29], logistic regression [27], and gradient boosting [43].
[6] Unstructured data, such as image and timeseries, is a collection of many varied types that maintains their native form, while structured data is aggregated from original (raw) data and is usually stored in a tabular form.
[7] *Expressivity* describes the classes of functions a model can approximate, and *capacity* measures how much "brute force" the model has to fit the data.

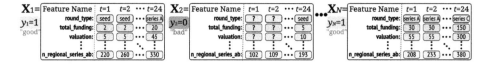

Fig. 1. An illustration of multivariate time series dataset (N samples) to train models for VC and GC sourcing.

3 The Approach

Our approach tackles the problem of identifying good investment targets for VC and GC by framing it as a multivariate time series classification task. Specifically, each potential investment target (i.e. candidate company) is represented by a multivariate time series $\mathbf{X} \in \mathbb{R}^{T \times K}$, as shown in Fig. 1. $\mathbf{X}$ consists of T observations, each containing K variables that describe different aspects (e.g., funding, revenue, etc.) of the corresponding company. Formally, each sample $\mathbf{X}$ is a sequence of T feature vectors: $\mathbf{X} = [\mathbf{x}_1, \mathbf{x}_2, \ldots, \mathbf{x}_t, \ldots, \mathbf{x}_T]$, where $\mathbf{x}_t \in \mathbb{R}^K$. At each time step t, we collect K numerical or categorical features about the company to form the vector $\mathbf{x}_t$, which captures a multi-view snapshot of the company at that time point. The last vector $\mathbf{x}_T$ represents the most recent state of the company. Depending on the data available, the time interval between two adjacent time points, t and $t+1$, can be set to a month, a quarter, or any other length of choice. By adopting this representation, we can model a multi-view evolution of each company over time and make informed predictions about their future success.

We collect a set of N samples, each corresponding to a company, denoted by $\mathbf{X}_1, \mathbf{X}_2, \ldots, \mathbf{X}_n, \ldots, \mathbf{X}_N$. For each sample $\mathbf{X}_n$, we have a binary ground truth label $y_n \in \{0, 1\}$ indicating a "bad" or "good" investment target according to some criteria. Details of how we define and collect these labels are explained in Sect. 3.3. We construct a dataset $\mathfrak{U}$ from these samples and labels as $\mathfrak{U} = \{(\mathbf{X}_1, y_1), (\mathbf{X}_2, y_2), \ldots, (\mathbf{X}_N, y_N)\}$, where $n \in \mathbb{Z} \cap [1, N]$. Our objective then is to **train a model on $\mathfrak{U}$ to accurately predict the ground truth labels y_n using $\mathbf{X}_n$**. We use $\hat{y}_n \in [0, 1]$ to denote the predicted probability of future success of the company represented by $\mathbf{X}_n$, in order to distinguish it from the ground truth label y_n. For the sake of brevity, we use general terms $\mathbf{X}$, y, and $\hat{y}$ to denote $\mathbf{X}_n$, y_n, and $\hat{y}_n$, respectively.

3.1 Time Series Features

We define the input time series features $\mathbf{X}$ by constructing 16 time series that fall into 6 feature categories, as summarized in [9]. These categories are ① **funding**, ② **founder/owner**, ③ **team**, ④ **investor**, ⑤ **web**, and ⑥ **context**, and below we will introduce the selected features under each category. Each time series feature contains precisely T values corresponding to the T time steps. For a concrete example of $\mathbf{X}$, see Fig. 1. All time series features are numerical, with the exception of the first one (round_type), which is categorical. Each time step corresponds to a calendar month, and the steps are aligned monthly.

① **Funding** category contains statistics of historical funding received by the company, showing recognition from investors.

- round_type indicates the latest funding round type that a company has received up to time t, such as *Seed* or *Series A*, providing insights into its funding stage and maturity. It is a categorical feature with 60 unique values.
- total_funding is the cumulative amount of funding in USD that the company has received up to time t, indicating the amount of capital it has been able to attract. The value range is from 1 to approximately 2×10^{11}.
- valuation: the estimated USD valuation of the company immediately after its latest funding round and is included to provide insight into a company's overall financial value. It is a numerical feature with values ranging from 1 to about 1×10^{12}.

② **Founder/Owner**: this category captures attributes of the founding team, which are critical to a company's short-term success and long-term survival [21].

- n_founder shows the number of a company's founders still with the company at time t. The value ranges from 0 to 38.

③ **Team**: this category captures the statistics of the employees of the company.

- n_employee: the number of employees at time t, implying the company's growth trajectory. The feature has a value range of 1 to 113,757.

④ **Investor** category captures the statistics of investors who have funded the company, indicating its early attractiveness.

- n_investor represents the total number of unique investors who have provided funding to the company up to time t. This feature provides insights into the diversification of the company's investment sources. The value ranges from 1 to 240.
- growth_investor_rate is the ratio of unique GC investors[8] among the company's unique investors up to time t. This feature indicates the investors' beliefs in the company's future growth potential.
- average_cagr is the average Compound Annual Growth Rate (CAGR)[9] of all exited deals made by the company's investors up to time t. This feature is meant to demonstrate the past investment performance of the involved investors.
- 2x_cagr_rate is calculated as the ratio of investment deals up to time t with a CAGR ≥ 2 among all exited deals made by the company's investors. This feature reflects the proportion of investors with a history of impressive returns who are currently invested in the company.

---

[8] GC investors are defined as those who have participated in a funding round of 50 million USD or valuation above 200 million USD.

[9] CAGR $= (EV/SV)^{1/Y} - 1$ is calculated for each deal the investor has exited, where SV and EV stand for the starting and exiting value of the investment, respectively; Y is the number of holding years (from investment till divestment) of the invested asset.

⑤ **Web**: this category covers any feature extracted from web pages that are related to the company in focus.

- cu_popularity describes the company's domain name popularity rank at time t. This rank is determined based on the domain's network traffic as measured by Cisco Umbrella (CU)[10].
- sw_global_rank describes at time t the monthly unique visitors and pageviews of the company website(s). The higher this sum, the higher the site's rank. This feature is obtained from SimilarWeb[11].
- n_desktop_visitor and n_mobile_visitor are two features indicating the number of unique visitors to the company's website utilizing a desktop and mobile device, respectively. Both are sourced from SimilarWeb.
- n_news counts the number of times a company is mentioned across approximately 3,700 news websites to a time point t, reflecting its media visibility and recognition. The value range of the dataset is 1 to 389.

⑥ **Context**: this category captures extrinsic factors[12] that may be (but are not limited to) competition, regional, environmental, cultural or economical based.

- n_regional_seed_round represents the number of seed funding rounds in the company's region[13] between adjacent time points $t–1$ and t. This feature offers context on the company's performance relative to regional competitors and financial conditions, highlighting potential company success even if regional investments are low.
- n_regional_series_ab: same as the previous one except that it is counting the Series A and B rounds instead.

3.2 TMTSC Architecture

As illustrated in Fig. 2, TMTSC learns to predict $\hat{y}_n$ using time series input $\mathbf{X}$. At the t-th time step, each input feature vector $\mathbf{x}_t$ consists of a numerical part (often normalized), denoted as $\mathbf{u}_t$, and a categorical part, denoted as $\mathbf{v}_t$. Thus, $\mathbf{x}_t = [\mathbf{u}_t; \mathbf{v}_t]$, where ";" represents a vector concatenation operation. To convert the categorical features $\mathbf{v}_t$ into dense embeddings, we utilize embedding layers, which can be collectively represented by a learnable function $\mathcal{E}$. The embedded categorical features are then given by $\mathbf{v}'_t = \mathcal{E}(\mathbf{v}_t)$. As a result, the K-dimensional vector $\mathbf{x}_t$ is transformed to a new numerical vector $\mathbf{x}'_t$ that has K' ($K' > K$) dimensions:

$$\mathbf{x}'_t = [\mathbf{u}_t; \mathcal{E}(\mathbf{v}_t)] \in \mathbb{R}^{K'} \text{ and } \mathbf{x}_t = [\mathbf{u}_t; \mathbf{v}_t] \in \mathbb{R}^K. \qquad (1)$$

Then, $\mathbf{x}'_t$ is linearly projected onto a D-dimensional vector space, where D is the dimension of the Transformer model sequence element representations:

---

[10] http://s3-us-west-1.amazonaws.com/umbrella-static/index.html.
[11] https://support.similarweb.com/hc/en-us/articles/213452305-Rank.
[12] While intrinsic features act from within a company, extrinsic ones wield their influence from the outside. The company may impact the former, yet not the latter.
[13] A region is a collection of countries such as *Great Britain, DACH, France Benelux, Southern Europe, Nordics, South Asia, South East Asia*, and so on.

$$\mathbf{h}_t = \mathbf{W}\mathbf{x}'_t + \mathbf{b}, \qquad (2)$$

where $\mathbf{W} \in \mathbb{R}^{D \times K'}$ and $\mathbf{b} \in \mathbb{R}^D$ are learnable parameters and $\mathbf{h}_t \in \mathbb{R}^D$, $t \in \mathbb{Z} \cap [1, T]$ are the input vectors to the Transformer model. Although Eqs. (1) and (2) show the operation for a single time step for clarity, all raw input vectors $\mathbf{x}_t$, $t \in \mathbb{Z} \cap [1, T]$ are embedded in the same way concurrently. It is worth mentioning that the above formulation can also accommodate univariate time series (i.e., $K = 1$), though in the scope of this work, we will only evaluate the approach on multivariate time series.

It is important to note that the Transformer is a feed-forward architecture that does not inherently account for the order of input elements. To address the sequential nature of time series data, we incorporate positional encodings, denoted as $\mathbf{P} = [\mathbf{p}_1, \mathbf{p}_2, \ldots, \mathbf{p}_T] \in \mathbb{R}^{T \times D}$, to the input vectors $\mathbf{H} = [\mathbf{h}_1, \mathbf{h}_2, \ldots, \mathbf{h}_T] \in \mathbb{R}^{T \times D}$, resulting in the final input $\mathbf{H}'$:

$$\mathbf{H}' = \mathbf{H} + \mathbf{P} = [\mathbf{h}'_1, \mathbf{h}'_2, \ldots, \mathbf{h}'_T] \in \mathbb{R}^{T \times D}, \qquad (3)$$

where $\mathbf{h}'_t \in \mathbb{R}^D = \mathbf{h}_t + \mathbf{p}_t$. Closely following the approach in [44], we employ fully learnable positional encodings, as they have been reported to yield better performance compared to deterministic sinusoidal encodings [40] for multivariate time series classification tasks. We also utilize batch normalization (rather than layer normalization), as it is considered effective in mitigating the impact of outlier values in time series data, an issue that does not arise for textual inputs.

The Transformer-based model architecture depicted in Fig. 2 generates T output vectors z_t corresponding to the T input time steps. These output vectors are concatenated to form a single output matrix $\mathbf{Z} = [\mathbf{z}_1; \mathbf{z}_2; \ldots; \mathbf{z}_T]$, which serves as the input for a linear layer. As shown in Eq. (4), the linear layer is parameterized by $\mathbf{W}_{\text{out}} \in \mathbb{R}^{C \times (T \cdot D)}$ and $\mathbf{b}_{\text{out}} \in \mathbb{R}^C$, where C denotes the number of classes to be predicted.

$$\hat{\mathbf{y}} = Softmax(\mathbf{W}_{\text{out}}\mathbf{Z} + \mathbf{b}_{\text{out}}). \quad (4)$$

3.3 Optimization Target

In the absence of a universally agreed-upon definition of "true success" of startups and scaleups, most existing definitions tend to focus on "growth", which can be measured from various perspectives, such as funding, revenue, employee count,

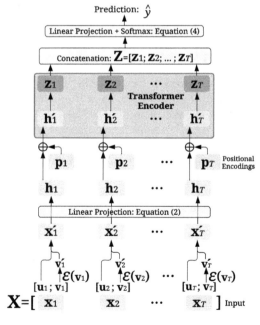

Fig. 2. TMTSC architecture: $\mathbf{u}_t$ and $\mathbf{v}_t$ are numerical and categorical part respectively.

Table 1. Specification of tasks/datasets: split with an investor-centric strategy [9].

Dataset	#feat.	#sample	#time step	#class	#train	#val.	#test
VC	16	86,886	24	2	63,562	11,178	12,146
GC	16	21,163	24	2	16,275	–	4,888

and valuation, among others [9]. As a well-established investment firm, we have access to a large volume of expert evaluations (akin to [4,28]) that represent quantified assessments from human experts. These evaluations encompass multiple categories/terms, such as "inbound", "reviewing", "reach-out", "follow", "negotiating", and "out-of-scope"[14], which are updated periodically by investment professionals for companies in the context of VC and GC. To further simplify the prediction task, we assign each evaluation term to either a good ("1") or bad ("0") binary bucket denoted by $\mathbf{y}_n$ in Fig. 1, implying $C = 2$ in Eq. (5). In this way, each company is annotated with two ground-truth binary labels – one for VC and the other for GC; and the loss function $\mathcal{L}$ is

$$\mathcal{L} = -\frac{1}{N}\sum_{n=1}^{N}\left[\mathbf{y}_n \log(\hat{\mathbf{y}}_n) + (1 - \mathbf{y}_n)\log(1 - \hat{\mathbf{y}}_n)\right]. \quad (5)$$

4 Experiments on Real-World Investment Tasks

Following the details introduced in the previous section, we prepare two real-world proprietary datasets: "**VC**" for the VC context and "**GC**" for the GC context, performing data augmentation to obtain monthly time steps. To eliminate overly sparse time series, we discard the samples whose time series features are all shorter than six months. Missing `valuation` values are approximated with the cumulative funding received up to that point. Missing `total_funding` values are filled by taking the value of the previous month (if available) or 0 otherwise. For the time steps where the values are still missing, we fill them with "−1". Finally, we pad all time series to the same length of 24 months. As for scaling, we empirically apply log-transform to 13 numerical features (excluding `cu_popularity` and `n_employee`). The specification is presented in Table 1. It is worth noting that we also experimented with two public TSC (time series classification) benchmark datasets[15]: Ethanol [30] and PEMS-SF [17]. Since they do not directly relate to the investment business domain, we chose to leave them outside the scope of this paper. For in-depth information about experiments on public datasets, we recommend reading [25].

---

[14] The complete evaluation framework is withheld as it is proprietary.
[15] The overall performance can be found on Motherbrain's blog post: https://motherbrain.ai/applying-transformers-to-score-potentially-successful-startups-7893284efb01.

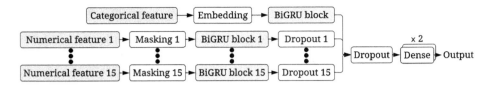

Fig. 3. U-GRU: each univariate time series is modeled by a BiGRU block.

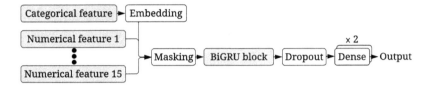

Fig. 4. M-GRU: all time series features are modeled by one single BiGRU block.

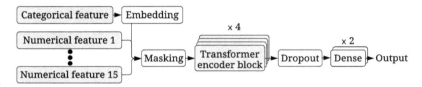

Fig. 5. TE architecture with 4 Transformer encoder blocks.

4.1 Baselines and Hyper-parameters

GRU (Gated Recurrent Unit) [16] is a highly relevant baseline for comparison due to its ability to model sequential dependencies and capture long-term dependencies in time series data. We experiment both U-GRU (Univariate GRU) and M-GRU (Multivariate GRU), whose architectures are illustrated in Figs. 3 and 4 respectively. We adopt the implementation of BiGRU (Bidirectional GRU) [15], masking, embedding, dropout, and dense layers from Keras[16]. In both architectures, the masking layer is added to inform the model to ignore any values marked as missing in its computation. As a close relative and foundation of TMTSC, the Transformer Encoder (TE) [40] is also selected as a baseline. As shown in Fig. 5, we adopt the same layers as the original implementation [40]. For comparability, the input features are ingested, embedded and concatenated in the same way as M-GRU as shown in Fig. 4. The hyper-parameters are selected based on the highest AUC-ROC ("Area Under the Curve" of the Receiver Operating Characteristic curve) score on the validation split of the dataset. Refer to [25] for the searched and selected hyper-parameter values.

---

[16] Keras Layer documentation: https://keras.io/api/layers.

4.2 Overall Performance: A Precision-Centric Comparison

When interpreting the results in Table 2, the costs of different types of prediction errors must be considered. The outcome from false negatives (failing to identify a successful company) is that investors are simply not made aware of a successful company and therefore no action is taken. In that regard, there is an upside loss in terms of lost profit but no detriment in terms of time or money invested. False positives (incorrectly predicting a company will be successful), on the other hand, can lead to wasted time spent on due diligence, or, in the worst case, an investment that loses money. For that reason, it is more important to evaluate a model with respect to its precision, or the number of its positive predictions that are actually positive. Observing the precision scores in Table 2, TMTSC outperforms all other methods, achieving scores of 0.86 and 0.83 for VC and GC scenario, respectively. Additionally, Fig. 6 provides a more balanced and comprehensive view using AUC-ROC metric. TMTSC clearly outperforms on the VC task, achieving an average score of 0.92, 12% better than M-GRU, the next

Table 2. Overall performance comparison.

Task	Metric	U-GRU	M-GRU	TE	TMTSC
VC	Accuracy ± STDEV	0.548 ± 0.013	0.731 ± 0.026	0.655 ± 0.081	**0.863 ± 0.015**
	Precision ± STDEV	0.704 ± **0.015**	0.740 ± 0.044	0.699 ± 0.062	**0.864 ± 0.016**
	AUC-ROC ± STDEV	0.628 ± 0.015	0.819 ± 0.020	0.780 ± 0.081	**0.924 ± 0.009**
GC	Accuracy ± STDEV	0.934 ± 0.011	0.924 ± 0.027	0.933 ± 0.021	**0.956 ± 0.004**
	Precision ± STDEV	0.701 ± 0.058	0.794 ± 0.101	0.765 ± 0.108	**0.831 ± 0.026**
	AUC-ROC ± STDEV	**0.977 ± 0.008**	0.939 ± 0.002	0.971 ± 0.002	0.971 ± **0.001**

Table 3. The comparison of training efficiency. Underlined values indicate shortest time per step for each dataset.

Task	Method	Sec./Step	Relative Time
VC	U-GRU	2.000	83.3 ×
	M-GRU	<u>0.024</u>	1.0 ×
	TE	0.057	2.4 ×
	TMTSC	0.100	4.2 ×
GC	U-GRU	1.232	46.6 ×
	M-GRU	<u>0.026</u>	1.0 ×
	TE	0.056	2.1 ×
	TMTSC	0.101	3.8 ×

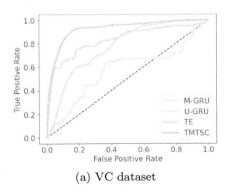

(a) VC dataset

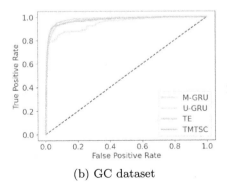

(b) GC dataset

Fig. 6. ROC (Receiver Operating Characteristic) curves.

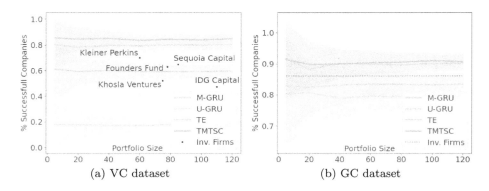

Fig. 7. Portfolio simulation: success rate vs. portfolio size.

best method. All methods perform extremely well on GC task, and TMTSC's AUC-ROC score of 0.97 is less than 1% lower than the winner U-GRU.

4.3 Training Stability and Efficiency

The standard deviation values (STDEV) in Table 2 indicate that the TMTSC training is relatively stable in both the VC and GC datasets, as the values are relatively low compared to the other baselines. To measure training efficiency, we record per-step training time for each dataset and method using the same batch size (=512) and hardware configurations. The results are presented in Table 3, where the "Relative Time" column shows how the time consumption for the corresponding method relates to the fastest method (i.e., "1.0 ×"). Take the VC task for example, "2.4 ×" for TE would therefore mean TE took over twice as long as M-GRU. It is evident that M-GRU requires the least amount of training time, largely due to its design, which favors simplicity. TMTSC and TE take only a small amount of extra time to train, despite their increased complexity. This is likely due to their multi-head architecture, allowing parallelization of self-attention computations.

4.4 Portfolio Simulation

To further evaluate the model in the context of the real-world investment scenario, portfolio simulations are executed and visualized in Fig. 7. Concretely, we assemble a set by isolating the companies confirmed to be potentially good investment targets in the VC or GC datasets (i.e., that are positively labeled). From this set, we randomly sample i companies to simulate forming VC/GC investment portfolios of size i and calculate the percentage of companies each model predicts to be successful within the sample. To address the stochasticity of this process, we perform each simulation 100 times. Different portfolio sizes (i.e., values of i) are simulated; and for each i (X-axis), the mean and standard deviations are plotted (Y-axis), resulting in a colored line with a shaded area in

Fig. 7. In VC and GC contexts, we can see that (1) TMTSC performs the best among all methods, (2) performance becomes less variable as simulated portfolio size increases, and (3) the models evaluated performed more variably on the VC dataset than the GC dataset.

To roughly compare these methods against real-world VC and GC fund performance, Fig. 7(a), includes the portfolio size and performance of five VC funds [42], showing a performance largely on par with M-GRU and inferior to TMTSC and TE. For the GC simulation, a horizontal line representing the real-world GC success rate of 86.3% [33] is included in Fig. 7(a). Here, the real-world GC success rate is outperformed by M-GRU and TMTSC. It is important to note that investment firms are much more constrained than the simulation: they cannot invest in every attractive company they encounter due to factors like founders' preference, portfolio conflict, investment focus, and available funds.

5 Conclusion and Future Work

In this work, we propose using a Transformer-based Multivariate Time Series Classifier (TMTSC) to facilitate sourcing investment targets for Venture Capital (VC) and Growth Capital (GC). Specifically, TMTSC utilizes multivariate time series as input to predict the probability that any candidate company will succeed in the context of a VC or GC fund. We formally define the sourcing problem as a multivariate time series classification task, and introduce the key components of our implementation, including input features, model architecture, and optimization target. Our extensive experiments on two proprietary datasets (collected from real-world VC and GC contexts) demonstrate the effectiveness, stability, and efficiency of our approach compared with three popular baselines. To further evaluate the model in the context of the real-world investment scenario, portfolio simulations are executed, showing TMTSC's high success rate in both VC and GC sourcing. The main future work includes (1) incorporating global features along with time series input, (2) and learning generic and condensed representations for multivariate time series for varies downstream prediction tasks.

References

1. Allu, R., Padmanabhuni, V.N.R.: Predicting the success rate of a start-up using LSTM with a swish activation function. J. Control Decis. **9**(3), 355–363 (2022)
2. Ang, Y.Q., Chia, A., Saghafian, S.: Using machine learning to demystify startups' funding, post-money valuation, and success. In: Innovative Technology at the Interface of Finance and Operations, pp. 271–296. Springer, Cham (2022)
3. Arroyo, J., Corea, F., Jimenez-Diaz, G., Recio-Garcia, J.A.: Assessment of machine learning performance for decision support in venture capital investments. IEEE Access **7**, 124233–124243 (2019)
4. Bai, S., Zhao, Y.: Startup investment decision support: application of venture capital scorecards using machine learning approaches. Systems **9**(3), 55 (2021)

5. Blank, S.: Why the lean start-up changes everything. Harvard Bus. Rev. **91**(5), 63–72 (2013)
6. Block, J., Fisch, C., Vismara, S., Andres, R.: PE investment criteria: an experimental conjoint analysis of venture capital, business angels, and family offices. J. Corp. Finan. **58**, 329–352 (2019)
7. Bonaventura, M., Ciotti, V., Panzarasa, P., Liverani, S., Lacasa, L., Latora, V.: Predicting success in the worldwide start-up network. Sci. Rep. **10**(1), 1–6 (2020)
8. Cao, L., et al.: A scalable and adaptive system to infer the industry sectors of companies: prompt + model tuning of generative language models. In: Proceedings of the IJCAI Workshop on Financial Technology and Natural Language Processing (FinNLP), pp. 1–11, August 2023
9. Cao, L., von Ehrenheim, V., Krakowski, S., Li, X., Lutz, A.: Using deep learning to find the next unicorn: a practical synthesis. arXiv preprint arXiv:2210.14195 (2022)
10. Cao, L., Horn, S., von Ehrenheim, V., Stahl, R.A., Landgren, H.: Simulation-informed revenue extrapolation with confidence estimate for scaleup companies using scarce time series data. In: Proceedings of the 31st ACM International Conference on Information and Knowledge Management (CIKM 2022), 17–21 October 2022, Atlanta, GA, USA, p. 12. Association for Computing Machinery (ACM), New York, NY, USA, October 2022
11. Cavallo, A., Ghezzi, A., Dell'Era, C., Pellizzoni, E.: Fostering digital entrepreneurship from startup to scaleup: the role of venture capital funds and angel groups. Technol. Forecast. Soc. Chang. **145**, 24–35 (2019)
12. Chen, M., et al.: A trend-aware investment target recommendation system with heterogeneous graph. In: International Joint Conference on Neural Networks, pp. 1–8 (2021)
13. Cheng, C., Tan, F., Hou, X., Wei, Z.: Success prediction on crowdfunding with multimodal deep learning. In: IJCAI, pp. 2158–2164 (2019)
14. Chernenko, S., Lerner, J., Zeng, Y.: Mutual funds as venture capitalists? Evidence from unicorns. Rev. Finan. Stud. **34**(5), 2362–2410 (2021)
15. Cho, K., et al.: Learning phrase representations using RNN encoder–decoder for statistical machine translation. In: Proceedings of the Conference on Empirical Methods in Natural Language Processing (EMNLP), pp. 1724–1734. Association for Computational Linguistics, Doha, Qatar, October 2014
16. Chung, J., Gulcehre, C., Cho, K., Bengio, Y.: Empirical evaluation of gated recurrent neural networks on sequence modeling. In: NIPS 2014 Workshop on Deep Learning, December 2014 (2014)
17. Cuturi, M.: Fast global alignment kernels. In: Proceedings of the 28th International Conference on Machine Learning (ICML-2011), pp. 929–936 (2011)
18. Dellermann, D., Lipusch, N., Ebel, P., Popp, K.M., Leimeister, J.M.: Finding the unicorn: predicting early stage startup success through a hybrid intelligence method. In: International Conference on Information Systems (2021)
19. Garkavenko, M., Gaussier, E., Mirisaee, H., Lagnier, C., Guerraz, A.: Where do you want to invest? Predicting startup funding from freely, publicly available web info. arXiv:2204.06479 (2022)
20. Gastaud, C., Carniel, T., Dalle, J.M.: The varying importance of extrinsic factors in the success of startup fundraising: competition at early-stage and networks at growth-stage. arXiv:1906.03210 (2019)
21. Ghassemi, M., Song, C., Alhanai, T.: The automated venture capitalist: data and methods to predict the fate of startup ventures. In: AAAI Workshop on Knowledge Discovery from Unstructured Data in Financial Services (2020)

22. Gompers, P.A., Gornall, W., Kaplan, S.N., Strebulaev, I.A.: How do venture capitalists make decisions? J. Financ. Econ. **135**(1), 169–190 (2020)
23. Goodfellow, I., Bengio, Y., Courville, A.: Deep Learning. MIT Press, Cambridge (2016)
24. Guerzoni, M., Nava, C.R., Nuccio, M.: The survival of start-ups in time of crisis: a ML approach to measure innovation. arXiv preprint arXiv:1911.01073 (2019)
25. Halvardsson, G.: A Transformer-based scoring approach for startup success prediction. Master's thesis, KTH Royal Institute of Technology (2023)
26. Horn, S.: Deep learning models as decision support in venture capital investments: temporal representations in employee growth forecasting of startup companies. Master's thesis, KTH Royal Institute of Technology & EQT Partners (2021)
27. Kaiser, U., Kuhn, J.M.: The value of publicly available, textual and non-textual information for startup performance prediction. J. Bus. Ventur. Insights **14**, e00179 (2020)
28. Kinne, J., Lenz, D.: Predicting innovative firms using web mining and deep learning. PLoS ONE **16**(4), e0249071 (2021)
29. Krishna, A., Agrawal, A., Choudhary, A.: Predicting the outcome of startups: less failure, more success. In: International Conference on Data Mining Workshop, pp. 798–805 (2016)
30. Large, J., Kemsley, E.K., Wellner, N., Goodall, I., Bagnall, A.: Detecting forged alcohol non-invasively through vibrational spectroscopy and machine learning. In: Phung, D., Tseng, V.S., Webb, G.I., Ho, B., Ganji, M., Rashidi, L. (eds.) PAKDD 2018. LNCS (LNAI), vol. 10937, pp. 298–309. Springer, Cham (2018). https://doi.org/10.1007/978-3-319-93034-3_24
31. Lyu, S., et al.: Graph neural network based VC investment success prediction. arXiv preprint arXiv:2105.11537 (2021)
32. Malmström, M., Voitkane, A., Johansson, J., Wincent, J.: What do they think and what do they say? Gender bias, entrepreneurial attitude in writing and venture capitalists' funding decisions. J. Bus. Ventur. Insights **13**, e00154 (2020)
33. Mooradian, P., Auerback, A., Slotsky, C., Gilfix, J.: Growth equity: turns out, it's all about the growth (2019)
34. Raghu, M., Poole, B., Kleinberg, J., Ganguli, S., Sohl-Dickstein, J.: On the expressive power of deep neural networks. In: International Conference on Machine Learning, pp. 2847–2854. PMLR (2017)
35. Santisteban, J., Mauricio, D., Cachay, O., et al.: Critical success factors for technology-based startups. Int. J. Entrep. Small Bus. **42**(4), 397–421 (2021)
36. Shi, J., Yang, K., Xu, W., Wang, M.: Leveraging deep learning with audio analytics to predict the success of crowdfunding projects. J. Supercomput. **77**(7), 7833–7853 (2021)
37. Skawińska, E., Zalewski, R.I.: Success factors of startups in the EU - a comparative study. Sustainability **12**(19), 8200 (2020)
38. Stahl, R.H.A.: Leveraging time-series signals for multi-stage startup success prediction. Master's thesis, ETH Zurich & EQT Partners (2021)
39. Tang, Z., Yang, Y., Li, W., Lian, D., Duan, L.: Deep cross-attention network for crowdfunding success prediction. IEEE Trans. Multimedia **25**, 1306–1319 (2022)
40. Vaswani, A., et al.: Attention is all you need. In: Advances in Neural Information Processing Systems, vol. 30 (2017)
41. Williamson, K.: Research Methods for Students, Academics and Professionals: Information Management and Systems. Elsevier (2002)

42. Yin, D., Li, J., Wu, G.: Solving the data sparsity problem in predicting the success of the startups with machine learning methods. arXiv preprint arXiv:2112.07985 (2021). https://arxiv.org/abs/2112.07985
43. Żbikowski, K., Antosiuk, P.: A machine learning, bias-free approach for predicting business success using Crunchbase data. Inf. Process. Manage. **58**(4), 102555 (2021)
44. Zerveas, G., Jayaraman, S., Patel, D., Bhamidipaty, A., Eickhoff, C.: A transformer-based framework for multivariate time series representation learning. In: Proceedings of the 27th ACM SIGKDD Conference on Knowledge Discovery & Data Mining, pp. 2114–2124 (2021)
45. Zhang, S., Zhong, H., Yuan, Z., Xiong, H.: Scalable heterogeneous graph neural networks for predicting high-potential early-stage startups. In: ACM SIGKDD Conference on Knowledge Discovery and Data Mining, pp. 2202–2211 (2021)

MSIF: Multi-source Information Fusion for Financial Question Answering

Man Lin[1], Delong Zeng[1], Jiarui Ouyang[1], and Ying Shen[1,2,3(✉)]

[1] Sun Yat-Sen University, Shenzhen 518107, China
sheny76@mail.sysu.edu.cn
[2] Pazhou Lab, Guangzhou 510005, China
[3] Guangdong Provincial Key Laboratory of Fire Science and Intelligent Emergency Technology, Guangzhou 510006, China

Abstract. Research on hybrid data, which combines tabular and textual content, has garnered significant interest in financial question answering. Recent approaches mainly focus on the encoding of tables and texts to facilitate model generation. However, there is still room for improvement in the question reasoning type and question semantic topic. Therefore, we propose **MSIF**, a novel model based on **M**ulti-**S**ource **I**nformation **F**usion which integrates additional information, encompassing question reasoning type and question semantic topic, from multiple sources for the original question. Specifically, we first identify the question reasoning type that helps discrete reasoning from the existing corpus TAT-QA. We then obtain the question semantic topic that helps analyze the question by leveraging the large-scale generative language model ChatGPT. Finally, we introduce an information fusion module to integrate additional information into the context representation. Experiments on the FinQA dataset show the effectiveness of our model in financial question answering, which outperforms most baselines on Execution Accuracy and Program Accuracy.

Keywords: Financial Question Answering · Data Augmentation · Numerical reasoning

1 Introduction

Financial documents contain a hybrid of tabular and textual content, requiring complex quantitative analysis. Recently, Financial Question Answering (FQA) has attracted increased attention in Natural Language Processing (NLP), aiming to answer queries from users by extracting pertinent information from financial documents. FQA finds applications in diverse domains, such as financial customer service for answering account-related inquiries [2], investment advisory systems for offering market insights [13], and legal counsel services for transactions [16].

Contemporary FQA systems face the challenge of handling hybrid data. In recent years, numerical reasoning over hybrid data has attracted much attention [5,24]. The current methods mainly focus on the encoding of tables and

(...abbreviate...) A reconciliation of the provision for income taxes, with the amount computed by applying the statutory federal income tax rate 35% in 2006, 2005, and 2004 to income before provision for income taxes, is as follows :

Year	Computed expected tax	...	Provision for income taxes	Effective tax rate
2006	$987	...	$829	29%
2005	$633	...	$480	27%
2004	$129	...	$104	28%

(...abbreviate...) the net tax benefits from employee stock option transactions were $ 419 million, $ 428 million, and $ 83 million in 2006, 2005, and 2004, respectively (...abbreviate...)

Question: What was the 2006 tax expense?
Gold Facts: the effective tax rate of 2006 is 29%; the effective tax rate of 2005 is 27%; the effective tax rate of 2004 is 28%;
the provision for income taxes of 2006 is $ 829; the provision for income taxes of 2005 is $ 480; the provision for income taxes of 2004 is $ 104;
Golden Program: Multiply(829, 29%)
Golden Answer: 240.41

Fig. 1. An example of Financial Question Answering: the system needs to leverage the supporting facts from hybrid data and generate the reasoning program to obtain the answer. The relevant values are highlighted in red. (Color figure online)

texts, trying to find the most relevant part to the question from either tables or sentences [10]. The exploration of question analysis in FQA has been limited, potentially leading to incorrect results. For instance, as shown in Fig. 1, we should analyze the question, then extract the keywords "2006" and "tax expense", and use them to locate needed information from tables and paragraphs. The needed information will serve as supporting facts for subsequent program generation. Failure to accurately analyze the question may result in identifying incorrect supporting facts and selecting erroneous mathematical symbols.

To enhance the question analysis capability for FQA, we introduce **MSIF**, a novel model based on Multi-Source Information Fusion, which integrates additional information, including question reasoning types and question semantic topics from various sources, into the question. More specifically, the question reasoning type is derived from the existing corpus TAT-QA by similarity calculation, while the question semantic topic is obtained from the topic model and large language model. Furthermore, we introduce an information fusion module to integrate comprehensive information into the context representation. Our model surpasses some baselines in different pre-trained models.

The main contribution of this work can be summarized as follows:

- We propose a novel augmentation approach to obtain additional information, encompassing question reasoning type and question semantic topic, from different sources, including existing corpus TAT-QA and the large language model ChatGPT.

- We introduce an information fusion module that helps capture the relevance between the additional information and the context representation, thereby enhancing answer generation.
- The experiment results outperform the most recent baselines on Execution Accuracy and Program Accuracy, showing the effectiveness of our model.

2 Related Work

Financial NLP in question answering has attracted much attention recently, and some studies have focused on hybrid data, which requires numerical reasoning to obtain an answer.

Previous work mainly utilized traditional machine learning techniques such as deep learning [12] and Latent Dirichlet Allocation (LDA) [8]. Chen et al. [3] use augmented financial datasets to train deep reinforcement learning without access to real financial data. Poria et al. [15] propose Sentic LDA to shift clustering from a syntactic to a semantic level. Recently, pre-trained language models such as LLaMA and ChatGPT have been used for numerical reasoning by post-training [14] or prompt-training [20] over texts. These models are pre-trained at scale to produce semantically rich texts and can guide the generation process to meet specific task requirements. Suadaa et al. [18] investigate template-based table representations for numerical reasoning in a table-to-text pre-trained model. Deng et al. [7] handle numerical reasoning over hybrid text to understand heterogeneous representations. Chen et al. [6] propose the CONVFINQA to investigate the chain of numerical reasoning in conversational question answering. Yang et al. [21] propose a system paradigm MM-REACT using ChatGPT to achieve multi-modal reasoning and action. In addition, semantic analysis of QA pairs in the financial domain has recently gained attention, which can provide valid contextual information for numerical reasoning [19], thus helping to obtain accurate responses. Chen et al. [5] propose the FinQA dataset that contains tabular and textual content and then use a retriever to classify the facts whose semantics are relevant to the questions. Zhu et al. [24] propose the TAT-QA dataset that contains more complex data and then adopts sequence tagging to extract relevant cells from the table along with relevant spans from the text to infer their semantics.

3 Methodology

3.1 Task Definition

Given a question q and a long document consisting of tabular content T and textual content E, the financial question answering aims to generate the reasoning program G consisting of mathematical operations that can be executed to get the answer A:

$$P(A|q,T,E) = \sum P(G|q,T,E). \qquad (1)$$

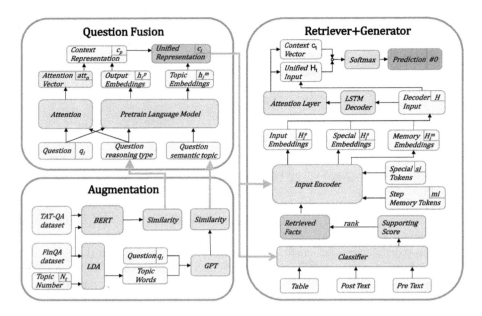

Fig. 2. The overall architecture of our MSIF model. The model contains an augmentation module to obtain additional information from different sources, a fusion encoder to fuse questions with their additional information, a retriever, and a program generator to retrieve facts and generate programs.

We use Domain Specific Language, which consists of mathematical and table operations, to construct the reasoning program.

The overall architecture of our model is shown in Fig. 2, and we will show the details in the next sections. In Sect. 3.2, we introduce the augmentation approach, which leverages the existing corpus TAT-QA, the LDA model, and the large language model ChatGPT to construct the additional information. In Sect. 3.3, we introduce the information fusion module, which combines questions with their additional information to promote numerical reasoning.

3.2 Augmentation

As shown in Algorithm 1, We obtain the additional information for questions in two steps. We first obtain the question reasoning types by calculating the similarity over the corpus TAT-QA, annotated with question reasoning types. Then, we obtain the semantic topics of questions by employing the LDA model, which analyzes the topic structure of the question, and ChatGPT, which generates sufficient semantic descriptions.

Question Reasoning Type. Considering high consistency as appropriateness, we utilize the corpus TAT-QA to infer the reasoning types for questions in the

Algorithm 1. Augmentation Procedure

Input: The experimental QA dataset $\mathbf{D} = \{(q_i, t_i, e_i, a_i) | i = 1, 2, ..., n\}$;
An existing corpus $\mathbf{B} = \{(s_j, type_j) | j = 1, 2, ..., m\}$;
Generate language model $\mathbf{G}_\theta$ like ChatGPT;
Number of topics $\mathbf{N_t}$;
Number of topic words in each topic $\mathbf{N_w}$;
Number of question semantic topics for each question $\mathbf{N_m}$;
Output: QA dataset $\widetilde{\mathbf{D}}$ with question reasoning types and question semantic topics

1: Train a LDA model $\mathbf{L}$ with hyperparameters $\mathbf{N_t}$ and $\mathbf{N_w}$ on $\{q_1, ..., q_n\}$
2: Get topic words $topic_k = \{w_{k1}, w_{k2}, ..., w_{kN_w}\}, k = 1, 2, ..., N_t \leftarrow \mathbf{L}$
3: Initialize the topic descriptions dictionary $\mathbf{R}$
4: **for** i = 1,2,..., n **do**
5:     Get $topic_k$ of q_i by the LDA model
6:     Construction the instruction $x_i \leftarrow Prompt(topic_k, q_i)$
7:     Get the question topic descriptions for the question $\{r_{i1}, ..., r_{iN_m}\} \leftarrow G_\theta(x_i)$
8:     $R.q_i = \{r_{i1}, ..., r_{iN_m}\}$
9: **end for**
10: Initialize an empty dataset $\widetilde{\mathbf{D}}$
11: **for** i = 1,2,...,n **do**
12:     // Get the question reasoning type
13:     Select the most similar sentences in $\mathbf{B}$, $j^* \leftarrow \arg\max_{j=1,...,m} \mathbf{similarity}(q_i, s_j)$
14:     // Get the question semantic topic
15:     Search question topic descriptions $R.q_i = \{r_{i1}, ..., r_{iN_m}\}$
16:     Select the most relevant part $r* \leftarrow \arg\max_{R.q_i = \{r_{i1}, ..., r_{iN_m}\}} \mathbf{similarity}(q_i, R.q_i)$
17:     $\widetilde{\mathbf{D}} \leftarrow \widetilde{\mathbf{D}} \bigcup (q_i, type_{j*}, r*, t_i, e_i, a_i)$
18: **end for**

experimental dataset. The question reasoning types are categorized into four types: Span, Multiple Spans, Counting, and Arithmetic. We first constitute set $B = \{(s_j, type_j) | j = 1, 2, ..., m\}$ consisting of question s_j and its question reasoning type $type_j$ from the corpus TAT-QA. We then constitute set $D = \{(q_i, t_i, e_i, a_i) | i = 1, 2, ..., n\}$ consisting of question q_i, table content t_i, text content e_i and answer a_i from the experimental dataset. After that, we convert the questions in D and B into feature vectors using the BERT model and obtain the feature sets X and Y. Within Y, we calculate a representative vector by averaging the feature vectors associated with the same reasoning type. These representative vectors are then reassembled into Y. Finally, we perform a cosine similarity calculation between representative vectors in Y and each feature vector in X:

$$similarity\ \ score = \frac{x_i \cdot y_j}{||x_i|| * ||y_j||}. \tag{2}$$

The pair exhibiting the highest similarity score is identified, and the reasoning type $type_{j*}$ of the corresponding representation vector is subsequently assigned to the question q_i as its question reasoning type.

> Given the questions and the words under the corresponding topics, please use as few phrases to indicate the semantic topics of the questions.
>
> Input: question: *What is the net change in net revenue in 2007 compare to 2006?*
>
>   topic words: *Value Total Cash Fair Millions Percentage December Flow Average Change*
>
> Output: *Ask about changes in values ; Inquire about total cash; Calculate millions of dollars*
>
> ....
>
> Input: question: { *the target question* }
>
>   topic words:{ *the topics words* }

Fig. 3. The prompt used for generating the semantic topic descriptions from ChatGPT. It consists of instructions, contextual examples, and input.

Question Semantic Topic. We derive the semantic topics for the questions in D using both the LDA model and ChatGPT. First, we preprocess the questions by removing numbers and stopwords to prepare the text for further analysis. Then we utilize the LDA model with the gensim API [17] to process the preprocessed questions. We evaluate perplexity and consistency to determine the optimal number of topics N_t. We first compute the perplexity when the topic number N_t varies from one to twenty:

$$p(w_i) = \sum_{j=1}^{J} p(d_j)p(w_i|d_j) = \sum_{j=1}^{J} p(d_j) \sum_{k=1}^{N_t} p(w_i|z_k)p(z_k|d_j), \qquad (3)$$

$$perplexity = e^{\frac{-\sum_{i=1}^{N} log(p(w_i))}{N}}, \qquad (4)$$

where N is the total number of words, $p(w_i)$ is the probability of word i appearing in the LDA model, $p(z_k|d_j)$ and $p(w_i|z_k)$ are topic and word distributions, respectively.

Then, we compute the consistency of each topic by computing the relatedness between topic words using Pointwise Mutual Information (PMI):

$$consistency_k = \frac{2}{N_w * (N_w - 1)} \sum_{i=2}^{N_w} \sum_{j=1}^{i-1} log \frac{p(w_i, w_j)}{p(w_i)p(w_j)}, \qquad (5)$$

where $\mathbf{N_w}$ is the total number of topic words in each topic, $p(w_i, w_j)$ refers to the probability that the word i and the word j appear in the sentence at the same time. Then we average the consistency scores for all topics:

$$consistency = \frac{\sum_{k=1}^{N_t} consistency_k}{N_t}. \qquad (6)$$

Low perplexity and high consistency suggest fewer uncertainties. By calculating the perplexity and consistency scores under different numbers of topics, we determine the topic number and obtain the topics for D.

For each question q_i, we determine its corresponding topic $topic_k$ and concatenate the topic words under $topic_k$ into a sequence. Then, the sequence and the question will be processed by the prompt in Fig. 3 and used as input to the large language model ChatGPT. While receiving text descriptions returned from ChatGPT using in-context learning, we perform denoising by removing stopwords and standardizing the format to create topic descriptions for the question, denoted as $R.q_i = \{r_{i1}, ..., r_{iN_m}\}$. We select the most relevant part of $R.q_i$ as the semantic topic $r*$ for q_i through the cosine similarity calculation. After obtaining the reasoning types and semantic topics of the questions, we construct the augmented experimental dataset, denoted as $\widetilde{D} = \{(q_i, type_{j*}, r*, t_i, e_i, a_i) | i = 1, 2, ..., n\}$.

3.3 Information Fusion Module

We propose an information fusion module consisting of a fusion encoder to fuse questions and additional information, a retriever to extract supporting facts from the hybrid data, and a program generator to create the reasoning program for answer calculation.

Fusion Encoder. We employ hierarchical fusion to merge questions with their additional information to obtain unified representations. First, we use a pre-trained language model to encode question q_i concatenated with its reasoning type and denote the output embeddings as h_i^p. Meanwhile, we calculate the attention vector att_p, capturing the relationship between the question and its reasoning type. A context vector c_p consolidates all the semantic information:

$$c_p = W_p[att_p; h_i^p]. \tag{7}$$

Then, we encode the semantic topic and denote the topic embedding as h_i^m which is applied to obtain the unified representation c_i:

$$c_i = c_p + \rho W_m h_i^m, \tag{8}$$

where ρ is the adjustment parameter used to balance original information and additional information. The unified representation c_i is used as input in the following fact retriever and program generator.

Retriever. To comply with the token number requirements, we retrieve the supporting facts from the financial report. The lower right part of Fig. 2 illustrates the retrieving process applied to both input tables and text. Tables are converted into row-based sentences. We use BERT to process each fact to obtain the embedding f_i. Then, we concatenate each embedding f_i with the unified representation c_i to the classifier. We use a pooling layer followed by a linear layer to obtain the supporting score s_i:

$$s_i = W_r classifier([f_i, c_i]). \tag{9}$$

We take the top n retrieved facts according to their supporting scores. These retrieved facts are then reordered to match their appearance in the financial report. Finally, the retrieved facts f_i^n serve as input to the program generator.

Program Generator. The program generator uses the unified representation c_i from the fusion encoder and the supporting facts f_i^n from the retriever to formulate the reasoning program required to answer the question. First, we concatenate the unified representation c_i and the supporting facts f_i^n to generate input embeddings H_i^e:

$$H_i^e = Encoder([c_i, f_i^n]). \quad (10)$$

Then, we denote the embeddings of special tokens s_i as H_i^s and the embeddings of step memory tokens m_i as H_i^m. The input H of the LSTM decoder is obtained as:

$$H = [H_i^e; H_i^s; H_i^m]. \quad (11)$$

As the decoder output h_t is generated, we utilize it to compute the attention vector att_H and att_h over the input and the decoding history, respectively. Subsequently, we apply the attention vector att_H to the input embeddings H:

$$H_t = W_h[H; H \cdot att_H]. \quad (12)$$

Finally, we derive the context vector and the prediction of the program operation at step t:

$$c_t = W_c[att_H; att_h; h_t], \quad (13)$$

$$p_t = softmax(H_t \cdot c_t). \quad (14)$$

At the end of each step, we update the embeddings of the step memory via the attention vectors and the decoder output. Following the program generation, we execute the program to obtain the predicted answer.

4 Experiments

To the best of our knowledge, two datasets require generating arithmetic expressions for numerical reasoning over tabular and textual data: FinQA [5] and MultiHiertt [23]. However, we did not use MultiHiertt due to its novelty, and it is yet to be further polished after contacting its authors. We use FinQA in our experiments. Initially, we conduct a statistical analysis of the augmented FinQA and evaluate its data quality via human evaluation. Then, we validate the importance of additional information in the FinQA dataset for the financial question-answering task. The overall results and ablation results demonstrate the strong performance of our model.

4.1 Baselines and Metrics

To evaluate the impact of including additional information, we choose baselines that also employ retriever-generator frameworks without any additional modules. We consider the following models:

- **Retriever + Seq2seq** [1]: The model adopts a Seq2seq architecture for the generator, which includes a bidirectional LSTM for encoding and an LSTM for decoding.
- **Retriever + NeRd** [4]: The Neural Symbolic Reader (NeRd) is a pointer-generator model that learns the program with a nested format.
- **FinQANet** [5]: FinQANet is a retriever-generator model. The retriever uses a BERT classifier to identify relevant facts, while the generator generates arithmetic expressions.
- **ELASTIC** [22]: ELASTIC comprises the RoBERTa as the encoder and a compiler that uses an adapted program solver to separate the generation of operators and operands. It demonstrates domain agnosticism by facilitating the integration of operators.
- **G-Eval** [11]: This model leverages large-scale generative language models to generate new contexts for the origin question.
- **DyRRen** [9]: DyRRen is a retriever-reranker-generator framework. A dynamic reranking of retrieved sentences enhances each generation step.
- **Human Expert and General Crowd** [5]: Human Expert assessments represent the expert performance, while General Crowd assessments represent the layman performance.

In line with prior research, we use Execution Accuracy and Program Accuracy to evaluate our model. Execution Accuracy measures the accuracy between the final answer and the golden answer. Program Accuracy measures the accuracy between the predicted program and the golden program.

4.2 Implementation Details

We conduct experiments on NVIDIA GeForce RTX 3090 GPUs and Ubuntu 18.04, using Torch 1.11.1. For the LDA model, we set the topic number N_t as 7 and the word number of each topic N_w as 10. For the large language model ChatGPT, we set the semantic topic number N_m as 5. For the fusion encoder, we use the BERT-base to add additional information hierarchically and set the adjustment parameter ρ as 0.2. For the retriever, we utilize the BERT-base as the classifier with 12 hidden layers and 768 hidden units. We use the Adam optimizer with a learning rate = 2e−5. Our training setup includes epoch = 50, seed = 1234, and batch size = 16. We select the top 3 supporting facts as the retriever results. For the program generator, we employ four popular large pre-trained models as encoders: BERT-base, BERT-large, RoBERTa-base, and RoBERTa-large. We set the maximum sequence length to 256 and used the Adam optimizer with a learning rate of 2e−5. Training is conducted over 30 epochs with seed = 1234. The batch size is set to 32, while BERT-large and RoBERTa-large are set to 16 due to GPU memory limitations.

Table 1. Comparison of statistics between augmented FinQA and the original FinQA.

Dataset	FinQA	Augmented FinQA
Examples (Q&A pairs)	8,281	8,281
Avg. # tokens in question	16.63	16.63
Avg. # tokens in question reasoning type	–	1.07
Avg. # tokens in question semantic topic	–	4.89
Avg. # tokens in additional information	–	5.96
Avg. # tokens in all inputs (text&table)	687.53	687.53

4.3 Dataset Analysis

The FinQA dataset is divided into three sets: train (6,251), validation (883), and test (1,147), following a 75%/10%/15% split. Table 1 shows the general statistics of the original FinQA and the augmented FinQA. Compared to the original FinQA, the augmented FinQA extends questions with reasoning types and semantic topics. The total number of tokens in the question reasoning type and question semantic topic accompanying each question averages 5.96. The additional information is not long compared to the total input tokens, although it provides contextual information for the questions, enabling a more accurate understanding of the question, and facilitating numerical reasoning and program generation.

To evaluate the quality of additional information, we randomly select 200 samples and solicit evaluations from three expert annotators and three layman annotators for both question reasoning types and question semantic topics. Specifically, we select the metric *Agreement*, which assesses the semantic match between the question and its corresponding additional information. The expert annotators achieve agreement rates of 86% and 90% for question reasoning types and question semantic topics, respectively, while the layman annotators achieve rates of 92% and 96%. The agreement rates exceeding 85% indicate the high quality of additional information.

4.4 Main Results

On the test set, as shown in Table 2, our model outperforms most baselines in Execution Accuracy and Program Accuracy. MSIF performs better than the Seq2seq and NeRd baselines, which do not use pre-trained models for program generation, indicating that large language models can better understand semantic and contextual information.

Compared with FinQANet, MSIF further improves performance. Specifically, on the best-performing RoBERTa-large model, the Execution Accuracy improves by 1.97 points, while the Program Accuracy improves by 2.47 points. On the remaining pre-trained models, improvements are also observed to varying degrees.

Table 2. The experiment results of Execution Accuracy and Program Accuracy in the test set. The original papers revise the results of the baselines. The best results are highlighted in bold, and the second-best results are underlined.

Baseline	Pre-trained Models	Execution Accuracy	Program Accuracy
Retriever+Seq2seq [1]	–	19.71	18.38
Retriever+Nerd [4]	BERT-base	48.57	46.76
FinQANet (EMNLP2021) [5]	BERT-base	50.00	48.00
	BERT-large	53.52	51.62
	RoBERTa-base	56.10	54.38
	RoBERTa-large	61.24	58.86
ELASTIC (NeurIPS2022) [22]	RoBERTa-large	62.16	57.54
G-Eval (NLP4ConvAI2023) [11]	BERT-base	49.76	48.47
	BERT-large	54.70	52.05
DyRRen (AAAI2023) [9]	RoBERTa-large	**63.30**	<u>61.29</u>
MSIF (Ours)	BERT-base	51.35	49.78
	BERT-large	54.93	52.83
	RoBERTa-base	58.50	56.67
	RoBERTa-large	<u>63.21</u>	**61.33**
Human Expert [5]	–	91.16	87.49
General Crowd [5]	–	50.68	48.17

Compared with G-Eval, which augments the questions with a large language model GPT4, our model performs better. It may be because the additional information of the question can provide more accurate information to the retriever, aid in fact localization, and facilitate the answer generation.

Compared with ELASTIC and DyRRen, our model demonstrates further improvements in performance, particularly in Program Accuracy. This enhancement could be attributed to the integration of the question reasoning type, which guides the process of the program generation. However, additional information reduces the model's emphasis on the semantic information of the original input, resulting in a slightly lower Execution Accuracy compared to the DyRRen, which rearranges the input at each generation step. Nevertheless, our model still achieves the second-best performance in Execution Accuracy and consistently outperforms the baselines in Program Accuracy.

Compared with human performance, while MSIF is not comparable with Human Expert, it exceeds General Crowd using different pre-trained models. This implies that our model can discover complex relationships in hybrid data and learn features that are difficult for people to identify.

4.5 Ablation Study

To evaluate the effectiveness of additional information, we compare MSIF with several variants:

Table 3. Study for evaluating the contribution of the question reasoning type, the question semantic topic, the fusion encoder, and the attention layer respectively.

Baseline	Pre-trained Models	Execution Accuracy	Program Accuracy
MSIF	BERT-base	51.35	49.78
	BERT-large	54.93	52.83
	RoBERTa-base	58.50	56.67
	RoBERTa-large	**63.21**	**61.33**
MSIF	BERT-base	50.61	49.15
	BERT-large	54.21	52.31
(w/o question reasoning type)	RoBERTa-base	57.38	55.62
	RoBERTa-large	61.92	60.25
MSIF	BERT-base	51.15	49.32
	BERT-large	54.84	52.51
(w/o question semantic topic)	RoBERTa-base	57.54	55.75
	RoBERTa-large	61.81	60.12
MSIF	BERT-base	49.52	48.11
	BERT-large	54.14	52.12
(w/o fusion encoder)	RoBERTa-base	57.04	55.23
	RoBERTa-large	61.05	59.80
MSIF	BERT-base	50.72	49.18
	BERT-large	54.62	52.36
(w/o attention layer)	RoBERTa-base	57.56	55.78
	RoBERTa-large	62.20	60.57

- MSIF(w/o question reasoning type), the MSIF model removes the question reasoning type.
- MSIF(w/o question semantic topic), the MSIF model removes the question semantic topic.
- MSIF(w/o the fusion encoder), the MSIF model removes the fusion encoder and uses summation to fuse additional information and the questions.
- MSIF(w/o the attention layer), the MSIF model removes the attention layer in the program generator.

The results of the ablation experiment are presented in Table 3. It can be seen that the performance of our MSIF model broadly decreases when a component is removed from each, indicating that each component contributes to facilitating the model to generate better programs and answers. Compared to the other three components, the fusion encoder plays a more important role in the financial question answering task. When it is removed, all metrics decrease the most, proving its role in fusing additional information with questions to help capture inner semantics.

Table 4. Case study sampled from FinQANet and MSIF. The keywords of the question are highlighted in red.

Question: What was the change in millions of operating income from 2016 to 2017?
Gold Program: subtract(11503, 10815) **Gold Result:** 688.0

FinQANet

Supporting Scores: T1: 4.23; T2: 4.17; T3: 3.31

Retrieved Sentences:

T1: the cost of sales for the year ended December 31, 2017 is $ 10432; the cost of sales for the year ended December 31, 2016 is $ 9391... (from table)

T2: we are continuing to evaluate the impact that the tax cuts and jobs act will have on our tax liability. (from text)

T3: we are regularly examined by tax authorities around the world, and we are currently under examination in several jurisdictions. (from text)

Predicted Program: add(10432, 9391)
Calculation: 10432 + 9391 = 19823.0 **Result:** 19823.0

MSIF

Additional Information: Arithmetic/Ask about changes in incomes
Supporting Scores: T1: 4.12; T2: 3.68; T3: 3.36

Retrieved Sentences:

T1: the operating income for the year ended December 31, 2017 is 11503; the operating income for the year ended December 31, 2016 is 10815... (from table)

T2: the cost of sales for the year ended December 31, 2017 is $ 10432; the cost of sales for the year ended December 31, 2016 is $ 9391... (from table)

T3: 2022 price increases $1.4 billion, partly offset by 2022 higher marketing, and research costs $570 million and 2022 unfavorable currency $ 157 million. (from text)

Predicted Program: subtract(11503, 10815)
Calculation: 11503 − 10815 = 688.0 **Result:** 688.0

4.6 Case Study

As shown in Table 4, we list a control group of samples generated by the Fin-QANet model and our MSIF model, respectively. It is clearly shown that the generation process of FinQANet faces some problems. (1) The retrieved sentences should be about "operating income" rather than "cost", which shows the weakness of FinQANet in question semantic analysis. (2) The question asks about "change", so the mathematical symbol should be "subtract" rather than "add", which shows that the judgment on the question reasoning type is not accurate enough. At the same time, our MSIF model successfully retrieves the sentences relevant to the question, and the answer is contained in the top sentence, which is about the annual operating income from 2015 to 2017. The relevant sentences are input to the program generator using the RoBERTa-large pre-trained model.

It can be seen that MSIF generates the correct operator "subtract" and successfully extracts the arguments "11503" and "10815", finally obtains the correct answer "688.0". The results above affirm the important role of additional information about the question in facilitating information localization and program generation.

5 Conclusion

In this paper, we propose a novel model named MSIF for financial question answering. The model obtains and fuses the additional information, encompassing question reasoning type and question semantic topic, from different sources. Specifically, we first obtain question reasoning type based on the existing corpus TAT-QA and then obtain question semantic topic from the large language model ChatGPT. Furthermore, we propose the information fusion module to amalgamate the additional information into the context representation. The results of extensive experiments in the FinQA dataset indicate that incorporating additional information enhances Execution Accuracy and Program Accuracy.

References

1. Amini, A., Gabriel, S., Lin, S., Koncel-Kedziorski, R., Choi, Y., Hajishirzi, H.: MathQA: towards interpretable math word problem solving with operation-based formalisms. In: Proceedings of the 2019 Conference of the North American Chapter of the Association for Computational Linguistics: Human Language Technologies, vol. 1 (Long and Short Papers), pp. 2357–2367. Association for Computational Linguistics, Minneapolis, Minnesota, June 2019. https://doi.org/10.18653/v1/N19-1245, https://aclanthology.org/N19-1245
2. Carmel, D., Lewin-Eytan, L., Maarek, Y.: Product question answering using customer generated content-research challenges. In: The 41st International ACM SIGIR Conference on Research & Development in Information Retrieval, pp. 1349–1350 (2018)
3. Chen, P., Zhong, J., Zhu, Y., et al.: Intelligent question answering system by deep convolutional neural network in finance and economics teaching. Comput. Intell. Neurosci. **2022**, 5755327 (2022)
4. Chen, X., Liang, C., Yu, A.W., Zhou, D., Song, D., Le, Q.V.: Neural symbolic reader: scalable integration of distributed and symbolic representations for reading comprehension. In: International Conference on Learning Representations (2019)
5. Chen, Z., et al.: FinQA: a dataset of numerical reasoning over financial data. In: Proceedings of the 2021 Conference on Empirical Methods in Natural Language Processing, pp. 3697–3711 (2021)
6. Chen, Z., Li, S., Smiley, C., Ma, Z., Shah, S., Wang, W.Y.: ConvFinQA: exploring the chain of numerical reasoning in conversational finance question answering. In: Proceedings of the 2022 Conference on Empirical Methods in Natural Language Processing, pp. 6279–6292 (2022)
7. Deng, Y., Lei, W., Zhang, W., Lam, W., Chua, T.S.: Pacific: towards proactive conversational question answering over tabular and textual data in finance. In: Proceedings of the 2022 Conference on Empirical Methods in Natural Language Processing, pp. 6970–6984 (2022)

8. Jelodar, H., et al.: Latent Dirichlet Allocation (LDA) and topic modeling: models, applications, a survey. Multimedia Tools Appl. **78**, 15169–15211 (2019)
9. Li, X., Zhu, Y., Liu, S., Ju, J., Qu, Y., Cheng, G.: DyRRen: a dynamic retriever-reranker-generator model for numerical reasoning over tabular and textual data. In: Proceedings of the AAAI Conference on Artificial Intelligence, vol. 37, pp. 13139–13147 (2023)
10. Liang, C., Berant, J., Le, Q., Forbus, K.D., Lao, N.: Neural symbolic machines: Learning semantic parsers on freebase with weak supervision. arXiv preprint arXiv:1611.00020 (2016)
11. Lin, Y.T., Chen, Y.N.: LLM-Eval: unified multi-dimensional automatic evaluation for open-domain conversations with large language models. In: Chen, Y.N., Rastogi, A. (eds.) Proceedings of the 5th Workshop on NLP for Conversational AI (NLP4ConvAI 2023), pp. 47–58. Association for Computational Linguistics, Toronto, Canada, July 2023. https://doi.org/10.18653/v1/2023.nlp4convai-1.5, https://aclanthology.org/2023.nlp4convai-1.5
12. Liu, C., Ventre, C., Polukarov, M.: Synthetic data augmentation for deep reinforcement learning in financial trading. In: Proceedings of the Third ACM International Conference on AI in Finance, pp. 343–351 (2022)
13. Maybury, M.: Toward a question answering roadmap, pp. 8–11 (2003)
14. Pi, X., et al.: Reasoning like program executors. arXiv preprint arXiv:2201.11473 (2022)
15. Poria, S., Chaturvedi, I., Cambria, E., Bisio, F.: Sentic LDA: improving on LDA with semantic similarity for aspect-based sentiment analysis. In: 2016 International Joint Conference on Neural Networks (IJCNN), pp. 4465–4473 (2016). https://doi.org/10.1109/IJCNN.2016.7727784
16. Pundge, A.M., Khillare, S., Mahender, C.N.: Question answering system, approaches and techniques: a review. Int. J. Comput. Appl. **141**(3), 0975–8887 (2016)
17. Řehůřek, R., Sojka, P., et al.: Gensim-statistical semantics in Python. Retrieved from genism.org (2011)
18. Suadaa, L.H., Kamigaito, H., Funakoshi, K., Okumura, M., Takamura, H.: Towards table-to-text generation with numerical reasoning. In: Proceedings of the 59th Annual Meeting of the Association for Computational Linguistics and the 11th International Joint Conference on Natural Language Processing (Volume 1: Long Papers), pp. 1451–1465. Association for Computational Linguistics, Online, August 2021. https://doi.org/10.18653/v1/2021.acl-long.115, https://aclanthology.org/2021.acl-long.115
19. Tanwar, A., Zhang, J., Ive, J., Gupta, V., Guo, Y.: Unsupervised numerical reasoning to extract phenotypes from clinical text by leveraging external knowledge. In: Shaban-Nejad, A., Michalowski, M., Bianco, S. (eds.) Multimodal AI in Healthcare. Studies in Computational Intelligence, vol. 1060, pp. 11–28. Springer, Cham (2023). https://doi.org/10.1007/978-3-031-14771-5_2
20. Wang, X., et al.: Self-consistency improves chain of thought reasoning in language models. arXiv preprint arXiv:2203.11171 (2022)
21. Yang, Z., et al.: MM-React: prompting ChatGPT for multimodal reasoning and action. arXiv preprint arXiv:2303.11381 (2023)
22. Zhang, J., Moshfeghi, Y.: Elastic: numerical reasoning with adaptive symbolic compiler. In: Koyejo, S., Mohamed, S., Agarwal, A., Belgrave, D., Cho, K., Oh, A. (eds.) Advances in Neural Information Processing Systems, vol. 35, pp. 12647–12661. Curran Associates, Inc. (2022). https://proceedings.neurips.cc/paper_files/paper/2022/file/522ef98b1e52f5918e5abc868651175d-Paper-Conference.pdf

23. Zhao, Y., Li, Y., Li, C., Zhang, R.: MultiHiertt: numerical reasoning over multi hierarchical tabular and textual data. arXiv preprint arXiv:2206.01347 (2022)
24. Zhu, F., et al.: TAT-QA: a question answering benchmark on a hybrid of tabular and textual content in finance. In: Proceedings of the 59th Annual Meeting of the Association for Computational Linguistics and the 11th International Joint Conference on Natural Language Processing (Volume 1: Long Papers), pp. 3277–3287 (2021)

Artificial Intelligence in Education

A Temporal-Enhanced Model for Knowledge Tracing

Shaoguo Cui, Mingyang Wang(✉), and Song Xu

College of Computer and Information Science, Chongqing Normal University, Chongqing 401331, China
csg@cqnu.edu.cn, wangmingyang0221@163.com

Abstract. Knowledge Tracing (KT) aims to predict students' future practice performance through their historical interaction with Intelligent Tutoring Systems (ITS). This method plays an important role in computer-assisted education and adaptive learning research. In the learner's learning process, as the learning time increases, the time distance between the learner's historical records continues to increase, resulting in a long-term dependency problem when capturing the correlation of knowledge concepts in exercises. In addition, the learner's learning status is affected by time. How to accurately judge the learner's knowledge status at different time levels is also a challenge. To tackle the above problems, we propose A Temporal-Enhanced Model for Knowledge Tracing (TEKT). On the one hand, the problem of long-term dependence is solved by using the Probabilistic Sparse Attention mechanism; On the other hand, the Fine-grained Temporal Features are embedded to capture learner's knowledge status at different time granularities. The method proposed in this paper has been fully experimented and verified on three datasets. The experimental results show that the proposed method demonstrates an improvement in Accuracy (ACC) and Area Under The Receiver Operating Characteristic Curve (AUC) evaluation metrics, which compared to the existing KT methods. Thus, it proves the effectiveness of the proposed method.

Keywords: Knowledge Tracing · Probabilistic Sparse Attention Mechanism · Fine-grained Temporal Feature

1 Introduction

With the rapid development of the internet, the shortcomings of traditional teaching have become increasingly prominent, such as the lack of personalization, adaptability, and flexibility in teaching methods. Knowledge Tracing is a well-known and effective method for estimating student proficiency [6]. It uses students' historical behavioral data and combines them with knowledge concepts in exercises to trace students' current knowledge status. This approach not only improves students' learning efficiency, but plays an important role in students' adaptive learning and personalised recommendation of exercises. After decades of development, KT has been widely used in various fields. Researchers have proposed a number of KT models and demonstrated their effectiveness under certain conditions [2]. These models analysis students' behavioral data like the accuracy of question responses and the frequency of problem-solving attempts, so that to deduce a students' knowledge state and progress in learning. This provides educators with valuable information, allowing them to provide customized instruction based on students' individual characteristics and learning needs.

In early KT studies, the traditional method was used to model students' knowledge proficiency. For example, Corbett et al. proposed Bayesian Knowledge Tracing (BKT) was the first time elicited the concept of knowledge tracing [6]. BKT models the learning process as a Markov chain parameterized by guessing, sliding, acquisition and initial learning. It uses hidden variables in a Hidden Markov Model (HMM) [18], which derives the probability that students will answer questions correctly in the future. However, one assumption of the BKT model is that once students master certain knowledge concepts, they will never forget it, which is not true in reality. For traditional knowledge tracing in addition to the BKT model there is the Georg Rasch et al. proposed Item Response Theory (IRT) [11] model, Albert Corbett et al. proposed Dynamic Bayesian Knowledge Tracing (DBKT) [10] model, John R. Anderson et al. proposed Additive Factor Model (AFM) [3]. Although traditional KT models can be based on learners' answer patterns and performance, conduct personalized diagnosis and assessment of their knowledge status, ability level and learning progress, they still struggle to process and utilize large-scale data. What's more, certain modeling assumptions don't align perfectly with real-world situations, and it is difficult to cover the complexity of the learning process. Hence, there is a need for further enhancement and optimization of these models.

With the rapid development of deep learning, inspired by deep learning, these Deep Learning-based Knowledge Tracing (DLKT) models [7] have demonstrated good effects and generalization capabilities. Chris Piech et al. proposed Deep Knowledge Tracing (DKT) which is the first model to utilize deep learning methods in addressing the KT problem [17]. DKT uses Recurrent Neural Network (RNN) to model students' proficiency in knowledge concepts, and it achieves better results compared to traditional KT methods. Moreover, within deep neural networks, some KT methods based on the Transformer architecture [22] exhibit superior performance. Alberto Garcia-Mendoza et al. proposed Self-Attentive Knowledge Tracing (SAKT) [14] is the first model to incorporate the self-attention mechanism in knowledge tracing, which can automatically capture the attention weights of students towards each knowledge concepts, enabling accurate prediction of future performance on exercises. Compared to the DKT model, the SAKT model performs better.

However, with the increase of student interaction data and the extension of recording time, existing KT methods encounter difficulties in capturing students' forgetting behaviors. Besides, due to the complexity of the learning process, it is difficult to accurately predict students' mastery of knowledge concepts solely based on simple time features. To solve the above problems, this paper proposes an enhanced knowledge tracing method based on probabilistic sparse attention mechanism and fused with fine-grained temporal features. Initially, a method utilizing fine-grained temporal features is employed, aiming to overcome the limitations associated with using simple time features to model changes in students' knowledge status on the temporal scale. Subsequently, in response to the issue of capturing the forgetting of students' knowledge related to distant temporal intervals, this study introduces a multi-head probabilistic sparse attention mechanism. This advanced method adeptly captures students' forgetful behavior while effectively addressing long-term dependency. Through the above methods, the learner's knowledge status and cognitive development over a long time sequence can be further captured.

The main contributions of this paper are as follows:

- A probabilistic sparse attention mechanism is proposed, which can effectively deal with the problem of capturing the correlation of knowledge concepts over long temporal distances.
- A fine-grained temporal feature is proposed to enable the model to better capture timing information at different granularities, effectively solving the problem which simple time features are difficult to obtain more timing information.
- This paper conducts extensive experiments on three real datasets, demonstrating the superior performance of the proposed model compared to related knowledge tracing methods.

2 Related Work

Knowledge tracing, as a fundamental area of study in the field of Artificial Intelligence (AI) education, has garnered significant interests and scrutiny due to its role in enabling adaptive learning and personalized exercise recommendations in smart education [4]. Existing models for KT can be broadly classified into traditional probabilistic models and deep learning-based methods [20].

Classical traditional methods of KT are usually divided into two categories: Bayesian Knowledge Tracing (BKT) and Factor Analysis Models (FAM) [25]. The BKT model generally uses the probabilistic graphical model, which is based on the HMM to trace the changing knowledge status of students in the process of addressing problems [16]. The FAM method, based on the principles of Item Response Theory (IRT), enables the inference of students' latent traits or abilities from their assessment performances, thereby modeling the diversity and correlations of student abilities. Traditional KT methods offer a solid foundation for preliminary analyses and predictions of changes in students' knowledge status. However, they are usually based on predefined model structures and assumptions that require a large number of model parameters, which limits the expressive power of the models.

With the continuous advancement of deep learning, knowledge tracing methods based on deep learning have emerged. DKT is the first KT method that combines knowledge tracing with deep learning. It achieves this integration by inputting the student's learning sequence[1] into an RNN or LSTM networks [9]. Based on DKT, a variety of KT models have been proposed. For example, Cheng et al. proposed Dynamic Key-Value Memory Network (DKVMN) [26], which stores student knowledge status through the key memory matrix, and the value memory matrix contains the concept of exercises to enhance DKT. Abdelrahman G. et al. proposed Sequential Key-Value Memory Network (SKVMN) [1], which addresses the issue of the knowledge concepts in previous answered questions not necessarily being relevant to the current questions. Furthermore, with the introduction of the Transformer architecture, scholars have begun to explore the application of attention mechanisms in the realm of KT tasks, constituting a domain of great significance and ripe for further exploration. For instance, Jiani Zhang et al. proposed Attentive Knowledge Tracing (AKT) [8], which employs a monotonic attention mechanism and exponential weight decay. Choi Y et al. proposed Separated Self-Attentive Neural Knowledge Tracing (SAINT) [5], which is the first model to employ the complete Transformer architecture and utilize an encoder-decoder structure. For Pandey S et al. proposed Relation-Aware Self-Attention for Knowledge Tracing (RKT) [15], which utilizes the textual information of questions to represent them and estimate the relationships between questions in past interaction sequences. However, due to the existence of various relationship structures in interactive data, some KT models have begun to try to use graph representation learning technology to solve related problems. Likely, Yang Y et al. proposed Graph-based Interaction Knowledge Tracing (GIKT) [24], it assumes that a single knowledge concept may be related to multiple exercises, and an exercise may correspond to multiple knowledge concept. Therefore, expressing the relationship between exercises and knowledge concepts in the form of a graph for embedding of answer prediction, which can further improve the performance and effect of the KT task. Zhao S et al. proposed Transition-Aware Multi-Activity Knowledge Tracing (TAMAKT) [27], which actively learns knowledge transfer by activating and learning a set of knowledge transition matrices.

Although the above methods have achieved good results to some extent, it still faces challenges in dealing with capturing relevant knowledge concepts over long temporal distances. Additionally, the evolution of students' knowledge states over time poses a problem as simple time features cannot capture more detailed temporal information during the learning process. To address the aforementioned problems, we propose the model of TEKT. This method integrates fine-grained temporal features and probabilistic sparse attention to tackle the above two problems separately.

3 Problem Formulation

In this section, the paper formalizes the learning process of students and introduces the definition of knowledge tracing.

In the scenario where students respond to a series of questions provided by an online learning system. Assuming there is a group of students $S =$

---
[1] https://sites.google.com/site/assistmentsdata/.

$\{s_1, s_2, s_3, \ldots, s_u\}$, a corresponding set of exercises $E = \{e_1, e_2, e_3, \ldots, e_m\}$, $R = \{r_1, r_2, r_3, \ldots, r_h\}$ indicates whether the answer to the exercise is correct or not, and the knowledge concepts involved in the question $K = \{k_1, k_2, k_3, \ldots, k_n\}$. Each exercise is associated with its own corresponding concept, so the relationship between exercises and concepts is represented by a matrix $Q \in m \times n$ composed of 0 and 1 [23]. If $Q_{m,n} = 1$, it indicates that the exercise is related to the corresponding knowledge concept; Otherwise, $Q_{m,n} = 0$. For fine-grained time, it is assumed to be $T = \{t_1, t_2, t_3, \ldots, t_q\}$, When a group of students' learning sequence $X = \{(e_1, r_1, t_1), (e_2, r_2, t_2), (e_3, r_3, t_3), \ldots, (e_t, r_t, t_t)\}$, Where X represents a triplet $\{E, R, T\}$.

Based on the above assumptions, we can understand that knowledge tracing aims to input historical interaction information X that integrates multiple features into the model, monitor students' changing knowledge status during the learning process, and predict their performance at $t+1$. This can be further used to develop personalized learning strategies and improve learning efficiency.

4 Model

In this section, the TEKT model structure is described in detail, as is shown in Fig. 1. It consists of four parts: a timing processing layer, an input layer, an encoding layer and a decoding layer. Firstly, the sequences of exercises and knowledge i.e. (E, K) as well as the position information are input into the encoder. In the encoder module, knowledge concepts that are highly relevant to the current exercise in the whole sequence are searched by multi-head probabilistic sparse

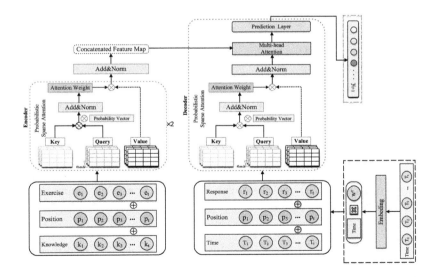

Fig. 1. The architecture of the TEKT model

self-attention. Then, it is output to the multi-head self-attention in the decoder through the feature mapping layer. In the decoder, temporal features are processed at a fine-grained level prior to input. Then, the exercise responses R and position information are fused with the encoded temporal features and input to the decoder. The fusion of different features can provide a more comprehensive description of the student's learning behavior and knowledge state. Whereas for multi-level temporal information can be obtained by fine-grained time granularity. Subsequently, the output of the encoder and the input of the first layer of the decoder are jointly input to the multi-head attention. Finally, predictions are generated through the prediction layer. This method has a positive impact on improving the performance and effectiveness of KT models and providing more accurate learner modeling and instructional intervention strategies for personalised education.

4.1 Fine-Grained Temporal Features

In this paper, the simple temporal features are processed at a fine-grained granularity before being input to the model. Unlike many KT models which simply use the time interval between two exercise units or the time spent on an exercise as inputs. The paper processes the temporal features into five granularities: year, month, day, hour, minute, and discretizes them into a certain range, as shown in Fig. 2.

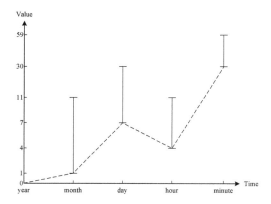

Fig. 2. The graph represents discrete time ranges. In the case of hours, which can be discretised into intervals such as 1–11 h. Using 4 h as a lower bound, knowledge forgetting increases when the time exceeds 4 h.

Besides the method of processing fine-grained temporal features can more accurately capture the temporal information in the learning process and provide a more refined temporal representation for the model. At the same time, the processing of fine-grained temporal features can better reflect the changes in students' learning behaviors at different time periods. After fine-graining the timing features, the feature diversity can be enriched as well as more timing information obtained. Meanwhile, the impact of changes at different time levels on students' knowledge proficiency can also be captured through fine-grained temporal features. As shown in Fig. 3.

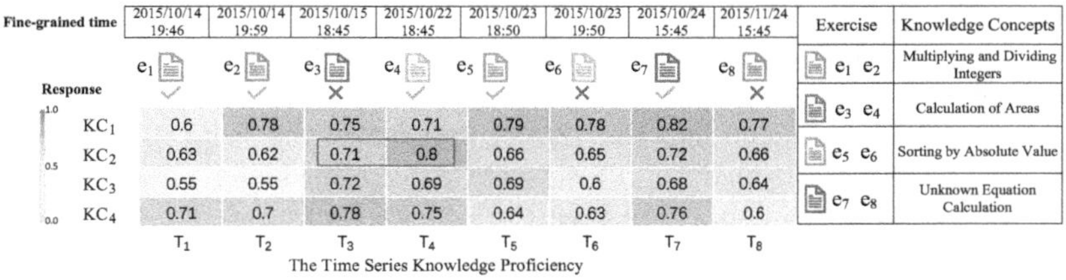

Fine-grained time	2015/10/14 19:46	2015/10/14 19:59	2015/10/15 18:45	2015/10/22 18:45	2015/10/23 18:50	2015/10/23 19:50	2015/10/24 15:45	2015/11/24 15:45	Exercise	Knowledge Concepts
	e_1	e_2	e_3	e_4	e_5	e_6	e_7	e_8	e_1 e_2	Multiplying and Dividing Integers
Response	✓	✓	✗	✓	✓	✗	✓	✗	e_3 e_4	Calculation of Areas
KC_1	0.6	0.78	0.75	0.71	0.79	0.78	0.82	0.77		
KC_2	0.63	0.62	0.71	0.8	0.66	0.65	0.72	0.66	e_5 e_6	Sorting by Absolute Value
KC_3	0.55	0.55	0.72	0.69	0.69	0.6	0.68	0.64	e_7 e_8	Unknown Equation Calculation
KC_4	0.71	0.7	0.78	0.75	0.64	0.63	0.76	0.6		
	T_1	T_2	T_3	T_4	T_5	T_6	T_7	T_8		

The Time Series Knowledge Proficiency

Fig. 3. The evolution process of students' knowledge state is traced by TEKT. The top part represents his/her fine-grained time at each time step, and the bottom indicates the student's performance. The right side is the knowledge concept involved in each topic.

Using the red box as a reference, let's explain and illustrate the graph. In KC_2, exercises e_3 and e_4 were involved. When students answered e_3 incorrectly, their proficiency in the corresponding knowledge decreased. However, since there was a one-week interval between e_3 and e_4, the increase in proficiency was slightly lower in KC_2 compared to KC_1 (with a thirteen-minute interval). It is worth noting that proficiency varies with time granularity. In e_7 and e_8, which represent monthly temporal granularity, we can observe a faster decay of knowledge compared to the other exercises.

Specifically, by mapping each granularity of the time feature to an integer value within a certain range, each granularity of the temporal feature can be processed using a global temporal encoding. For each granularity assume T_i, the time at each granularity is spliced into a set of temporal feature matrices, each row represents a discrete feature encoding of a sample, then the feature matrix $T_{\{u,i\}}$ of a single sample is:

$$T = [T_{\{u,1\}}, T_{\{u,2\}}, T_{\{u,3\}}, \ldots, T_{\{u,i\}}] \tag{1}$$

$$\tilde{T} = T * W^{d_k} \tag{2}$$

The expression $\tilde{T} \in \mathbb{R}^{d_u \times k}$ represents a multi-granularity temporal feature vector, where T represents the temporal feature matrix. The $T_{\{u,i\}}$ represents the integer encoding of the temporal features at the i-th granularity for the u-th sample, where i belongs to the set of integers $(1, 5)$. Moreover, the encoded feature matrix is projected onto a low-dimensional vector space using the embedding matrix $W \in \mathbb{R}^{d_k}$, where each granular feature is represented as a dense vector. This enables the extraction of more temporal information.

4.2 Multi-feature Fusion Input Layer

In the input module, the fine-grained temporal feature matrix undergoes positional encoding and global time encoding to effectively capture the temporal data sequence. The encoded temporal features represent the changes which occur in time at different levels of granularity. In the encoder input block, K_t and E_t are used as inputs. In the decoder input block, (R_t, T_t) is used as input, where T_t represents the encoded fine-grained temporal features. By combining the fine-grained temporal features with the responses, a better understanding of the student's learning state and knowledge evolution can be obtained. By integrating E_t and K_t in the encoder, the model can learn the relationship between excises and knowledge concepts. This integration allows the model to capture the dynamics and dependencies between practice and knowledge, enhancing its ability to simulate students' learning state and knowledge evolution.

4.3 Probabilistic Sparse Attention Layer

The traditional attention mechanism takes three input matrices: Query, Key, and Value [22]. These input matrices are multiplied by three different weight vectors, W_i^Q, W_i^K, and W_i^V, respectively, using the dot product formula. The dot product formula is as follows:

$$head_i = Softmax\left(Mask\left(\frac{Q_i K_i^T}{\sqrt{d}}\right)\right) V_i \qquad (3)$$

where d is the dimension of the input vector matrix. Then, the scaled dot-product attention is used to calculate each head matrix, and the final output is obtained by concatenating the linear transformations of each head matrix to form the result of the multi-head attention.

$$Multihead = Concat\left(head_i, head_2, \ldots, head_h\right) W_o \qquad (4)$$

where W_o is the weight matrix of linear transformation, and $Concat()$ is the splicing function.

From the visualization results of the scaling dot product attention, it can be analyzed that it obeys the probability of long tail distribution [28]. Concretely, from Eq. 5, the attention of the ith query to all keys is defined as the probability $P(k_j|q_i)$, the output is a combination of values V. Specifically, it is the process of calculating the attention distribution of each query and all keys according to the degree of association between the query and the key, and then combining the values. It can be found that the dominant dot product pair q, its attention probability distribution is far from uniform distribution.

$$\Gamma(q_i, K, V) = \sum_j \frac{k(q_i, k_j)}{\sum_l k(q_i, k_j)} v_j \qquad (5)$$

The function $P(k_j|q_i)$ measures the correlation between the query q_i and the key k_j. This function can be computed using an asymmetric exponential kernel, where the kernel is defined as $exp\left(\frac{(q_i^T*k_j)}{\sqrt{d}}\right)$, with q_i^T representing the transpose of the query q_i, and d denoting the dimension of the features.

Furthermore, the values v_j are combined together using self-attention mechanism and weighted by the calculated probabilities $P(k_j|q_i)$. The weighted values are then summed up to generate the output, making that the keys k_j with higher relevance to the query q_i are assigned greater weights in the output. Thus, sparse probability attention is designed to find crucial q and capture knowledge concepts that exhibit high relevance to the current exercise, and the Kullback-Leibler [21] divergence is employed as a measure of the relevance of historical knowledge concepts to acquire the queries. The computation formula is provided below:

$$\mathcal{M}(Q_i, K) = \ln \sum_{j=1}^{L_k} e^{\frac{Q_i k_j^T}{\sqrt{d}}} - \frac{1}{L_k} \sum_{j=1}^{L_k} \frac{Q_i k_j^T}{\sqrt{d}} \tag{6}$$

The first term is the logarithm and expression of q_i on all keys, and the second term is their arithmetic mean. Moreover, in the multi-head sparse probability attention mechanism, unique sparse query-key pairs are generated for each head, addressing the issue of significant information loss. Specifically, self-attention is achieved by assigning each key to handle only u subset of explicit queries. Furthermore, the probability sparse attention mechanism selects the keys that exhibit the highest relevance to the queries, thereby focusing on the essential information that is pertinent to the current learning objective. This is illustrated in Eq. 7:

$$MultiheadSP_i(Q_i, K, V) = Softmax\left(\frac{Q_i K^T}{\sqrt{d}}\right) V \tag{7}$$

where $MultiheadSP$ represents multi-head probabilistic sparse attention, Q is a sparse matrix of the same size as q, which contains only $Top-u$ queries under the sparse metric $M(q, K)$.

4.4 Encoder Layer

In the encoder block, we introduce positional encoding to the encoded exercises and knowledge features, enabling the model to capture positional information within the input sequence. This helps the model understand the relevance of different positions in the sequence. After that, the feature vectors are concatenated with the positional encoding, and then inputted into two layers of multi-head sparse probabilistic attention, a feed-forward neural network, and a normalization layer in the encoding stage. Finally, the encoded feature representation is extracted through the feature mapping layer to obtain more expressive and discriminative feature representations. Then, these feature representations are outputted to the decoder block. The specific process is described by the following equation:

$$X_e = K \oplus E \oplus P \tag{8}$$

$$FN = W_2\left(ReLU\left(W_1 X + b_1\right)\right) + b_2 \tag{9}$$

$$E = FN\left(LN\left(\text{MultiheadSP}(X_e)\right)\right) \quad (10)$$

$$E_{out} = FM(E) \quad (11)$$

where $\oplus$ means the concatenation operation, X_e denotes the fused features, FN represents the feed-forward neural network, LN denotes the layer normalization, and FM represents the feature mapping layer.

4.5 Decoder Layer

In the decoder block, this study takes fine-grained temporal features and exercise responses as inputs, while also incorporating positional encoding. After features concatenation, the input goes through multi-head sparse probabilistic attention and layer normalization. Then, the output from the encoding side and the first-layer output from the decoding side are jointly used as inputs for multi-head self-attention. Finally, the prediction layer is used to output the prediction results.

$$D_e = R \oplus T \oplus P \quad (12)$$

$$D = LN\left(MMultiheadSP\left(D_e\right)\right) \quad (13)$$

$$D_1 = LN\left(MMultihead\left(D, E_{out}\right)\right) \quad (14)$$

$$PL = \sigma\left(WX + b\right) \quad (15)$$

$$y = PL\left(D_1\right) \quad (16)$$

where D_e denotes the fused features, $MMultiheadSP$ represents multi-head sparse probability attention, D refers to the output of the first layer, $MMultihead$ denotes multi-head self-attention, D_1 represents the output of the multi-head self-attention, PL represents the prediction layer, and y represents the output of the decoder.

5 Experiment

5.1 Datesets

In this section, we conduct extensive experiments to evaluate the performance of our model on three public online education datasets. The following is a detailed description of the three datasets:

- **ASSITments2012-2013:** The data was collected from students using the ASSISTments online learning system from September 2012 to September 2013. The students participated in various learning activities, such as classroom exercises and homework assignments. The data consists of 5,801,073 instances and 64 features, summarizing the learning behavior of students when answering individual questions on the ASSISTments platform.

- **ASSITments2015:** This dataset is collected based on the ASSISTments educational platform in 2015. Unlike the dataset from 2009, this dataset utilizes records in an interactive mode and includes all of these records for analysis. During the preprocessing process, records with correct values other than 1 or 0 were removed. The final dataset consists of 19,840 students, 100 exercise labels, and 683,801 interaction records.
- **ASSITments2017:** This dataset is derived from the ASSISTments 2017 Data Mining Competition and comprises a series of data related to student learning. The datasets covers the learning records of 1,709 students, including 942,816 question-answer interactions on the platform. In addition, the dataset provides 102 exercise labels that describe the different types of exercises in which students participated.

5.2 Baseline Model

In this section, we compare our proposed model with five KT models: DKT, IEKT, SAINT+, SAKT, and AT-DKT. To ensure fairness, all models are implemented using PyTorch. Here is a brief description of the five KT models:

- **DKT** [17]: DKT is a simple model based on RNN which is used to model the knowledge state of students. The model utilizes exercise and response as input features and represents the student's knowledge state as an RNN hidden state vector.
- **SAKT** [14]: SAKT is the first KT models that utilizes the Transformer self-attention architecture in the decoder to perform sequence tasks.
- **IEKT** [13]: Tracing Knowledge State with Individual Cognition and Acquisition Estimation (IEKT) is a model that estimates students' cognitive states for questions before making answer predictions. It also evaluates their knowledge acquisition sensitivity for questions before updating the knowledge state.
- **SAINT+** [19]: SAINT + is a knowledge tracing model based on Transformer. Two time features are introduced on the basis of SAINT: time passing (time for students to answer questions) and lag time (time interval between adjacent learning activities).
- **AT-DKT** [12]: Enhancing Deep Knowledge Tracing with Auxiliary Tasks (AT-DKT), this model uses two auxiliary learning tasks, namely Question Tag (QT) prediction task and Personalized Prior Knowledge (PK) prediction task.

5.3 Model Evaluation

To evaluate the performance of the model, this study divided the dataset into training, validation, and test sets in a ratio of 7 : 1 : 2. For the hyperparameter settings in the experiments, which used the Adam optimizer to train the model for a maximum of 200 epochs. Early stopping was employed when the performance did not improve after 10 epochs. Moreover, the loss function utilized for calculating the loss was Binary Cross Entropy With Logits Loss (BCEWithLogitsLoss). The learning rate was set to 0.001, the batch size was 64, and the model consisted of 4 layers with 8 attention heads. In this experiment, ACC and AUC were used as evaluation indicators. To be more specific, the ACC metric measures the proportion of correctly classified samples in the model's predictions. It ranges from 0 to 1, with a value closer to 1 indicating a higher level of accuracy. AUC is a metric which measures the quality of the model's classification results at different thresholds. The AUC value also ranges between 0 and 1, with a value closer to 1 indicating better model performance.

5.4 Main Result

In this section, we primarily present the experimental evaluation of our model on three datasets. The performance comparison between our proposed TEKT model and other baseline KT models is shown in Table 1. It can be seen from the experimental results that compared with the baseline model, the ACC and AUC of TEKT are better than the baseline model on all datasets, thus the effectiveness of this model is improved.

Table 1. Comparison of ACC and AUC between TEKT and the baseline model.

Model	ASSIT2012-2013		ASSIT2015		ASSIT2017	
	ACC	AUC	ACC	AUC	ACC	AUC
DKT	0.6979	0.7128	0.7553	0.7311	0.6876	0.7187
SAKT	0.7611	0.7623	0.7718	0.7415	0.7021	0.7042
SAINT+	0.7693	0.7703	0.7637	0.7899	0.7692	0.7901
IEKT	0.7787	0.7863	0.7714	0.7921	0.7899	0.7818
AT-DKT	0.7586	0.7687	0.7661	0.8013	0.7678	0.7934
TEKT	**0.7814**	**0.7865**	**0.7718**	**0.8017**	**0.7912**	**0.8183**

5.5 Ablation Experiments

In this section, ablation experiments were conducted to further demonstrate the impact of fine-grained temporal features and probability sparse attention on the final results of the model. To achieve this, the paper divided the complete model into two variants, where each variant corresponds to removing one module from the complete model. Specifically, TEKT refers to the complete model, TEKT/Q refers to the variant without Probability Sparse Attention, without considering the forgetting effect. TEKT/MT refers to the variant where Fine-grained Temporal Features are removed, and only simple timestamps are used. Table 2 presents the comparative performance of each variant with the complete model, highlighting important findings from the results.

Firstly, forgetting is a common phenomenon in the learning process, and it plays a crucial role throughout the learning process. Neglecting the student forgetting scenario would lead to a significant decrease in the prediction performance. Furthermore, temporal features play an indispensable and essential role in the learning process. From the experimental results, it is evident that finer-grained temporal features help us understand the variations in knowledge acquisition across different time granularities during the learning process. On the other hand, replacing simple timestamps with finer-grained temporal features can enhance the model's performance. Additionally, it was observed in this study that capturing the relevance of knowledge concepts that are distant from the current practice becomes challenging for long temporal sequences. Many KT models tend to disregard sequence information that is far from the current item. From the perspective of learners' cognitive processes, handling long sequences involves assigning higher weights to closely related knowledge concept and lower weights to distantly related ones. This approach better aligns with the consistency of learners' learning processes and serves as evidence for the rationality and effectiveness of the experimental findings in this paper.

Table 2. Ablation experimental results on three datasets

Model	ASSIT2012		ASSIT2015		ASSIT2017	
	ACC	AUC	ACC	AUC	ACC	AUC
TEKT	0.7814	0.7865	0.7718	0.8017	0.7912	0.8183
TEKT/**Q**	0.7790	0.7818	0.7665	0.7957	0.7815	0.8079
TEKT/**MT**	0.7802	0.7798	0.7688	0.7995	0.7896	0.8120

6 Conclusion and Future Work

In this paper, fine-grained embedding of temporal features with feature fusion is proposed. By this method, information about learners' knowledge state and practice sequence at different time levels can be obtained, which is consistent with the consistency of learners' learning process. In addition, multi-head probabilistic sparse attention is utilised to capture the long-term dependency and the characteristic of student forgetting. Finally, the effectiveness of the proposed model is verified on three public education datasets.

In future studies, it is crucial to take into account not only learners' behaviors on exercises but also their contextual features, such as extracurricular tutoring and supplementary video materials etc. Furthermore, considering the varying levels of knowledge relevance, leveraging knowledge graphs and graph structures can be utilized to establish relationships between knowledge concepts.

Acknowledgements. Supported by Chongqing Social Science Planning Project (2022NDYB119), the Humanities and Social Sciences Project of Chongqing Municipal Education Commission(23SKGH072), Chongqing Normal University Graduate Research Innovation Project (YZH23006).

References

1. Abdelrahman, G., Wang, Q.: Knowledge tracing with sequential key-value memory networks. In: Proceedings of the 42nd International ACM SIGIR Conference on Research and Development in Information Retrieval, pp. 175–184 (2019)
2. Abdelrahman, G., Wang, Q., Nunes, B.: Knowledge tracing: a survey. ACM Comput. Surv. **55**(11), 1–37 (2023)
3. Cen, H., Koedinger, K., Junker, B.: Comparing two IRT models for conjunctive skills. In: Woolf, B.P., Aïmeur, E., Nkambou, R., Lajoie, S. (eds.) Intelligent Tutoring Systems: 9th International Conference, ITS 2008, Montreal, Canada, 23–27 June 2008 Proceedings 9, pp. 796–798. Springer, Cham (2008). https://doi.org/10.1007/978-3-540-69132-7_111
4. Chen, L., Chen, P., Lin, Z.: Artificial intelligence in education: a review. IEEE Access **8**, 75264–75278 (2020)
5. Choi, Y., et al.: Towards an appropriate query, key, and value computation for knowledge tracing. In: Proceedings of the Seventh ACM Conference on Learning@ Scale, pp. 341–344 (2020)
6. Corbett, A.T., Anderson, J.R.: Knowledge tracing: modeling the acquisition of procedural knowledge. User Model. User-Adap. Inter. **4**, 253–278 (1994)
7. Gabriella, C., Grilli, L., Limone, P., Domenico, S., Daniele, S., et al.: Deep learning for knowledge tracing in learning analytics: an overview. In: CEUR Workshop Proceedings, vol. 2817, pp. 1–10. CEUR-WS (2021)
8. Ghosh, A., Heffernan, N., Lan, A.S.: Context-aware attentive knowledge tracing. In: Proceedings of the 26th ACM SIGKDD International Conference on Knowledge Discovery & Data Mining, pp. 2330–2339 (2020)
9. Hochreiter, S., Schmidhuber, J.: Long short-term memory. Neural Comput. **9**(8), 1735–1780 (1997)
10. Käser, T., Klingler, S., Schwing, A.G., Gross, M.: Dynamic Bayesian networks for student modeling. IEEE Trans. Learn. Technol. **10**(4), 450–462 (2017)
11. Kean, J., Bisson, E.F., Brodke, D.S., Biber, J., Gross, P.H.: An introduction to item response theory and Rasch analysis: application using the eating assessment tool (EAT-10). Brain Impairment **19**(1), 91–102 (2018)
12. Liu, Z., et al.: Enhancing deep knowledge tracing with auxiliary tasks. In: Proceedings of the ACM Web Conference 2023, pp. 4178–4187 (2023)
13. Long, T., Liu, Y., Shen, J., Zhang, W., Yu, Y.: Tracing knowledge state with individual cognition and acquisition estimation. In: Proceedings of the 44th International ACM SIGIR Conference on Research and Development in Information Retrieval, pp. 173–182 (2021)
14. Pandey, S., Karypis, G.: A self-attentive model for knowledge tracing. arXiv preprint arXiv:1907.06837 (2019)
15. Pandey, S., Srivastava, J.: RKT: relation-aware self-attention for knowledge tracing. In: Proceedings of the 29th ACM International Conference on Information & Knowledge Management, pp. 1205–1214 (2020)
16. Pelánek, R.: Bayesian knowledge tracing, logistic models, and beyond: an overview of learner modeling techniques. User Model. User-Adap. Inter. **27**, 313–350 (2017)

17. Piech, C., et al.: Deep knowledge tracing. In: Advances in Neural Information Processing Systems, vol. 28 (2015)
18. Rabiner, L.R.: A tutorial on hidden Markov models and selected applications in speech recognition. Proc. IEEE **77**(2), 257–286 (1989)
19. Shin, D., Shim, Y., Yu, H., Lee, S., Kim, B., Choi, Y.: SAINT+: integrating temporal features for EdNet correctness prediction. In: LAK21: 11th International Learning Analytics and Knowledge Conference, pp. 490–496 (2021)
20. Song, X., Li, J., Cai, T., Yang, S., Yang, T., Liu, C.: A survey on deep learning based knowledge tracing. Knowl.-Based Syst. **258**, 110036 (2022)
21. Van Erven, T., Harremos, P.: Rényi divergence and Kullback-Leibler divergence. IEEE Trans. Inf. Theory **60**(7), 3797–3820 (2014)
22. Vaswani, A., et al.: Attention is all you need. In: Advances in Neural Information Processing Systems, vol. 30 (2017)
23. Wang, F., et al.: Neural cognitive diagnosis for intelligent education systems. In: Proceedings of the AAAI Conference on Artificial Intelligence, vol. 34, pp. 6153–6161 (2020)
24. Yang, Y., et al.: GIKT: a graph-based interaction model for knowledge tracing. In: Hutter, F., Kersting, K., Lijffijt, J., Valera, I. (eds.) Machine Learning and Knowledge Discovery in Databases: European Conference, ECML PKDD 2020, Ghent, Belgium, 14–18 September 2020, Proceedings, Part I, pp. 299–315. Springer, Cham (2021). https://doi.org/10.1007/978-3-030-67658-2_18
25. Yong, A.G., Pearce, S., et al.: A beginner's guide to factor analysis: focusing on exploratory factor analysis. Tutorials Quant. Methods Psychol. **9**(2), 79–94 (2013)
26. Zhang, J., Shi, X., King, I., Yeung, D.Y.: Dynamic key-value memory networks for knowledge tracing. In: Proceedings of the 26th International Conference on World Wide Web, pp. 765–774 (2017)
27. Zhao, S., Wang, C., Sahebi, S.: Transition-aware multi-activity knowledge tracing. In: 2022 IEEE International Conference on Big Data (Big Data), pp. 1760–1769. IEEE (2022)
28. Zhou, H., et al.: Informer: beyond efficient transformer for long sequence time-series forecasting. In: Proceedings of the AAAI Conference on Artificial Intelligence, vol. 35, pp. 11106–11115 (2021)

Social Network Analysis

Position and Type Aware Anchor Link Prediction Across Social Networks

Dongwei Zhu[1,2], Yongxiu Xu[1,2(✉)], Hongbo Xu[1], Hao Xu[1,2], Qi Wang[1,2], and Wenhao Zhu[1,2]

[1] Institute of Information Engineering, Chinese Academy of Sciences, Beijing, China
{zhudongwei,xuyongxiu,hbxu,xuhao,wangqi2022,zhuwenhao}@iie.ac.cn
[2] School of Cyber Security, University of Chinese Academy of Sciences, Beijing, China

Abstract. Anchor link prediction (ALP) aims to align the accounts of the same natural person on different social networks, which is essential for cross-platform recommendations and comprehensive characterization of user characteristics. In recent years, the method based on network embedding has become the mainstream method for cross-network anchor link prediction. The main goal of network embedding is to learn high-quality node characterization vectors by paying attention to the attribute information of nodes and the connection relationship between nodes, so as to make nodes more personalized. However, for heterogeneous networks, traditional methods cannot make full use of the heterostructure information of nodes. To address the challenge, we propose a novel cross-heterogeneous social network anchor link prediction model (PT-ALP). Specifically, PT-ALP obtains the position information of each neighbor by maximizing the mutual information between the central node and its neighbors at all levels, and then uses a two-layer graph attention architecture (GAT) to obtain the embedding representation that contains both position and type aware information. Finally, to verify the effectiveness of our proposed method, we conduct extensive experiments on several real-world datasets. Experimental results show that the proposed model achieves better performance than state-of-the-art methods.

Keywords: Anchor link prediction · Mutual information estimation · Neural network · Network alignment · Social networks

1 Introduction

With the deepening of participation in social networks, people's demand for social network services has gradually increased, and a single social network can

---

Supported by the National Key R&D Plan project of China (2021YFB3100600), the Youth Innovation Promotion Association, Chinese Academy of Sciences (No. 2020163) and the Strategic Pilot Science and Technology Project of Chinese Academy of Sciences (XDC02040400).

no longer meet the needs of users for different network services [1]. In order to enjoy the colorful and personalized services provided by different social networks, people often participate in multiple social networks at the same time [2], such as using Twitter to express personal political views and posting fun life videos to Facebook. Although accounts in different social networks may belong to the same person, for security and anonymity reasons, these accounts scattered on different social platforms are usually isolated from each other [3]. Anchor link prediction (ALP) aims to recognize the accounts of the same natural person across different networks, and the links between these accounts are anchor links, as shown in Fig. 1. Specifically, ALP will help many downstream tasks, such as link recommendation and community analysis [4]. For example, in Foursquare network, social connections and activities of new users can be very sparse. The friend and position recommendations for such users are very hard using only one network. However, if we also know the user's Twitter account, his/her social connections and position data in Twitter network can also be used to improve the recommendation performances in the Foursquare network.

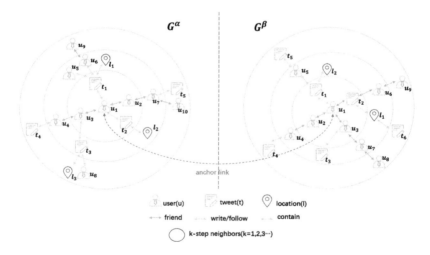

Fig. 1. An example of anchor link prediction across social networks.

However, different social networks often contain different node attributes and different social network structure information, which brings difficulties to cross-network user alignment. How to identify and mark the accounts of the same person on different social networks has become an urgent problem to be solved, so the anchor link prediction task came into being. Virtual links between accounts belonging to the same person in different social networks are like anchor chains that connect different social networks together, and each account is similar to an anchor. The anchor link prediction task aims to identify potential anchor chains between different social networks to align users across social networks.

With the rise of network embedding, the anchor link prediction method based on network embedding representation has become the mainstream trend [5]. The purpose of network embedding is to obtain the low-dimensional characterization vector of each node in each network through the method of network embedding, which will make the network nodes more distinguishable from a fine-grained perspective. For heterogeneous networks, the existing methods only simply integrate the various nodes associated with the user node to obtain the fusion vector, and then predict the anchor link based on the similarity between the fusion vectors corresponding to the user node. Although this crude method can achieve a certain alignment effect, the rich node information in heterogeneous networks will not be fully utilized, because the fusion vector ignores the impact of local type information on the alignment of user nodes [6].

In addition, existing methods usually only focus on how to model the neighbor information of a node, and few methods pay attention to the position information of the node in the network [7]. The position information of nodes often contains a wider sensory field and more comprehensive structural context information. This kind of information is essential in some data mining tasks related to structural analysis [8]. Some researchers proved that by paying attention to node type information or position information in the network, the quality of embedded representation of user nodes can be improved [6,9], thereby improving the accuracy of cross-social network anchor link prediction. However, there is no work to integrate the above two kinds of information into the node characterization vector at the same time, so the above two kinds of information cannot be fully utilized. Therefore, we designed experiments to explore the importance of the above two information to the alignment of cross-social network users separately, and proposed a novel model to comprehensively use the two information to further improve the accuracy of anchor link prediction.

In order to achieve the above purposes, we have designed a network embedding method based on position and type perception, which is convenient to integrate the information of the two into the node characterization at the same time, and then obtain richer characterization information of the node. On this basis, we propose a new cross-heterogeneous social network anchor link prediction model (PT-ALP). PT-ALP first obtains the position information of each neighbor node by maximizing the mutual information between the central node and its neighbors of all levels, and then solves the network embedding and position & type-aware alignment problems at the same time under the unified optimization framework based on the two-layer graph attention architecture. The main contributions of this paper can be summarized as follows:

- We use a mutual information estimation method based on K-step neighbors and combine GCN to obtain the initial embedding vector of nodes containing global and local structural features.
- We use a two-layer graph attention architecture to obtain the location and type-aware embedding of user nodes and the fusion vector of the two respectively.

- We propose a method that comprehensively utilizes node position information and type information, and further improving the prediction accuracy of anchor links by using both of them at the same time.
- To verify the effectiveness of the method, we conduct experiments on real-world data. The experimental results show that our method is effective compared to state-of-the-art algorithms.

2 Related Work

Depending on the information used, the models used for anchor link prediction can be divided into three categories: feature-based model, social relations-based model and feature & social relations-based model. Section 2.1 introduces feature-based models. Section 2.2 presents social relations-based models. Section 2.3 introduces the model in which the feature and social relations are used at the same time.

2.1 Feature-Based Models

In the process of joining and using new social networks, users generally bind some basic attributes with distinguishing characteristics to their accounts, such as name, gender, age, address, and content of posts. Obviously, using these basic attributes of users to identify and predict accounts on different social platforms is the easiest and most effective way. Therefore, most of the early anchor link prediction work uses the basic attributes of users to predict. Riederer et al. [2] realized cross-network user identification by modeling user location information. Malhotra et al. [3] used the user's digital footprint information on social networks to achieve a better cross-network anchor link prediction effect. Carmagnola et al. [10] combined the user profile and the content characteristics of the user's posts to make anchor link prediction, and jointly completed the anchor link prediction task by comparing the similarity of the user profile and the content of the posts in different social networks. In the above research on user characteristics, UserID and name have been confirmed to be the most discriminating features for disambiguating user profiles [10], and have been widely used in the research of anchor link prediction. However, this method can achieve better results if the user name is not repeated, but many platforms do not set the user name to be non-repeatable [11], so the above method has inevitable defects.

2.2 Social Relations-Based Models

When the user's attribute information was not available, researchers began to try to get breakthroughs from the user's social relationships. Compared with the user's attribute characteristics, the user's social relationship is easier to obtain and has higher credibility. In recent years, more and more researchers have tended to use users' social relationships to predict anchor links. Narayanan et al. [12] first proved that when user characteristics are lacking, anchor link prediction can be achieved only through the user's social relationships. Zhou et al. [13] conducted in-depth research on the basis of the above work. They believe that a two-way relationship is more stable than a one-way relationship, and it is more likely to be a real friend relationship. Based on this, they proposed an anchor link prediction model based on the number of friends. Inspired by the above work, Tang et al. [14] proposed a model based on the degree penalty mechanism. The model not only considers the number of friends, but also assigns different weights to each friend to measure its contribution to similarity. Liu et al. [15] combined network embedding and anchor link prediction in a unified framework and proposed a network-based attention embedding model.

2.3 Feature and Social Relations-Based Models

In order to further improve the prediction accuracy of anchor links. Zhang et al. [16] used both user attribute information and network structure information to solve the problem of anchor link prediction. Using an unsupervised approach, they proved that better anchor link prediction performance can be obtained by combining user attribute information and network structure information at the same time. Based on a semi-supervised method, Zhang et al. [17] proposed an effective anchor link prediction method with the help of user attribute information and network structure information. Specifically, they leverage potentially useful information possessed by the existing anchor link, and then develop a local expansion model to identify new social links, which are taken as a generated anchor link to be used for iteratively identifying additional new social link. Wang et al. [18] introduced the heterogeneous social network features and proposed a graph embedding method LHNE for anchor link prediction which collectively leveraged structural and content information in a unified framework to learn comprehensive representations of users. Inspired by the above work, in our paper, the user's basic attribute characteristics and social relationship characteristics are all utilized.

3 Proposed Model

Given that both the location information and type information of the node can help to obtain a better node embedding, in this paper, both of them are considered by us for anchor link prediction. As shown in Fig. 2, our method consists of three parts, including a node position perception module, a two-layer graph attention mechanism module, and an anchor link prediction module. In the node position perception module, we use the K-step mutual information estimation method to learn the initial representation of the position information contained in the node. The two-layer attention mechanism module is composed of two-layer attention networks. The work of the first layer is to integrate embedding vectors belonging to the same type, and obtain a local representation of the user node on this type of information, which is called the position & type aware embedding. The second layer aims to merge the different types of perceptual embedding vectors of user nodes to obtain a global embedding vector, which is called the position & type fusion embedding. By integrating the above two embeddings, the n-tuple representation of the node is finally obtained by us. In the anchor link prediction module, we jointly guide the anchor link prediction by synergistically comparing the similarity of each element of the n-tuple representation of the node.

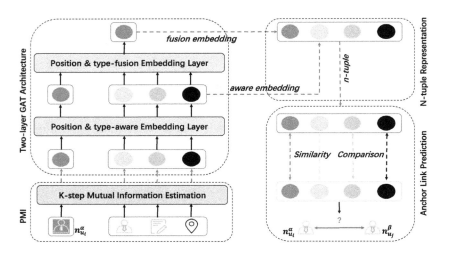

Fig. 2. Overview of model framework.

3.1 Problem Formulation

Homogeneous and Heterogeneous Network. A homogeneous network refers to a network that contains only one type of node and one type of edge. It can be expressed as $G = \{N, E\}$, where N represents the set of nodes and E represents the set of edges. A network containing different types of nodes and/or multiple types of edges is defined as a heterogeneous network, formally $G = \{N, E, T\}$, where T represents a set of node types. As shown in Fig. 1, there are two heterogeneous networks, in which the source network and the target network are formalized as G^α and G^β respectively. Take G^α as an example, $G^\alpha = \{N^\alpha, E^\alpha, T^\alpha\}$, $N^\alpha = \{n^\alpha_{u1}..., n^\alpha_{t1}..., n^\alpha_{l1}...\}$, $E^\alpha = \{(n^\alpha_{u1}, n^\alpha_{u2}), (n^\alpha_{u1}, n^\alpha_{t1}), (n^\alpha_{u1}, n^\alpha_{l1})...\}$, $T^\alpha = \{u, t, l\}$.

K-step Mutual Information Estimation. Mutual information is a measurement method based on fragrant entropy. It captures the nonlinear dependence between variables and can usually be used to measure the true dependence between two variables. The mutual information estimation of the two variables P and Q is equivalent to calculating the K-L divergence of the joint distribution J and the marginal distribution M of the two, The formula is as follows:

$$I(P;Q) = KL(J|M) \tag{1}$$

K-step mutual information estimation is the use of mutual information between nodes to make the central node have the ability to perceive the location of neighbors of each step.

Position and Type-Aware Embedding and Fusion Embedding. Given a user node $n_{u_i}^\alpha \in G^\alpha$, and its set of neighbor nodes of each level $\mathcal{N}_{n_{u_i}^\alpha}^{t,k}$ ($t \in T^\alpha$, $k = 1, 2, 3$, represents the level of the neighbor). For each node $n_{u_j}^\alpha \in \mathcal{N}_{n_{u_i}^\alpha}^{t,k}$, its embedding vector is denoted as $\boldsymbol{n}_{u_j}^\alpha$, integrating each embedding vector $\boldsymbol{n}_{u_j}^\alpha$ ($j = 1, 2, \ldots \left|\mathcal{N}_{n_{u_i}^\alpha}^{t,k}\right|$) of node in $\mathcal{N}_{n_{u_i}^\alpha}^{t,k}$ can obtain the position & type-aware embedding vector of $n_{u_i}^\alpha$, denoted as $\vec{n}_{u_i}^{\alpha,t}$. The position & type-fusion embedding is to integrate each position & type-aware embedding vector $\vec{n}_{u_i}^{\alpha,t}$ associated with $\boldsymbol{n}_{u_i}^\alpha$, denoted as $\vec{f}_{u_i}^\alpha$.

Position and Type-Aware Anchor Link Prediction. Given two heterogeneous networks G^α and G^β. If node $n_{ui}^\alpha \in G^\alpha$ and node $n_{uj}^\beta \in G^\beta$, and they belong to the same person, then $(n_{ui}^\alpha, n_{uj}^\beta)$ is defined as an anchor link. PT-ALP predict the unobserved anchor links by matching position & type-aware embedding and fusion embedding representation vectors between each pair of user nodes across G^α and G^β.

3.2 N-Tuple Representation

In order to discuss the details of our model, in this section, we will introduce the initial vector acquisition method of each node and how each element in the n-tuple representation is obtained.

Initial Vector Representation. To solve the problem of formula (1) limited application scenarios, [19] using DV (Donsker-Varadhan) to characterize the next generation of derivation formula:

$$I(P;Q) \geq E_J[D_M(p,q)] - \log E_M[e^{D_M(p,q)}] \quad (2)$$

Where $D_M(\cdot,\cdot)$ is a bilinear function with parameter M. On this basis, [20] applied the neural estimation of mutual information to unsupervised image characterization learning for the first time, and proposed DIM model. The model replaces the K-L divergence with the J-S divergence, and uses adversarial training to maximize the mutual information between positive samples:

$$E_p[-sp(-D_M(p,q))] - E_{p \times \tilde{p}}[sp(D_M(r,q))] \quad (3)$$

Among them, $sp(x) = \log(1 + e^x)$ is the softplus activation function, and the variable r is the sampled negative sample. Inspired by the above work, we apply mutual information estimation to the characterization and learning of social network nodes:

$$I(N_i^k; n_{u_i}) = E_p[-sp(-D_{M^k}(y_i^k, x_i))] - \sum_{v_z \sim \tilde{p}}^{n_z} E_{\tilde{p}}[sp(D_{M^k}(y_z^k, x_i))] \quad (4)$$

Among them, $N_i^k = \{n_{u_{i_1}}^k, n_{t_{i_1}}^k, n_{l_{i_1}}^k ...\}$ is the set of k-step neighbors of node n_{u_i}, and $y_i^k = R(\{x_{i_1}^k, x_{i_2}^k, ...\})$ is the characteristic of all k-step neighbors determined by the Readout function. Node v_z is a negative sample node randomly sampled according to the distribution $\tilde{P}$, the distribution obeys a uniform distribution and has $P = \tilde{P}$, while n_z represents the number of negative samples collected for each central node n_{u_i}, M^k represents the parameter matrix of the bilinear function corresponding to the k-th neighbor. Formula (4) encourages each central node n_{u_i} to perceive its k-step neighbor information, so as to achieve the purpose of allowing the characterization of node n_{u_i} to carry its k-step neighbor information.

Position and Type-Aware Embedding. Similar to [6], we also use a GAT with two layers of attention. As shown in Fig. 2, the role of the first layer is to learn the position and type perception vectors, and the purpose of the second layer is to obtain the fusion embedding of the position and type perception vectors. For a node $n_{u_i}^\alpha$ in G^α and each of its neighbors $n_{u_j}^\alpha$ belongs to $\mathcal{N}_{n_{u_i}^\alpha}^{t,k}$, we formally define its initialization embedding representation as $\boldsymbol{n}_{u_i}^\alpha$ and $\boldsymbol{n}_{u_j}^\alpha$ respectively, and the embedding representation is obtained from the previous step. Then, we feed these initialized representations containing position information into the first layer of GAT to obtain embedded representations based on position and type perception. Specifically, we use the masking attention method [21] to output the adjacency matrix of G^α to the attention mechanism:

$$\alpha_{i,j}^t = \frac{exp(LeakyRelu(l_{i,j}^t))}{\sum_{k=1}^{\left|N_{n_{u_i}^\alpha}^t\right|} exp(LeakyReLU(l_{i,k}^t))} \quad (5)$$

In which, $l_{i,j}^t$ is the importance score of $n_{u_i}^\alpha$'s arbitrary neighbor $n_{u_j}^\alpha$ to $n_{u_i}^\alpha$. The t-th position and type aware vector $\boldsymbol{n}_{u_i}^{\alpha,t}$ of $n_{u_i}^\alpha$ from step k can be calculated as follows:

$$\boldsymbol{n}_{u_i}^{\alpha,t} = \Theta(\sum_{j=1}^{\left|N_{v_{u_i}^\alpha}^{k,t}\right|} \alpha_{i,j}^t W^t \overrightarrow{x}_j) \quad (6)$$

Where Θ is the Elu activation function, and $W^t = \mathbf{R}^{D' \times D}$ is the linear transformation weight matrix, which transfer the initial features into higher-level features.

In order to stabilize the learning process of self-attention, we employ multi-head attention [6] to calculate the position and type perception embedding vector:

$$\boldsymbol{n}_{u_i}^{\alpha,t} = \Theta(\frac{1}{\Phi}\sum_{\phi=1}^{\Phi} \sum_{j=1}^{\left|N_{v_{u_i}^\alpha}^{kt}\right|} \alpha_{i,j}^{\phi,t} W^{\phi,t} \overrightarrow{x}_j) \quad (7)$$

In which, Φ is the number of attention mechanisms, and $\alpha_{i,j}^{\phi,t}$ is the normalized attention coefficient calculated by the ϕ-th attention mechanism, and $W^{\phi,t}$ represents the weighting matrix of the ϕ-th attention mechanism.

Position and Type-Fusion Embedding. In order to obtain the position & type fusion-aware embedding vector of $n_{u_i}^{\alpha}$, we feed all the embedded representations $\vec{n}_{u_i}^{\alpha,t}$ ($t \in T$) and the initial representations of the central node obtained in the previous step into the second layer of GAT, and then aggregate these vectors with attention coefficients. In order to solve the problem of different dimensions of each vector pair $\vec{x}_i$ and $\vec{n}_{u_i}^{\alpha,t}$, we adopt additive-attention [22] to compute attention coefficient between them, and then the fusion embedding $\vec{f}_{u_i}^{\alpha}$ is calculated. Based on the above steps, all the constituent elements of the n-tuple have been obtained by us.

3.3 Position and Type-Aware Anchor Link Prediction

In the anchor link prediction stage, if there is an anchor link between the two nodes, the closer their corresponding n-tuple embedding representations are, otherwise the distance between their n-tuple embedding representations should be as far as possible. In other words, the distance between aligned user nodes should be minimized, while the distance between unaligned user nodes should be maximized.

3.4 Loss Function

Based on the above theory, there are two parts of the loss that need to be calculated, namely: the position & type aware loss ($\mathcal{L}_{aware}$) and the position & type fusion loss ($\mathcal{L}_{fusion}$). We define the position & type aware loss as follows:

$$\mathcal{L}_{aware} = \sum_{t \in T} \lambda^t [d(\vec{n}_{u_i}^{\alpha,t}, \vec{n}_{u_j}^{\beta,t}) + \xi - d(\vec{n}_{u_i}^{'\alpha,t}, \vec{n}_{u_j}^{'\beta,t})] \qquad (8)$$

where $(\vec{n}_{u_i}^{\alpha,t}, \vec{n}_{u_j}^{\beta,t})$ refers to the embedding of an anchor link, $(\vec{n}_{u_i}^{'\alpha,t}, \vec{n}_{u_j}^{'\beta,t})$ refers to the embedding of a non-anchor link. $d(,)$ is a distance formula. T is the number of node types and ξ is a margin hyper-parameter separating anchor links and unanchored links. For the position & type fusion loss, we define it as follows:

$$\mathcal{L}_{fusion} = \sum [d(\vec{f}_{u_i}^{\alpha}, \vec{f}_{u_j}^{\beta}) + \xi - d(\vec{f}_{u_i}^{'\alpha}, \vec{f}_{u_j}^{'\beta})] \qquad (9)$$

where $(\vec{f}_{u_i}^{\alpha}, \vec{f}_{u_j}^{\beta})$ refers to the fusion embedding of an anchor link, $(\vec{f}_{u_i}^{'\alpha}, \vec{f}_{u_j}^{'\beta})$ refers to the embedding of a non-anchor link. ξ is a hyper-parameter shared with ξ in $\mathcal{L}_{aware}$. Finally, we obtain the final loss function:

$$\mathcal{L} = \mathcal{L}_{aware} + \omega \mathcal{L}_{fusion} \qquad (10)$$

where ω is the hyper-parameter, which is used to balance the importance between position & type-aware similarity and position & type-fusion similarity for anchor link prediction.

4 Experiments

4.1 Datasets and Evaluation

We evaluate our method on two real-world social network datasets collected from Foursquare and Twitter [23]. Among them, Twitter contains 5,220 users and 164,916 edges, Foursquare contains 5,315 users and 76,972 edges, and there are 3,148 anchor links between the two networks, as shown in Table 1.

Table 1. Statistics of the Datasets.

Networks	Nodes	#Nodes	Rel.	#Rel.	#Anc.
Twitter	User	5220	U-U	164916	
	Tweet	9490707	U-T	9490707	3148
	Location	297183	U-L	615515	
Foursquare	User	5315	U-U	76972	
	Tweet	48755	U-T	48756	3148
	Location	38921	U-L	48756	

During the experiment, all the hyperparameters of PT-ALP were adjusted to perform best on the test set. In detail, in the stage of learning position perception characterization based on k-step mutual information estimation, we set the maximum step $k = 3$ and the number of negative samples $n_z = 5$. In the stage of learning to characterize based on type perception, all our hyperparameter settings are consistent with [6]. To evaluate the effectiveness of PT-ALP and other models, both the $Precision@k(P@k)$ [24] and the Mean Average Precision (MAP) [25] are used.

4.2 Baselines

We compare our proposed model TP-ALP with the following classic and state-of-the-art models as follows:

- **DeepLink** [25]: DeepLink is an ALP method that can process heterogeneous network data. It uses unbiased random walks to generate embedding, and then uses MLP to map users. Since the source code of DeepLink is not disclosed, and the method uses the same experimental data set as our model, we directly copy the experimental results reported in DeepLink to compare with our method.
- **HAN** [26]: As a powerful network embedding model based on GAT, Han can use the attention of the vertex level and the attention of the semantic level to learn the importance of the vertex and meta-path respectively, so as to capture the complex structural information and rich semantic information of heterogeneous graphs. Semantic information.

- **PME** [27]: PME constructs the embedding of nodes and relationships in the node space and the relationship space respectively, rather than mapping the embedding of nodes and relationships into the same space. Moreover, it projects various types of links into different subspaces, and finally obtains the overall embedding vector of each node.
- **HHNE** [28]: HHNE is an active learning method, which can use simple active learning to obtain the embedding vector of each node in a heterogeneous network, and then use the obtained embedding vector to make anchor link prediction.
- **PMI** [7]: PMI is an embedded learning method based on position perception. The model can learn the comprehensive embedding of nodes by maximizing the mutual information between each central node and its neighbors of all levels.
- **TALP** [6]: As one of the best models for anchor link prediction, TALP not only focuses on the comprehensive embedded representation of nodes, but also on the type characteristics of nodes' neighbors, and proposes an anchor link prediction method based on type perception. Users are mapped together by comparing the comprehensive embedding of nodes and the neighbor type-aware embedding.

Table 2. Performance comparison on anchor link prediction.

Metrics (%)	P@1	P@5	P@9	P@21	P@30	MPA@30
DeepLink	34.37	59.42	66.09	70.00	70.48	47.78
HAN	38.69	60.38	71.16	75.49	78.33	50.22
PME	40.51	59.89	73.18	78.54	80.95	52.42
HHNE	38.72	60.45	69.92	75.96	79.13	51.28
PMI	40.91	61.27	78.84	89.42	92.81	54.25
TALP	43.79	72.73	93.22	95.20	98.69	59.33
PT-ALP	**45.52**	**74.06**	**94.87**	**96.32**	**99.10**	**61.88**

4.3 Overall Evaluation Results

Table 2 shows the experimental results of each model in the cross-social network anchor link prediction task. The results show that our model's performance on the Twitter-Foursquare dataset exceeds all baseline methods and has brought almost consistent improvements in different indicators. In addition, to better present the $P@K$ changes of each model, we show it in Fig. 3.

PT-ALP can achieve better performance. The reason is that the previous method was only based on the pairwise similarity of the fusion vector to measure whether there is an anchor link between the two nodes. PT-ALP aligns the anchor user nodes by simultaneously fusing the node position perception vector, the type perception vector, and the fusion vector. The experimental results show that the location information and type information of nodes in the network are beneficial to improve the accuracy of anchor link prediction.

PT-ALP performed better than PMI and TALP. Since the key components of PT-ALP are PMI and TALP, these two methods can be regarded as ablation experiments for PT-ALP. The difference between PMI and PT-ALP is that it only focuses on the location information of the node. Obviously, adding node type-aware information can improve the performance of anchor user node alignment. For TALP and PT-ALP, the only difference is whether to pay attention to the location information of the node in the network. From the experimental results, we can see that the addition of node location perception information can also improve the performance of the model.

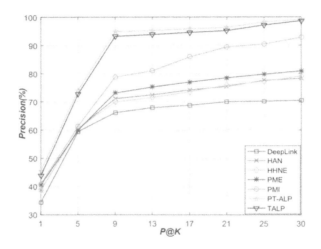

Fig. 3. Performance of each model on different $P@K$.

TALP's performance surpassed PMI. One reason is that the graph attention architecture can better model the fusion vector of user nodes. Another reason may be that for anchor link prediction tasks, the importance of node type perception information is greater than that of node position perception information. In other words, in the task of improving the anchor link prediction model, node type perception information dominates. In addition, HAN, PME, and HHNE are better than the homogeneous network anchor link prediction method DeepLink, which shows that the embedding vector of user nodes in heterogeneous networks contains richer information than that of homogeneous networks.

4.4 Ablation Study

We conduct ablation studies to explore the effectiveness of each component in our method. To be specific, we consider four settings in the ablation study.

- w/o Two-layer GAT Architecture is the method that we completely ignore the type-aware information of the node. In this case, we only use the position information of the node to align users according to the number of neighbors of different steps of the node.
- w/o Position & type-aware Embedding is a method which only uses position & type-aware fusion embedding vectors to align user nodes across heterogeneous networks.
- w/o Position & type-fusion Embedding is a method which only uses position and type-aware embedding vectors to align user nodes across heterogeneous networks.
- w/o PMI is the method that we ignore the position information of nodes. In this case, like TALP, we randomly initialize the vector of the node, and then only use the type-aware vector and the fusion vector for cross-network anchor link prediction.

Table 3. The performance of model variants at different $Precision@K$ (Results of ablation experiment).

Precision@K	1	5	9	13	17	21	25	30
Ours w/o Two-layer GAT Architecture	40.91	61.25	78.84	81.04	85.94	89.42	90.47	92.81
Ours w/o Position & type-aware Embedding	42.66	62.32	88.50	89.81	90.22	90.40	92.44	93.79
Ours w/o Position & type-fusion Embedding	43.37	69.82	89.96	91.48	92.29	92.57	93.71	94.27
Ours w/o PMI	43.79	72.73	93.22	93.82	94.57	95.20	97.16	98.96
Ours (PT-ALP)	**45.52**	**74.06**	**94.87**	**95.22**	**95.86**	**96.32**	**98.27**	**99.10**

As shown in Table 3, the ablation experiment results show that all modules contribute to the performance of the final model, and any deletion of any module will lead to a decrease in the performance of the model. From ablation experiment 1(w/o Two-layer GAT Architecture) and ablation experiment 4(w/o PMI), it is not difficult to see that it is much better to pay attention to the type information of the node through the two-layer GAT architecture than to use only the location information of the node. This proves to a certain extent that the type information of the node is more important than the location information of the node. In addition, the results of ablation experiment 3(w/o Position & type-fusion Embedding) exceed the results of ablation experiment 2(w/o Position & type-aware Embedding), which indicates that the Position & type-aware information is more efficient than the information of Position & type-fusion for anchor links prediction across heterogeneous networks.

5 Conclusion

In this paper, we propose a novel cross-heterogeneous social network anchor link prediction model (PT-ALP). PT-ALP obtains the location information of each neighbor node by maximizing the mutual information between the central node and its neighbors of all levels, and then completes the overall embedding process of the node under the unified optimization framework based on the two-layer graph attention architecture. PT-ALP can simultaneously perceive the location information of the node in the network and the type information of the node, and integrate the two perceptual information into the node characterization, thereby obtaining richer characterization information of the node. A large number of experiments on real social network data sets have shown that the proposed method is effective and efficient compared with several state-of-the-art methods.

References

1. Fan, S., et al.: Metapath-guided heterogeneous graph neural network for intent recommendation. In: Proceedings of the 25th ACM SIGKDD Conference on Knowledge Discovery and Data Mining (KDD 2019), Anchorage, AK, USA. ACM (2019)
2. Riederer, C., Kim, Y., Chaintreau, A., Korula, N., Lattanzi, S.: Linking users across domains with location data: theory and validation. In: Proceedings of the 25th International Conference on World Wide Web, pp. 707–719 (2016)
3. Malhotra, A., Totti, L., Meira Jr., W., Kumaraguru, P., Almeida, V.: Studying user footprints in different online social networks. In: 2012 IEEE/ACM International Conference on Advances in Social Networks Analysis and Mining, pp. 1065–1070. IEEE (2012)
4. Yin, H., Zou, L., Nguyen, Q.V.H., Huang, Z., Zhou, X.: Joint event-partner recommendation in event-based social networks. In: 2018 IEEE 34th International Conference on Data Engineering (ICDE), pp. 929–940. IEEE (2018)
5. Song, X., Li, J., Lei, Q., Zhao, W., Chen, Y., Mian, A.: Bi-CLKT: bi-graph contrastive learning based knowledge tracing. Knowl.-Based Syst. **241**, 108274 (2022)
6. Li, X., Shang, Y., Cao, Y., Li, Y., Tan, J., Liu, Y.: Type-aware anchor link prediction across heterogeneous networks based on graph attention network. Proc. AAAI Conf. Artif. Intell. **34**, 147–155 (2020)
7. Chu, X., Fan, X., Yao, D., Zhu, Z., Huang, J., Bi, J.: Cross-network embedding for multi-network alignment. In: The World Wide Web Conference, WWW 2019, pp. 273–284, New York, NY, USA. Association for Computing Machinery (2019)
8. Liu, D., Chang, Z., Yang, G., Chen, E.: Community hiding using a graph autoencoder. Knowl.-Based Syst. **253**, 109495 (2022)
9. Bi, X., Chu, J., Fan, X.: Characterization learning of location-aware networks based on k-order mutual information estimation. Comput. Res. Dev. **34**(1) (2021)
10. Carmagnola, F., Cena, F.: User identification for cross-system personalisation. Inf. Sci. **179**(1), 16–32 (2009)
11. Yang, Y., Wang, L., Liu, D.: Anchor link prediction across social networks based on multiple consistency. Knowl.-Based Syst. **257**, 109939 (2022)
12. Narayanan, A., Shmatikov, V.: De-anonymizing social networks. In: 2009 30th IEEE Symposium on Security and Privacy, pp. 173–187. IEEE (2009)

13. Zhou, X., Liang, X., Zhang, H., Ma, Y.: Cross-platform identification of anonymous identical users in multiple social media networks. IEEE Trans. Knowl. Data Eng. **28**(2), 411–424 (2015)
14. Tang, R., Jiang, S., Chen, X., Wang, H., Wang, W., Wang, W.: Interlayer link prediction in multiplex social networks: an iterative degree penalty algorithm. Knowl.-Based Syst. **194**, 105598 (2020)
15. Liu, L., Zhang, Y., Shun, F., Zhong, F., Jun, H., Zhang, P.: ABNE: an attention-based network embedding for user alignment across social networks. IEEE Access **7**, 23595–23605 (2019)
16. Zhang, Y., Tang, J., Yang, Z., Pei, J., Yu, P.S.: COSNET: connecting heterogeneous social networks with local and global consistency. In: Proceedings of the 21th ACM SIGKDD International Conference on Knowledge Discovery and Data Mining, pp. 1485–1494 (2015)
17. Zhang, Y., Wang, L., Li, X., Xiao, C.: Social identity link across incomplete social information sources using anchor link expansion. In: Bailey, J., Khan, L., Washio, T., Dobbie, G., Huang, J., Wang, R. (eds.) Advances in Knowledge Discovery and Data Mining: 20th Pacific-Asia Conference, PAKDD 2016, Auckland, New Zealand, 19–22 April 2016, Proceedings, Part I 20, pp. 395–408. Springer, Cham (2016). https://doi.org/10.1007/978-3-319-31753-3_32
18. Wang, Y., Feng, C., Chen, L., Yin, H., Guo, C., Chu, Y.: User identity linkage across social networks via linked heterogeneous network embedding. World Wide Web **22**, 2611–2632 (2019)
19. Hjelm, R.D., et al.: Learning deep representations by mutual information estimation and maximization. arXiv preprint arXiv:1808.06670 (2018)
20. Veličković, P., Fedus, W., Hamilton, W.L., Liò, P., Bengio, Y., Hjelm, R.D.: Deep graph infomax. arXiv preprint arXiv:1809.10341 (2018)
21. Veličković, P., Cucurull, G., Casanova, A., Romero, A., Lio, P., Bengio, Y.: Graph attention networks. arXiv preprint arXiv:1710.10903 (2017)
22. Bahdanau, D., Cho, K., Bengio,Y.: Neural machine translation by jointly learning to align and translate. Comput. Sci. (2014)
23. Zhang, J., Philip, S.Y.: Integrated anchor and social link predictions across social networks. In: Twenty-Fourth International Joint Conference on Artificial Intelligence (2015)
24. Liu, H., Zafarani, R., Shu, K., Tang, J., Wang, S.: User identity linkage across online social networks: a review. SIGKDD Explor. **18**, 5–17 (2016)
25. Zhou, F., Liu, L., Zhang, K., Trajcevski, G., Wu, J., Zhong, T.: DeepLink: a deep learning approach for user identity linkage. In: IEEE INFOCOM 2018 - IEEE Conference on Computer Communications, pp. 1313–1321 (2018)
26. Wang, X., et al.: Heterogeneous graph attention network. In: The World Wide Web Conference, pp. 2022–2032 (2019)
27. Chen, H., Yin, H., Wang, W., Wang, H., Nguyen, Q.V.H., Li, X.: PME: projected metric embedding on heterogeneous networks for link prediction. SIGKDD Explor. **22**(Udisk) (2018)
28. Wang, X., Zhang, Y., Shi, C.: Hyperbolic heterogeneous information network embedding. Proc. AAAI Conf Artif. Intell. **33**, 5337–5344 (2019)

Artificial Intelligence and Music

LSTM-MorA: Melody-Accompaniment Classification of MIDI Tracks

Hui Liu[✉][⑩], Leon Flaack, Shiyao Zhang[⑩], and Tanja Schultz[⑩]

University of Bremen, 28358 Bremen, Germany
hui.liu@uni-bremen.de

Abstract. Many studies based on symbolic music signals require retaining only melody tracks or accompaniment tracks from musical instrument digital interface (MIDI) files. However, this seemingly simple setting often becomes a stumbling block in the first step because the MIDI format does not have any mandatory regulations for the track numbers of melody/accompaniment tracks. This study delves into the classification of melody and accompaniment parts within MIDI files, pioneering the use of long-short-term memory (LSTM) for this purpose. An LSTM network is trained to classify multivariate time series of varying lengths, representing the tracks within MIDI files as either melody or accompaniment. Experimental results of over 0.91 accuracy, precision, recall and F score reveal that our proposed methodology, LSTM-MorA (Melody or Accompaniment), could be one of the solutions for MIDI melody-accompaniment classification.

Keywords: MIDI · Melody · Accompaniment · LSTM · MorA

1 Introduction

Since the inception of musical instrument digital interface (MIDI) in 1982, its standard has served as a cornerstone for transcribing played music into a symbolic format and composing new music. As a result, a vast repository of MIDI files is now accessible online. Many studies based on symbolic music signals require the retention of only melody tracks or accompaniment tracks from MIDI files. The former involves research on automatic harmonization [3], music information retrieval (MIR) [8,9], and automatic orchestration [12], among others, while the latter is of great significance for tasks such as karaoke audio rendering [10]. However, except for datasets/bases with deliberately formulated unified rules or rigorous annotation, the melody tracks and accompaniment tracks of any MIDI file are often black boxes to users prior to the careful listening along tracks. In studies that require processing large amounts of MIDI files, such an uncertainty leads to a surge in upfront work.

Successfully classifying melody and accompaniment tracks of a MIDI file is therefore a pivotal task within the field of MIR, unlocking the potential of leveraging the abundance of MIDI files available online, thereby providing significant benefits for further scientific endeavors, including machine learning. Approaches exist in the realm of melody extraction or selection from polyphonic MIDI files.

Uitdenbogerd and Zobel's algorithms [18,19] extract melody tracks employing different strategies, involving aggregating the highest note at each time point, as well as calculating the entropy within each track and selecting the one with the highest entropy. Regardless of accuracy, this method obviously needs to know in advance how many melody tracks there are in the MIDI file to be extracted (although multiple melody-track situations are not necessarily common). In other words, applying this method still requires a priori knowledge of the music file, even prelistening.

In the research conducted by Tang, Yip, and Kao [17], five different criteria-based approaches are explored, named AvgVel, PMRatio, SilenceRatio, TrackName, and Range. Subsequently, they considered the first n tracks to be melody. Still, n is here a preset parameter that makes the algorithm not suitable for the usage of a black box or huge amounts of data.

Criteria such as note duration and number to discard unpromising tracks were proposed by Velusamy, Thoshkahna, and Ramakrishnan [20]. This method is again based on the ranking of the weighted sum of calculated features; thus, the number of melody tracks is certainly a needed variable.

The approach of Jiang and Dannenberg [5] begins by dividing the notes of the tracks into measures. These measures are then classified using a Bayesian maximum likelihood model.

Regarding deep learning, Li, Yang, and Chen [7] computed various features for each track to train a neural network to select the melody tracks.

It can be found that the reviewed literature is all early. Along with attention, the availability of data has faded. Unfortunately, many datasets applied in the above-mentioned works or other peer MIDI researches are either untraceable or no longer accessible, which makes it difficult to compare the approach investigated in this paper under the same conditions. Additionally, some early publications did not disclose their model parameters transparently, further complicating a fair comparison with this study.

This study aims to innovate and assess the effectiveness of a machine learning approach based on long short-term memory (LSTM) technology [4] in tackling the melody-accompaniment classification of MIDI tracks, while also pinpointing potential challenges. The paper details the methodology, the modeling configurations, the experimental evaluation, and the comparison of the results with existing methods. The peer comparison is conducted using a standardized dataset and the metrics outlined in the methodology section, providing a coherent framework for analysis and interpretation.

2 Dataset, MIDI Processing, and Data Filtering

To effectively train the LSTM-based recurrent neural network (RNN), an open source available online dataset comprising well-labeled input sequences, specifying whether they represent melody or accompaniment, was applied, sourced from the *Clean MIDI* subset of the *Lakh MIDI* dataset [14,15]. It encompasses 17,256 MIDI files, featuring a range of popular and classical music. Notably, many musical compositions appear multiple times, each represented by transcriptions that may exhibit varying degrees of dissimilarity, which should be considered as different musical works.

The extraction of individual monophonic note sequences from polyphonic MIDI files was facilitated using the *Python* library *pretty_midi*. Each sequence corresponds to a distinct track within the files. To identify the melody we inspected each file for any track labeled as "melody" via the *Instrumentname* event previously described. Subsequently, the files bearing such a "melody" label underwent additional processing (see Sect. 5) to prepare them for training and testing the LSTM network.

In order to ensure that the tracks labeled as "melody" exclusively contain melody, a more stringent criterion was applied. Specifically, only MIDI files containing exactly one track marked as "melody" were retained. This criterion has the following advantages:

- The melody remains confined within a single track and does not transition across multiple tracks or instruments, as observed in certain musical compositions, which means no track contains the whole melody information of the music work;
- The labeling as a melody is ensured to be intentional and not a result of idiosyncrasies within the MIDI transcription program used.

As a validation of our mechanism, we manually checked 80 random samples from the set of MIDI files containing only one track labeled as a melody. All 80 examined files were correctly labeled.

Figure 1 illustrates the distribution of the number of voices labeled as melody per file in the dataset. In particular, 279 files were not possible to process due to corruption, whose invalid data exceeds 127 bytes.

3 Methodology

This section elucidates our proposed approach, LSTM-MorA (Melody or Accompaniment), to classifying MIDI tracks into melody and accompaniment.

3.1 Feature Extraction

Feature extraction is essential for machine learning study applying feature engineering, such as subsequence search [2] and self-similarity matrix [16]. Instead of using a universal feature extraction library such as TSFEL [1], we applied the music-specific *Python* library *pretty_midi* for the calculation of the features, where the *note* object contains the absolute time of the Note On message (in seconds), the absolute time of the Note Off message (in seconds), and the pitch value of the Note On message, among others.

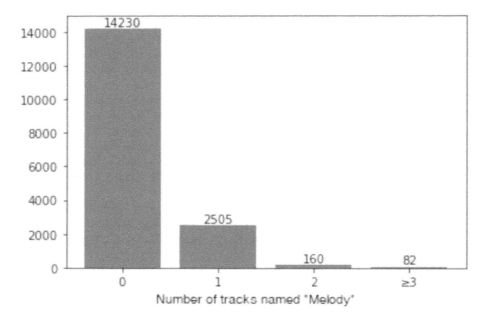

Fig. 1. Distribution of how many "melody" tracks each MIDI file has.

The resulting input sets (see Sect. 4.1) for feeding the trainer and tester include a series of four features per note for each track in the applied data corpus, calculated by the attributes mentioned above.

- the absolute time of the `Note off` message
- the note duration (computed through the `Note On` and `Note off` messages)
- the pitch
- the strength of the attack.

This paper is the beginning of verifying LSTM as one of the solutions of the research topic, where the basic features are chosen. More sophisticated representations of MIDI have already been explored and could potentially increase performance [11], which will be one of the essential further studies.

3.2 Architecture

The network was developed and trained using the *PyTorch* library. It comprises an LSTM layer, a linear layer, and a concatenated sigmoid function to facilitate the classification task.

Since *PyTorch* tensors mandate entries matching the tensor's size in each dimension, yet the length of individual sequences of each track varies, they are padded with the value -1 in the collate function. This choice is made because -1 cannot be represented by any of the features.

LSTM Layer. A schematic illustration of the LSTM unit applied in this work is presented in Fig. 2.

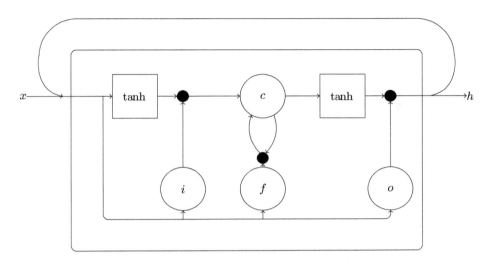

Fig. 2. The selected LSTM architecture.

Each input x_t is a vector with a length of four features. To achieve accurate estimation results, the number of elements in the cell-to-cell (CEC) and the output tensor play a crucial role. A larger number of entries enables the consideration of more aspects from inputs of earlier time steps. However, choosing an excessively high size can lead to inefficiencies, as some elements may become redundant without providing any additional benefit. For example, certain elements may contain overlapping or redundant information, but are still calculated in each step.

To determine an appropriate size, several networks were trained with identical parameters, varying only the size of the CEC and the output of the LSTM layer (`hidden_size`), whose results and analysis are presented in Sect. 4. The findings revealed that a `hidden_size` of 64 yielded satisfactory performance and was therefore selected to build the LSTM layer.

The output of the LSTM layer takes the form (`hidden_size`), resulting in a one-dimensional tensor. To calculate this, two distinct approaches, labeled "Last" and "Average" (the arithmetic mean), were explored, detailed in Sect. 4.

Linear Layer. The linear layer takes the output of the LSTM layer as input and applies a linear transformation to each element. It contains learnable weights and bias values for the connections from the LSTM layer, represented as a weight matrix and a bias vector. These parameters combine the input vector, with a size of `in_features = 64`, into an output vector with a size of out_features = 1. The output vector of length 1 represents the value that is passed on to the next layer.

Sigmoid Layer. A threshold value of 0.5 was chosen, which facilitates rounding (up/down) of the output to make the prediction.

Binary Cross-Entropy and Parameter Training. The evaluation of the network output is based on the (binary) cross entropy as the chosen cost function.

We applied Adam [6], a gradient-based stochastic optimization method for the learnable parameters of artificial neural networks based on gradient descent.

3.3 Training Strategies

As introduced in Sect. 3.2, the Adam optimization method was applied. We divided the available data into separate training and validation sets. Within the training set, further subdivision into mini-batches occurs, which are then collectively processed by the network. For each element within these mini-batches, costs are calculated based on their labels, and the average cost is computed. Subsequently, the network parameters are adjusted accordingly. It is essential to acknowledge that larger mini-batch sizes may also have associated drawbacks, as discussed in Sect. 3.4.

To avoid overfitting, early stop, as proposed by Prechelt [13], is used. In this approach, the network undergoes standard training, but its performance is periodically evaluated every k epochs using separate validation data. Training continues until a predetermined stopping criterion is met. At this point, the training stops and the parameter values yielding the lowest average costs during validation are selected.

By using distinct validation data, the network's tendency to overfit to the training data is mitigated. Additionally, by carefully choosing the stopping criterion, training persists until the point where the performance on unknown data starts to decline. Given that certain fluctuations in the validation data costs are inevitable, not solely attributed to overfitting, a minor decline in estimation quality over a few epochs is typically accepted to avoid prematurely ending training. It is recognized that a meaningful decrease could still occur even after a temporary increase in costs.

The stopped criterion chosen, denoted UP_s by $s = 5$ and $k = 5$, dictates that training stops when the validation data costs have exceeded those of the last validation for consecutive instances of $s = 5$, which occur every k epochs. The termination transpired after 195 epochs in this training instance, revealing that the parameter values obtained after epoch 156 yielded the lowest observed costs. Consequently, the network with these parameters is identified as the most effective among those examined in classifying the validation data, ideally providing superior classifications for unknown data. This network will be further evaluated in the following sections.

3.4 Hyperparameter Tuning

Learning Rate and `hidden_size`. We conducted multiple experiments of learning rate and `hidden_size`. For comparison of results, we computed the F score with the target class "melody" after each epoch using the validation data. We examined the learning rates of 0.01, 0.001, and 0.0001, as well as the `hidden_sizes` of 16, 32, 64, and 128 (see Figs. 3 and 4). In particular, with a `hidden_size` of 16, we can observe the impact of the size of the learning rate on learning success: a relatively high learning rate of 0.01 led to rapid but unstable increases in the F score, while a lower rate of 0.0001 showed slower but more stable improvements. An average learning rate of 0.001 offered a quickly converging, yet relatively stable compromise, yielding the best validation results in most epochs.

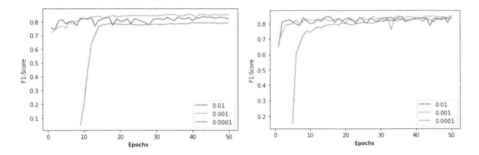

Fig. 3. Left: f1-score of experiments on learning rates of 0.01 (blue), 0.001 (orange), 0.0001 (green) over 50 epochs and `hidden_size` of 16. Right: f1-score of experiments on learning rates of 0.01 (blue), 0.001 (orange), 0.0001 (green) over 50 epochs and `hidden_size` of 32. (Color figure online)

The learning rate of 0.01 and `hidden_sizes` 16 and 32 were excluded from consideration. This decision was based on several factors. First, the learning rate of 0.01 resulted in unstable training and produced inferior results, especially with a larger `hidden_size`. Additionally, the F scores for `hidden_sizes` 16 and 32 across all learning rates seemed to plateau around 0.8. In contrast, the larger `hidden_sizes` showed F scores above 0.85, with indications of further improvement even after 50 epochs.

To investigate the effects of the remaining learning rates-0.001 and 0.0001-as well as `hidden_sizes` of 64 and 128, we conducted a parallel experiment spanning 250 epochs (see Fig. 5).

Upon analyzing the F scores, we consistently observed superior performance from networks trained with a learning rate of 0.001. As a result, we decided to adopt this learning rate.

Furthermore, we observed strikingly similar learning behavior between 64 and 128 of the `hidden_sizes`. Therefore, we decided to stick with the less resource

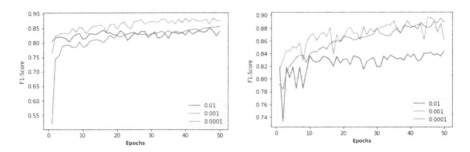

Fig. 4. Left: f1-score of experiments on learning rates of 0.01 (blue), 0.001 (orange), 0.0001 (green) over 50 epochs and **hidden_size** of 64. Right: f1-score of experiments on learning rates of 0.01 (blue), 0.001 (orange), 0.0001 (green) over 50 epochs and **hidden_size** of 128. (Color figure online)

intensive **hidden_size** of 64, under the assumption that further increases beyond this value would not significantly enhance the quality of the estimation.

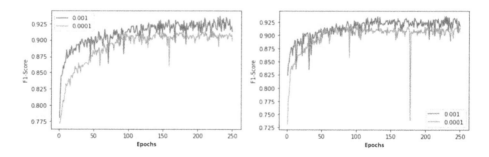

Fig. 5. Left: f1-score of experiments on learning rates of 0.001 (blue), 0.0001 (orange) over 250 epochs and **hidden_size** of 64. Right: f1-score of experiments on learning rates of 0.001 (blue), 0.0001 (orange) over 250 epochs and **hidden_size** of 128. (Color figure online)

Note Duration Versus Note Ends. While the number of features per time step is inherent to the properties of the notes, the choice between representing a feature as either the note duration or the note end could impact the quality of classification. The note end represents the absolute time in seconds at which the `Note Off` message should occur, while the note duration indicates the duration of the note in seconds, calculated from the difference between the note end and the note start.

To determine the preferred feature for classification, we prepared two subsets. Each set contained known time series with four features per element: one with note durations and the other with note ends. We then trained networks with identical hyperparameters, except for the learning rate, over 250 epochs. Specifically, we used the last considered learning rate values of 0.001 and 0.0001 to assess the influence of feature choice under varied conditions. The results of this experiment are depicted in Fig. 6.

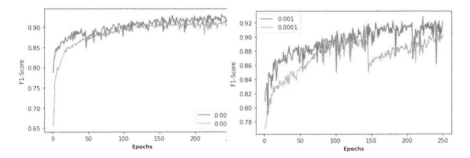

Fig. 6. Experiments on note duration versus note end: grade duration (left) and end of grade (right) over 250 epochs.

Given that opting for the feature "note duration" appears to yield a more stable learning process without compromising classification quality, we preferred the set containing note duration as a feature.

Mini-batchsize. We experimentally explored three values to identify a suitable batch size: 32, 64, and 128. Larger values were not viable due to technical constraints. The results are depicted in Fig. 7.

Obviously, the larger mini-batch sizes appear to perform slightly better than smaller ones. For the purpose of this work, the mini-batch size of 128 is therefore assumed.

Optimization Process. The selection of the optimization method plays a crucial role in parameter optimization. We trained the same network using gradient descents and Adam over 250 epochs to determine the most suitable method. Subsequently, we examined the trend of the mean cost of the validation data using binary cross-entropy (see Fig. 8).

Obviously, Adam leads to a faster decrease in costs, which is why it was chosen as the optimization method.

Output of the LSTM Layer. In Sect. 3.2, we discussed two approaches, "Average" and "Last", for computing the output from the LSTM layer. In each case, this results in a one-dimensional tensor of size hidden_size, with its elements determined by different methods. We trained two networks over 250 epochs, each implementing one of these approaches (see Fig. 9).

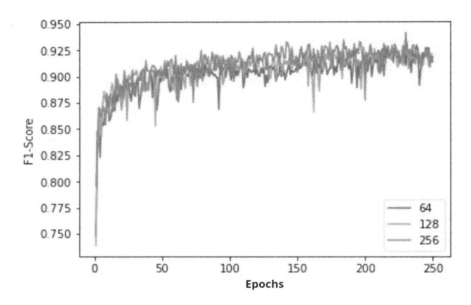

Fig. 7. Experiments on mini-batch sizes

As the strategies for the learning process appear to be equivalent, the approach "Last" was chosen as the strategy for determining the output of the LSTM layer, as this eliminates the need to calculate the mean value.

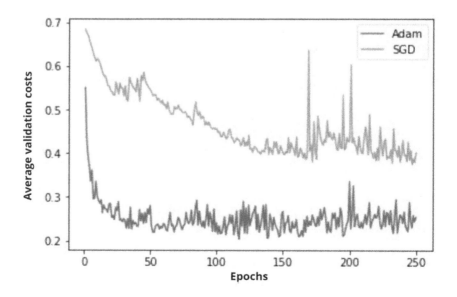

Fig. 8. Average validation costs after optimization process

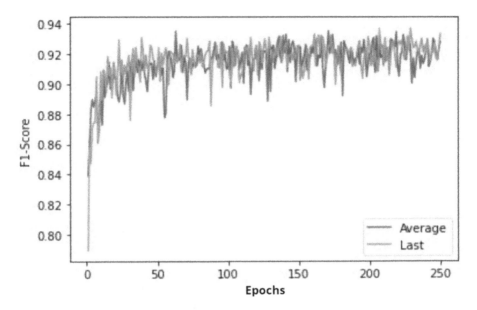

Fig. 9. Learning success according to LSTM output strategy over 250 epochs

4 Results and Discussion

4.1 Data Partitioning

As concluded in Fig. 1, the data corpus for this work consists of 2,505 MIDI files, each containing a single melody track. Initially, 20% of the data, that is, 501 randomly selected files, were set aside as a test data set to ensure their separation from the rest for an unbiased evaluation. This segregation is crucial to prevent the test data from influencing the network's training, thus enabling an impartial assessment of estimate quality.

For training and validation, data were drawn from the remaining 2,004 files. All tracks from these files were extracted, resulting in a combined dataset comprising 2,004 melody tracks and 2,004 randomly selected accompaniment tracks to ensure the balance of training resources. This deliberate measure was taken because, understandably, there are dramatically more accompaniment tracks (18,694) than melody tracks.

4.2 Metrics and Results

We applied accuracy, precision, recall, F score, Youden-index [21], and confusion matrix, as the evaluation metrics, of which the not quite common applied Youden-index is calculated as

$$J = \frac{tp \cdot tn - fn \cdot fp}{(tp+fn)(tn+fp)} = \frac{tp}{tp+fn} + \frac{tn}{tn+fp} - 1 \qquad (1)$$

This index falls between -1 and 1, with the network classifying better the further the index is from 0 (negative values actually mean worse classification, but provide a classifier whose index is better than 0 by selecting the class that is not estimated).

The results of the metrics applied on the 501-file test set are as follows:

- Accuracy: 91.22%
- Precision: 0.9105
- Recall: 0.9142
- F score: 0.9124
- Youden-index: 0.8244

Figure 10 provides the confusion matrix.

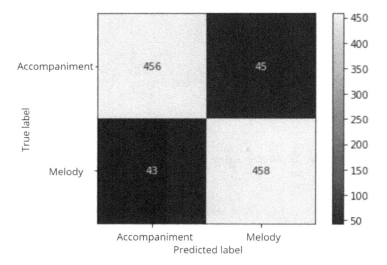

Fig. 10. Confusion matrix of the experiments

Another verification method that could be performed is cross-validation. If the dataset division is sequentially ordered, we could, of course, use five-fold cross-validation for 20% of the test set, which also offers full reproducibility; since we partitioned the data randomly, it is of little significance to list multiple experiments; in fact, there were no significant differences between the results of repeated runs.

Observation. Given the assumption of reasonably representative test data, the evaluation metrics illustrate the model's quality: accuracy, precision, recall, and F score all exceed 0.91. Furthermore, the Youden index rates the classification at 0.8244, indicating robust performance even for negative cases. This alignment is reinforced by the similarity in quality between positive and negative classes, as depicted in the confusion matrix.

4.3 Strategies for Peer Comparison and Results

Considering that numerous related studies primarily address the extraction of melodic tracks from MIDI files, while fewer delve into the classification of individual tracks, the following comparison evaluates the quality of melodic track selection in MIDI files. The evaluation contrasts estimates generated by the developed network with approaches utilized by predecessors. The comparison highlights the importance of accurately classifying melody tracks. To tackle this, a strategy is proposed wherein the desired track is selected based on the network's estimates.

Melody Track Selection Strategy. When dealing with a MIDI file containing a melody track along with multiple accompaniment tracks, the mesh estimates the class of each track, producing values between 0 and 1 due to the sigmoid layer.

Following this, the tracks can be arranged based on their estimated values, allowing the selection of the track with the highest estimation as the melody.

Alternatively, it could be opted to select the first n tracks and apply the metrics of "track precision" and "track recall" as outlined in [17] for comparative analysis with other methodologies.

Test Set. The test set for this comparison has been established in previous chapters. However, in this instance, the melody and accompaniment segments are not extracted, as the objective is to select the melody segment directly from the files themselves. Consequently, the test dataset comprises 501 MIDI files, each containing a melody segment and multiple accompaniment segments. The trained network was then tested using this dataset and the melody track selection strategy. This testing process yielded the values presented in Table 1.

Table 1. Results of the melody track selection

	Selecting the correct track (tp)	Selecting the false track (fp)
#	423	78

From these values, the precision of the network for the test data and the selection strategy is 0.8443, which implies that in 84.43% of the test files the chosen track accurately corresponds to a melody track.

Comparison. The upcoming comparison contrasts the melody selection results obtained from the network with those derived from two methodologies detailed by Tang, Yip, and Kao [17], acclaimed for delivering optimal outcomes, with the exception of one additional strategy. Both approaches involve selecting n candidate tracks presumed to contain the melody.

The two approaches, along with the network trained in the preceding chapters, were evaluated using the test set. This assessment yielded the values shown in Table 2 for the metrics (where $n = 3$ for the approaches of Tang et al.). Evidently, our proposed LSTM-MorA excels in significantly selecting the melody track compared to the considered approaches.

Table 2. Track precision and track recall according to Tang et al. [17]

	Track precision	Track recall
(1) AvgVel	0.3134	0.3393
(2) PMRatio	0.1816	0.2814
LSTM-MorA ($n = 1$)	0.8443	0.8443

Further Existing Approaches. Evaluation of many existing approaches becomes challenging due to the unavailability of implementation details, hindering straightforward re-implementation. Furthermore, the datasets used are often unspecified or difficult to reconstruct. As a result, comparing with other existing approaches is limited to a potentially biased assessment based solely on metrics for separate test data.

Velusamy et al. [20] claimed a precision of 97%, which was not achieved after implementation using the provided test data. This discrepancy could arise from inadequate post-implementation of the approach.

Li et al. [7] reported a precision of 84.2% using deep learning, comparable to the 84.43% achieved by the LSTM-MorA proposed in this study. However, conducting a direct comparison is challenging due to the inability to retrieve the learned parameters of the network and the lack of clear labeling.

4.4 Challenges and Possible Solutions

A notable challenge encountered during the evaluation of the trained network is its difficulty in classifying and selecting melodic tracks, especially those that diverge from the conventional patterns found within the dataset. These instances may include melodies featuring prolonged simultaneous note playing, which deviates from typical melodic characteristics, as well as melodies with structures that are underrepresented in the dataset, such as music from diverse cultural backgrounds or with unconventional interpretations of music theory. This limitation could have influenced the comparison with the melody track selection approach of Velusamy et al., particularly since their approach was tested on Hindi music. To address this issue, one potential solution could involve training the network with more diverse datasets. Additionally, the use of human musicological knowledge during the classification process can help identify and classify such musical nuances.

Another aspect to consider is the current strategy of simply selecting the track with the highest probability of being a melody, which overlooks the harmonic structure of the music piece and fails to utilize potentially valuable information. Implementing a more refined strategy that accounts for these factors could further improve the quality of melody selection.

An overarching challenge that has emerged during this research is the absence of a standardized dataset for uniformly evaluating different approaches to the problems investigated here and similar ones. This lack of standardization makes it difficult to compare these approaches effectively, hindering the assessment of the value of new or modified methodologies. Although developing and providing such a dataset can involve substantial effort, it could offer substantial benefits in facilitating fair and comprehensive evaluations.

5 Conclusion and Future Work

This study delved into the problem of classifying melody tracks, with the aim of determining the suitability of an LSTM network-based approach. To accomplish this, we applied labeled training and test data from a freely available dataset of MIDI files. We then constructed an LSTM network, called LSTM-MorA, and trained it on these data, meticulously optimizing hyperparameters through selection and experimental adjustment.

Following training, we evaluated the LSTM network using various metrics, revealing high classification quality on the test data, with accuracy, precision, recall and F score exceeding 91%. Moreover, we assessed the LSTM network alongside a straightforward selection strategy in one of its potential applications: melody selection from MIDI files. Remarkably, in the test dataset, it performed comparably to, if not better than, existing specialized approaches, showcasing its efficacy in this application context.

The study innovates the application of LSTM networks in classifying and selecting melody tracks from MIDI files by treating them as multivariate time series, and demonstrated its effectiveness.

References

1. Barandas, M., et al.: TSFEL: time series feature extraction library. SoftwareX **11**, 100456 (2020). https://doi.org/10.1016/j.softx.2020.100456
2. Folgado, D., et al.: TSSEARCH: time series subsequence search library. SoftwareX **18**, 101049 (2022). https://doi.org/10.1016/j.softx.2022.101049
3. Groves, R.: Automatic harmonization using a hidden semi-Markov model. In: Proceedings of the AAAI Conference on Artificial Intelligence and Interactive Digital Entertainment, vol. 9, pp. 48–54 (2013)
4. Hochreiter, S., Schmidhuber, J.: Long short-term memory. Neural Comput. **9**(8), 1735–1780 (1997)
5. Jiang, Z., Dannenberg, R.B.: Melody identification in standard midi files (2019). https://api.semanticscholar.org/CorpusID:204766730
6. Kingma, D.P., Ba, J.: Adam: A Method for Stochastic Optimization (2017)

7. Li, J., Yang, X., Chen, Q.: Midi melody extraction based on improved neural network. In: 2009 International Conference on Machine Learning and Cybernetics, July 2009. https://doi.org/10.1109/icmlc.2009.5212378
8. Liu, H., Jiang, K., Gamboa, H., Xue, T., Schultz, T.: Bell shape embodying Zhongyong: the pitch histogram of traditional chinese anhemitonic pentatonic folk songs. Appl. Sci. **12**(16) (2022).https://doi.org/10.3390/app12168343
9. Liu, H., Xue, T., Schultz, T.: Merged pitch histograms and pitch-duration histograms. In: Proceedings of the 19th International Conference on Signal Processing and Multimedia Applications (SIGMAP 2022), pp. 32–39. INSTICC, SCITEPRESS - Science and Technology Publications (2022). https://doi.org/10.5220/0011310300003289
10. Mohapatra, A., Sethy, P.K.: International Journal of Advanced Research in Information and Communication Engineering. ISSN, pp. 2321–8762 (2015)
11. Oore, S., Simon, I., Dieleman, S., Eck, D., Simonyan, K.: This time with feeling: learning expressive musical performance. Neural Comput. Appl. **32**, 955–967 (2020)
12. Pachet, F.: A joyful ode to automatic orchestration. ACM Trans. Intell. Syst. Technol. (TIST) **8**(2), 1–13 (2016)
13. Prechelt, L.: Early stopping - but when? In: Orr, G.B., Müller, K.-R. (eds.) Neural Networks: Tricks of the Trade. LNCS, vol. 1524, pp. 55–69. Springer, Heidelberg (1998). https://doi.org/10.1007/3-540-49430-8_3
14. Raffel, C.: The lakh midi dataset v0.1 (2016). https://colinraffel.com/projects/lmd/
15. Raffel, C.: Learning-based methods for comparing sequences, with applications to audio-to-midi-alignment and matching. Dissertation, Columbia University (2016)
16. Rodrigues, J., Liu, H., Folgado, D., Belo, D., Schultz, T., Gamboa, H.: Feature-based information retrieval of multimodal biosignals with a self-similarity matrix: Focus on automatic segmentation. Biosensors **12**(12) (2022). https://doi.org/10.3390/bios12121182
17. Tang, M., Yip, C.L., Kao, B.: Selection of melody lines for music databases. In: Proceedings 24th Annual International Computer Software and Applications Conference, COMPSAC 2000 (2000). https://doi.org/10.1109/cmpsac.2000.884725
18. Uitdenbogerd, A., Zobel, J.: Melodic matching techniques for large music databases. In: Proceedings of the Seventh ACM International Conference on Multimedia (Part 1), October 1999. https://doi.org/10.1145/319463.319470
19. Uitdenbogerd, A.L., Zobel, J.: Manipulation of music for melody matching. In: Proceedings of the Sixth ACM International Conference on Multimedia - MULTIMEDIA '98 (1998). https://doi.org/10.1145/290747.290776
20. Velusamy, S., Thoshkahna, B., Ramakrishnan, K.R.: A novel melody line identification algorithm for polyphonic MIDI music. In: Cham, T.-J., Cai, J., Dorai, C., Rajan, D., Chua, T.-S., Chia, L.-T. (eds) MMM 2007. LNCS, vol. 4352, pp. 248–257. Springer, Heidelberg (2006). https://doi.org/10.1007/978-3-540-69429-8_25
21. Youden, W.J.: Index for rating diagnostic tests. Cancer **3**(1), 32–35 (1950)

Software Security

Ch4os: Discretized Generative Adversarial Network for Functionality-Preserving Evasive Modification on Malware

Christopher Molloy[✉], Furkan Alaca, and Steven H. H. Ding

Queen's University, Kingston, ON K7L 3N6, Canada
{chris.molloy,furkan.alaca,steven.ding}@queensu.ca

Abstract. Rapid advancements in Artificial Intelligence (AI) have led to applications in varying domains. Due to the exponential growth of cyberspace in recent years, the domain of cybersecurity has seen substantial integrations of AI to aid in handling large amounts of data. The discipline of malware analysis within cybersecurity has leveraged AI to develop advanced analysis techniques. Within malware analysis, AI has been applied to both malware detection and evasive malware generation. Adversarial Learning on Malware (ALM) is the study of evasive modifications that is focused on AI-based detection tools. Most of the existing ALM evasive modification methods produce samples that are not valid executables. Solutions that produce effective valid executables are limited to injecting random code from a finite set of benign samples. Instead of using known code, we aim to optimize the injected bytes to increase evasion probability through adversarial learning. We propose Ch4os, a malware modification system trained in a Generative Adversarial Network setting. We introduce the Valid Machine Code Execution (VaME) activation function, guaranteeing functionality of modified malware samples while preserving differentiability of the learning process. As well, to address the challenge of learning efficiency and stability, we introduce the Binary Copier Pre-training (BCP) method. We conduct experiments on a dataset of chronologically separated malware for a simulated real-world detection scenario and show Ch4os can generate 152% more evasive samples compared to the state-of-the-art.

Keywords: Machine Learning · Adversarial Learning · Malware Analysis

1 Introduction

Due to notable advancements, Artificial Intelligence (AI) has garnered widespread attention and adoption worldwide. One area that has seen a substantial integration of AI is cybersecurity. With the current number of internet-connected devices sharing data with one another, human-based analysis for malware detection is now impractical. Notably, the production of novel malware has reached unprecedented levels in recent years. According to the AVTest institute, there have been more than 79 million new unique malware samples indexed in 2022 that attack Microsoft systems[1]. At this scale, it is impracticable to assume human investigation can suffice. The conventional signature-based method is also unable to cope with the dynamic and versatile characteristics of malware variants. To augment human investigation at scale, multiple AI approaches have been proposed for different challenges within cybersecurity.

One research area within the cybersecurity space is attacking AI-based anti-malware tools. Given some malware and an AI-based anti-malware tool, attackers modify incoming samples or feature vectors of malware with the intent of having the sample classified as benign. This area of research, as well as the practice of creating AI methods that are more challenging to evade, is known as Adversarial Learning on Malware (ALM). Multi-agent systems have been used for ALM in recent works to better simulate real-world defender-attacker environments [13].

Generative Adversarial Networks (GANs) are multi-agent neural networks designed to create approximations to some data space with higher resolution than conventional generative networks. GANs have been applied to the cybersecurity space in both malware generation and detection [6,8,11,14,15]. [6] used a GAN to generate adversarial feature vectors for malware samples against a black box system. A drawback of this work is the GAN was only trained to produce a modified feature vector of the sample. Reinforcement Learning (RL) is also used for adversarial sample generation. Current RL functional adversarial sample generation methods train against AI-based anti-malware engines with a finite set of actions for malware sample modification [1,13]. Molloy et al.. show that a two-party game environment improves the evasiveness of generated samples, as well as the detection ability of an AI-based anti-malware tool [13]. Molloy et al.. [13] found that the most effective method for producing functional adversarial samples was appending benign machine code to the malicious executable. One drawback of this method is that it relies on a repository of pre-written benign machine code that can be sampled from to modify malicious executable samples.

We propose a different route, Ch4os, a functionality-preserving adversarial sample generation tool. Ch4os generates and directly optimizes functionality-preserving machine code injected into a malicious file's overlay, augmenting its evasiveness against detection tools. Ch4os is trained in a GAN architecture against a pre-trained state-of-the-art Deep Learning based anti-malware tool for adversarial machine code optimization. Unlike the previous methods that have proposed GAN architectures for malware generation, we introduce the VaME activation function, a discretized function which maps continuous values to machine code. To address the learning efficiency and stability challenge of adversarial machine code optimization from conventionally initialized network weights, we propose a pre-training method that prepares the network for adversarial byte generation by optimizing to learn the data distribution of benign machine code. Our system has shown successful evasive sample genera-

---

[1] https://portal.av-atlas.org/.

tion against a state-of-the-art anti-malware engine trained on a research standard dataset. The main contributions of this paper are as follows:

1. We propose the first GAN for functional evasive malware modification with the novel VaME activation function, enabling differentiability of the complete learning process.
2. To address the challenge of learning efficiency and stability, we propose the Benign Copier Pre-training (BCP) method, which accelerates the initial convergence problem.
3. We compare our training method against contemporary solutions using a real-world dataset of malware samples and show our method can produce 152% more evasive samples than the state-of-the-art.

2 Related Work

GANs are a neural network architecture proposed by Goodfellow *et al.* GANs use two networks, a generator and a discriminator, to train off one another for better results. GANs have been designed against adversarial attacks, attacks in which data samples are modified to be misclassified by the network [10]. This idea has extended to cybersecurity in the domains of ALM and zero-day malware detection [8,11,14,15]. Kim *et al.* proposed the tGAN (transferred GAN) for classifying malware [8]. The proposed system used an autoencoder to generate images of malware from rescaled assembly code visualizations. The discriminator is then transferred to a malware detector with a family classification accuracy of 96.39%. Moti *et al.* proposed a GAN that simulated potential zero-day malware Binary header information [14]. The simulated headers were incorporated into a malware dataset and trained through multiple classifiers. Training the classifier with the generated data showed an improvement in malware classification accuracy from 97.12% to 98.14%. Lu *et al.* proposed a method of increasing a malware dataset size by simulating malware samples with a GAN [11]. Lu *et al.* showed that training a network with the simulated data could raise the classification accuracy by 6%.

Hu *et al.* proposed MalGAN, a GAN architecture for generating adversarial samples against a black box malware detector [6]. The system proposed by Hu *et al.* was able to achieve a true positive percent of zero in some test datasets. The generator in this work modifies each malware feature vector and does not generate a functional piece of obfuscated malware. Kargaad *et al.* built on top of this system by proposing a malware detection network that is trained on the resulting data from MalGAN [7]. Nazari *et al.* proposed a CGAN-based upsampling method to produce software data samples for balancing malware datasets [16]. Trehan *et al.* showed that multiple different GAN architectures can be used to generate malware mnemonic opcode sequences for model training [20].

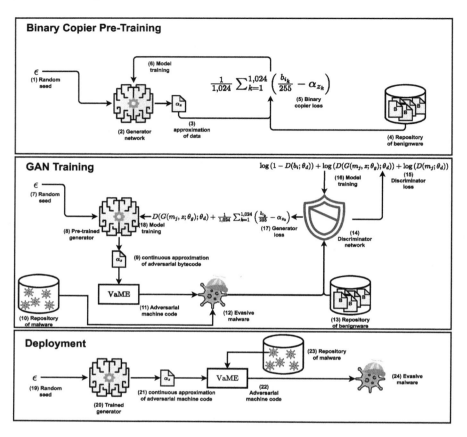

Fig. 1. An overview of the training and deployment process of Ch4os. Nodes represent an action, and edges represent the next action that requires such action. (1) Generate a random vector from prior distribution. (2) Pass vector as input to generator network. (3) Generate an approximation of data b. (4) Retrieve a sample of benignware. (5) Generate benign copier loss from benign sample and approximate data. (6) Train generator on loss. (7) Generate a random vector from prior distribution. (8) Pass random vector into generator network. (9) Generate continuous approximation of adversarial machine code. (10) Retrieve a sample of malware. (11) Generate adversarial machine code. (12) Generate evasive malware. (13) Retrieve a sample of benignware. (14) Evaluate samples with discriminator network. (15) Generate discriminator loss. (16) Train the discriminator network on discriminator loss. (17) Calculate generator loss. (18) Train generator network on generator loss. (19) Generate a random vector from prior distribution. (20) Pass random vector into trained generator network. (21) Generate continuous approximation of adversarial machine code. (22) Generate adversarial machine code. (23) Retrieve a sample of malware. (24) Generate evasive malware.

3 Methodology

The Ch4os system is a deep generative neural network for adversarial machine code generation. An overview of the Ch4os system can be seen in Fig. 1. Given a random vector input, a kilobyte of adversarial machine code is generated to be repeatedly appended to the end of the malware machine code. The resulting malware sample contains machine code that has been optimized to evade static analysis-based anti-malware engines. The adversarial machine code is generated from the generator network and goes through two steps of training optimization. The first step of training is optimizing the generator network to create benign machine code from a random vector. The second step of training is optimizing the generator to create evasive bytes in a multi-agent environment against a pre-trained malware detection network.

3.1 Data

We define data b as benign Microsoft PE machine code. The PE file format is the most used executable file format for Microsoft Windows operating systems. PE Executables are written in machine code and are comprised of various headers and sections [17]. Headers contain important information such as the intended machine type, the number of sections, and the number of symbols. PE sections contain code blocks for PE-specific execution. Also, a PE Executable file may contain an overlay, optional code machine code added to the end of the file that is not mapped to anywhere in memory [5].

Our data m is malicious Microsoft PE machine code. Malware is defined as any piece of code that has been designed to cause harm or subvert a computer's intended function when executed [12]. Malicious PE files have the same format as benign files but are designed to perform different actions on a user's computer than benign code.

3.2 Adversarial Networks

We build on the work of Goodfellow *et al.* for defining our adversarial network [4]. In this work, we optimize the distribution p_g of the generator network G over data b. We define data m as input to the generator. We also define a prior noise variable z from distribution p_z as input to the generator. We define $G(m, z; \theta_g)$ as a mapping from the space of data m and the noise variable z to the b data space. G is a deep neural network with parameters θ_g. We define $D(\cdot; \theta_d)$ that outputs a single scalar between 0 and 1. $D(\cdot; \theta_d)$ represents the probability that the input data came from p_g rather than b. This can be interpreted as if the input data is untrue to the data b. We train D to maximize correct assignment of probability to both training samples from b and data generated from G. While D is trained, we concurrently train G to minimize the correct assignment from D for data generated from G. It follows that G and D play a two-player minimax game with the value function $V(G, D)$:

$$\min_G \max_D V(G, D) = \mathbb{E}_{b \sim p_{data}(b)}\left[1 - D(b)\right] + \mathbb{E}_{z \sim p_z(z)}\left[D(G(z))\right] \quad (1)$$

3.3 Adversarial Attack

The target of the Ch4os system is deep learning models trained in malware detection [18]. Given an input sample executable, some preprocessing may be required for feature set generation, and some models perform inference directly from the software binary code [18]. These networks have a sigmoid activation head for malware prediction. Malicious software has a true positive value of 1, and benign software has a true positive value of 0. Based on experimentation results, a classification threshold $t \in 0 \leq \mathbb{R} \leq 1$ is chosen for the network to optimize the area under the ROC curve [19]. The Ch4os system aims to modify malware binaries at the machine code level to reduce the classification prediction below the target detection network's threshold without implicit knowledge of the threshold. This is done by adding machine code generated by $G(m, z; \theta_g)$ to the overlay of the attacking malware sample. As is the case with benign PE executables, PE malware executables can also have machine code added to the overlay that does not disrupt intended functionality. The adversarial attack is conducted as follows:

1. A sequence of bytes fitted to data b is generated by the generator network $G(m, z; \theta_g)$.
2. The sequence of generated bytes is concatenated to the end of a malware sample executable.
3. The concatenation is repeated with the same machine code until the modified sample is over the maximum input size of the target anti-malware engine.

It is intended that the adversarial machine code at the end of the malware sample reduces the malicious probability under the detection threshold against the target anti-malware engine.

Table 1. Structure of generator network.

Layer	Size-in	Size-out	Kernel	Params
random sample		1		0
dense1a	1	16		32
dense1b	16	256		4352
conv1a	256	256	1	65,792
lnorma	256	256		512
conv1b	256	512	1	131,584
lnormb	512	512		1024
conv1c	512	1024	1	525,312
lnormc	1024	1024		2048
flatten	1024	1024		0
Total				730,656

3.4 Generator

A challenge for designing the generator network $G(m, z; \theta_g)$ is mapping output from a continuous differentiable neural network to the discrete space of machine code. Deep Learning optimization and training require a continuous and differentiable network for backpropagation. A challenge in this context arises from the limitation that machine code is exclusively comprised of integers, but we require the real-valued output of a continuous activation head for model optimization. Due to this, we propose the Valid Machine Code Execution (VaME) activation function. We define VaME : $0 \leq \mathbb{R}^n \leq 1 \Rightarrow 0 \leq \mathbb{Z}^n \leq 255$, where $n \in \mathbb{N}$ is arbitrary, as

$$\text{VaME}(x) = \lfloor 255x \rfloor. \qquad (2)$$

The VaME activation head is selectively employed in the generation of a functional adversarial sample, while its utilization is absent in the computation of the generator network's loss. This utilization technique poses no challenge to model efficacy, as the VaME activation function has no trainable parameters.

We will now describe the generator network $G(m, z; \theta_g)$. The input to $G(m, z; \theta_g)$ is some malware machine code m and a randomly generated noise z. We sample z from the standard normal distribution $p_z(z) \sim N(0, 1)$. The random noise is upsampled by two fully connected neuron layers to a vector of length 256. The vector is then upsampled further through three deep convolution blocks of one-dimensional convolution and layer normalization. The result of the final convolution block is then flattened to a vector, α_z, of length 1,024. The vector α_z is then passed through a sigmoid activation function

$$\sigma(x) = \frac{1}{1 + e^{-x}}. \qquad (3)$$

The sigmoid activation function transforms α_z such that $0 \leq \alpha_z \leq 1 \in \mathbb{R}^{1,024}$.

The vector α_z is used for model loss calculation as well as adversarial sample generation. We will now explain adversarial sample generation. Given the input dimension of the discriminator function, q, we concatenate the executable code generated by α_z, to the input malware sample m with starting length r, until the length of the malware sample m is larger q. We denote this vector as

$$\beta = m \,\|_{i=0}^{\left\lceil \frac{q-r}{1,024} \right\rceil} \text{VaME}(\alpha_z). \qquad (4)$$

We then remove the trailing bytes until the length of β is the correct input length for the discriminator. The resulting machine code is the output of the generator network, thus

$$G(m, z; \theta_g) = (\beta_i)_{1 \leq i \leq q}. \qquad (5)$$

The network architecture can be seen in Table 1.

Table 2. Structure of discriminator network.

Layer	Size-in	Size-out	Kernel	Params
embedding	1,048,576	1,048,576 × 8		2,056
conv1a	1,048,576 × 8	2097 × 128	500	512,128
conv1b	1,048,576 × 8	2097 × 128	500	512,128
matrix multiply	(2097 × 128, 2097 × 128)	2097 × 128		0
mpool	2,097 × 128	128		0
dense1a	128	128		16,512
dense1b	128	1		129
Total				1,042,953

3.5 Discriminator

We require a pre-trained anti-malware network to act as the discriminator of the Ch4os system. For the discriminator, we train with the modified MalConv detection network trained on the EMBER dataset [2,18]. This pre-trained network was tested in [2], resulting in an ROC of 0.998 on the testing set.

MalConv is a gated deep convolution network for malware detection proposed by Raff et al. [18]. Table 2 shows the architecture of the discriminator model. The input to the discriminator is a bytestring of the first megabyte of the executable's machine code. The bytestring is a vector $b \in \mathbb{Z}^{1,048,576}, 0 \leq b \leq 256$. There are a total of 257 possible values for each byte value due to the use of 256 as a padding character if the software sample is less than one megabyte in length. The bytestring is first embedded into a space that represents each byte in the byte string as a vector of length eight. This space is represented by $D^{|257| \times 8}$ due to the vocabulary of bytes in the executable being 257.

The input to the gated convolution unit is the embedded bytestring. The gated convolution unit takes the input of the embedding and passes it through two parallel one-dimensional convolution units. One of the convolution units acts as an attention layer, and the other a filter layer [18]. Both convolution layers have the same hyper-parameters of a filter size of 128, a kernel size of 500, and a stride of 500, but it is important to note there is a difference in layer activation. One layer uses the rectified linear unit (ReLU) activation function, and the other layer uses a sigmoid activation. The output of the two convolution layers are multiplied with one another to yield the gated unit result. The sigmoid activation is used on one of the convolution layers to create an attention scalar, which is multiplied by the second layer. The output matrix of the attention is max pooled on the embedded dimension. The output of the pooling is sent through two fully connected neuron layers to reduce with output dimensions of 128 and one, respectively. The final neuron has a sigmoid activation function, making the output of the MalConv network a probability. The output of the MalConv network is the probability that the input is malicious. The network architecture can be seen in Table 2.

The most significant difference between the MalConv network originally proposed by Raff et al. and the modified version used in [2] and this work is the input size. The input size to the original MalConv network was the first two megabytes of machine code from the software executable, whereas the input size to the modified MalConv is the first single megabyte of the software sample. This input size was chosen to accommodate the memory capacity of state-of-the-art Graphics Processing Units [2].

We use Transfer learning for the pre-trained discriminator. As previously discussed, the original training task of the MalConv network is classifying if an incoming sample is malicious. Due to software being only benign or malicious, it follows that the MalConv network is trained to classify if an incoming sample is not benign. This can be understood as assigning the probability to if the sample is not from data b. In this study, we consider non-functional software benign due to the inability of non-functional software to cause harm to a user's computer. We transfer this to the task of giving the probability that the incoming information is not from our data b but originates from $G(m, z; \theta_g)$.

3.6 GAN Loss

We will now describe the loss of the discriminator D. We optimize D to maximize the probability of correct assignment of training samples from b and data generated from G. As well we also optimize D to maximize the correct assignment of samples from m. This additional optimization procedure is implemented with the aim of ensuring discriminator D retains the capacity to assign whether a given sample originates from data b based on malicious machine code. For a single train step, we require some random vector z, some data b_i sampled from b, and some data m_j sampled from m. First, we find the probability that sample b_i is from data b. This is done by finding the log loss of assigning b_i to data b calculated as

$$L_1 = \log\left(1 - D(b_i; \theta_d)\right). \tag{6}$$

We then find the probability that our adversarial sample $G(m_j, z : \theta_g)$ is from data b. This is done by finding the log loss of assigning $G(m_j, z : \theta_g)$ to data b calculated as

$$L_2 = \log\left(D(G(m_j, z; \theta_g); \theta_d)\right). \tag{7}$$

Subsequently, we calculate the probability that m_j from data m is assigned to data b. Similarly to the above, this is done by finding the log loss of assigning m_i to data b. This is calculated as

$$L_3 = \log\left(D(m_j; \theta_d)\right). \tag{8}$$

Finally we compute the loss for the discriminator as the sum as the previously discussed losses as $L_D = L_1 + L_2 + L_3$.

We will now describe the loss of generator G. Following the same train step in calculating the loss for the discriminator D, the following is how the generator loss is calculated for the same z, b_i, and m_j. As described in the minimax game, we are optimizing the generator to minimize the correct assignment from the discriminator generated from G. To incorporate this into the generator loss, we create the reward r, where

$$r = 1 - D(G(m_j, z; \theta_g); \theta_d). \tag{9}$$

The reward is 1 if the discriminator assigns the adversarial sample to not in data b with probability 0, and the reward is 0 if the discriminator assigns the adversarial sample to not in data b with probability 1.

With the discriminator correct assignment minimization loss, we also find the mean squared error of the adversarial bytes α_z against the data b_i to further optimize the generator in mapping z to the data b. For the mean squared error calculation, we multiply all bytes in data b_i by $\frac{1}{255}$ to map the machine code to the range of α_z. This is calculated as

$$L_4 = \frac{1}{1,024} \sum_{k=1}^{1,024} \left(\frac{b_{i_k}}{255} - \alpha_{z_k} \right). \tag{10}$$

Finally, the calculation of the total loss of the generator network is performed as $L_G = 1 - r + L_4$.

In experiments, we perform GAN training with the Adam Stochastic Gradient Descent method [9] for both the generator and discriminator. Both models are trained with a learning rate of 10^{-9} for 1,000 epochs.

3.7 Benign Copier Pre-training

We use the weight initialization method proposed by Glorot and Bengio [3] as the method of conventional weight initialization. When initialized with random weights, $G(m, z; \theta_g)$ proves inept at generating an adversarial β. To prepare the generator $G(m, z; \theta_g)$ for adversarial sample generation, the network is pre-trained in a benign machine code generation task. We refer to this training process as Benign Copier Pre-training (BCP).

The input to the generator network $G(m, z; \theta_g)$ is some malware machine code m and a randomly generated noise z. As well, we require a set of benign machine code as a training set. Similar to GAN training, the noise z is sampled from the standard normal distribution. For BCP, malware machine code is not used for training, so the input is some arbitrary machine code m. We optimize the network to generate machine code that is similar to samples from data b. Given some random noise z, we compare the generated machine code VaME(α_z) to a vector of benign machine code of the same length from our training set. The advantage of this pre-training is that before any training in the GAN environment, our network can already generate machine code that is similar in distribution to data b. In BCP training, we optimize the generator to generate machine code that is similar to data b whereas in GAN training, we optimize the generator to generate an adversarial malware sample that is similar to data b.

3.8 Benign Copier Loss

We will now describe the loss of the BCP. In BCP, we optimize the generator $G(m, z; \theta_g)$ in generating machine code, α_z, that is similar to data b. For a single train step, we require a random noise z and some benign machine code b_i sampled from data b. First, we find the continuous output of the generator for the random noise z as α_z. We then optimize our network on the comparison of each continuous value of α_z to each value in b_i that is mapped from the machine code space in the same continuous space of α_z. This mapping is simply done by dividing each byte of b_i by 255. We compare these bytes by finding the mean squared error for each value in the two vectors α_z and b_i. We treat b_i as the true positive value and α_z as the predicted value. We calculate this loss as

$$L_{BCP} = \frac{1}{1{,}024} \sum_{k=1}^{1{,}024} \left(\frac{b_{i_k}}{255} - \alpha_{z_k} \right). \tag{11}$$

The loss L_{BCP} has the same form as L_4. The loss derived from L_{BCP} is then propagated through the network to optimize for benign machine code generation.

In experiments, we perform the BCP with the Adam Stochastic Gradient Descent method [9]. We train the model with a learning rate of 10^{-5} for $10{,}000$ epochs.

4 Experiments

For our experiments, we required four datasets. All samples used for training and testing were Microsoft PE Executable files. All samples used were collected from various online repositories.

The first dataset was used to conduct the BCP generator training. For BCP training, we used a dataset of 1,000 benignware samples upsampled from a set of 915 unique benign software binaries through sampling with replacement. For the BCP training, we used a train-validation split of $0.8 - 0.2$.

The second dataset used was for GAN training. For GAN training, we used a dataset of 15,000 malicious-benign pairs. Within the entire dataset, there were 15,000 unique malware samples first identified in 2021 from 173 unique families and 4,873 unique benign samples that were up-sampled to the required 15,000 pairs.

The third dataset was required for testing the evasive ability of the Ch4os system. This dataset was referred to as the Detection Testing Set, and was used to determine an optimal detection threshold of the target detection network. This dataset was comprised of 6,800 malware and benignware samples with 3,413 malicious and 3,387 benign. All malware samples in the Detection Testing Set were first identified in 2021.

The fourth dataset was a holdout set of malware for novel generation. The holdout set was comprised of 6,000 unique malware samples first identified in 2022 from 63 unique families.

We conducted two experiments to demonstrate the efficacy of the Ch4os system. The first experiment validated the BCP method. We performed GAN training with a generator that had weights initialized by the Glorot and Bengio method [3] as well as a generator that was first optimized with the BCP method. As described in Sect. 3.6, for GAN training, both models trained for 1,000 epochs on the GAN training dataset. Prior to GAN training, the generator that was optimized with the BCP method was trained for 10,000 epochs on the BCP training dataset. We refer to the model that was pre-trained with BCP as Ch4os and the model that was not pre-trained as GAN-NO-BCP.

For our first experiment, we used four metrics from the training and validation sets to measure the results of training the GAN-NO-BCP model compared to the Ch4os model. The first metric used was the generator loss. This was the loss of the generator network on the final epoch of training. The optimal generator loss value was 0. The second metric was the discriminator loss. This was the loss of the discriminator function on the final epoch of training. The higher the discriminator loss, the better the performance of the generator. The third metric used was the generator reward. The optimal value for the generator reward was 1. The fourth metric used was the detection accuracy of the generator given based on a threshold of 0.5. The closer the training reward was to 0 indicated a higher performance in the generator.

Table 3. Results of training models with different weight initialization in the GAN architecture. We refer to training loss as T-Loss and validation loss as V-Loss. We refer to training reward as T-Reward and validation reward as V-Reward. We refer to training accuracy as T-Accuracy and validation accuracy as V-Accuracy.

Model	Gen. T-Loss	Disc. T-Loss	Gen. T-Reward	Disc. T-Accuracy
GAN-No-BCP	0.6611	4.2006	0.5875	0.7015
Ch4os	**0.1332**	**6.8089**	**0.8916**	**0.6016**
Model	Gen. V-Loss	Disc. V-Loss	Gen. V-Reward	Disc. V-Accuracy
GAN-No-BCP	0.6544	4.2718	0.594	0.6966
Ch4os	**0.1158**	**7.0781**	**0.909**	**0.5956**

The second experiment validated the evasive ability of the Ch4os system. We validated the evasive ability of the Ch4os system by generating an evasive set of samples using the holdout set of malware. We then evaluated the classification accuracy of the MalConv network trained on the EMBER dataset against the evasive samples [2,18]. To simulate a real-world malware triage environment, we used the Detection Testing Set to find the Optimal Threshold (OT) for the successful classification of the MalConv network. We determined the OT by maximizing the Youden's J statistic of the ROC curve. We then evaluated our evasive set using the found OT. We compared our results against two state-of-the-art RL-based function malware generation systems [13] and [1] with the same training and holdout sets as benchmarks. We also tested the evasive ability of randomly generated bytes and the unmodified malware for further benchmarks for the Ch4os system. Finally, we tested the evasive ability of a network only trained through BCP and the BCP-NO-GAN networks to further study the efficacy of Ch4os. Three metrics were used to measure the evasive experiment. The first metric used was accuracy. This accuracy refers to the accuracy of the MalConv network that was used as the target anti-malware engine. The lower the accuracy, the higher the evasive ability of the system. The second metric used was False Negative (FN). FN indicate the number of malware samples that were incorrectly classified as benign. This metric was used to measure the rise or fall in the number of evasive samples compared to the unmodified malware. As well, for measuring the MalConv network on the Detection testing set we used the metrics Area Under the ROC curve (AUC), Accuracy, F1, Precision, Recall, False Negatives, and Optimal Threshold.

Table 4. MalConv-Ember model on results on the Detection Testing Set.

Model	AUC	Accuracy	F1	Precision	Recall	FN	OT
MalConv-Ember [2]	0.9134	0.9134	0.9129	0.9217	0.9042	327	0.0007

As can be seen in Table 3, Ch4os outperformed GAN-No-BCP on both the training and validation sets. In both training and validation, the Ch4os generator had a significantly lower loss (82% decrease and 79% decrease, respectively). Ch4os also had a higher reward than the GAN-No-BCP generator on both sets. For the discriminator of the GAN, the loss was much higher against Ch4os compared to GAN-No-BCP, and the accuracy was much lower against the Ch4os generator.

Table 5. Results of MalConv-Ember network against different adversarial sets.

Evasion Method	Accuracy	FN	% Increase of FN	OT
Unmodified	0.8958	625	–	0.0007
Random	0.9705	177	28%	0.0007
Anderson et al.. [1]	0.8965	621	99%	0.0007
Molloy et al.. [13]	0.8903	658	105%	0.0007
GAN-NO-BCP	0.9677	194	31%	0.0007
BCP	0.7947	1232	197%	0.0007
Ch4os	**0.7323**	**1606**	**257%**	0.0007

The results of the MalConv-Ember [2] network can be seen in Table 4. From these results, we found that the optimal classification threshold was 0.0007. This was the OT used to compare Ch4os to other methods.

The results of the evasive experiment can be seen in Table 5. As described above, the OT from Table 4 is used as the OT for this experiment. Ch4os performed best against all other baselines in generating adversarial samples. These results showed that using a generative network to create adversarial bytes was more effective than the current state-of-the-art method of choosing benign bytes from a finite set.

5 Conclusion

In this work, we propose Ch4os, the first functionality-agnostic GAN system for problem-space evasive malware generation. A limitation of the Ch4os system is that it only appends generated bytes to the end of the malware sample. Future work for Ch4os includes testing the system against industry anti-malware engines and adding adversarial machine code throughout sample executables. Future work also includes testing the Ch4os system against different anti-malware ML systems with varying datasets of malware and benignware.

References

1. Anderson, H., Kharkar, A., Filar, B., Evans, D., Roth, P.: Learning to evade static PE machine learning malware models via reinforcement learning. CoRR, abs/1801.08917 (2018)
2. Anderson, H.S., Roth, P.: EMBER: an open dataset for training static PE malware machine learning models. CoRR, abs/1804.04637 (2018)
3. Glorot, X., Bengio, Y.: Understanding the difficulty of training deep feedforward neural networks. In: Proceedings of the Thirteenth International Conference on Artificial Intelligence and Statistics, AISTATS 2010, Chia Laguna Resort, Sardinia, Italy, 13–15 May 2010, vol. 9 of *JMLR Proceedings*, pp. 249–256. JMLR.org (2010)
4. Goodfellow, I.J., et al.: Generative adversarial nets. In: Advances in Neural Information Processing Systems: Annual Conference on Neural Information Processing Systems 2014, vol. 27, 8–13 December 2014, Montreal, Quebec, Canada, pp. 2672–2680 (2014)
5. Hahn, K., INM Register: Robust static analysis of portable executable malware. HTWK Leipzig, vol. 134 (2014)
6. Hu, W., Tan, Y.: Generating adversarial malware examples for black-box attacks based on GAN. In: Tan, Y., Shi, Y. (eds.) Data Mining and Big Data - 7th International Conference, DMBD 2022, Beijing, China, 21–24 November 2022, Proceedings, Part II, vol. 1745 of CCIS, pp. 409–423. Springer, Cham (2022). https://doi.org/10.1007/978-981-19-8991-9_29
7. Kargaard, J., Drange, T., Kor, A.-L., Twafik, H., Butterfield, E.: Defending it systems against intelligent malware. In: 2018 IEEE 9th International Conference on Dependable Systems, Services and Technologies (DESSERT), pp. 411–417 (2018)
8. Kim, J.-Y., Bu, S.-J., Cho, S.-B.: Malware detection using deep transferred generative adversarial networks. In: Liu, D., Xie, S., Li, Y., Zhao, D., El-Alfy, E.S. (eds.) Neural Information Processing - 24th International Conference, ICONIP 2017, Guangzhou, China, 14–18 November 2017, Proceedings, Part I. LNCS, vol. 10634, pp. 556–564. Springer, Cham (2017). https://doi.org/10.1007/978-3-319-70087-8_58

9. Kingma, D.P., Ba, J.: Adam: a method for stochastic optimization. arXiv preprint arXiv:1412.6980 (2014)
10. Liu, G., Khalil, I., Khreishah, A.: GanDef: a GAN based adversarial training defense for neural network classifier. In: Dhillon, G., Karlsson, F., Hedstrom, K., Zuquete, A. (eds.) ICT Systems Security and Privacy Protection - 34th IFIP TC 11 International Conference, SEC 2019, Lisbon, Portugal, 25–27 June 2019, Proceedings, IFIP Advances in Information and Communication Technology, vol. 562, pp. 19–32. Springer, Cham (2019). https://doi.org/10.1007/978-3-030-22312-0_2
11. Lu, Y., Li, J.: Generative adversarial network for improving deep learning based malware classification. In: 2019 Winter Simulation Conference, WSC 2019, National Harbor, MD, USA, 8–11 December 2019, pp. 584–593. IEEE (2019)
12. McGraw, G., Morrisett, G.: Attacking malicious code: a report to the Infosec research council. IEEE Softw. **17**(5), 33–41 (2000)
13. Molloy, C., Ding, S.H.H., Fung, B.C.M., Charland, P.: H4rm0ny: a competitive zero-sum two-player Markov game for multi-agent learning on evasive malware generation and detection. In: 2022 IEEE International Conference on Cyber Security and Resilience (CSR), pp. 22–29. Molloy-3 (2022)
14. Moti, Z., Hashemi, S., Namavar, A.: Discovering future malware variants by generating new malware samples using generative adversarial network. In: 2019 9th International Conference on Computer and Knowledge Engineering (ICCKE), pp. 319–324 (2019)
15. Nagaraju, R., Stamp, M.: Auxiliary-classifier GAN for malware analysis. CoRR, abs/2107.01620 (2021)
16. Nazari, E., Branco, P., Jourdan, G.-V.: Using CGAN to deal with class imbalance and small sample size in cybersecurity problems. In: 18th International Conference on Privacy, Security and Trust, PST 2021, Auckland, New Zealand, 13–15 December 2021, pp. 1–10. IEEE (2021)
17. Pietrek, M.: An in-depth look into the Win32 portable executable file format, Part 2. MSDN Mag. (2002)
18. Raff, E., Barker, J., Sylvester, J., Brandon, R., Catanzaro, B., Nicholas, C.K.: Malware detection by eating a whole EXE. In: The Workshops of the The Thirty-Second AAAI Conference on Artificial Intelligence, New Orleans, Louisiana, USA, 2–7 February 2018, vol. WS-18 of AAAI Workshops. AAAI Press (2018)
19. Tobiyama, S., Yamaguchi, Y., Shimada, H., Ikuse, T., Yagi, T.: Malware detection with deep neural network using process behavior. In: 40th IEEE Annual Computer Software and Applications Conference, COMPSAC Workshops 2016, Atlanta, GA, USA, 10–14 June 2016, pp. 577–582. IEEE Computer Society (2016)
20. Trehan, H., Di Troia, F.: Fake malware generation using HMM and GAN. In: Chang, SY., Bathen, L., Di Troia, F., Austin, T.H., Nelson, A.J. (eds.) Silicon Valley Cybersecurity Conference - Second Conference, SVCC 2021, San Jose, CA, USA, 2–3 December 2021, Revised Selected Papers. CCIS, vol. 1536, pp. 3–21. Springer, Cham (2021). https://doi.org/10.1007/978-3-030-96057-5_1

SSA-GAT: Graph-Based Self-supervised Learning for Network Intrusion Detection

Qian Liu[1], Hui Zhang[1(✉)], Youpeng Zhang[2], Lin Fan[1], and Xue Jin[1]

[1] China People's Police University, HeBei Province, China
{2021905004,zhanghui01,2021905002,2021905005}@cppu.edu.cn
[2] HuBei University, HuBei Province, China

Abstract. The attacks that derived from the advances of the Internet technology, despite sophistication, remain significantly outnumbered by benign traffic within networks. To deal with imbalance, traditional machine learning (ML) models rely heavily on resampling for data preprocessing, which changes the original data distribution and leads to suboptimal performance. Moreover, these ML algorithms operate on tabulated flows and thus ignore the network topology that is important to identify certain attacks. This paper proposes SSA-GAT, a Graph Attention Network (GAT)-based Network Intrusion Detection System (NIDS), which leverages a graph-based contrastive self-supervised learning framework. We propose four specific augmentation methods and a novel attentive readout function to facilitate the training process, utilizing the joint objective combining supervised and self-supervised losses to improve the performance of minority attack identification. Extensive experiments on the highly imbalanced CIC-IDS2017 dataset demonstrate that the proposed model has excellent performance. For extremely minority attacks, SSA-GAT achieves 62.5% and 66.7% F1 scores on infiltration (occurs only 36 times) and HeartBleed (occurs only 11 times), respectively.

Keywords: Graph Neural Networks · Self-Supervised Learning · Attention Mechanism

1 Introduction

As the Internet becomes ubiquitous in the daily life, recent years have witnessed a significant increase in malicious activities on the Internet; hence the need for Network Intrusion Detection Systems (NIDS). NIDS are cybersecurity tools that monitor network traffic for suspicious or unauthorized activities, safeguarding networks from malicious attacks and intrusions [16]. Traditional NIDS primarily rely on rule-based approaches, employing predefined patterns or signatures for identifying known threats [14]. The rapid evolution of malicious attacks poses a critical challenge to traditional NIDS, since rule-baed systems have limitations in detecting emerging and unknown threats.

Machine learning (ML) based NIDS have recently been deployed as viable solutions for detecting intrusions in computer networks [1]. ML-based NIDS create their own sets of rules based on the data they get trained on, without human supervision or intervention, and, as such, can be highly effective in detecting subtle and complex attack patterns that traditional rule-based approaches may not be able to capture.

Conventional ML-based NIDS classify each network flow as benign or malicious independently, by extracting prominent flow-level features that can characterize different attacks from network packets, while ignoring the topological patterns of network flows [31]. As a result, they can deliver high accuracy in detecting common attacks. However, certain malicious activities exhibit specific topological patterns. For example, DoS and Brute Force, apart from distinctive flow features such as abnormal packet sizes or connection durations, manifest topological features such as a large number of anomalous requests targeted at one or several nodes in a short period. Some patterns may also involve unauthorized nodes (Man-in-the-Middle) in communication links, abnormal redirection behaviors (Phishing), or a star-shaped topology indicating a one-to-many relationship (Botnet). Graph neural networks (GNNs) have been shown to achieve state-of-the-art performance on a variety of graph-based learning tasks [11–13,19,32]. Unsurprisingly, recent studies in NIDS have shifted their attention to using GNN to capture various topological structures in network flows [26].

Most existing NIDS are based on supervised learning, which generally requires large-scale labelled datasets. However, real-world NIDS datasets are highly imbalanced with scarcity of data in minority classes, since malicious cyberattacks are often hidden in large amounts of benign network traffic. Class imbalance poses a significant challenge, since most ML algorithms exhibit bias towards the prevalent classes and ignore minority classes. Resampling, which rebalances the class distribution for imbalanced datasets, has become the *de facto* approach to training NIDS. There are two principal methods for resampling: random oversampling and random undersampling, each with its own benefits and drawbacks. Oversampling works by generating new data instances, duplicated or synthetic, that belong to the minority classes. While oversampling increases the number of samples in the minority classes, it can lead to overfitting as the generated samples are either exact replications of the minority samples or synthesized samples with noises. Undersampling, on the other hand, removes samples from the majority classes to balance the class distribution [35]. However, undersampling inherently leads to information loss since it reduces the total number of samples on which the model is trained. This has motivated us to seek alternative solutions to data imbalance and scarcity.

Self-supervised learning (SSL) has emerged as a promising training paradigm to cope with data imbalance and data scarcity. SSL defines a series of auxiliary (pretext) tasks on abundant unlabelled data, which enables ML models to learn informative and intelligible representations; hence alleviating the need for large-scale labelled datasets. Therefore, SSL is particularly useful in scenarios where obtaining labelled data is difficult or expensive, e.g., medical imaging, fraud detection, and cybersecurity [7,25]. SSL has enjoyed tremendous success in domains such as computer vision (CV) [5,37,38], natural language processing

(NLP) [6], and recommender systems (RecSys) [22,23], while remaining largely unexplored by the NIDS community.

Inspired by the potential of graph neural networks in modeling the topological patterns in network flows [26], and by the success of SSL when dealing with imbalanced and scarce datasets, we propose a novel graph-based self-supervised learning framework for network intrusion detection. We begin by building a graph representation for network flows, where both hosts and flows are represented as nodes in the graph. Then we apply the graph attention networks (GAT) [32] to learn contextualized representations for flow nodes, which are subsequently used for network intrusion detection and classification. Furthermore, we design a set of data augmentation operators tailored to the task of NIDS, and employ the contrastive learning based SSL on the GAT backbone, which helps the model alleviate the data imbalance and scarcity issues commonly encountered in NIDS datasets.

The contribution of this paper can be summarized as follows: (1) We propose a novel NIDS approach based on graph self-supervised learning, which (i) captures the topological structures within the network flows, and (ii) addresses the data imbalance and scarcity issues through contrastive learning. (2) Experiments on CIC-IDS2017 [28], the widely used benchmark dataset for NIDS, show that our approach outperforms a number of state-of-the-art, baseline alternatives. In addition, we conduct an extensive ablation study, which further verifies the strengths of our model. (3) We demonstrate that interpretable explanations for predictions can be obtained by visualizing the attention scores in the model, which helps open the black-box of deep learning based NIDS.

2 Related Work

2.1 Conventional Approaches to NIDS

Existing NIDS methods rely primarily on random resampling to handle data imbalance. For example, SMOTE oversampling [4] has been applied in NIDS to synthesize minority classes [39]. LIO-IDS [9] proposes a two-stage approach [20, 35], where the first stage identifies intrusions from normal network flows, and the second stage classifies the detected intrusions into different attack classes, thus alleviating the bias towards majority classes. FCWGAN [24] relies on conditional WGAN [2] to synthesize minority classes, while [27] studies the effectiveness of SMOTE oversampling and random undersampling. Although simple and effective, resampling changes the data distribution, leading to sample selection bias.

2.2 Graph Neural Networks

Graph Neural Networks (GNN) leverage the message passing mechanism to learn contextualized node representations, which are subsequently used for downstream tasks such as node classification and link prediction [11,19]. Pujol et al. [26] argue that GNN can capture the rich topological features in network flows, making them a powerful tool for NIDS. Experiments on the CIC-IDS2017 dataset demonstrate that the GNN-based approach has improved robustness against adversarial attacks. BS-GAT [36] proposes a novel GNN algorithm based on graph attention networks (GAT) [32], where the graph is constructed using behavior similarity. ARGANIDS [34] proposes an adversarially regularized graph autoencoder (ARGA) algorithm for NIDS, and employs a line graph structure where each node represents a network flow and its features. GLD-Net [8] applies the long short-term memory (LSTM) after the graph attention network, thus mining the flow and topological features from both the time-series flow data and the network flow graph. Similarly, Tong et al. [30] introduce a GCN-BiLSTM-based abnormal behavior detection method, where a bidirectional long short-term memory network method with an integrated attention mechanism is applied after a graph convolutional neural network [11].

2.3 Self-supervised Learning

Self-supervised learning (SSL) is a novel representation learning paradigm that enables models to learn the latent features of the data. Chen et al. [5] introduce SimCLR, which emphasizes the importance of data augmentation combinations and introduces a learnable nonlinear transformations between learned representations and contrastive loss, yielding state-of-the-art performance. Liu et al. [17] categorize contrastive learning-based SSL frameworks into context-instance and instance-instance contrasts, noting their remarkable performance in downstream tasks, particularly in classification problems under linear protocols, where self-supervised learning's feature extraction capabilities are rapidly approaching the supervised alternatives. Anomal-E [3] is a self-supervised framework for NIDS. Anomal-E uses E-GraphSage [18] as the encoder and applies the modified deep graph infomax (DGI) [33] for SSL. Experiments on the undersampled CIC-IDS2018 dataset show that Anomal-E has better generalization ability and detection performance than baseline alternatives. Li et al. [15] combine self-supervised representations with traditional radiomics under a multi-task learning framework, resulting in significant improvements in brain tumor classification and lung cancer staging. Song et al. [29] propose a two-level contrastive learning scheme based on "exercise-to-exercise" (E2E) relational subgraphs, achieving superior predictive performance through contrastive learning at both the node level and the graph level.

3 Approach

3.1 Flow Graph Construction

To date, most GNN models follow the message-passing paradigm or its simplified variants, where the feature values of a node are obtained by aggregating features from its neighborhood nodes [32]. Thus, existing GNN models focus primarily on node features for node classification, without considering edge features. On

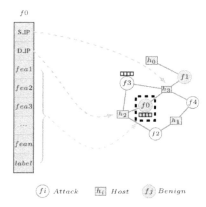

Fig. 1. Flow to Node: Red circles represent Attack flow nodes, orange circles represent Benign flow nodes, and blue rectangles denote Host nodes. The $1 \times n$ rectangular columns adjacent to each flow node denote the corresponding node features. (Color figure online)

the other hand, network flows are usually converted into graph structures by representing the source IP (s_ip) and destination IP (d_ip) of each flow as nodes, while modeling the flow as the edge connecting the nodes. Edges and their features, as such, play an essential role in graph-based NIDS.

Inspired by [26], we adopt an alternative flow graph construction approach, where we keep s_ip and d_ip as host nodes and transform the edges into flow nodes; see Fig. 1. Formally speaking, each flow f : (s_ip → d_ip), which is modeled orignally as an edge, is divided into two undirected edges: (s_ip → f) and (f → d_ip). In other words, we add specific nodes f representing each flow so that (1) flow representations can be learned directly as node embeddings, and (2) NIDS can be formulated as a node classification task. To differentiate between host nodes and flow nodes, we label host nodes with type = 0 and label flow nodes with type = 1.

3.2 Graph Attention Networks

To learn contextualized node representations from the constructed flow graph, we employ the graph attention networks (GAT) [32]. GAT introduces the multi-head attention mechanism, which enables each node to attentively aggregate information from its neighboring nodes. The key component in multi-head attention is the calculation of the attention score α_{ij}, which determines the relative importance of node j to node i among all node i's neighbor nodes:

$$\alpha_{ij} = \frac{\exp(\text{LeakyReLU}(\mathbf{a}^\top [\mathbf{W}_a \mathbf{h}_i \| \mathbf{W}_a \mathbf{h}_j]))}{\sum_{k \in \mathcal{N}(i)} \exp(\text{LeakyReLU}(\mathbf{a}^\top [\mathbf{W}_a \mathbf{h}_i \| \mathbf{W}_a \mathbf{h}_k]))} \quad (1)$$

where $\mathbf{W}_a \in \mathbb{R}^{d \times d}$ and $\mathbf{a} \in \mathbb{R}^{d \times 1}$ are learnable parameters, $\mathcal{N}(i)$ denotes the set of neighbors of node i, LeakyReLU is the extension of the ReLU activation function with negative input slope 0.2, $\|$ denotes the concatenation operation, and $\cdot^\top$ represents transposition. h_i and h_j are d-dimensional node features for node i and node j, respectively.

The attention scores are then used to compute a weighted sum of feature representations corresponding to all neighboring nodes, which, after applying the LeakyReLU nonlinearity, serves as the output features for node i:

$$\mathbf{h}'_i = \text{LeakyReLU}\left(\sum_{j \in \mathcal{N}(i)} \alpha_{ij} \mathbf{W}_u \mathbf{h}_j\right) \quad (2)$$

We employ a cascade of two GAT layers, where the first layer takes as input the original host and flow features, and the second layer takes the output from the first layer as node features. Both layers consist of 4 heads, where in the first layer we concatenate the node states from each head before feeding them to the second layer, and in the second layer we average the output from each head as the final output. After obtaining the contextualized node representations $\hat{\mathbf{h}}_i$ from the GAT, we apply a linear layer and a Softmax layer to all flow node (`type = 1`) representations:

$$\hat{\mathbf{y}}_i = \text{Softmax}(\mathbf{W}_o \hat{\mathbf{h}}_i + \mathbf{b}_o) \quad (3)$$

where $\hat{\mathbf{y}}_i \in \mathbb{R}^{C \times 1}$ denotes the predicted probabilities of the labels, C is the number of classes, and $\mathbf{W}_o \in \mathbb{R}^{C \times d}$ and $\mathbf{b}_o \in \mathbb{R}^{C \times 1}$ are learnable parameters.

The supervised learning objective is to minimize the cross-entropy loss:

$$\mathcal{L}_{\text{sup}} = \frac{1}{N} \sum_{i=1}^{N} \sum_{c=1}^{C} \mathbf{y}_{i,c} \log(\hat{\mathbf{y}}_{i,c}) \quad (4)$$

where $\mathbf{y}_i$ denotes the ground-truth label of the i-th network flow, in which $\mathbf{y}_{i,c} = 1$ if the c-th label is the true label for the i-th flow, and $\mathbf{y}_{i,c} = 0$ otherwise.

3.3 Self-supervised Learning

For self-supervised learning, we focus on contrastive learning, where representations are learned through a contrastive loss that encourages the model to distinguish between similar (positive) and dissimilar (negative) pairs of data instance [5,23]. Positive pairs are typically generated by constructing different views of the same data instance via augmentation, while negative pairs are generated from views of other data instances. To facilitate contrastive learning, we develop four graph-level data augmentation operators tailored specifically to the flow graph structures commonly encountered in NIDS: (1) node feature masking, (2) node resampling, (3) node feature shuffling, and (4) subgraph (Fig. 2).

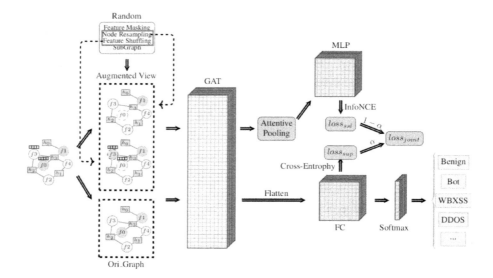

Fig. 2. The model architecture of SSA-GAT. Two randomly sampled augmentation operators are applied over Ori_Graph to construct the two augmented views. Both Ori_Graph and the two augmented views are then passed to a shared GAT for joint supervised and self-supervised learning.

Node Feature Masking. In this approach, we sample a binary mask tensor $M = [m_1, ..., m_i, ..., m_N]^T \in \{0,1\}^N$ from the Bernoulli distribution, where features of all nodes in the set of Selected Nodes (S) are set to zero. The nodes in S are determined by randomly selecting a proportion m_ratio of the number of benign nodes in the original graph (Ori_Graph). The masked features X are generated from the dot product of M and X.

$$t(X) = \tilde{g}(X, M) = M \circ X \tag{5}$$

Node Resampling. We combine the oversampling and undersampling techniques for resampling, applying oversampling at an add_ratio to attack nodes and undersampling at a del_ratio to the amount of benign nodes to balance the distribution of the view. However, alterations to the structure may result in nodes of degree zero. Therefore, we incorporate a pruning operation for the nodes where $degree(D) = 0$ to ensure that the view is devoid of nodes with $D = 0$.

Node Feature Shuffling. For node feature shuffling, which can also be recognized as graph permutation, a fraction of attack nodes, specified by ex_ratio, is randomly chosen from Ori_Graph. Their features are then swapped with an equal number of benign nodes within the same graph.

$$t(X) = \tilde{g}(X, M) = M_{r_i \leftrightarrow r_j} \circ X \tag{6}$$

Subgraph. For Subgraph generation, we begin by extracting all attack nodes from Ori_Graph to form a new initial view. After that, nodes and edges are complemented based on the N-order neighbors connected to these attack nodes within Ori_Graph, where N is a variable that can be freely adjusted. Finally, as alterations to the network structure may result in nodes with degree zero, a pruning operation for nodes with $D = 0$ is also incorporated.

Our contrastive learning approach focuses on learning graph-level discriminative representations. To extract the graph-level representation $\mathbf{h}_G$, we design an attentive readout function, which selectively aggregates node features to form the feature vector to summarize the whole graph representation:

$$\mathbf{h}_G = \sum_{i=1}^{n} \left(\mathbf{q}^\top \left(\mathbf{W}_1 \mathbf{h}_i + \mathbf{b} \right) \right) \mathbf{h}_i \tag{7}$$

where $\mathbf{q}, \mathbf{b} \in \mathbb{R}^d$ and $\mathbf{W}_1 \in \mathbb{R}^{d \times d}$ are all learnable parameters.

Given a batch of N flow graphs $\{\mathbf{x}_1, \mathbf{x}_2, ..., \mathbf{x}_l\}$, we generate two augmented views for graph by sampling separate augmentation operators. The views generated from the same instance are considered a positive pair, while the other $2(l-1)$ augmented views from the batch are taken as negative examples. Passing these views through the model produces $2l$ representations $\{(\mathbf{h}_1^1, \mathbf{h}_1^2), (\mathbf{h}_2^1, \mathbf{h}_2^2), ..., (\mathbf{h}_l^1, \mathbf{h}_l^2)\}$ where $(\mathbf{h}_i^1, \mathbf{h}_i^2)$ is the positive pair for the i-th input instance $\mathbf{x}_i$.

Inspired by SimCLR [5], we use an MLP with one hidden layer to obtain compact hidden representations:

$$\begin{aligned} \mathbf{z}_i^1 &= \max\left(0, \mathbf{h}_i^1 \cdot \mathbf{W}_1 + \mathbf{b}_1\right) \cdot \mathbf{W}_2 + \mathbf{b}_2 \\ \mathbf{z}_i^2 &= \max\left(0, \mathbf{h}_i^2 \cdot \mathbf{W}_1 + \mathbf{b}_1\right) \cdot \mathbf{W}_2 + \mathbf{b}_2 \end{aligned} \tag{8}$$

These representations are then passed to the contrastive loss [5]:

$$\mathcal{L}_{ssl} = -\frac{1}{l} \sum_{i=1}^{l} \log \left(\text{softmax} \left(\text{sim}(\mathbf{z}_i^1, \mathbf{z}_i^2)/\tau \right) \right) \tag{9}$$

where $\text{sim}(\boldsymbol{u}, \boldsymbol{v}) = \frac{\boldsymbol{u}^\top \boldsymbol{v}}{\|\boldsymbol{u}\| \|\boldsymbol{v}\|}$ is the cosine similarity, τ is the temperature, and softmax is normalized over all views in the batch. The contrastive objective aims to bring the views in each positive pair together and separate them from all other views in the batch.

3.4 Optimization Methodology

We formulate the optimization of our model as a multi-task learning problem [21, 40], where we use a joint objective combining supervised and self-supervised losses:

$$\mathcal{L}_{\text{joint}} = \alpha \cdot \mathcal{L}_{\text{sup}} + (1 - \alpha) \cdot \mathcal{L}_{\text{ssl}} \tag{10}$$

where $\alpha \in [0,1]$ is a hyperparameter that controls the relative importance of supervised and self-supervised losses. The model is trained for 50 epochs using the Adam optimizer with batch size 8, learning rate $1e-3$, and a weight decay of $1e-5$. Flow features are normalized to a mean of 0 and a standard deviation of 1 before feeding to the model.

4 Experiment

4.1 DataSet

Data Preprocessing. We conduct experiments on the CIC-IDS2017 [28] benchmark dataset. Table 1 represents the number of flows and the number of unique timestamps for each type of flow in the dataset. We can see that the dataset is highly imbalanced, with majority classes such as Benign and DoSs making up over 93% of the entire dataset. On the other hand, HeartBleed, the most infrequent class, occurs only 11 times. We clean the dataset by removing all duplicate, null, or infinite values, resulting in 2,827,575 valid records. Then we remove features with a standard deviation less than 0.01. The remaining 69 numerical features and 3 non-numerical features, s_ip, d_ip, and timestamp, are kept as input to the model.

Table 1. Statistics for the CIC-IDS2017 dataset

Label	# Flows	# Timestamps
BENIGN	2,271,021	27,444
DoSHulk	230,122	21
PortScan	158,804	27
DDoS	128,025	21
DoSGoldenEye	10,293	8
FTP-Patator	7,935	64
SSH-Patator	5,897	63
DoSslowloris	5,796	27
DoSSlowhttptest	5,499	19
Bot	1,956	163
WebAttackBruteForce	1,507	44
WebAttackXSS	652	21
Infiltration	36	28
WebAttackSqlInjection	21	3
Heartbleed	11	11

Data Split. Graph construction is based on the dataset's **timestamp** feature, partitioning by the second. [8] has used time windows for graph construction to capture prolonged connection features of typical attacks like DoSs and Patators. However, extending the time window may lead to an influx of Benign nodes, worsening the imbalance between categories and biasing metrics towards majority categories. Moreover, real-world network connections and disconnections are unpredictable, making it challenging to maintain consistent network topology over time. Therefore, for practicality and to mitigate imbalance, we chose to construct graphs based on seconds. The dataset is split into training, validation, and test sets using a ratio of 8:1:1 based on distinct timestamps. We tune hyper-parameters on the validation set and report results on the test set.

4.2 Comparative Experiments

Table 2. Comparison of SSA-GAT against three baseline models.

Model	SSA-GAT (ours)			MLP			Linear			LightGBM		
	F1	Pre	Recall	F1	Pre	Recall	F1	Pre	Recall	F1	Pre	Recall
Benign	0.9852	0.9975	0.9731	1.0000	1.0000	1.0000	0.9716	0.9447	1.0000	0.9953	0.9988	0.9918
DDOS	0.9130	0.8400	1.0000	1.0000	1.0000	1.0000	0.9902	0.9812	0.9993	0.9403	0.8929	0.9929
DOSHulk	0.9905	0.9812	1.0000	1.0000	1.0000	1.0000	0.9998	0.8664	0.9996	0.7645	0.9958	0.9960	0.9955
PortScan	0.9585	0.9905	0.9286	0.9316	0.8934	0.9732	0.9776	0.9820	0.9732	0.0835	0.0560	0.1637
SSHPatator	0.9947	0.9894	1.0000	0.9939	0.9818	1.0000	0.8342	0.7172	0.9967	0.0900	0.0486	0.5987
FTPPatator	0.8405	0.7763	0.9162	0.9953	0.9980	0.9926	0.9743	0.9567	0.9926	0.4922	0.6064	0.4142
DoSGoldenEye	0.9880	0.9763	1.0000	0.9960	0.9936	0.9984	0.8717	0.9960	0.7729	0.7515	0.7846	0.7211
DoSSlowHttpTest	0.9711	0.9728	0.9695	0.9072	0.8682	0.9500	0.8793	0.8491	0.9118	0.9970	0.9993	0.9946
DoSSolwLoris	0.7079	0.9950	0.5493	0.8560	0.9910	0.7534	0.3504	0.7961	0.2247	0.9900	0.9882	0.9917
WASqlInjection	0.0000	0.0000	0.0000	0.0000	0.0000	0.0000	0.0000	0.0000	0.0000	0.0000	0.0000	0.0000
WABruteForce	0.6009	0.5137	0.7238	0.6548	0.5108	0.9116	0.6047	0.4897	0.7901	0.0016	0.0008	0.1667
WAXSS	0.5185	0.4746	0.5714	0.0385	0.3333	0.0204	0.0000	0.0000	0.0000	0.0000	0.0000	0.0000
Infiltration	0.6250	1.0000	0.4545	0.0000	0.0000	0.0000	0.0000	0.0000	0.0000	0.1593	0.0865	1.0000
Bot	0.9914	0.9848	0.9981	0.9792	0.9610	0.9981	0.9564	0.9198	0.9961	0.8552	0.7658	0.9684
HeartBleed	0.6667	1.0000	0.5000	0.0000	0.0000	0.0000	0.0000	0.0000	0.0000	0.2695	0.1576	0.9312
Overall	0.7835	0.8328	0.7723	0.6902	0.7021	0.7065	0.6185	0.6421	0.6281	0.5081	0.4921	0.6620
LogLoss		0.0559			0.013			0.2089			0.2310	
Detection Count	14			12			11			13		

Category Performance. Table 2 compares our model with three baseline approaches, including simple linear regression, the multi-layer perceptron (MLP), and LightGBM [10]. We categorize attack classes in the dataset by counts: 0-1000 as extremely minority, 1000-11000 as minority, and more than 11000 as majority. We have several key findings: (*i*) Deep neural networks generally outperform traditional machine learning in detecting most attacks. In particular, MLP achieves near-perfect accuracy in detecting majority attacks but falls short in minority classes. (*ii*) In traditional machine learning, the LightGBM model shows commendable performance in identifying DOSs attacks but lags behind deep learning models in many other attacks. (*iii*) Our model consistently outperforms others in detecting minority attacks, surpassing MLP across most of the minority attacks, even exceeds comprehensive F1 scores of 60% in extremely minority attacks such as infiltration and HeartBleed.

Comprehensive Performance. In the bottom of Table 2, we report the micro-average of the LogLoss and the number of detected attack types for different models. We can see that SSA-GAT is able to detect more attack types than any other baseline model. MLP gets better LogLoss than SSA-GAT, yet underperforms in detecting minority attacks, skewing their aggregated F1 scores towards majority attacks and inflating overall results.

4.3 Ablation Experiments

We use supervised GAT as the baseline model and also experiment with GAT with mean pooling (SSM-GAT) as the readout function in the contrastive learning framework. These ablation experiments validate the effectiveness of contrastive learning and our attentive readout function.

Variation of Alpha Value. We vary α in Eq. 10 between $(1/11, ..., 11/11)$, and report the LogLoss of the model for each α; see Fig. 3. LogLoss is a preferred metric for this study, as it measures the uncertainty of predictions from a classifier. In other words, it measures how confident the classifier is in predicting the correct class for each instance. It is also known as cross-entropy loss; see Eq. 4. A lower LogLoss indicates a better classifier, while a higher LogLoss indicates a worse classifier. Recall that in Eq. 10, when $\alpha = 1.0$, the model degrades to the supervised-only setting. We can see that the model achieves the lowest LogLoss with $\alpha = 10/11$. This suggests that self-supervised learning leads to a better model.

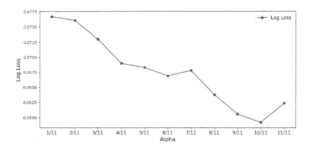

Fig. 3. The effect of varying α in Eq. 10 on the LogLoss of SSA-GAT. Lower LogLosses indicate better calibrated and more accurate probabilities.

Ablation Study. The attentive pooling mechanism in SSA-GAT further improves the detection accuracy of SSA-GAT. SSM-GAT uses mean pooling, which directly averages hidden representations of the GAT, and can overlook complex graph structures and node importance variations. In contrast, Attentive pooling dynamically weighs node importance, enhancing the model's adaptability to graphs of different sizes and node types, and aligns well with the internal attention mechanism of GAT, leading to the improved experimental results.

Table 3 shows: (i) For slow DOS attacks such as DosSlowhttptest and DosSlowloris, self-supervised learning achieves significantly higher F1 scores (97.11% and 70.79%) compared to supervised learning (82.62% and 0%). (ii) The infiltration attack's F1 score in self-supervised learning is 62.5%, significantly better than the supervised baseline. (iii) Extremely minority attacks such as WebAttacks have stable F1 scores ranging between 50% and 70%. (iv) Since WebAttackSqlInjection's timestamps occur only three times, all models fail to detect it.

Table 3. Ablation study on the SSA-GAT model.

Model	GAT			SSM-GAT			SSA-GAT (ours)		
	F1	Pre	Recall	F1	Pre	Recall	F1	Pre	Recall
Benign	0.9851	0.9974	0.9731	0.9852	0.9975	0.9731	0.9852	0.9975	0.9731
DDOS	0.913	0.84	1.0000	0.9130	0.84	1	0.913	0.84	1.0000
DOSHulk	0.9905	0.9812	1.0000	0.9905	0.9812	1	0.9905	0.9812	1.0000
PortScan	0.9585	0.9905	0.9286	0.9585	0.9905	0.9286	0.9585	0.9905	0.9286
SSHPatator	0.9947	0.9894	1.0000	0.9943	0.9886	1	0.9947	0.9894	1.0000
FTPPatator	0.8394	0.7759	0.9142	0.8394	0.7759	0.9142	**0.8405**	**0.7763**	**0.9162**
DoSGoldenEye	0.988	0.9763	1.0000	0.988	0.9763	1	0.988	0.9763	1.0000
Bot	0.9876	0.9773	0.9981	**0.9914**	**0.9829**	**1.0000**	**0.9914**	**0.9848**	**0.9981**
Infiltration	0.3077	1.0000	0.1818	**0.6250**	**1.0000**	**0.4545**	**0.6250**	**1.0000**	**0.4545**
HeartBleed	0.6667	1.0000	0.5000	0.6667	1.0000	0.5000	0.6667	1.0000	0.5000
DoSSlowhttptest	0.8262	0.7198	0.9695	**0.9711**	**0.9728**	**0.9695**	**0.9711**	**0.9728**	**0.9695**
DoSslowloris	0	0	0	**0.7079**	**0.9950**	**0.5493**	**0.7079**	**0.9950**	**0.5493**
WABruteForce	0.5365	0.4386	0.6906	0.5252	0.4237	0.6906	**0.6009**	**0.5137**	**0.7238**
WAXSS	0.3855	0.4706	0.3265	0.3855	0.4706	0.3265	**0.5185**	**0.4746**	**0.5714**
WASqlInjection	0	0	0	0	0	0	0	0	0
Overall	0.6920	0.7438	0.6988	0.7694	0.8263	0.7538	**0.7835**	**0.8328**	**0.7723**
LogLoss		0.0624			0.0592			**0.0559**	
Detection Count	13			14			14		

4.4 Visualization Results

Bot: Distinct Network Structure Features. Botnet exhibits a one-to-many infection pattern, involving communications between infected hosts for command execution, data synchronization or updation and distributed attacks. This pattern suggests a decentralized structure, implying greater autonomy among bots, enhancing resilience and stealth. The flow node graph Fig. 4a shows that, while two Bot hosts are mutually communicating, their corresponding flow nodes show in pairs.

Infiltration and HeartBleed: Fifth-Order Subgraph with Few Nodes.
Infiltration can be specifically recognized as abnormal behavior combined with kinds of attack. HeartBleed typically appears alongside DOSs attacks, but the attention mechanism can capture its attributes within the topology. In this dataset, the two attacks' fifth-order subgraph are with few nodes. Moreover, HeartBleed occurs predominantly on port 444. The attention scores are visualized in Fig. 4b.

DOSs and BruteForces and WebAttacks: Attributes and Topology.
DOSs, BruteForces and WebAttacks generally present similar network topologies, with a massive influx of traffic connected at one or a few hosts within a short period. However, the attention mechanism can further dissect their distinctly attribute features. For instance, the FTP-Patator (port 21) and SSH-Patator (port 22) attacks can be distinguished from port differences, DOSs and WebAttacks can be identified according to the magnitude and the attributes. Figure 4c illustrates the DoSSlowhttptest attack's attention score.

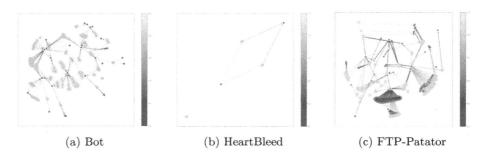

(a) Bot (b) HeartBleed (c) FTP-Patator

Fig. 4. Attention Visualization: Benign nodes and host nodes are colored same, while Benign nodes are with a bigger size. Attack nodes are painted in different colors from the others. The edges connected to attack nodes are colored according to the attention scores, a higher score with a lighter color.

5 Conclusions

We propose SSA-GAT, a graph-based contrastive self-supervised learning framework for network intrusion detection. We introduce four augmentation methods tailored to the network flow graph and a novel attentive readout function to facilitate contrastive self-supervised learning. Experimental results show that SSA-GAT is more effective in detecting minority attacks, when compared with a number of state-of-the-art baseline network intrusion detection models.

Acknowledgement. This work is funded by the Science and Technology Planning Project of the Ministry of Public Security of the People's Republic of China in 2022 (2022JSYJC24).

References

1. Ahmad, Z., Shahid Khan, A., Wai Shiang, C., Abdullah, J., Ahmad, F.: Network intrusion detection system: a systematic study of machine learning and deep learning approaches. Trans. Emerging Telecommun. Technol. **32**(1), e4150 (2021)
2. Arjovsky, M., Chintala, S., Bottou, L.: Wasserstein generative adversarial networks. In: International Conference on Machine Learning, pp. 214–223. PMLR (2017)
3. Caville, E., Lo, W.W., Layeghy, S., Portmann, M.: Anomal-e: a self-supervised network intrusion detection system based on graph neural networks. Knowl.-Based Syst. **258**, 110030 (2022)
4. Chawla, N.V., Bowyer, K.W., Hall, L.O., Kegelmeyer, W.P.: Smote: synthetic minority over-sampling technique. J. Artif. Intell. Res. **16**, 321–357 (2002)
5. Chen, T., Kornblith, S., Norouzi, M., Hinton, G.: A simple framework for contrastive learning of visual representations. In: International Conference on Machine Learning, pp. 1597–1607. PMLR (2020)
6. Devlin, J., Chang, M.W., Lee, K., Toutanova, K.: Bert: Pre-training of deep bidirectional transformers for language understanding. In: Proceedings of NAACL-HLT, pp. 4171–4186 (2019)
7. Eichenberger, C., et al.: Traffic4cast at neurips 2021-temporal and spatial few-shot transfer learning in gridded geo-spatial processes. In: NeurIPS 2021 Competitions and Demonstrations Track. PMLR (2022)
8. Guo, W., Qiu, H., Liu, Z., Zhu, J., Wang, Q.: Gld-net: Deep learning to detect ddos attack via topological and traffic feature fusion. Computational Intelligence and Neuroscience 2022 (2022)
9. Gupta, N., Jindal, V., Bedi, P.: Lio-ids: handling class imbalance using lstm and improved one-vs-one technique in intrusion detection system. Comput. Netw. **192**, 108076 (2021)
10. Ke, G., et al.: Lightgbm: a highly efficient gradient boosting decision tree. Advances in neural information processing systems **30** (2017)
11. Kipf, T.N., Welling, M.: Semi-supervised classification with graph convolutional networks. In: International Conference on Learning Representations (2016)
12. Krenn, M., et al.: Predicting the future of ai with ai: High-quality link prediction in an exponentially growing knowledge network. arXiv preprint arXiv:2210.00881 (2022)
13. Krenn, M., Buffoni, L., Coutinho, B., Eppel, S., Foster, J.G., Gritsevskiy, A., Lee, H., Lu, Y., Moutinho, J.P., Sanjabi, N., et al.: Forecasting the future of artificial intelligence with machine learning-based link prediction in an exponentially growing knowledge network. Nature Mach. Intell. **5**(11), 1326–1335 (2023)
14. Kumar, V., Sinha, D., Das, A.K., Pandey, S.C., Goswami, R.T.: An integrated rule based intrusion detection system: analysis on unsw-nb15 data set and the real time online dataset. Clust. Comput. **23**, 1397–1418 (2020)

15. Li, H., Xue, F.F., Chaitanya, K., Luo, S., Ezhov, I., Wiestler, B., Zhang, J., Menze, B.: Imbalance-aware self-supervised learning for 3d radiomic representations. In: Medical Image Computing and Computer Assisted Intervention–MICCAI 2021: 24th International Conference, Strasbourg, France, September 27–October 1, 2021, Proceedings, Part II 24, pp. 36–46. Springer (2021)
16. Liao, H.J., Lin, C.H.R., Lin, Y.C., Tung, K.Y.: Intrusion detection system: A comprehensive review. J. Network Comput. Appl. **36**(1) (2013)
17. Liu, X., Zhang, F., Hou, Z., Mian, L., Wang, Z., Zhang, J., Tang, J.: Self-supervised learning: generative or contrastive. IEEE Trans. Knowl. Data Eng. **35**(1), 857–876 (2021)
18. Lo, W.W., Layeghy, S., Sarhan, M., Gallagher, M., Portmann, M.: E-graphsage: A graph neural network based intrusion detection system for iot. In: NOMS 2022-2022 IEEE/IFIP Network Operations and Management Symposium, pp. 1–9. IEEE (2022)
19. Lu, Y.: Predicting research trends in artificial intelligence with gradient boosting decision trees and time-aware graph neural networks. In: 2021 IEEE International Conference on Big Data (Big Data), pp. 5809–5814. IEEE (2021)
20. Lu, Y.: An efficient two-stage gradient boosting framework for short-term traffic state estimation. arXiv preprint arXiv:2302.10400 (2023)
21. Lu, Y., Dong, R., Smyth, B.: Why i like it: multi-task learning for recommendation and explanation. In: Proceedings of the 12th ACM Conference on Recommender Systems, pp. 4–12 (2018)
22. Lu, Y., et al.: Session-based recommendation with transformers. In: Proceedings of the Recommender Systems Challenge 2022, pp. 29–33 (2022)
23. Lu, Y., Volkovs, M.: Robust user engagement modeling with transformers and self supervision. In: Proceedings of the Recommender Systems Challenge 2023, pp. 23–27 (2023)
24. Ma, Z., Li, J., Song, Y., Wu, X., Chen, C., et al.: Network intrusion detection method based on fcwgan and bilstm. Computational Intelligence and Neuroscience **2022** (2022)
25. Neun, M., et al.: Traffic4cast at neurips 2022–predict dynamics along graph edges from sparse node data: Whole city traffic and eta from stationary vehicle detectors. In: NeurIPS 2022 Competition Track, pp. 251–278. PMLR (2022)
26. Pujol-Perich, D., Suárez-Varela, J., Cabellos-Aparicio, A., Barlet-Ros, P.: Unveiling the potential of graph neural networks for robust intrusion detection. ACM SIGMETRICS Performance Evaluation Review **49**(4), 111–117 (2022)
27. Qazi, N., Raza, K.: Effect of feature selection, smote and under sampling on class imbalance classification. In: 2012 UKSim 14th International Conference on Computer Modelling and Simulation, pp. 145–150. IEEE (2012)
28. Sharafaldin, I., Lashkari, A.H., Ghorbani, A.A.: Toward generating a new intrusion detection dataset and intrusion traffic characterization. ICISSp **1**, 108–116 (2018)
29. Song, X., Li, J., Lei, Q., Zhao, W., Chen, Y., Mian, A.: Bi-clkt: Bi-graph contrastive learning based knowledge tracing. Knowl.-Based Syst. **241**, 108274 (2022)
30. Tong, X., Tan, X., Sun, X.: Abnormal behavior detection based on gcn-bilstm. In: Third International Conference on Machine Learning and Computer Application (ICMLCA 2022), vol. 12636, pp. 468–474. SPIE (2023)
31. Tsai, C.F., Hsu, Y.F., Lin, C.Y., Lin, W.Y.: Intrusion detection by machine learning: A review. Expert Syst. Appl. **36**(10), 11994–12000 (2009)
32. Veličković, P., Cucurull, G., Casanova, A., Romero, A., Liò, P., Bengio, Y.: Graph attention networks. In: International Conference on Learning Representations (2018)

33. Veličković, P., Fedus, W., Hamilton, W.L., Liò, P., Bengio, Y., Hjelm, R.D.: Deep graph infomax. In: International Conference on Learning Representations (2018)
34. Venturi, A., Ferrari, M., Marchetti, M., Colajanni, M.: Arganids: a novel network intrusion detection system based on adversarially regularized graph autoencoder. In: Proceedings of the 38th ACM/SIGAPP Symposium on Applied Computing. pp. 1540–1548 (2023)
35. Volkovs, M., Rai, H., Cheng, Z., Wu, G., Lu, Y., Sanner, S.: Two-stage model for automatic playlist continuation at scale. In: Proceedings of the ACM Recommender Systems Challenge 2018, pp. 1–6 (2018)
36. Wang, Y., Han, Z., Li, J., He, X.: Bs-gat behavior similarity based graph attention network for network intrusion detection. arXiv preprint arXiv:2304.07226 (2023)
37. Yu, W., Lu, Y., Easterbrook, S., Fidler, S.: Crevnet: Conditionally reversible video prediction. arXiv preprint arXiv:1910.11577 (2019)
38. Yu, W., Lu, Y., Easterbrook, S., Fidler, S.: Efficient and information-preserving future frame prediction and beyond. In: International Conference on Learning Representations (2019)
39. Zhang, H., Huang, L., Wu, C.Q., Li, Z.: An effective convolutional neural network based on smote and gaussian mixture model for intrusion detection in imbalanced dataset. Comput. Netw. **177**, 107315 (2020)
40. Zhang, Y., Yang, Q.: A survey on multi-task learning. IEEE Trans. Knowl. Data Eng. **34**(12), 5586–5609 (2021)

Author Index

A
Alaca, Furkan 461
An, Ning 171, 213
Artelt, André 155
Ashraf, Inaam 155

B
Brinkrolf, Johannes 155

C
Cao, Lele 373
Cardia, Marco 198
Chen, Hongwei 121
Chen, Jianxia 45
Chen, Rui 139
Chen, Yin 89
Chen, Zexi 121
Chessa, Stefano 198
Cui, Shaoguo 407

D
Delaney, Gary W. 325
Ding, Steven H. H. 461
Ding, Ying 139

E
Engsig-Karup, Allan P. 295

F
Fan, Chenyou 139
Fan, Lin 476
Fang, Yuchen 246
Feng, Dan 213
Flaack, Leon 443
Fu, Xianghua 89
Fukui, Ryota 33

G
Gambineri, Francesca 198
Gao, Lizhong 233

Gao, Weihao 233
Gu, Hengrui 74

H
Halvardsson, Gustaf 373
Hammer, Barbara 155
He, Xin 74
Herman, Pawel 373
Hinder, Fabian 155
Howard, David 325
Huang, Junshu 89
Huang, Yifan 171
Huang, Yubo 340

J
Ji, Peichen 105
Jiang, Chengling 357
Jiang, Gaohang 45
Jiang, Haiqi 139
Jiang, Shan 310
Jin, Beihong 60
Jin, Xue 476
Jiong, Wang 3

K
Kahandawa, Gayan 325

L
Lai, Xin 340
Le, Duy 325
Li, Beibei 60
Li, Xia 121
Li, Yi 340
Li, Yinmian 340
Lin, Man 389
Liu, Chaoqun 105
Liu, Diwen 171
Liu, Hui 443
Liu, Jie 3
Liu, Luanxuan 121

© The Editor(s) (if applicable) and The Author(s), under exclusive license to Springer Nature Switzerland AG 2024
M. Wand et al. (Eds.): ICANN 2024, LNCS 15024, pp. 493–495, 2024.
https://doi.org/10.1007/978-3-031-72356-8

Liu, Qian 476
Long, Zi 89
Lu, Mengyu 45
Luminare, Antonella Giuliana 198
Luo, Chenyang 340
Luo, Haiyong 246
lv, Lang 45

M

Ma, Jie 105
Martinů, Jan-Matyáš 186
Matsumoto, Yoshio 33
McCornack, Andrew 373
Meng, Yuan 310
Micheli, Alessio 198
Molloy, Christopher 461
Murshed, Manzur 325

N

Nakamura, Yutaka 19
Nguyen, Linh 325

O

Okadome, Yuya 19, 33
Ouyang, Jiarui 389

P

Peng, Xuemei 265
Phung, Truong 325
Plesner, Andreas 295

Q

Qi, Baozhen 357
Qiao, Chenbin 357
Qin, Jiwei 105

S

Schultz, Tanja 443
Shen, Ying 389
Shibasaki, Takahiro 19
Šimánek, Petr 186
Song, Zhina 45
Strotherm, Janine 155

T

Tan, Huailiang 280
True, Hans 295

V

Vaquet, Jonas 155
Vaquet, Valerie 155
von Ehrenheim, Vilhelm 373

W

Wang, Chenxing 246
Wang, Mingyang 407
Wang, Qi 425
Wang, Qiao 233
Wang, Ruobing 74
Wang, XiaoYi 3
Wang, Xin 74
Wang, Xinyan 280
Wang, Zixi 340
Wu, Yun 171, 213

X

Xiao, Liang 45
Xiong, Ao 357
Xiong, Haoyu 246
Xu, Hao 425
Xu, Hongbo 425
Xu, Qiuhan 265
Xu, Song 407
Xu, Yongxiu 425

Y

Yang, Jieming 171, 213
Yang, Xiaozong 280
Yang, Zhibin 105
Yao, Meihan 45
Ye, Muyang 340
Yuguchi, Akishige 33

Z

Zeng, Delong 389
Zhang, Donghao 105
Zhang, Fang 340
Zhang, Haichao 246
Zhang, Hui 476
Zhang, Ke 265
Zhang, Mingwei 233
Zhang, Shiyao 443
Zhang, Shuxi 45
Zhang, Youpeng 476
Zhao, Fang 246
Zhao, Rui 60

Author Index

Zhao, Wenhan 171
Zhao, Yongbin 171, 213
Zheng, Wei 213
Zheng, Yiyuan 60
Zhou, Danni 310

Zhou, Hao 265
Zhou, Jianshe 3
Zhu, Dandan 265
Zhu, Dongwei 425
Zhu, Wenhao 425

Printed in the USA
CPSIA information can be obtained
at www.ICGtesting.com
CBHW071423250924
14771CB00011B/105